Exploring drafting

basic fundamentals

by

JOHN R. WALKER

Bel Air High School
Bel Air, Maryland

South Holland, Illinois
THE GOODHEART-WILLCOX CO., Inc.
Publishers

Library of Congress Catalog Card Number 79–162277.
International Standard Book Number 0–87006–195–X.
23456789–75–9876

INTRODUCTION

A course in DRAFTING should be a part of your education. It will help you to develop the capacity to plan in an orderly fashion, to interpret the ideas of others, and to express in an understandable manner.

EXPLORING DRAFTING is a first course which teaches Drafting Fundamentals and Basic Constructions. As you proceed with the course, you will become familiar with methods and processes used by industry. You will make many of the Draftsman's skills your own.

Using selected material, EXPLORING DRAFTING may be used to present a six, nine, eighteen, or thirty six weeks drafting program.

Today, Drafting, the "Language of Industry," offers many opportunities for a lifetime of challenging and personally rewarding employment. Will you accept that challenge?

John R. Walker

Industry photo. Hundreds of drawings were required to plan and build the hotel in which this 53 ton, 70 ft. long truss was used. (Bethlehem Steel Corp.)

CONTENTS

Unit 1

WHY STUDY DRAFTING?

DRAFTING is the part of industry concerned with the preparation of drawings needed to develop and manufacture modern-day products.

Often drawings are the best means available to explain or show our ideas, Fig. 1-1. Drafting is frequently called a "universal language" because it communicates ideas in graphic or picture form. Like other languages, symbols (lines and figures) that have special and specific meanings are used to accurately describe the shape, size, type of material, finish and fabrication of an object. The symbols have been standardized over most of the world making it possible to interpret and understand drawings made in other countries, Fig. 1-2.

Fig. 1-1. Drawings are the best way to show how to construct something as complex as this helicopter. Can you imagine trying to explain how to make a helicopter using words only? (Sikorsky Aircraft)

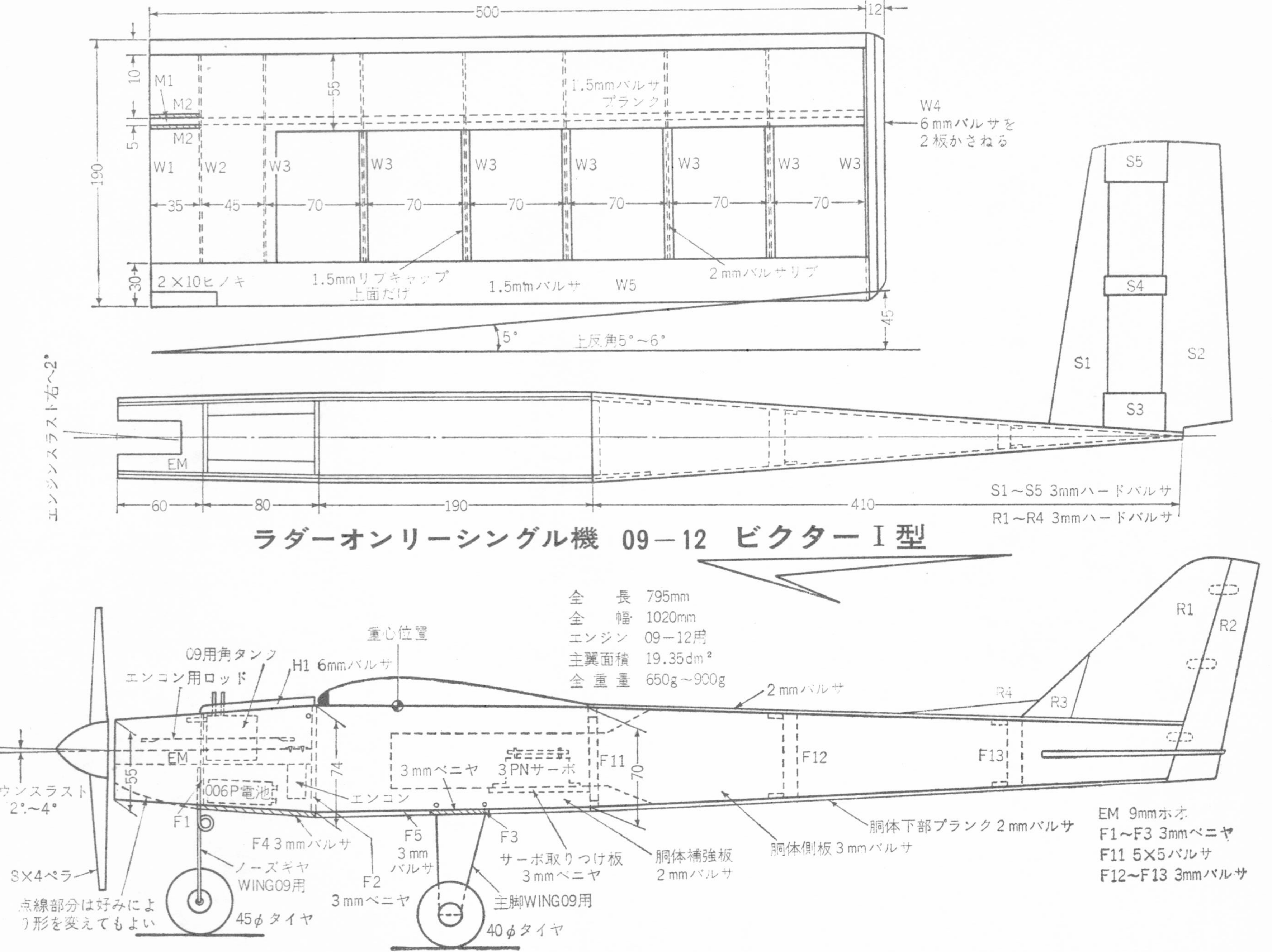

Fig. 1-2. Although the written instructions are in Japanese and the dimensions in the metric system it would be possible for you to use this plan to construct this radio controlled model airplane, if you understand drawings.

Fig. 1-3. Drafting is the "language of industry." Drawings show the craftsman what the designer had in mind. (American Motors)

Drafting is also the "language of industry." It is a precise language because drawings provide the craftsman with the information he needs to produce the product the designer had in mind, Fig. 1-3. It involves the recording of necessary production information on paper, film or tape, Fig. 1-4.

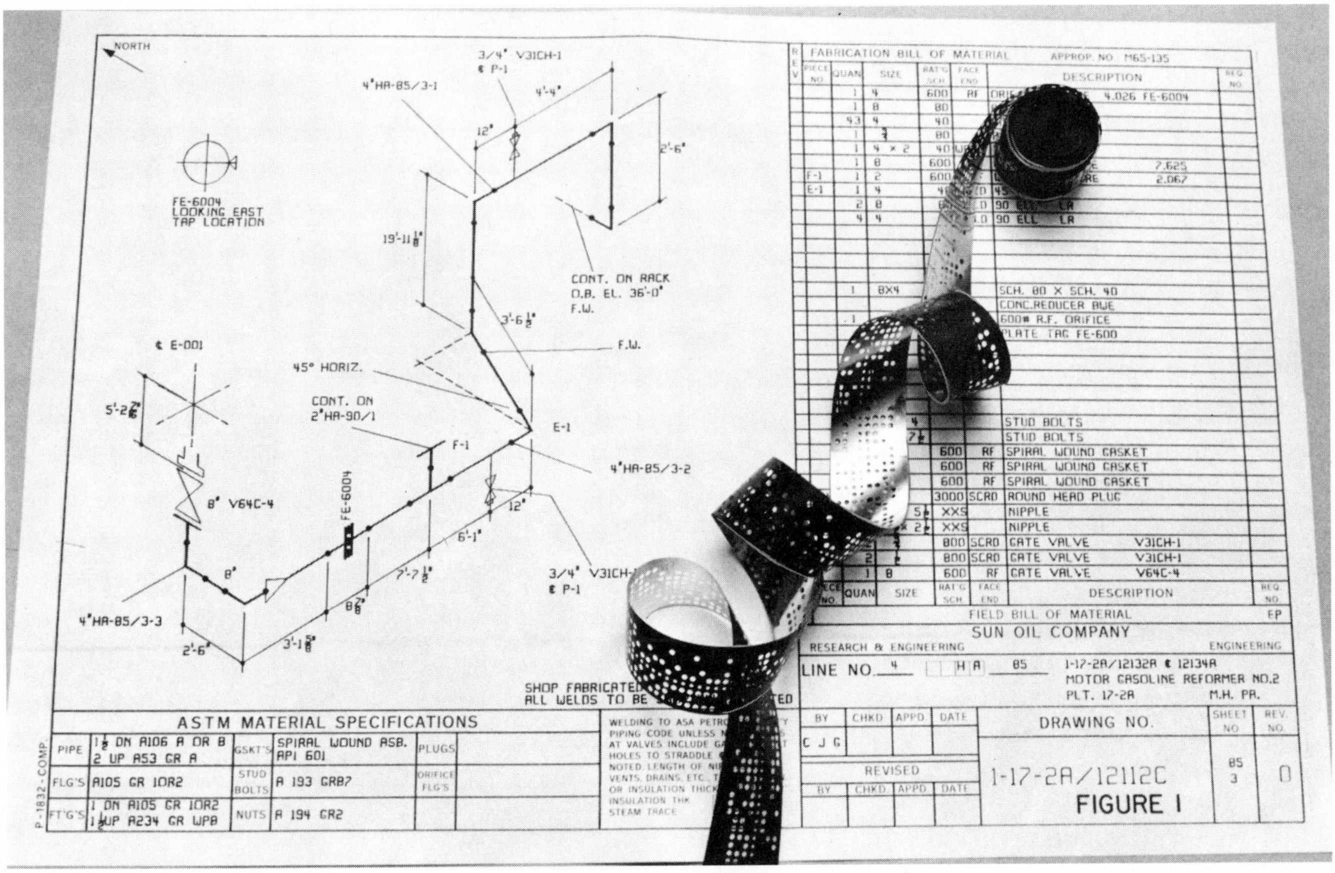

Fig. 1-4. Some specialized drafting can be done by machine. The required information is "punched" into a plastic tape. The tape is fed into a computer-like machine which interprets the hole sequence in the tape and prints it on a sheet of paper. This drawing of a piping diagram in an oil refinery was made in this manner. (Sun Oil Co.)

Fig. 1-5. You use drawings when you construct a model.

Drafting is very important to our modern world. It would be difficult to name an occupation that does not require the ability to read and understand drawings.

It is quite obvious that the craftsmen who build our homes, make our cars or produce our TV sets must use drawings. YOU use drawings when you assemble a model, Fig. 1-5.

Many specialized fields of drafting have been developed - aerospace, architectural, automotive styling, electronic and electrical drafting, structural drafting, engineering graphics, technical drawing and topographical drafting.

Basically, they use the same drafting equipment and employ similar drafting techniques. However, the type of work done varies greatly, Figs. 1-6 and 1-7.

TEST YOUR KNOWLEDGE - UNIT 1

(Write answers on a separate sheet of paper. Do not write in this book.)

1. What does the term drafting mean?
2. Drawings are often used because:
   a. They are easy to make.
   b. They are the best means available to explain or show many ideas.
   c. People who cannot read can understand them.
   d. All of the above.
   e. None of the above.
3. Drafting is also called:
   a. A universal language.
   b. The language of industry.
   c. A picture language.
   d. All of the above.
   e. None of the above.
4. In addition to the craftsmen who make the things we use every day, list several other people who make use of drawings.
5. List five specialized fields of drafting.

Fig. 1-6. The architectural draftsman must have the technical knowledge required to carry out the ideas of the architect. (Prestressed Concrete Institute)

Fig. 1-7. The fabrication and erection of steel structures like the Lamoille River (Vt.) Bridge requires many drawings. With drawings the draftsman conveys, in technical language, all the information required to fabricate the many structural members. (Bethlehem Steel)

OUTSIDE ACTIVITIES

1. Secure samples of drawings used by the following industries:
   a. Aerospace.
   b. Architecture.
   c. Automotive styling.
   d. Electrical and electronic.
   e. Structural.
   f. Manufacturing (technical drawings).
   g. Map making.
2. Make a collection of pictures (magazine clippings, photographs, etc.) that show products made by industries listed in No. 1.
3. Obtain copies of drawings made in a foreign country.
4. Visit a local architect who designs residences. After discussing a project he has in work, prepare a report on the steps he normally follows when he designs a home for a client. Prepare your questions carefully before your visit.
5. Visit a local surveyor and make a report on the work he does. Borrow samples of his work for a bulletin board display.

Unit 2
SHOP SKETCHING

SHOP SKETCHING is one of many drafting techniques. It is a convenient and rapid way of putting ideas into visual form. See Fig. 2-1.

A good sketch shows the shape of the object, and provides dimensions and special instructions on how the object is to be made and finished.

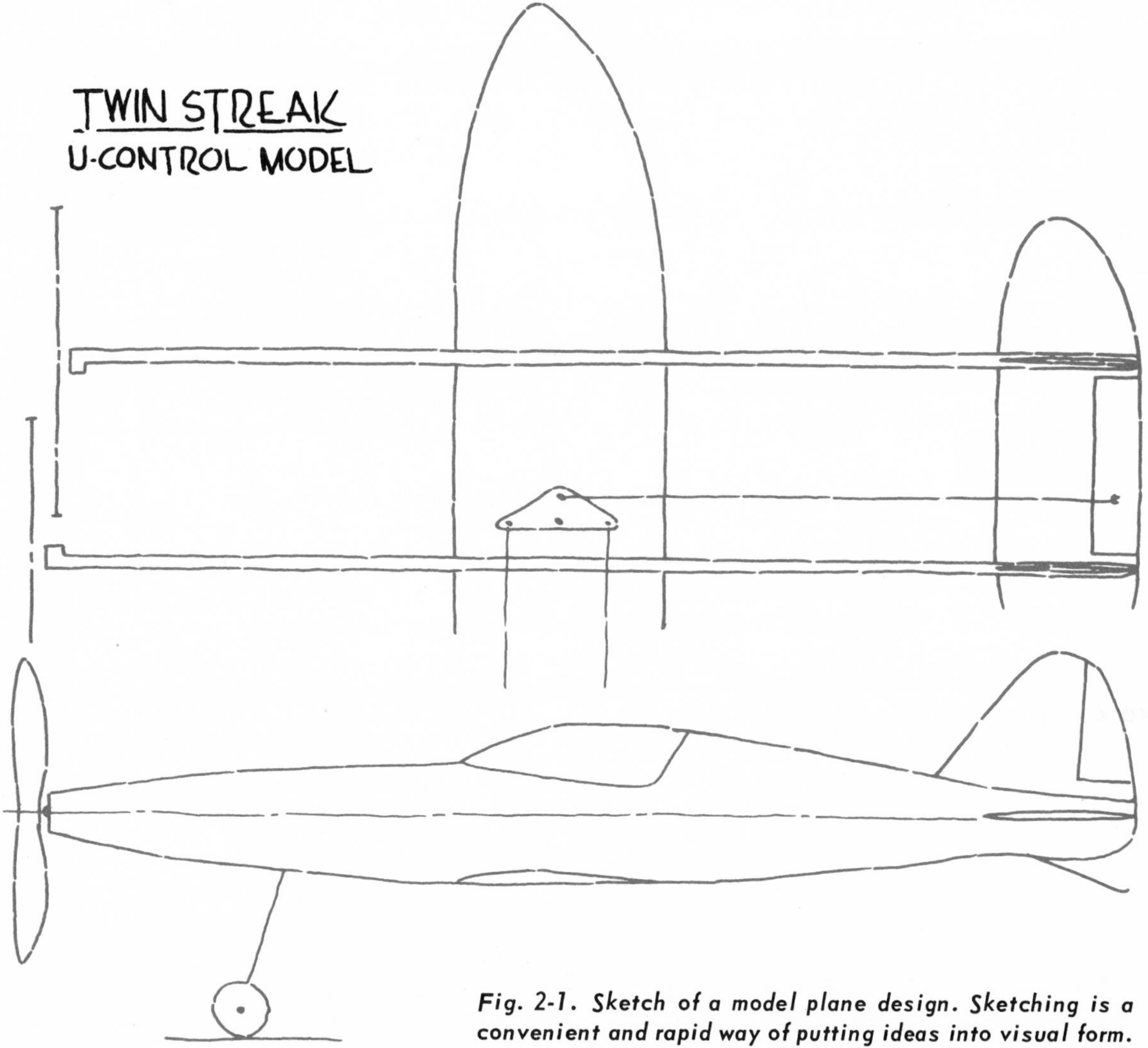

Fig. 2-1. Sketch of a model plane design. Sketching is a convenient and rapid way of putting ideas into visual form.

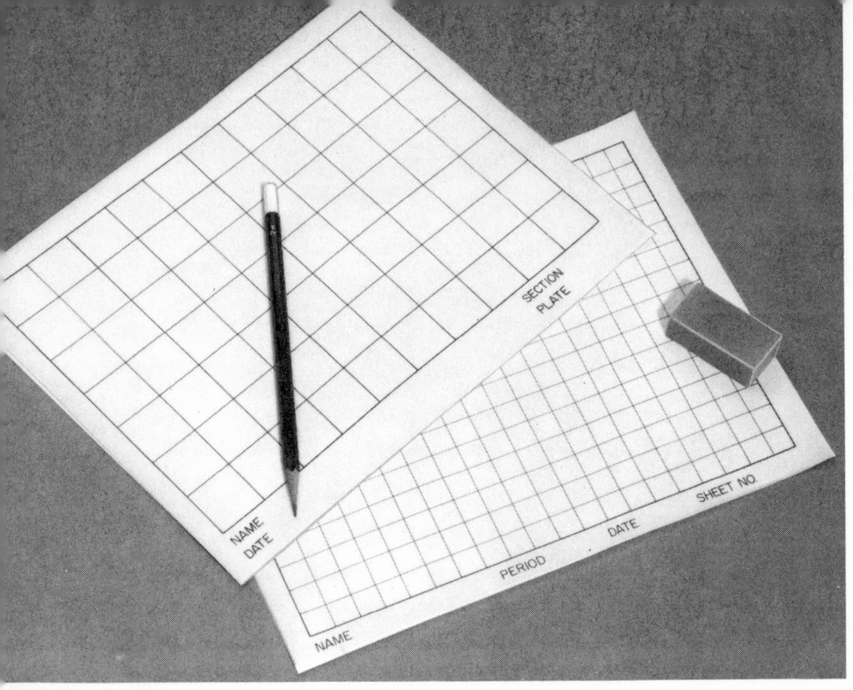

Fig. 2-2. Graph paper is useful in sketching. A good eraser is also a necessity.

Sketching does not require a great deal of equipment. Properly sharpened F, 2H or HB grade pencils and sheets of standard letter size paper (8 1/2 x 11 in.) are well suited for this purpose. See Fig. 2-2. The paper can be plain or cross sectioned (graph or squares). A good eraser is also needed.

In sketching, a line is drawn by making a series of short strokes, Fig. 2-3. It is recommended that light (thin) construction lines be drawn first--corrections are easier to make. Horizontal lines are drawn from left to right; vertical lines are drawn from the top down.

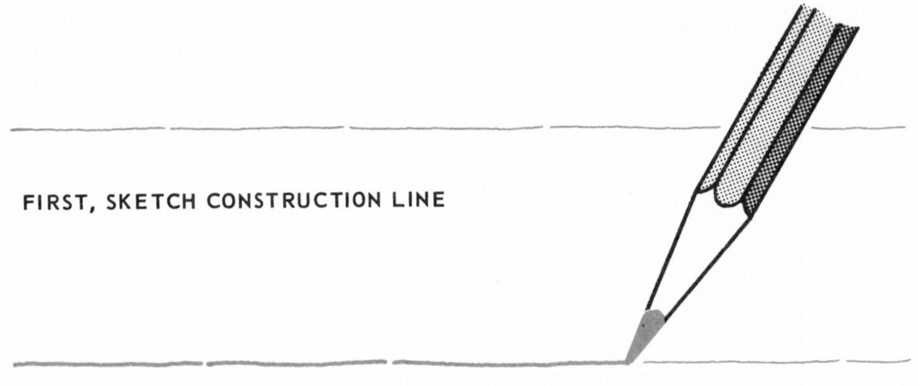

Fig. 2-3. When sketching, a line is drawn by making a series of short strokes.

The instructions that follow tell and show how to sketch basic geometric shapes. You will find that even the most complex drawings are made up of a combination of these basic geometric shapes.

ALPHABET OF LINES (SKETCHED)

A draftsman uses lines of various weights (thickness) to make a drawing. Each line has a special meaning, Fig. 2-4. Contrast between the various line weights or thicknesses helps to make a drawing easier to read. It is essential that you learn this ALPHABET OF LINES:

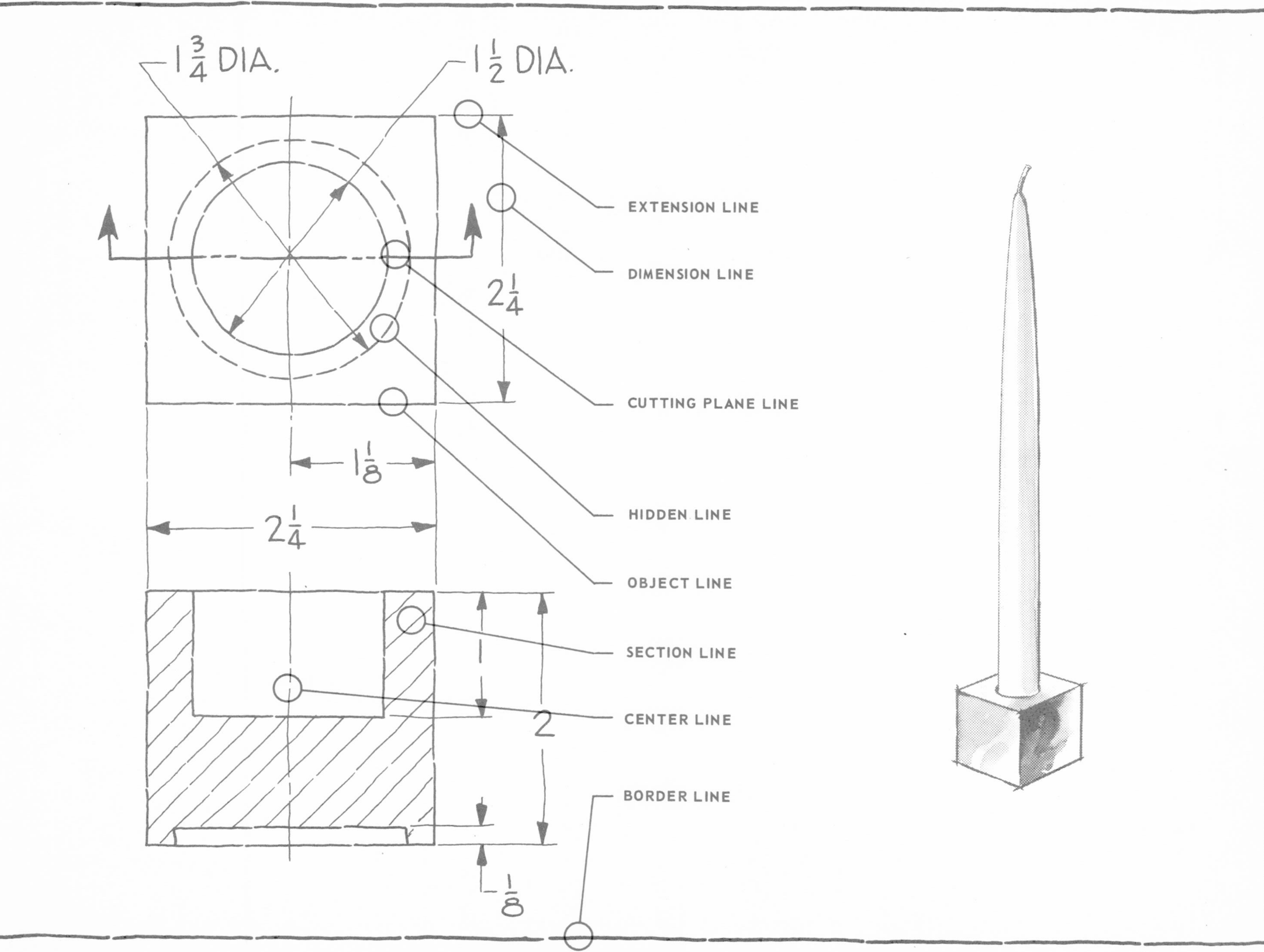

Fig. 2-4. ***ALPHABET OF LINES*** *as used in sketching a project.*

CONSTRUCTION AND GUIDE LINE

CONSTRUCTION LINES are used to lay out drawings. GUIDE LINES are used when lettering to help you keep the lettering uniform in height. These lines are drawn lightly using a pencil with the lead sharpened to a long conical point.

BORDER LINE

The BORDER LINE is the heaviest (thickest) line used in sketching. First draw light construction lines as a guide, then go over them using a pencil with a heavy rounded point to provide the border lines.

OBJECT LINE

The OBJECT LINE is a heavy line, but slightly less in thickness than the border line. The object line indicates visible edges. In sketching object lines, use a pencil with a medium lead and a rounded point.

HIDDEN LINE

HIDDEN LINES are used to indicate or show the hidden features of a part. The hidden line is made up of a series of dashes (1/8 in.) with spaces (1/16 in.) between the dashes.

EXTENSION LINE

EXTENSION LINES are the same weight as dimension lines. These lines indicate points from which the dimensions are given. The extension line begins 1/16 in. away from the view and extends 1/8 in. past the last dimension line.

2

DIMENSION LINE

DIMENSION LINES generally terminate in arrowheads at the ends. They are usually placed between two extension lines. A break is made, usually in the center, to place the dimension. The dimension line is placed from 1/4 in. to 1/2 in. away from the drawing. It is a fine line and is drawn using a pencil sharpened to a long conical point.

CENTER LINE

CENTER LINES are made up of alternate long (3/4 in. to 1 1/2 in.) and short (1/8 in.) dashes with 1/16 in. spaces between. These are drawn about the same weight as dimension and extension lines, and are used to locate centers of symmetrical objects.

CUTTING PLANE LINE

CUTTING PLANE LINES are used to indicate where an object has been cut to show interior features. A cutting plane line is made up by using a long dash (3/4 in. to 1 1/2 in.) followed by two short dashes (1/8 in.) with 1/16 in. spaces between. The cutting plane line is a heavy line that is drawn by using a pencil with a heavy rounded point.

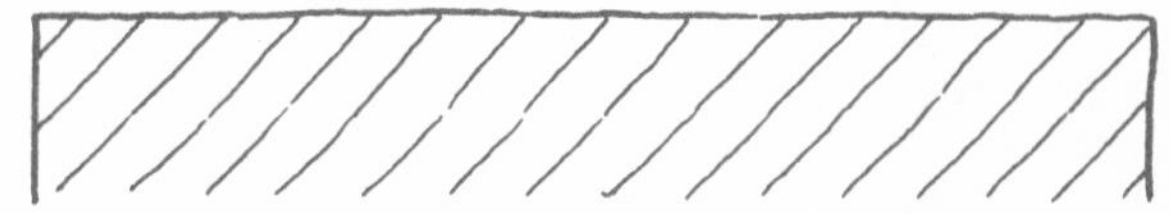

SECTION LINE

SECTION LINES are used when drawing the inside features of an object. They indicate the exposed surfaces cut by the cutting plane line. Section lines are also used to indicate general classifications of materials. These lines, light in weight, are also drawn using a pencil sharpened to a long conical point.

HOW TO SKETCH A HORIZONTAL LINE

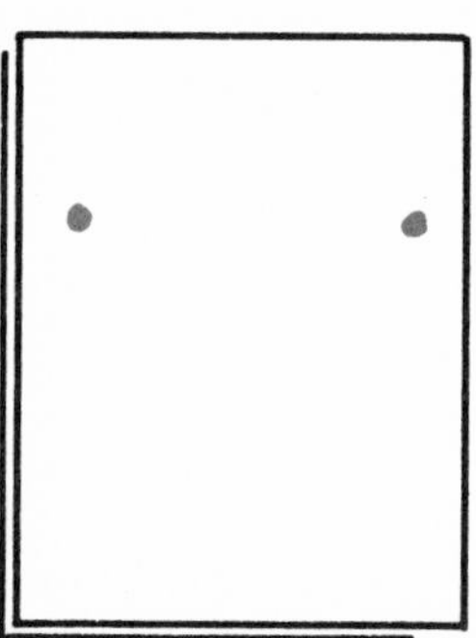

1. Mark off two points spaced a distance equal to the length of the line to be drawn. The points should be parallel to the top or bottom edge of your paper.

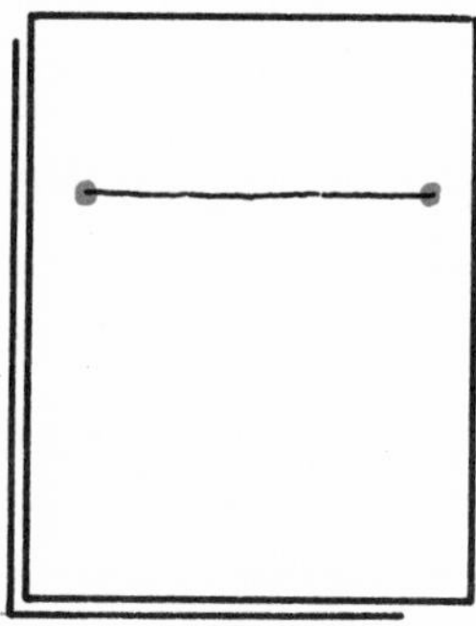

2. Move your pencil back and forth and connect these points with a construction line.

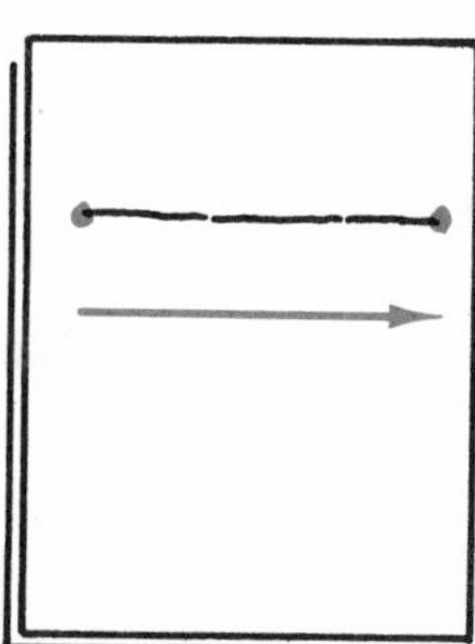

3. Start from the left point and sketch an object line to the right point. This line is sketched over the construction line.

HOW TO SKETCH A VERTICAL LINE

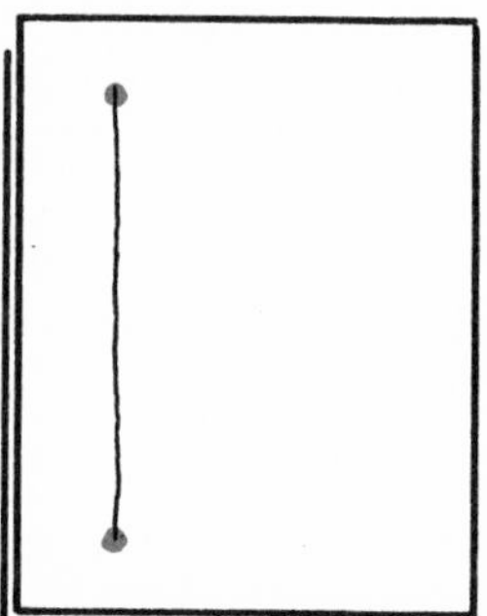

1. Mark off two points spaced a distance equal to the length of the line to be drawn. The two points should be parallel to the right or left edge of the sheet. Move your pencil back and forth and connect these points with a construction line.

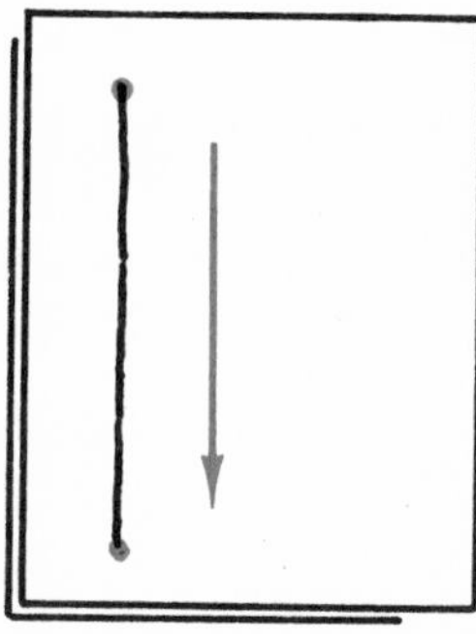

2. Start from the top point and sketch down and over the construction line to draw the desired line.

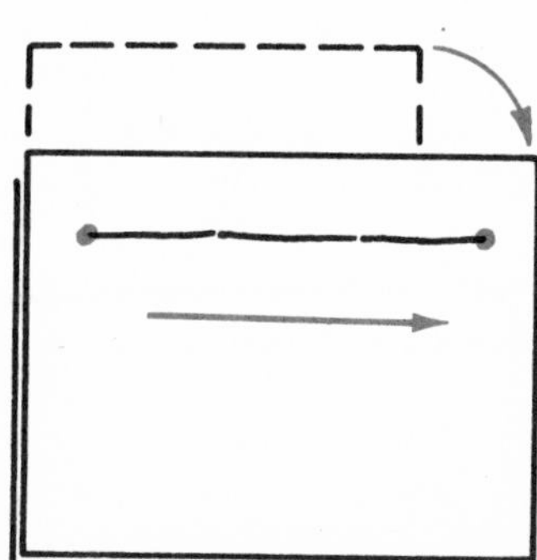

3. Vertical lines can also be sketched by rotating the paper into a horizontal position and proceeding as explained in HOW TO SKETCH A HORIZONTAL LINE.

HOW TO SKETCH AN INCLINED LINE

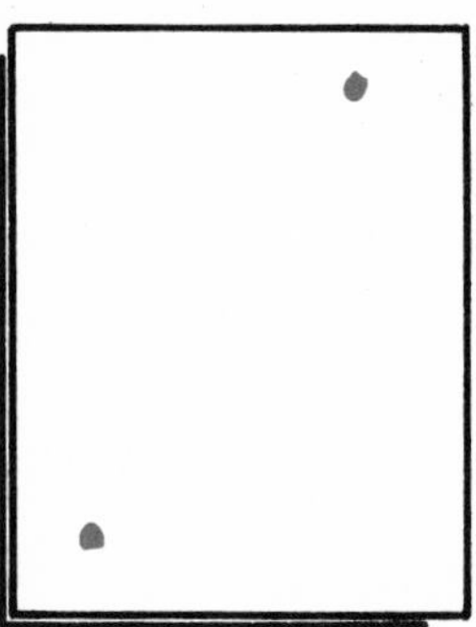

1. Mark off two points at the desired angle. Connect these points with a construction line.

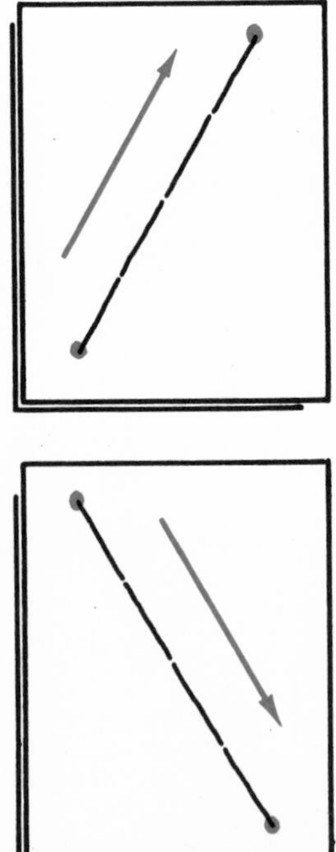

2. Sketch the desired weight line over the construction line. Sketch in the directions illustrated. Sketch up when the line inclines to the right. Sketch down when the line inclines to the left.

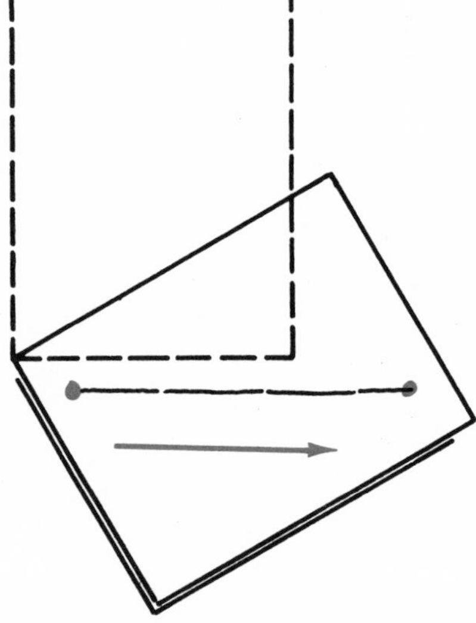

3. Inclined lines can also be sketched by rotating the sheet so the points are in a horizontal position. Sketch the line as previously described.

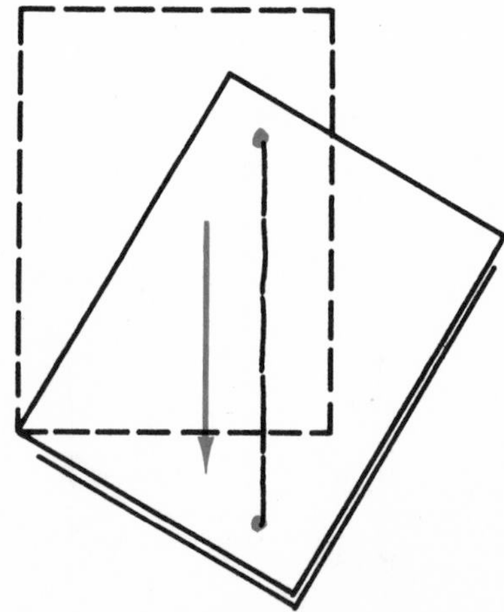

4. For some sketching problems, it may be easier to rotate the paper so the points are in a vertical position. Proceed as explained in HOW TO SKETCH A VERTICAL LINE.

HOW TO SKETCH SQUARES AND RECTANGLES

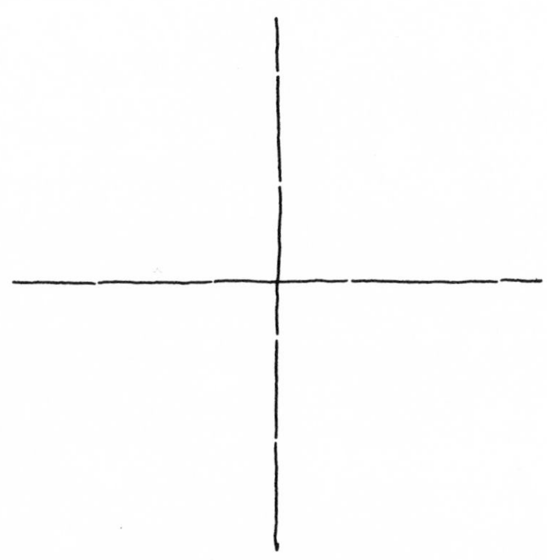

1. Sketch a horizontal line and a vertical line (axes).

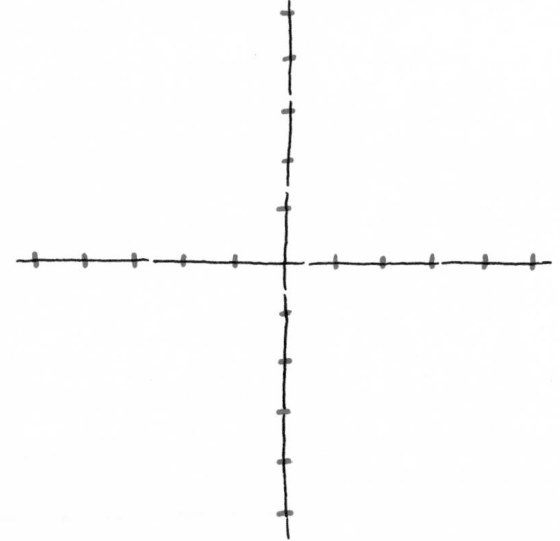

2. Begin at the intersection of these lines and lay out equal units on both lines in each direction. For example: If you want to draw a 2 in. square, you would estimate a unit of 1/4 in. and mark off four of these units on the vertical axis above and below the horizontal axis. Lay out the horizontal axis in the same manner.

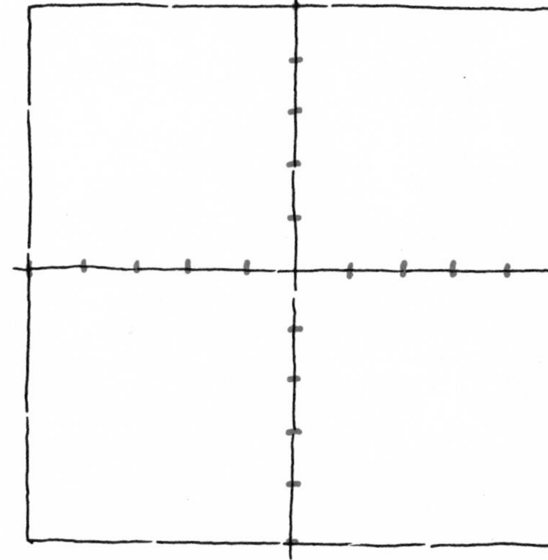

3. Sketch construction lines through the desired points.

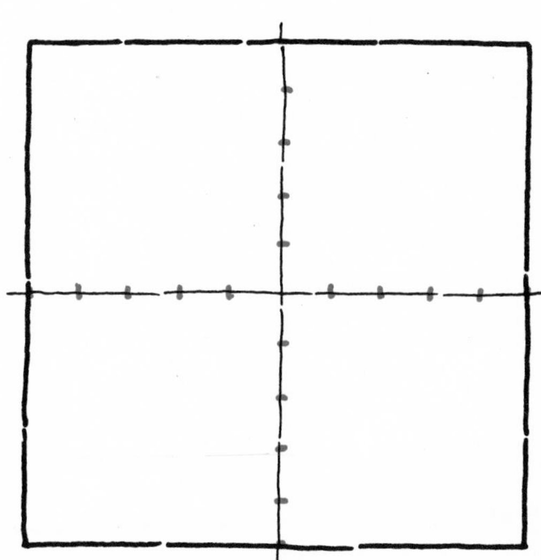

4. Go over the construction lines forming the square to produce the desired weight line.

5. Rectangles are sketched in the same way except that you will have more units on one axis (line) than the other axis (line).

HOW TO SKETCH ANGLES

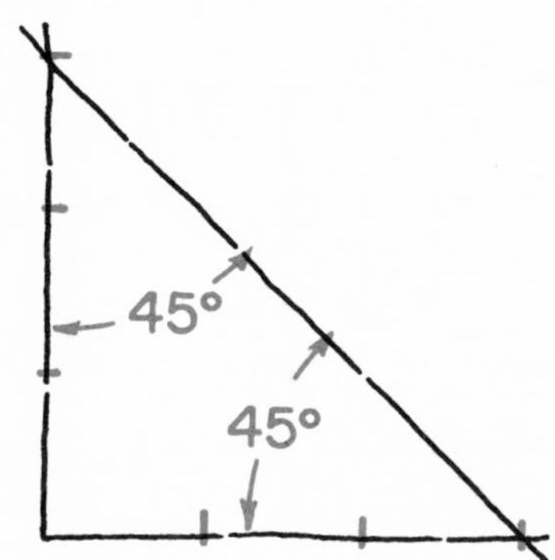

1. Sketch vertical and horizontal construction lines. These lines will form a 90 deg. or RIGHT ANGLE.

2. A 45 deg. angle is sketched by marking off an equal number of units on both lines. Connect the last unit of each line. This will form a 45 deg. angle with the vertical and the horizontal lines.

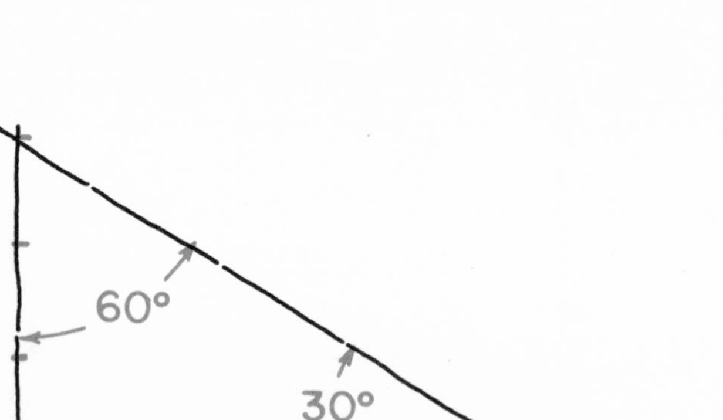

3. To sketch 30 and 60 deg. angles, mark off three units on one line and five units on the other line. Connecting the last unit on each line will give the required angles.

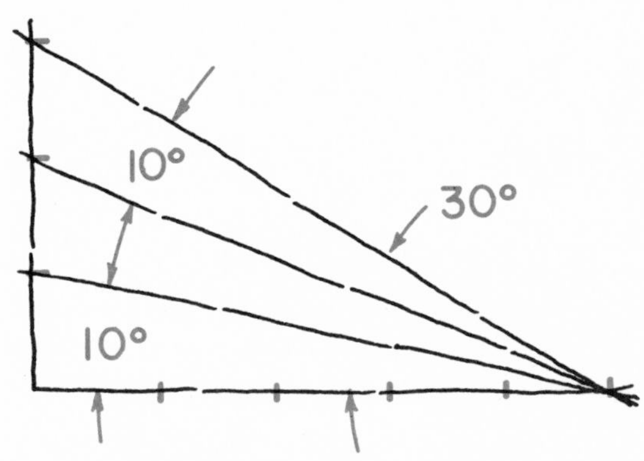

4. Other angles may be drawn by sketching an angle and subdividing this into the approximate number of degrees required. Example: Dividing a 30 deg. angle into thirds will give a 10 deg. angle.

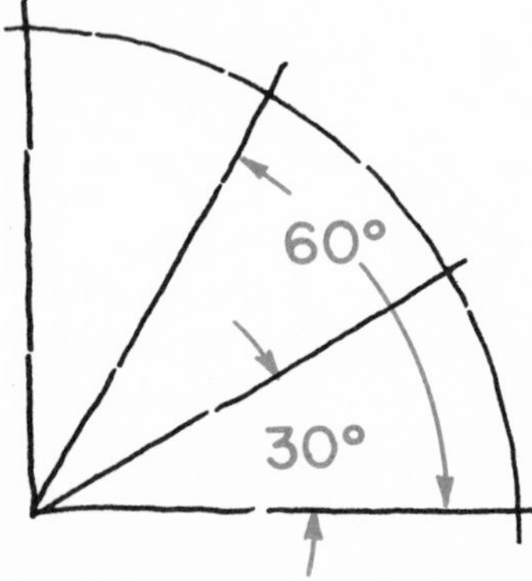

5. Another method used to develop angles in sketching, is to sketch a quarter circle and divide the resulting arc into the desired divisions. Example: Dividing the arc into three parts will give 30 and 60 deg. angles.

HOW TO SKETCH CIRCLES

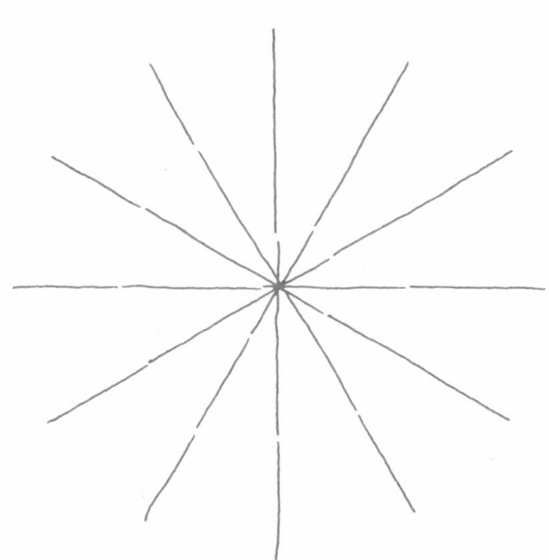

1. Sketch vertical, horizontal and inclined axes.

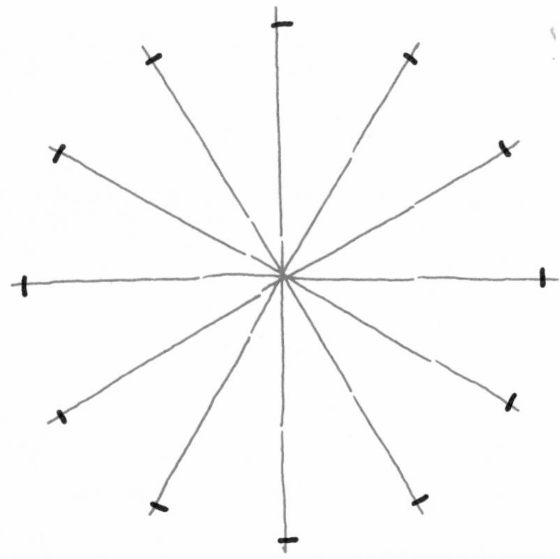

2. Mark off units equal to the radius of the required circle on each axis.

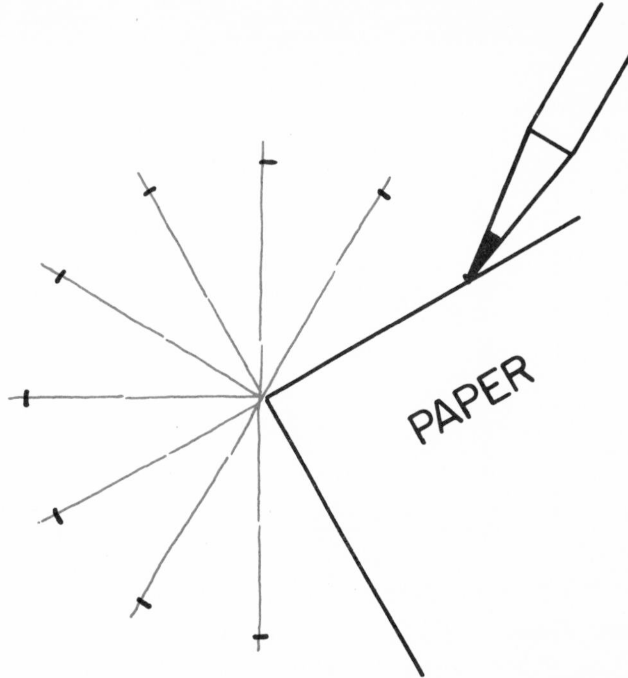

3. The radius units can be quickly and accurately located by marking off the desired radius on a piece of paper and using the paper as a measuring tool.

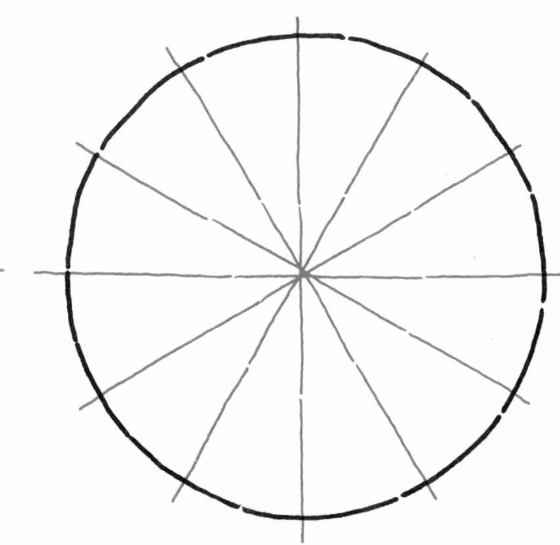

4. Sketch a construction line through the points. When satisfied with the construction line, fill it in with a line of the desired weight.

HOW TO SKETCH AN ARC

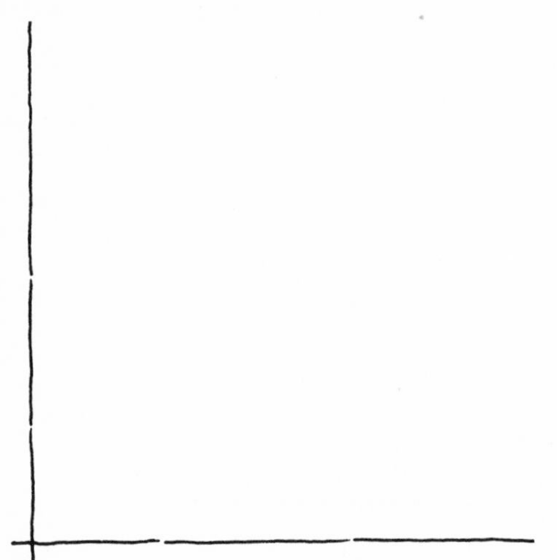

1. Sketch a right (90 deg.) angle. Use construction lines.

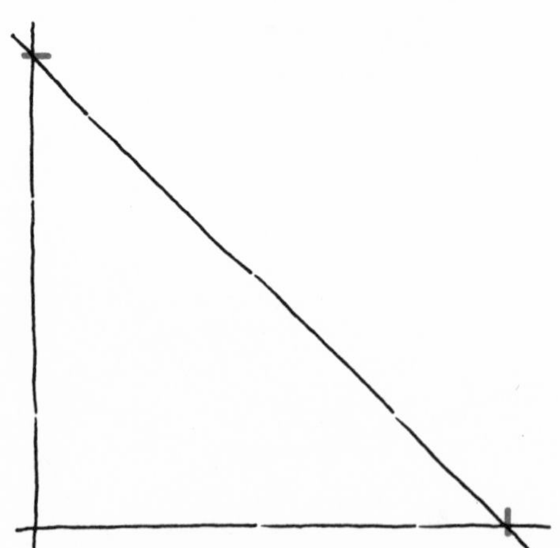

2. Units equal to the length of the desired radius are marked on each leg of the angle. Connect these points with a construction line.

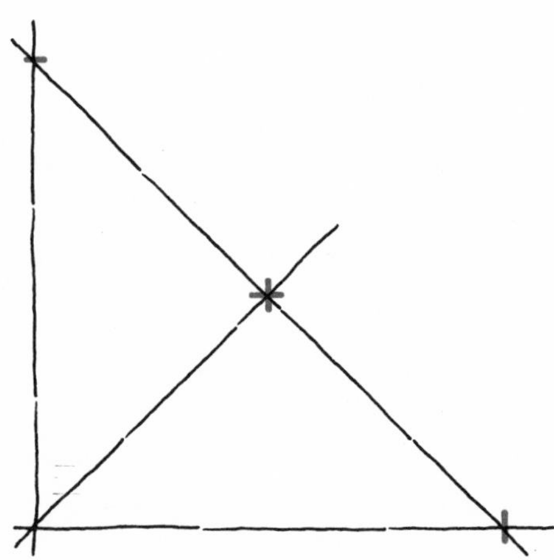

3. Divide this line into two equal parts. Starting from the point where the legs of the angle intersect, sketch a line through the dividing point of the diagonal line.

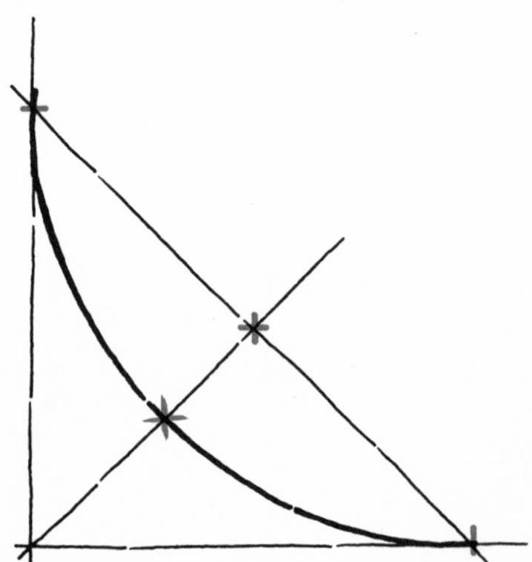

4. Mark off a point halfway between the diagonal line and the intersection of the legs of the angle. Sketch an arc through the three points as shown.

HOW TO SKETCH AN ELLIPSE

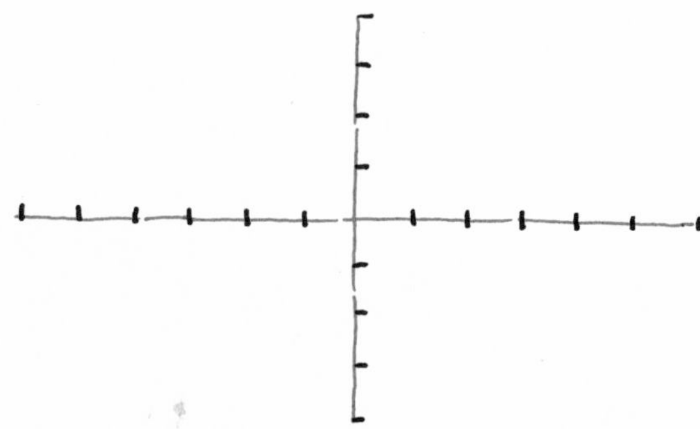

1. Sketch horizontal and vertical lines as shown. Mark off equal size units on the center lines to construct a rectangle with the dimensions equal to the major axis (the long axis) and the minor axis (the small axis) of the desired ellipse.

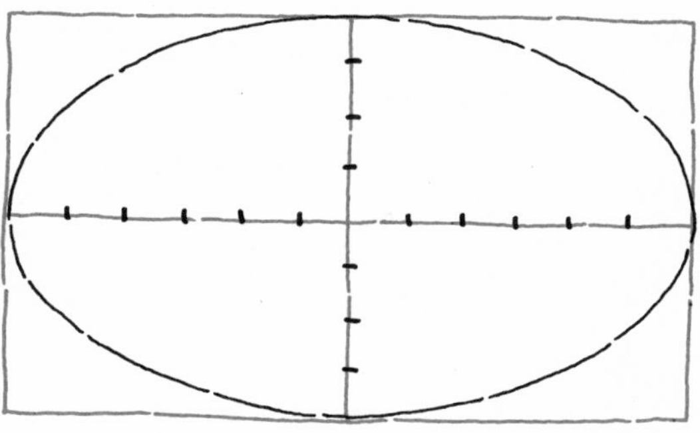

3. Lightly sketch arcs tangent to the lines that form the rectangle.

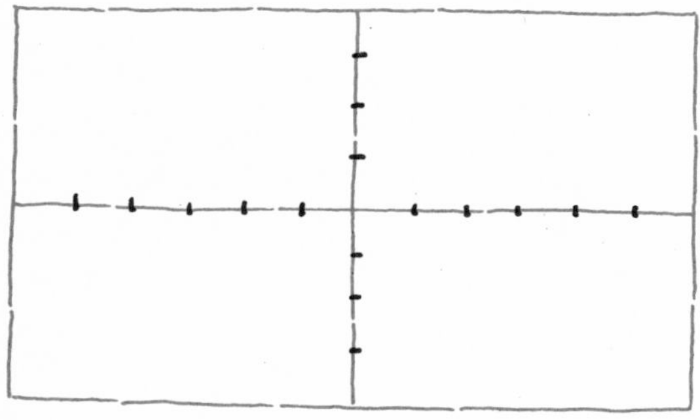

2. Construct the rectangle by sketching construction lines through the outer points.

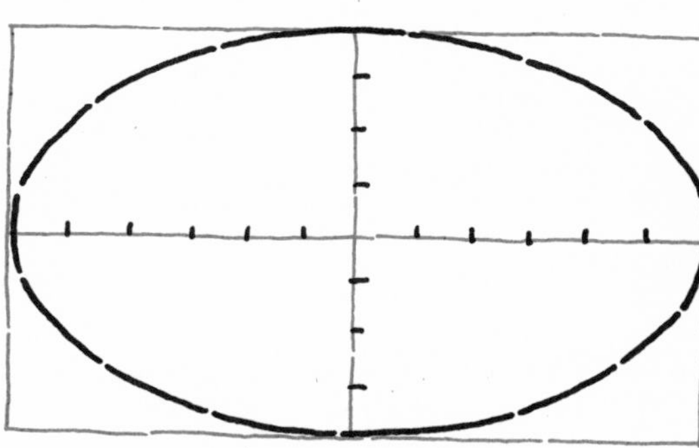

4. When you are satisfied with the shape of the ellipse, complete it by going over the construction lines with lines of the desired weight.

HOW TO SKETCH A HEXAGON

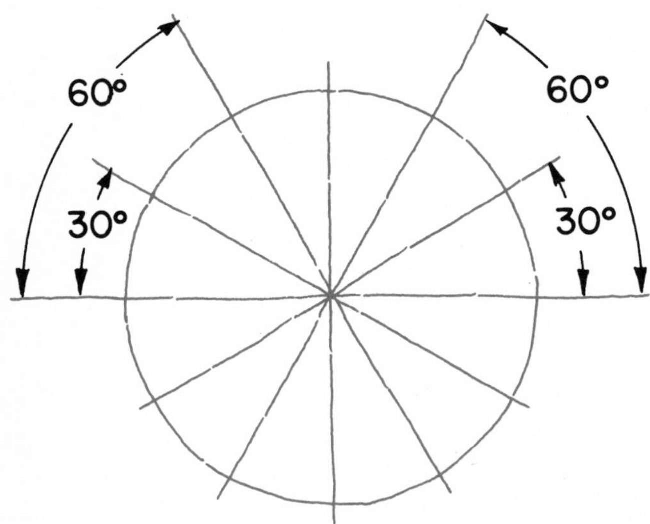

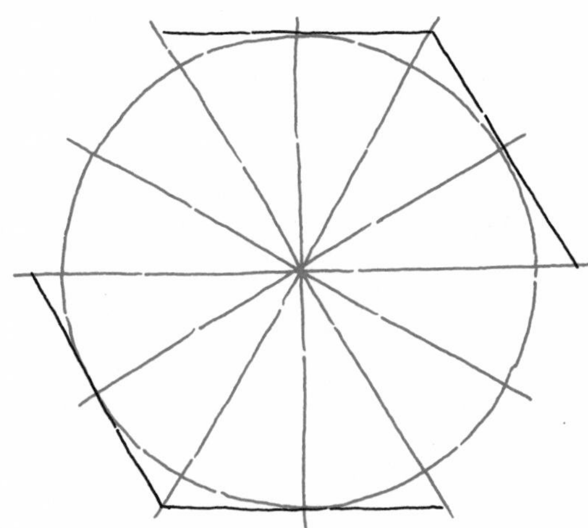

1. Sketch vertical and horizontal center lines, and inclined lines at 30 and 60 deg. Construct a circle with a diameter equal to the distance across the flats of the required hexagon. Use construction lines.

3. Sketch inclined parallel lines at 60 deg. and tangent to the circle at the point where the 30 deg. inclined line intersects the circle.

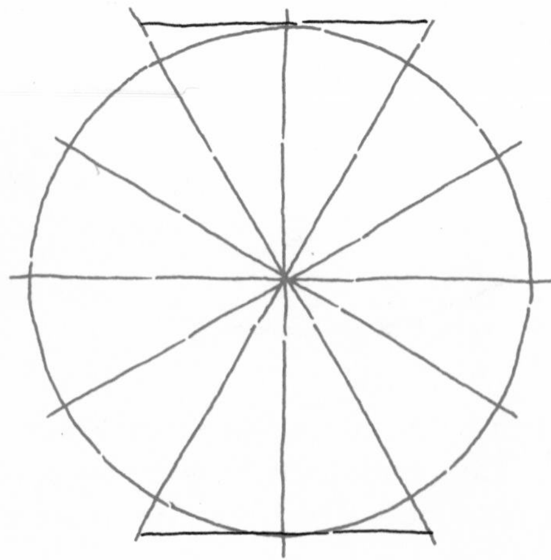

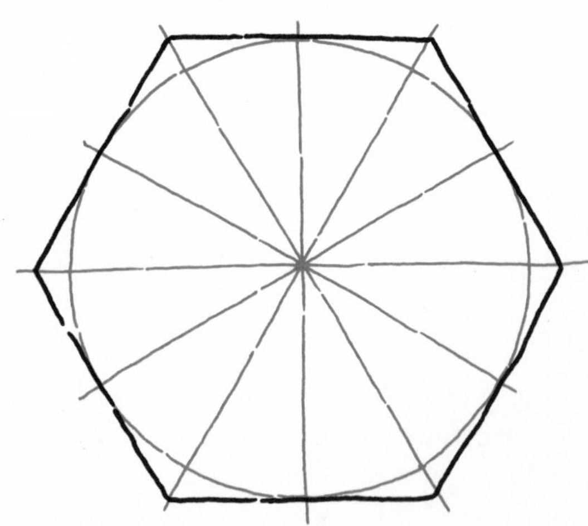

2. Sketch horizontal parallel lines at right angles (90 deg.) to the vertical center line. The lines are tangent to the circle at these points.

4. Complete the hexagon and go over the construction lines to produce the proper weight line.

HOW TO SKETCH AN OCTAGON

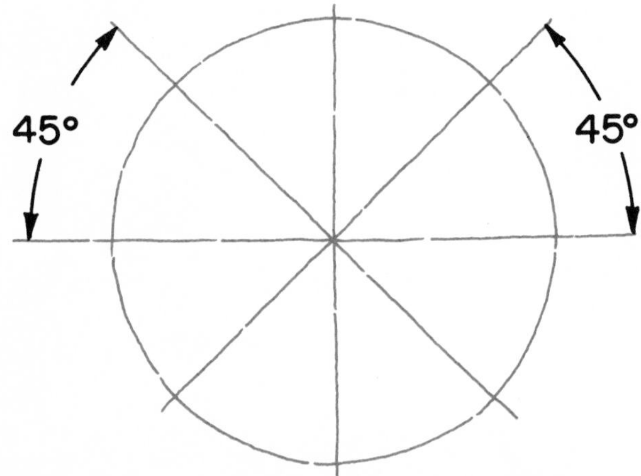

1. Sketch vertical and horizontal center lines and inclined lines at 45 deg. Construct a circle with a diameter equal to the distance across the flats of the required octagon. Use construction lines.

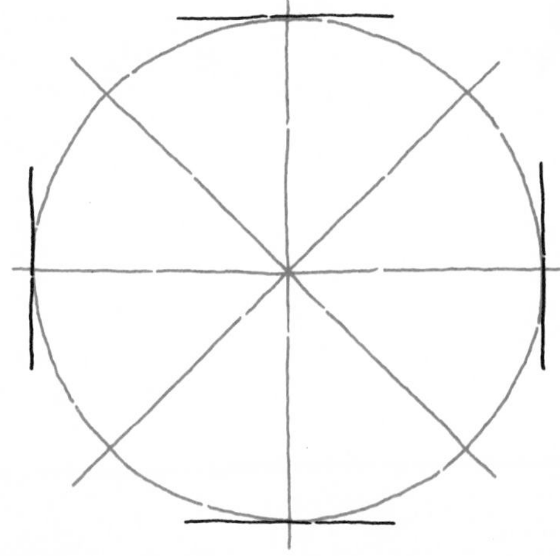

2. Sketch parallel lines tangent to the circle where the horizontal and vertical center lines intersect the circle.

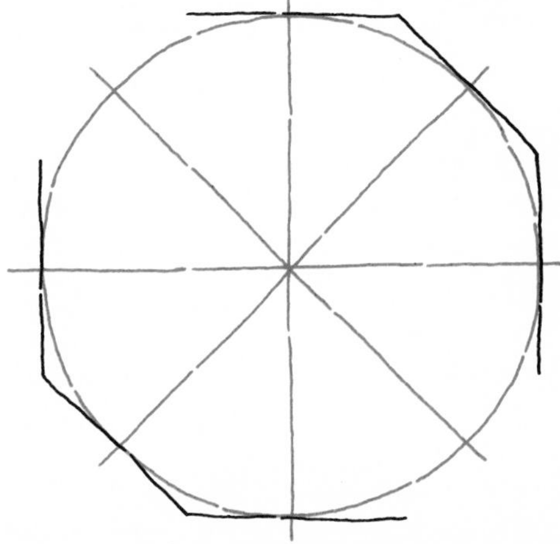

3. Sketch inclined parallel lines at 45 deg. and tangent to the circle at the point where the 45 deg. inclined lines intersect the circle.

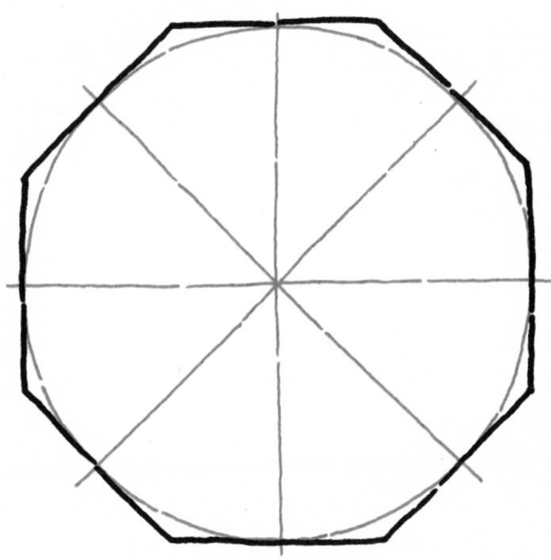

4. Complete the octagon and go over the construction lines to produce the desired weight line.

SHEET LAYOUT FOR SKETCHING

1. Sketch a 1/2 in. border around the edges of the paper. Use a construction line. The sheet should be 8 1/2 in. by 11 in. It may be plain or graph paper. Sketch in guide lines as shown in Fig. 2-5.

2. The edge of your drawing board or desk may be used as a guide in sketching the border and guide lines, Fig. 2-6. Place the pencil in a fixed position and move your fingers along the edge of the drawing board or desk.

3. Sketch a border line over the construction lines, letter in information as shown, Fig. 2-7, or as specified by your instructor.

4. Take your time and sketch in the border and information carefully and neatly.

Fig. 2-6. The edge of your drawing board can be used as a guide when sketching border lines. Note how the pencil is held.

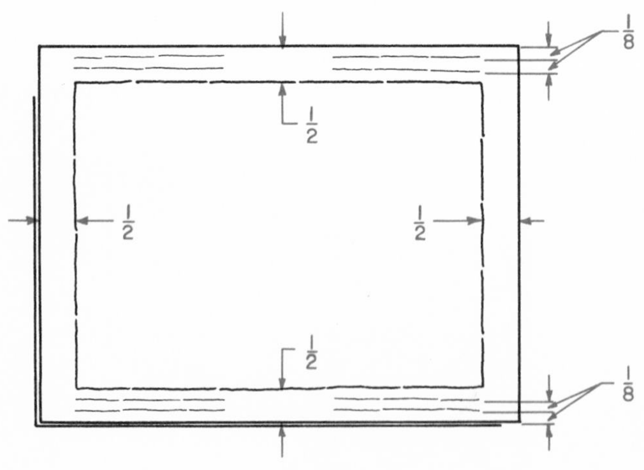

Fig. 2-5. Sketching in border and lettering guide lines.

Fig. 2-7. Lettering in the required information.

ENLARGING OR REDUCING BY THE GRAPH METHOD

Drawings can be reduced or enlarged easily and rapidly by using this technique. The original drawing is blocked off into squares, Fig. 2-8. After deciding how much larger or smaller the new drawing is to be made, draw squares of the new size on a blank sheet of paper. Using the design in the small squares as a guide, sketch the design into the larger squares.

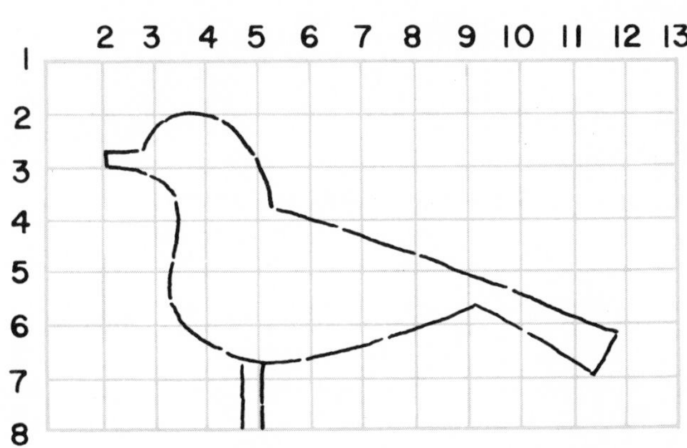

Fig. 2-8. Drawing to be enlarged has been blocked in.

EXAMPLE: A drawing of the design is to be enlarged to twice its original size. The original drawing is marked off into 1/4 in. squares (square size will vary depending on the size of the design). Number the squares starting at the upper left side and across the top. Another sheet is made up with squares that are twice the size of the 1/4 in. squares, or 1/2 in. squares. Number the large squares in the same manner as the small squares. Mark on the larger squares the points where the drawing crosses the squares. Then, sketch in the details freehand, Fig. 2-9.

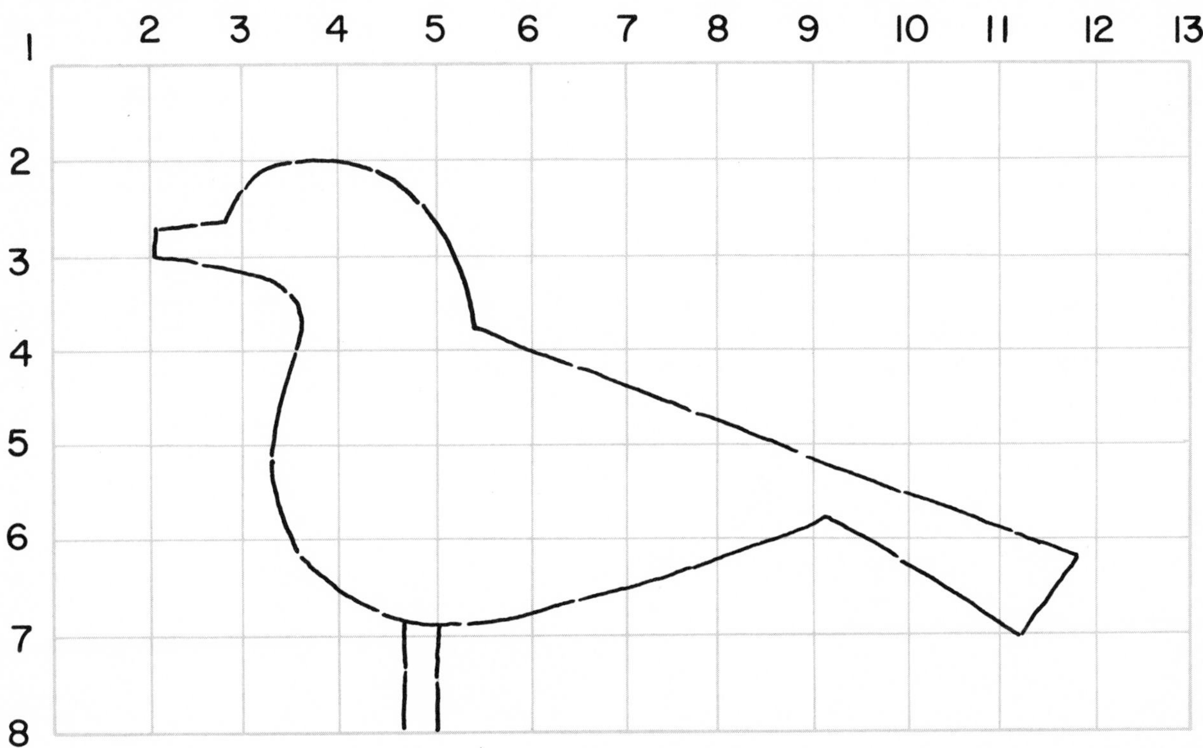

Fig. 2-9. Enlarged drawing.

TEST YOUR KNOWLEDGE - UNIT 2

1. In sketching, a line is drawn by making a series of ____________ ____________.
2. The heaviest line used in sketching is the ________________ line.
3. Drawings can be ______________ or ______________ easily by using the graph method.
4. When sketching an inclined line, sketch up when line inclines to ____________ and down when line inclines to ______________.
5. Extension lines are the same weight as __________________ lines.
6. In sketching, horizontal lines are drawn from ____________ ____________ ____________; vertical lines from the ______________ ______________.
7. Dimension lines generally terminate in __________________ at the ends.

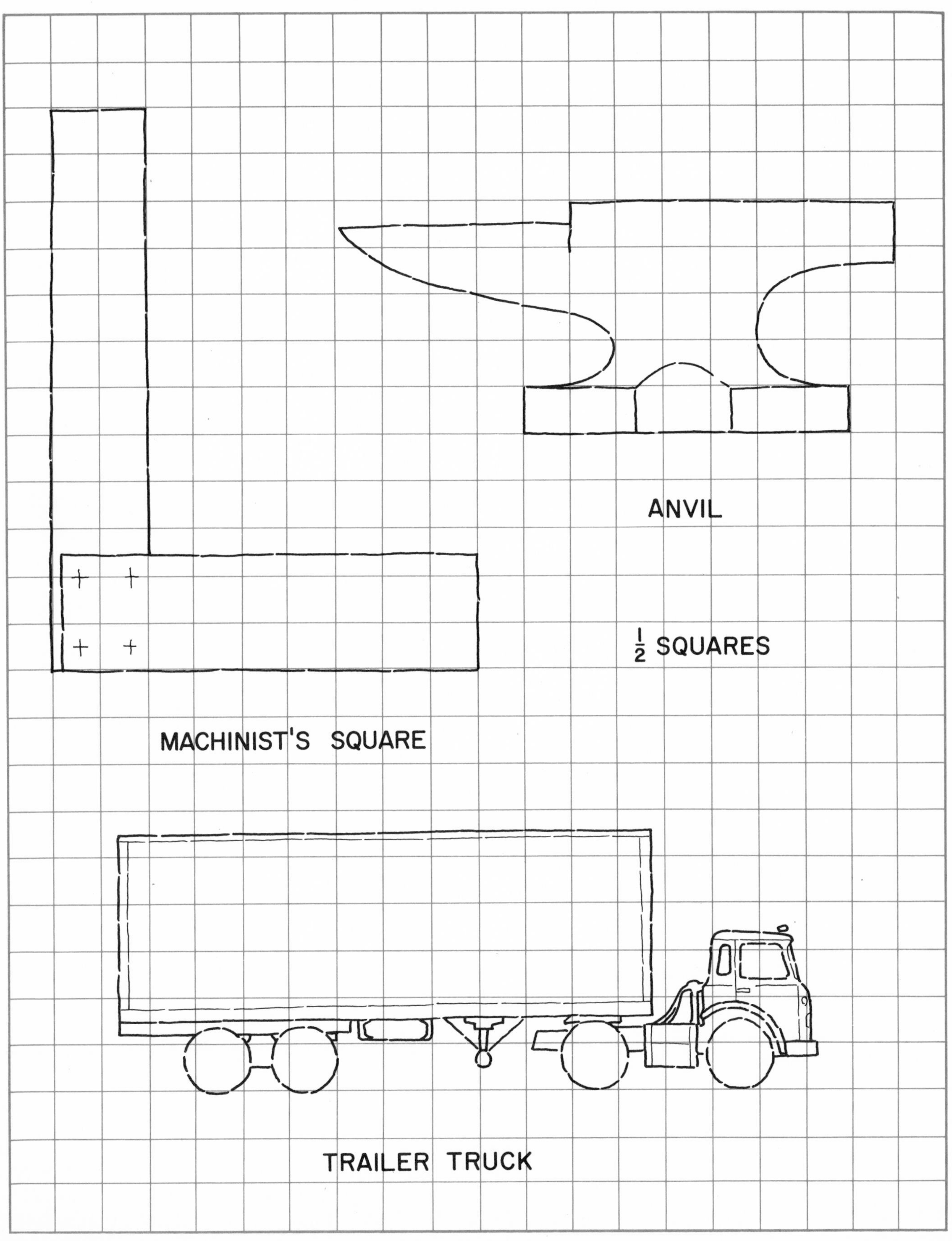
ANVIL
½ SQUARES
MACHINIST'S SQUARE
TRAILER TRUCK

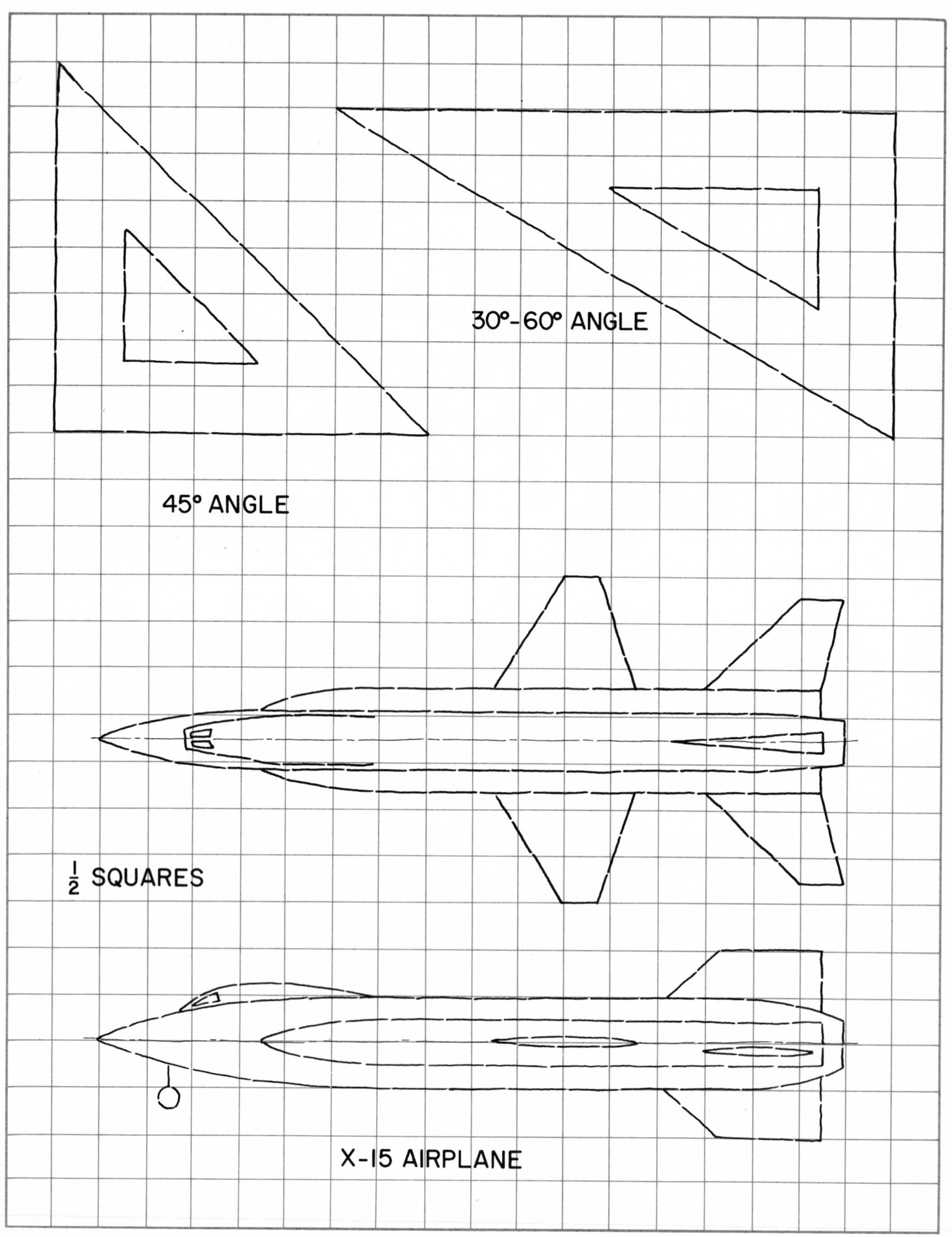
30°-60° ANGLE
45° ANGLE
½ SQUARES
X-15 AIRPLANE

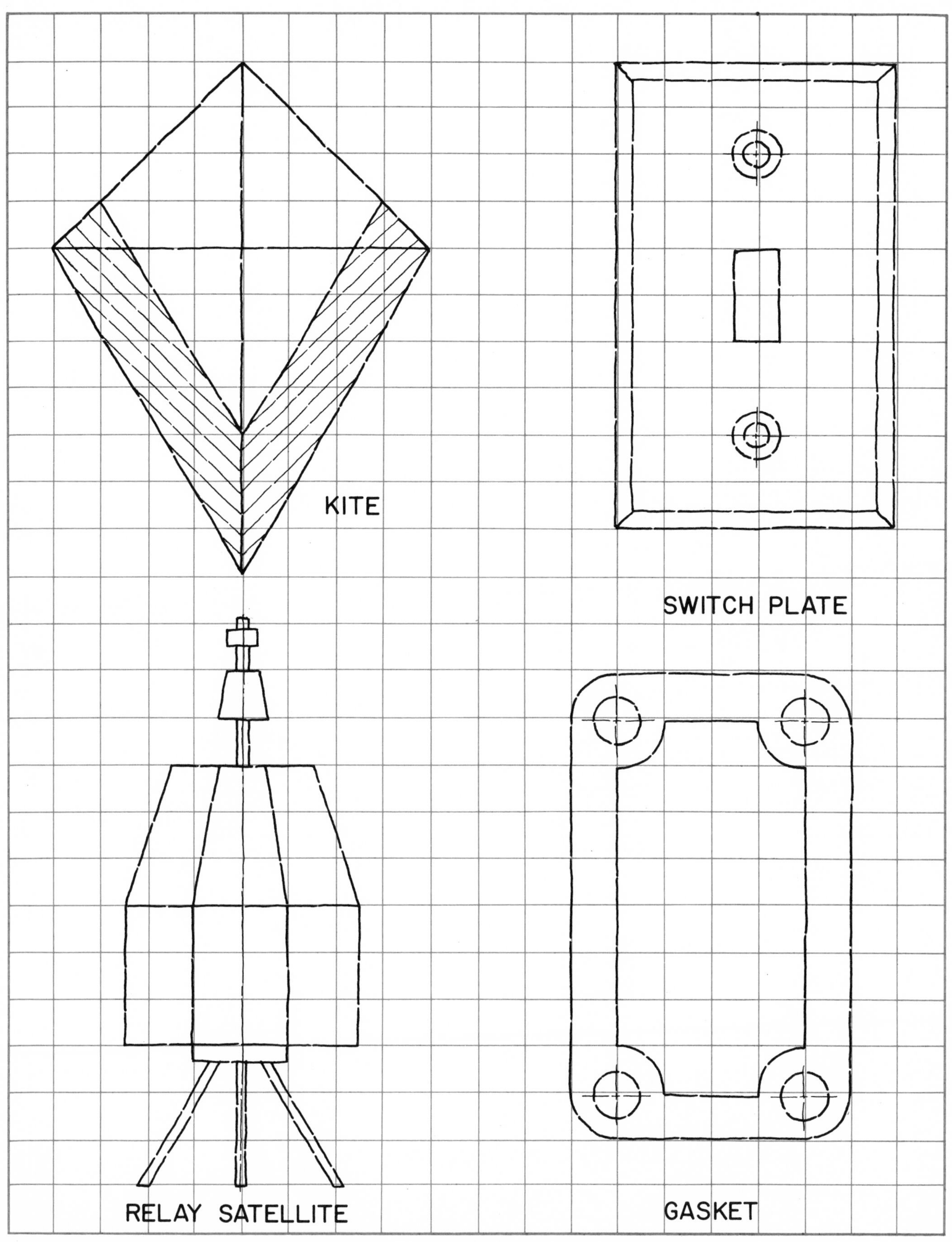
KITE
SWITCH PLATE
RELAY SATELLITE
GASKET

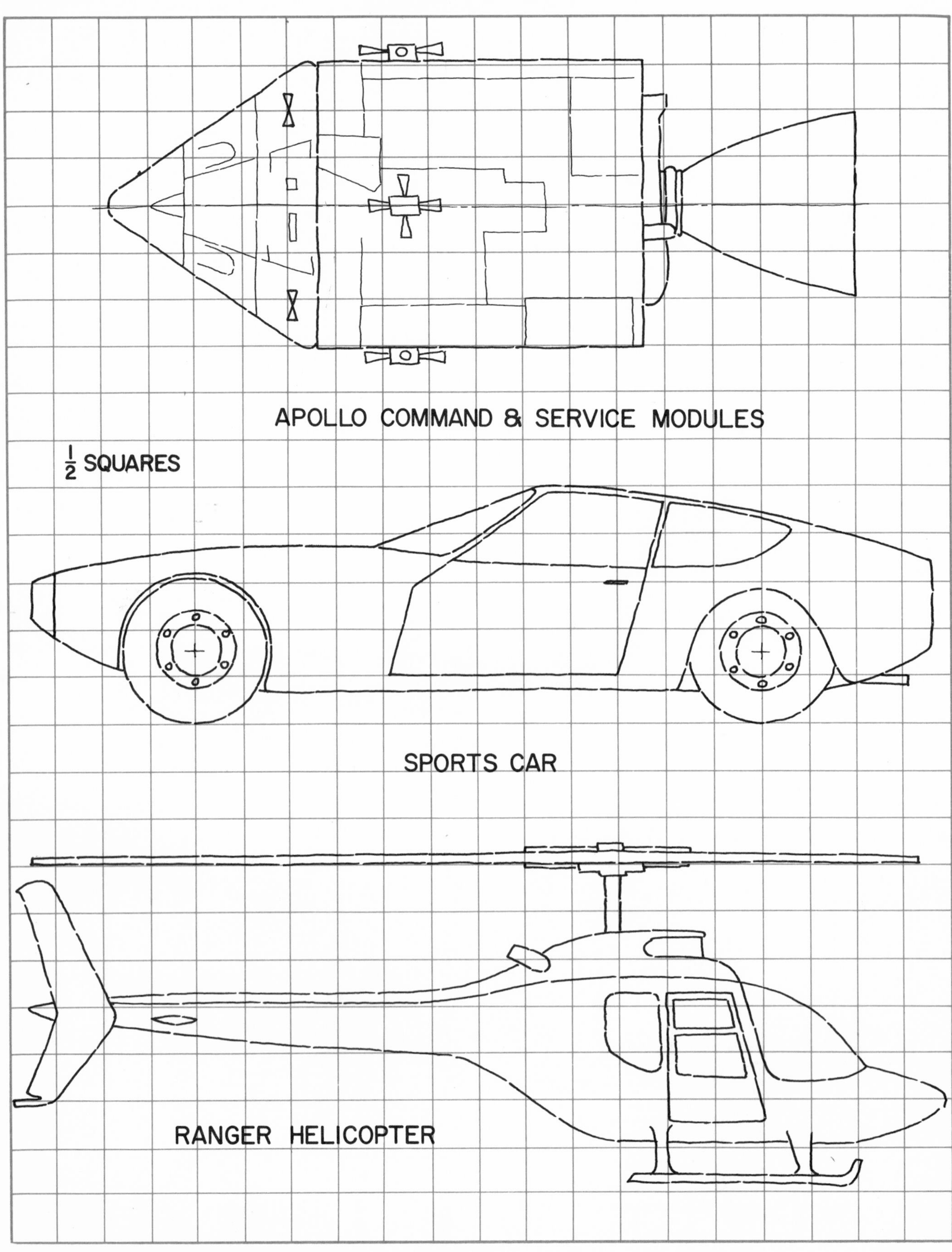
APOLLO COMMAND & SERVICE MODULES
$\frac{1}{2}$ SQUARES
SPORTS CAR
RANGER HELICOPTER

Unit 3
DRAFTING TOOLS

DRAWING BOARD

The DRAWING BOARD, Fig. 3-1, provides the smooth, flat surface needed for drafting. The tops of many drafting tables are designed for this purpose. Individual drawing boards are manufactured in a variety of sizes. The majority of them are made from selected, seasoned basswood.

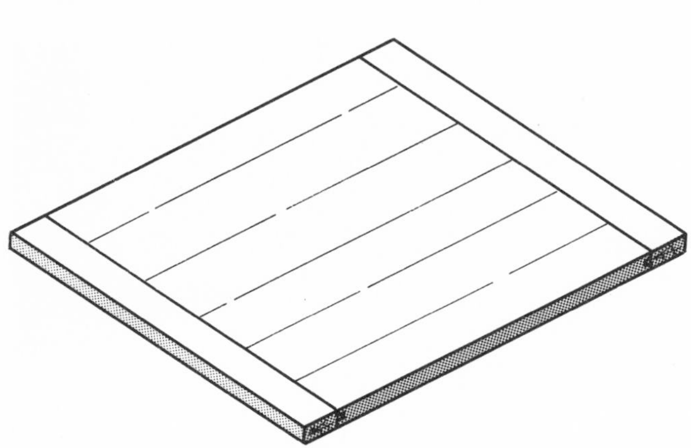

Fig. 3-1. Drawing board.

The right-handed draftsman will use the left edge of the board as the working edge; the left handed draftsman the right edge. The working edge should be checked periodically for straightness.

Draftsmen often tape a piece of heavy paper or a special vinyl board cover to the working face of the drawing board to protect its surface. The vinyl surface is easily cleaned.

T-SQUARE

Horizontal lines are drawn with the T-SQUARE, Fig. 3-2. It also supports triangles when they are used to draw vertical and inclined lines.

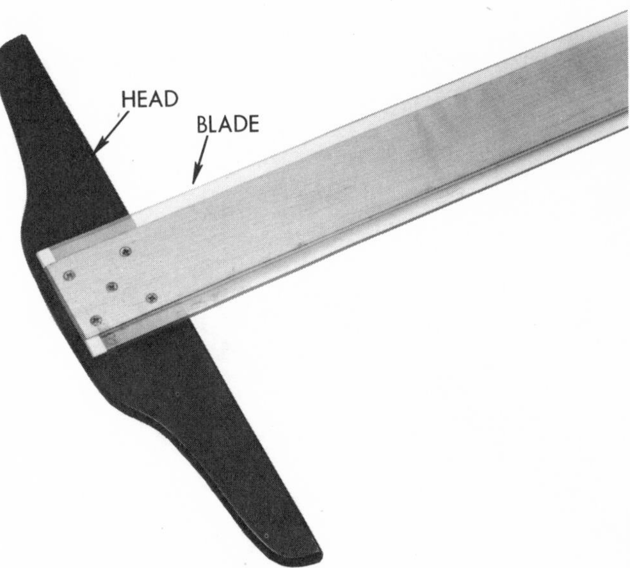

Fig. 3-2. T-square.

The T-square consists of two parts, the head and the blade or straight edge. The head is usually fixed solidly to the blade; however, a T-square with a protractor head and adjustable blade is also available.

Clear plastic strips inserted in the blade edge of some T-squares makes it easier to locate reference points and lines. The

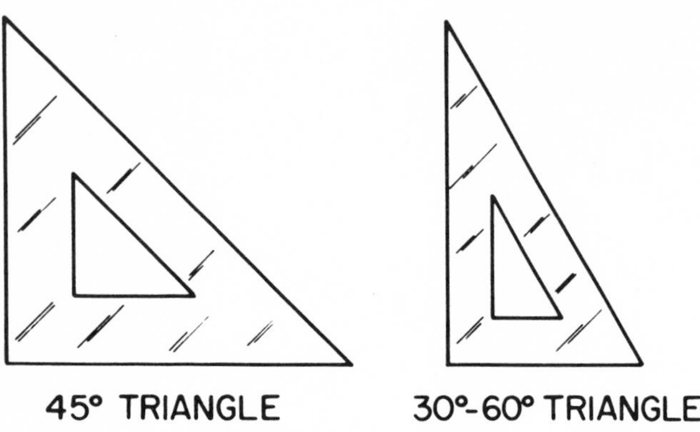

Fig. 3-3. Triangles.

blade must never be used as a guide for a knife or other cutting tool.

If accurate line work is to be done, it is essential that the head of the T-square be held firmly against the working edge of the board.

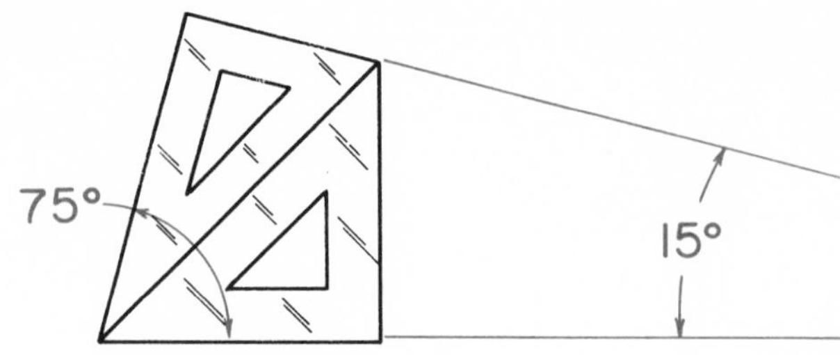

Fig. 3-4. Drawing 75 deg. and 15 deg. angles.

It is recommended that the blade be left flat on the board or suspended from the hole in its end. This will keep warping or bowing of the blade to a minimum.

TRIANGLES

When supported on the T-square blade, the 30-60 deg. and 45 deg. TRIANGLES, Fig. 3-3, are used to draw vertical and in-

Fig. 3-5. Drafting machine. (Post)

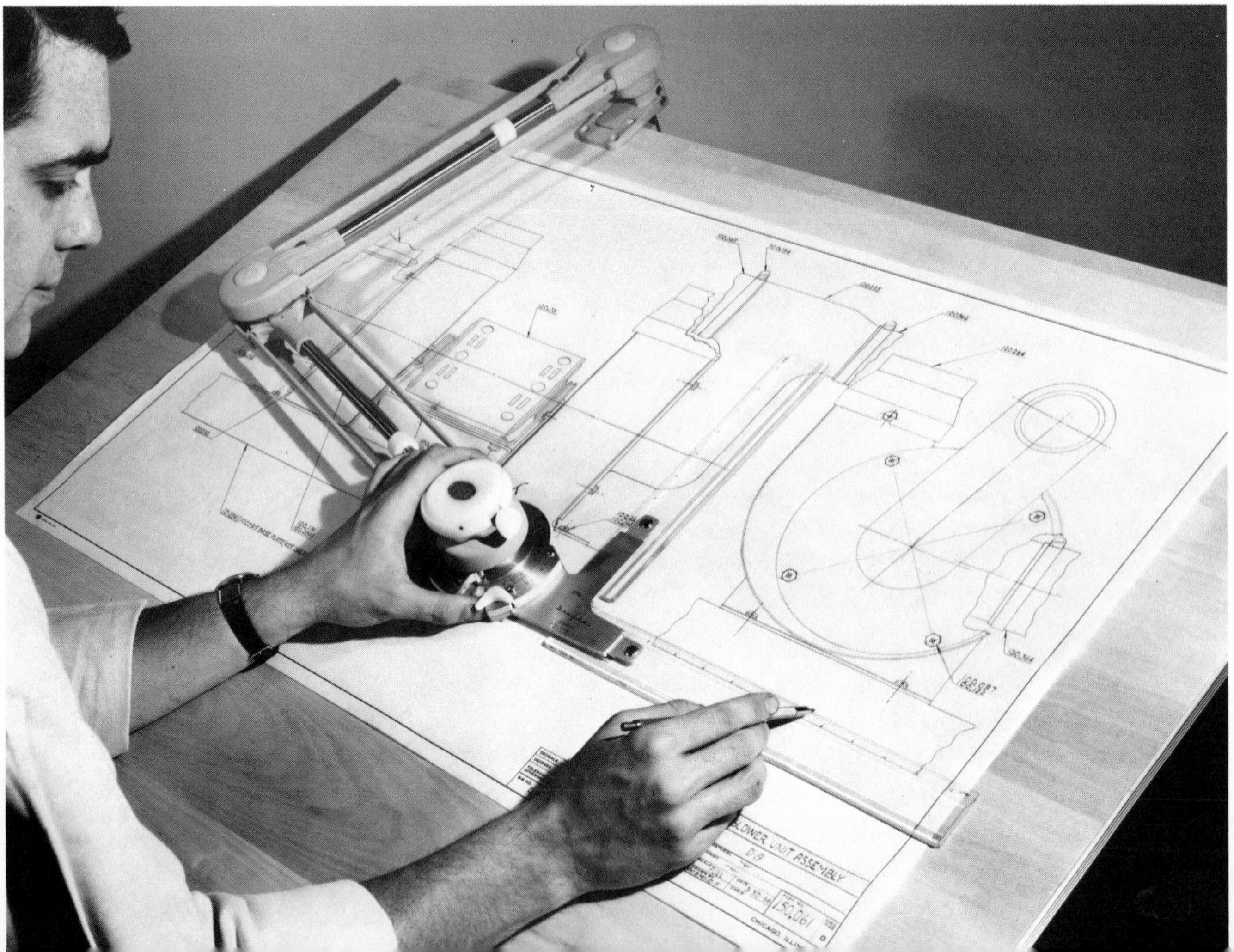

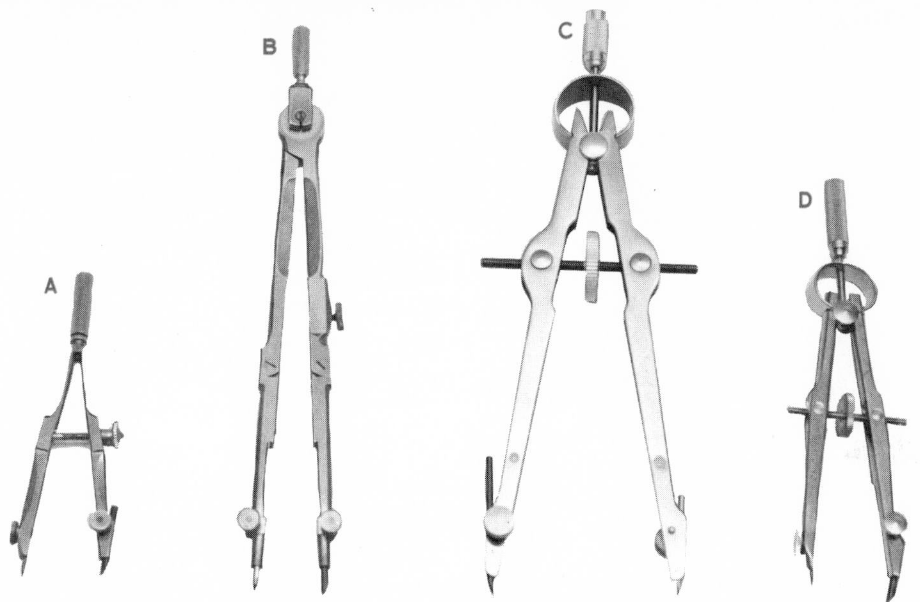

Fig. 3-6. Compasses. A–Spring bow. B–Friction type compass. C–Big bow compass. D–Small bow compass.

clined lines. They are made of transparent plastic and are available in a number of different sizes.

To prevent warping, the triangle should be left flat on the drawing board when not being used.

Angles of 15 and 75 deg. can be drawn by combining the triangles as shown in Fig. 3-4.

To draw vertical lines accurately, rest the triangle solidly on the T-square blade while holding the T-square head firmly against the working edge of the drawing board.

DRAFTING MACHINES

Industry makes considerable use of DRAFTING MACHINES, Fig. 3-5. This device replaces both the T-square and triangles. The straightedges can be adjusted to any angle. Drafting machines are gradually replacing T-squares and triangles in school drafting rooms.

COMPASS

In drafting, circles and arcs are drawn with a COMPASS, Fig. 3-6. The tool is held as shown in Fig. 3-7. For best re-

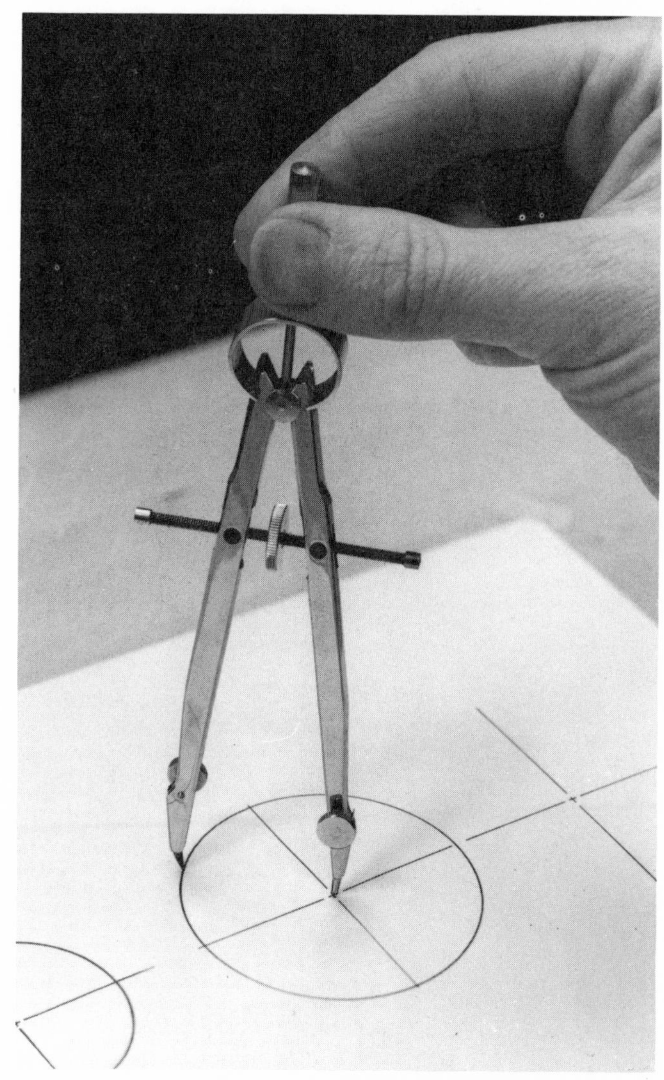

Fig. 3-7. The correct way to hold the compass when drawing a circle.

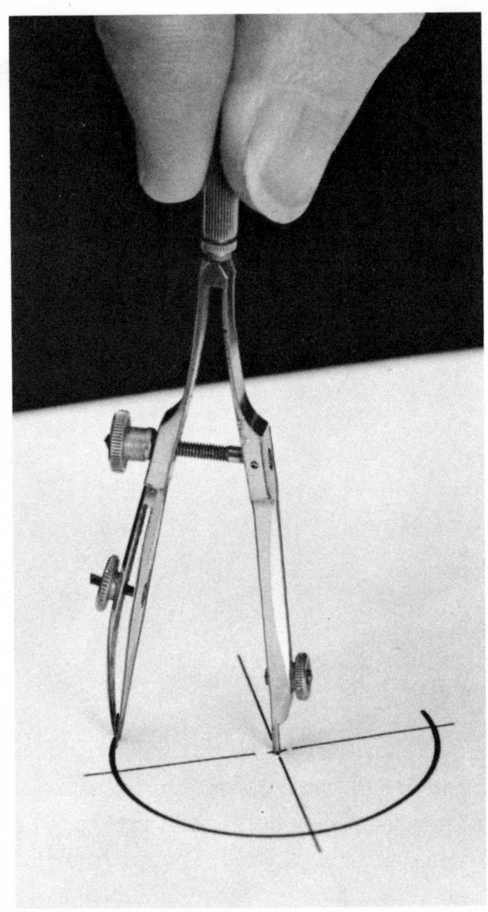

Fig. 3-8. Inking a circle.

sults the lead should be adjusted so that it is about 1/32 in. shorter than the needle. Both legs will be the same length when the needle penetrates the paper. Fit the compass with lead that is one grade softer than the pencil used to make the drawing. The lead must be kept sharp.

Several attachments are available for use with the compass. A pen is substituted when inking is to be done, Fig. 3-8. An extension makes it possible to draw a large circle, Fig. 3-9.

To set the compass to size, draw a line on a piece of clean scrap paper and measure off the required radius. Set the compass on this line. Avoid setting a compass on a scale. "Sticking" the compass needle into the scale will eventually destroy its accuracy.

DIVIDERS

Distances are subdivided and measurements are transferred with DIVIDERS, Fig. 3-10. Careful adjustment of the divider points is necessary.

Be careful where you place the dividers or compass after use. It is very painful to accidentally run the point into your hand.

PENCIL POINTER

It is not necessary to resharpen your drawing pencil every time it starts to dull.

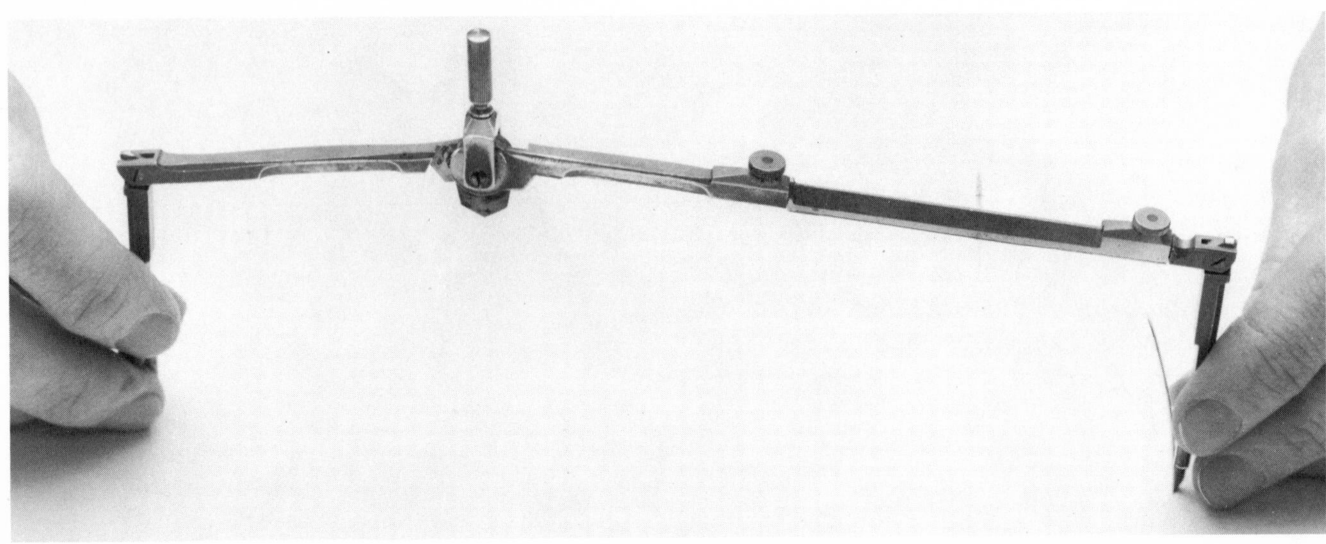

Fig. 3-9. Using an extension on the compass to draw large circles.

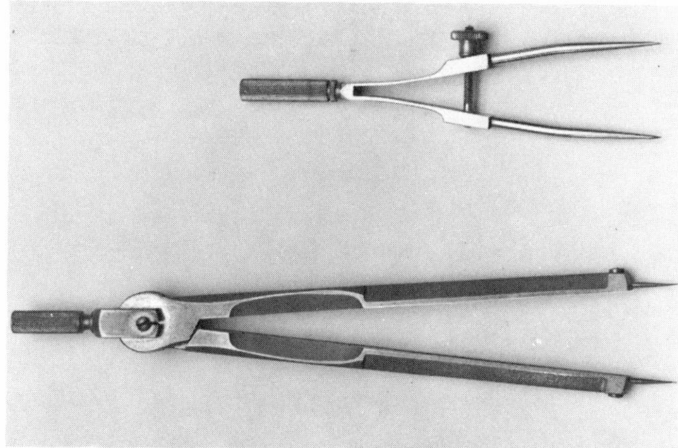

Fig. 3-10. Dividers.

It can be repointed quickly with a PENCIL POINTER, Fig. 3-11. Use the pencil sharpener only when the point becomes very blunt, or when it breaks.

Fig. 3-11. Mechanical pencil pointers.

Many commercial pencil pointers are available. The sand paper pad, Fig. 3-12, is most frequently found in the school drafting room. A piece of styrofoam cemented to the back of the pad is used to wipe graphite dust from the freshly pointed pencil.

Keep the pencil pointer clear of the drawing area when repointing your pencil. The graphite dust will smudge your paper when you attempt to wipe it off.

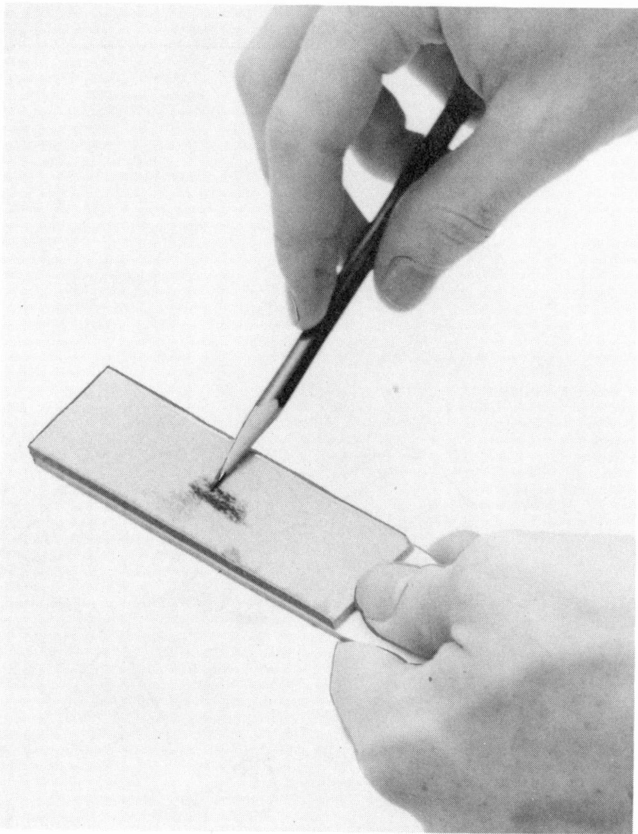

Fig. 3-12. Using a sandpaper pad to point a drafting pencil.

ERASERS

Many shapes and kinds of ERASERS, Fig. 3-13, are manufactured for use in the drafting room. The type of material being

Fig. 3-13. A few of the many different kinds of erasers used in the drafting room.

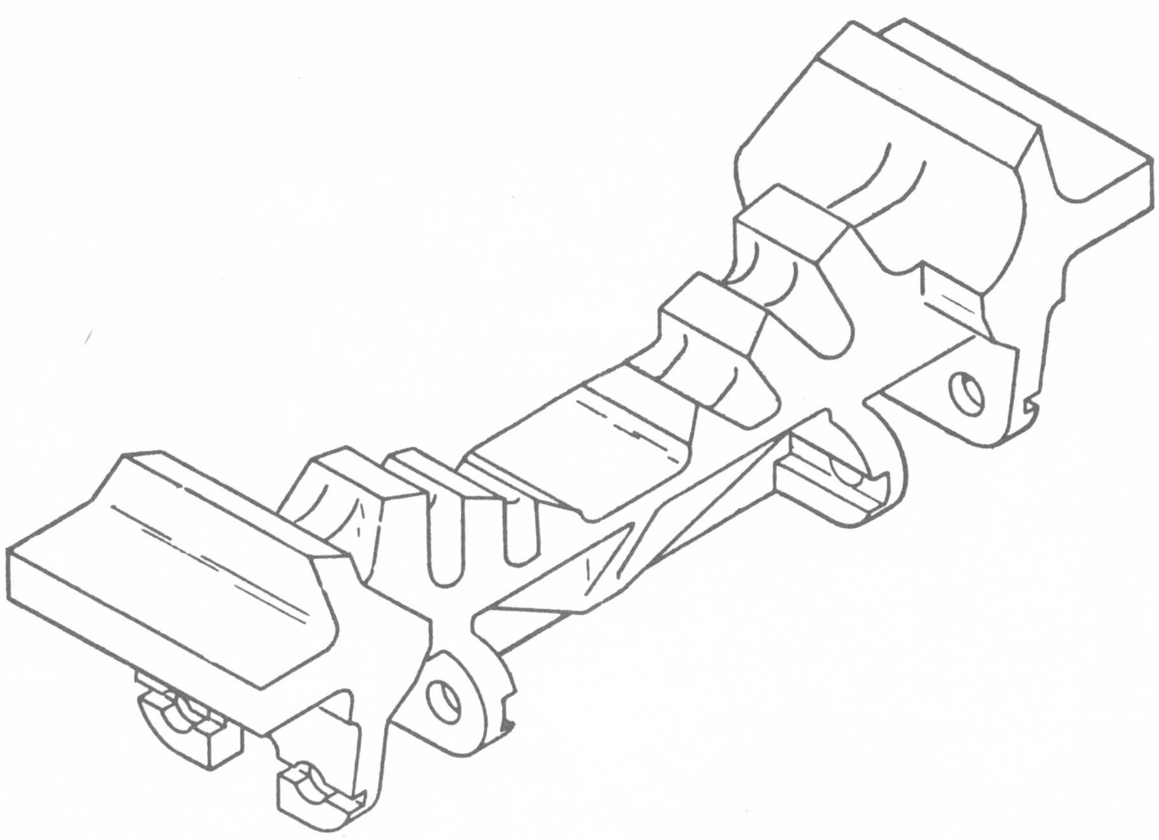

Above is an illustrator's isometric interpretation, of a complex casting used in logging equipment, which took 7.6 hours to prepare. Below is a drawing of the same casting which an automated drafting machine made in 3/4 of an hour. (Perspective Inc.)

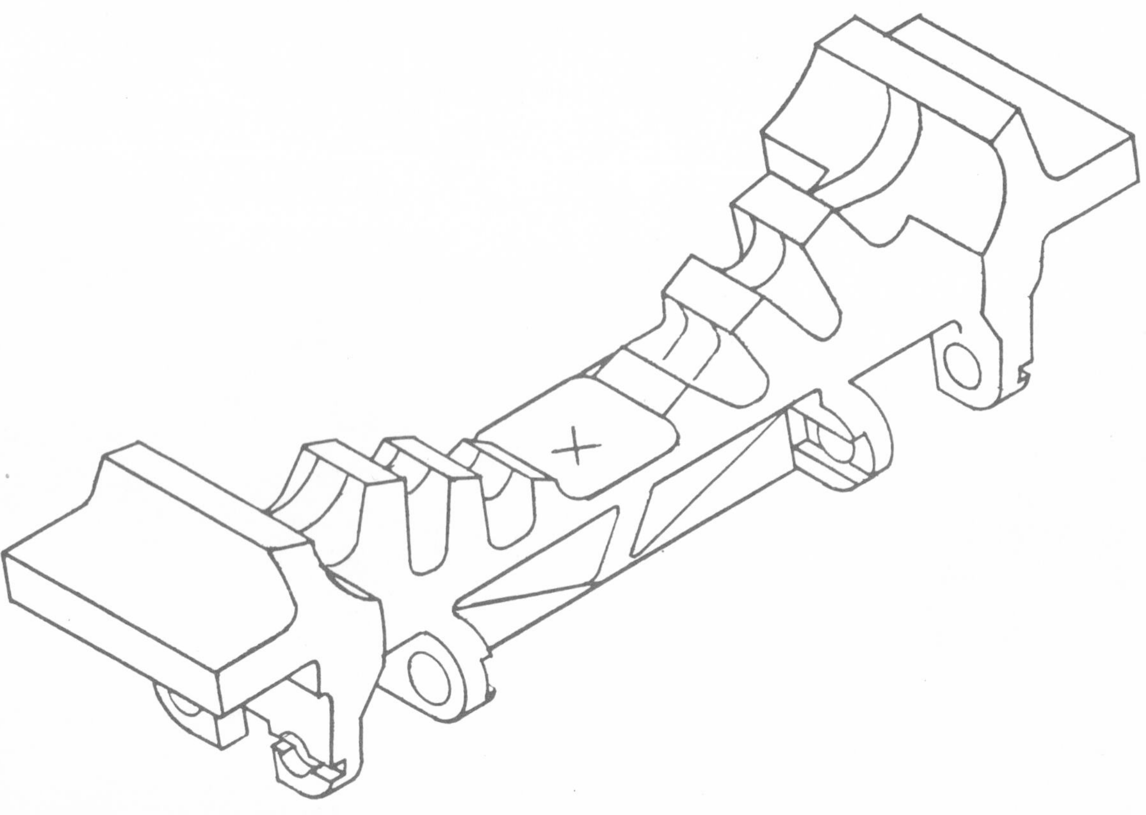

drawn upon - paper, film or vellum, will determine the kind of eraser to be used.

Brush away eraser crumbs before starting to draw again.

ERASING SHIELD

Small errors or drawing changes can be erased without soiling a large section of the drawing if an ERASING SHIELD is employed, Fig. 3-14. Place an opening in the shield of the proper shape and size over the area to be changed and then erase. The erasure is made without touching other parts of the drawing.

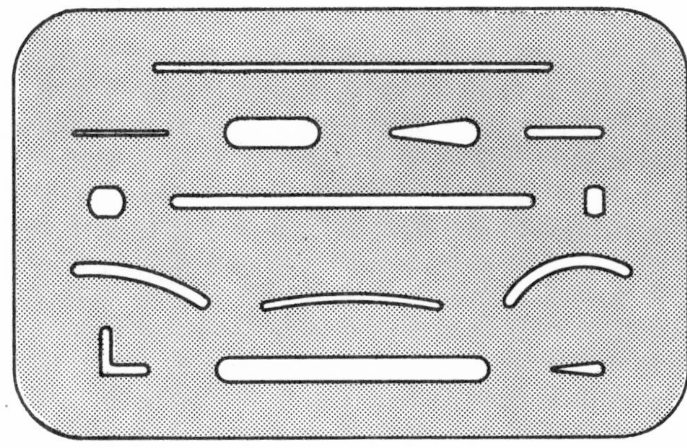

Fig. 3-14. Erasing shield.

A popular shield is made from stainless steel. This metal can be made very thin and still be strong. It is also wear resistant and does not stain or "smudge" the drawing.

Fig. 3-16. Removing erasure crumbs from the drawing with a dusting brush.

FASTENERS

Three methods of attaching paper to the drawing board are shown in Fig. 3-15. Tape is the most desirable of the methods. It does not puncture the paper or the drawing board.

Staples may be used but their continued use will tear up the working surface of the drawing board. They are often difficult to remove.

Thumb tacks are least recommended. They quickly destroy the smooth working surface of the board.

DUSTING BRUSH

No matter how careful you are, some erasing crumbs and dirt particles will collect on the drawing area. These should be removed by using a DUSTING BRUSH, Fig. 3-16, rather than your hands. Using your hands may cause smudges and streaks.

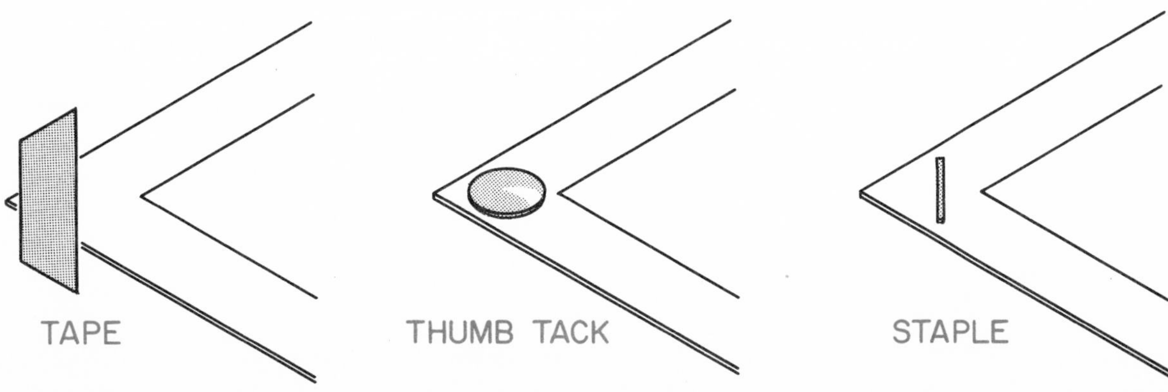

Fig. 3-15. Methods of attaching paper to the drawing board.

Dusting brushes are available in a number of sizes and with natural and man made bristles.

PROTRACTORS

A PROTRACTOR, Fig. 3-17, is used to measure and lay out angles on drawings. They are usually made of clear plastic and may be either circular or semicircular in shape. The degree graduations are scribed or engraved around the circumference of the protractor.

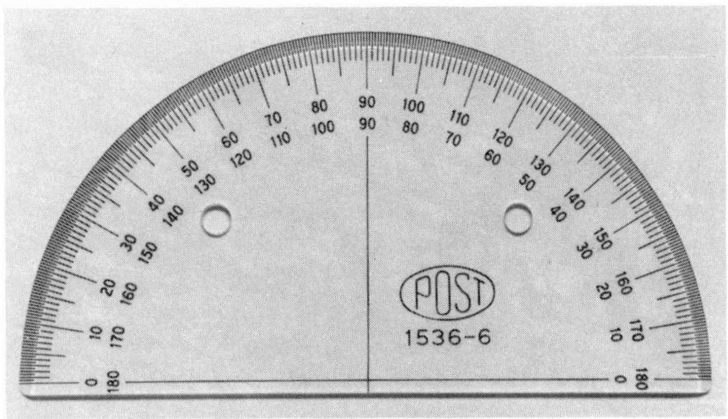

Fig. 3-17. Protractor.

When measuring or laying out an angle, place the center lines of the protractor at the point of the required angle as shown in Fig. 3-18. Read or mark the angle from the graduations on the circumference of the tool.

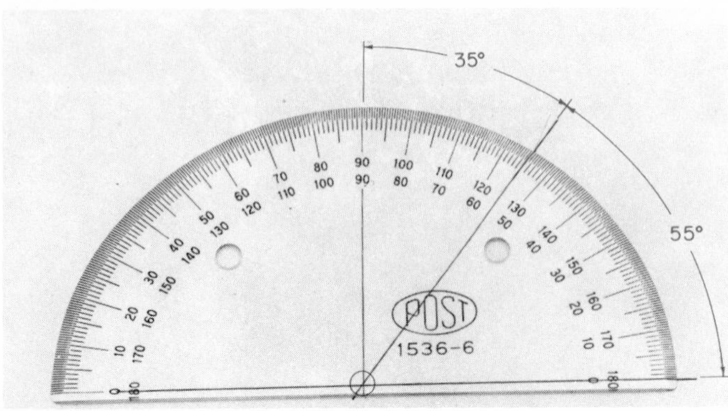

Fig. 3-18. Making a measurement with a protractor.

FRENCH CURVES

Curved lines that are not exactly circular in form are drawn with a FRENCH CURVE, Fig. 3-19. After the curved line is carefully plotted, it is drawn using a French Curve as shown in Fig. 12-6, page 152.

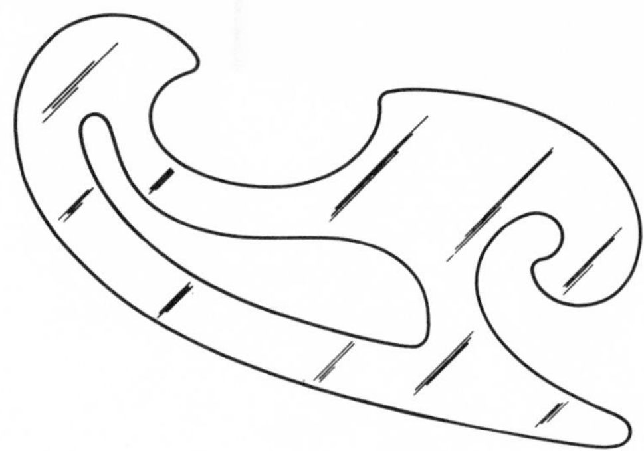

Fig. 3-19. French Curve.

The curves are made of transparent acrylic plastic. They range in size from a few inches to several feet in length and may be purchased individually or as a set.

TEMPLATES

Much time can be saved in drawing standard symbols and figures if TEMPLATES are used, Fig. 3-20. Made of thin plastic, these tools are available in a large number of styles and sizes.

DRAFTING MEDIA

Drawings are made on many different materials - paper, tracing vellum, film, etc. A heavyweight opaque paper that is white, buff or pale green in color is used in many school drafting rooms.

While this paper takes pencil lines well, it is difficult to erase because the pencil point makes a depression in the paper when a line is drawn. If this material is used, take special care to prevent mistakes.

Industry makes much use of tracing vellum and film because reproductions or prints must be made of all drawings.

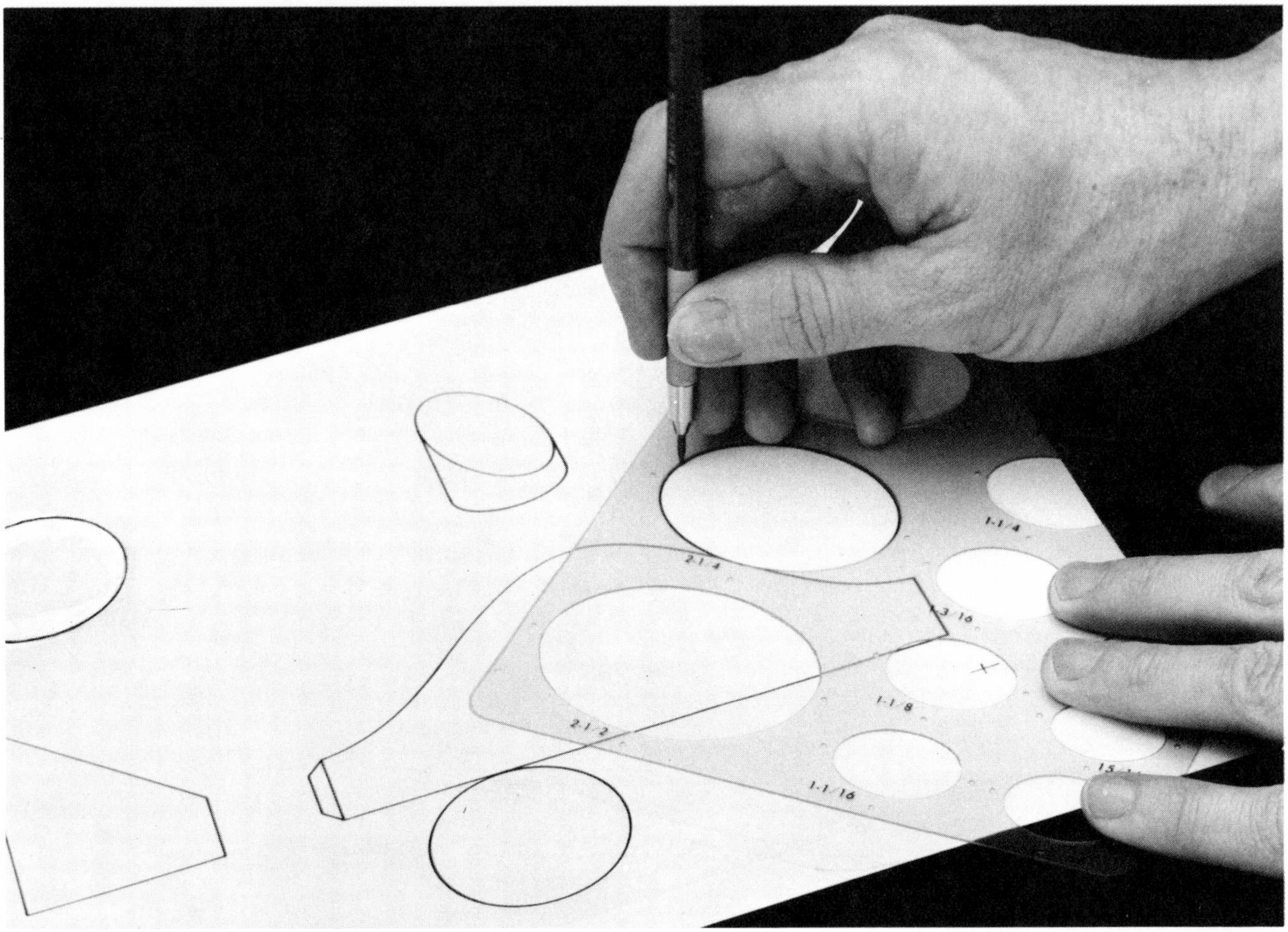

Fig. 3-20. Much time can be saved by using a template.

Drawing media is available in either sheet or roll form. Standard sheet sizes are identified by letter size:

A - 8 1/2 x 11 or 9 x 12
B - 11 x 17 or 12 x 18
C - 17 x 22 or 18 x 24
D - 22 x 34 or 24 x 36
E - 34 x 44 or 36 x 48

PENCILS

As most drawings are prepared with a pencil, it is important that the proper pencil be selected. The drawing media used will determine the type of pencil that should be used. The conventional graphite lead pencil is satisfactory with most papers and tracing vellums, while a pencil with plastic lead is necessary if the drawing is made on film.

The draftsman can select from 17 grades of pencils that range in hardness from 9H (very hard) to 6B (very soft), Fig. 3-21.

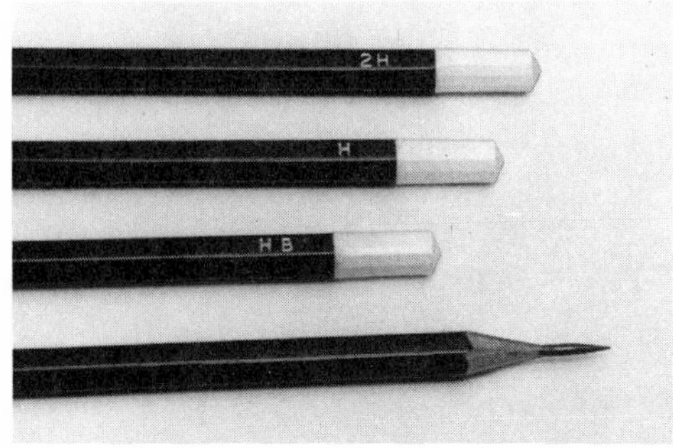

Fig. 3-21. Pencils used in drafting are available in a wide range of hardness.

Many draftsmen use a 4H or 5H pencil for lay out work and a H or 2H pencil to darken the lines and to letter. In general, use a pencil that will produce a sharp, dense black line because this type of line reproduces best on prints.

Avoid using a pencil that is too soft. It will wear rapidly, smear easily and soil your drawing. Also, the lines will be "fuzzy" and will not produce usable prints.

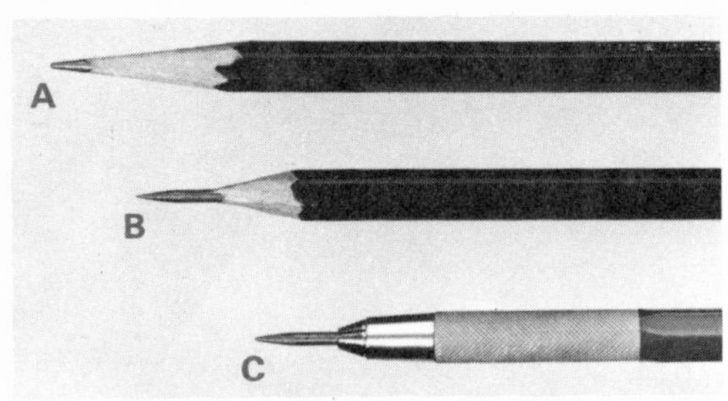

Fig. 3-22. Pencil points. A–Sharpened with regular pencil sharpener. B–Sharpened with a knife and sandpaper pad. C–Sharpened with mechanical pencil pointer.

A conical shaped pencil point is preferred for most general purpose drafting. To sharpen the pencil, cut the wood away from the unlettered end. Use a knife or mechanical sharpener and point the lead on a pencil pointer, Fig. 3-22.

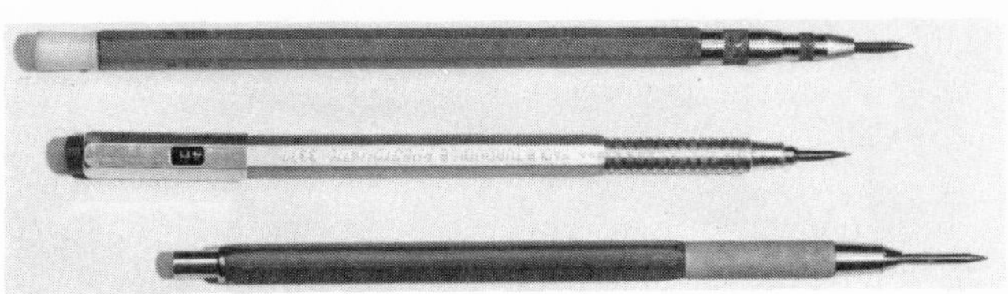

Fig. 3-23. A variety of semiautomatic drafting pencils.

A semiautomatic pencil, Fig. 3-23, is usually preferred to a wood pencil. With this type pencil it is not necessary to cut away the wood to expose the lead. The extended lead is shaped on a pencil pointer.

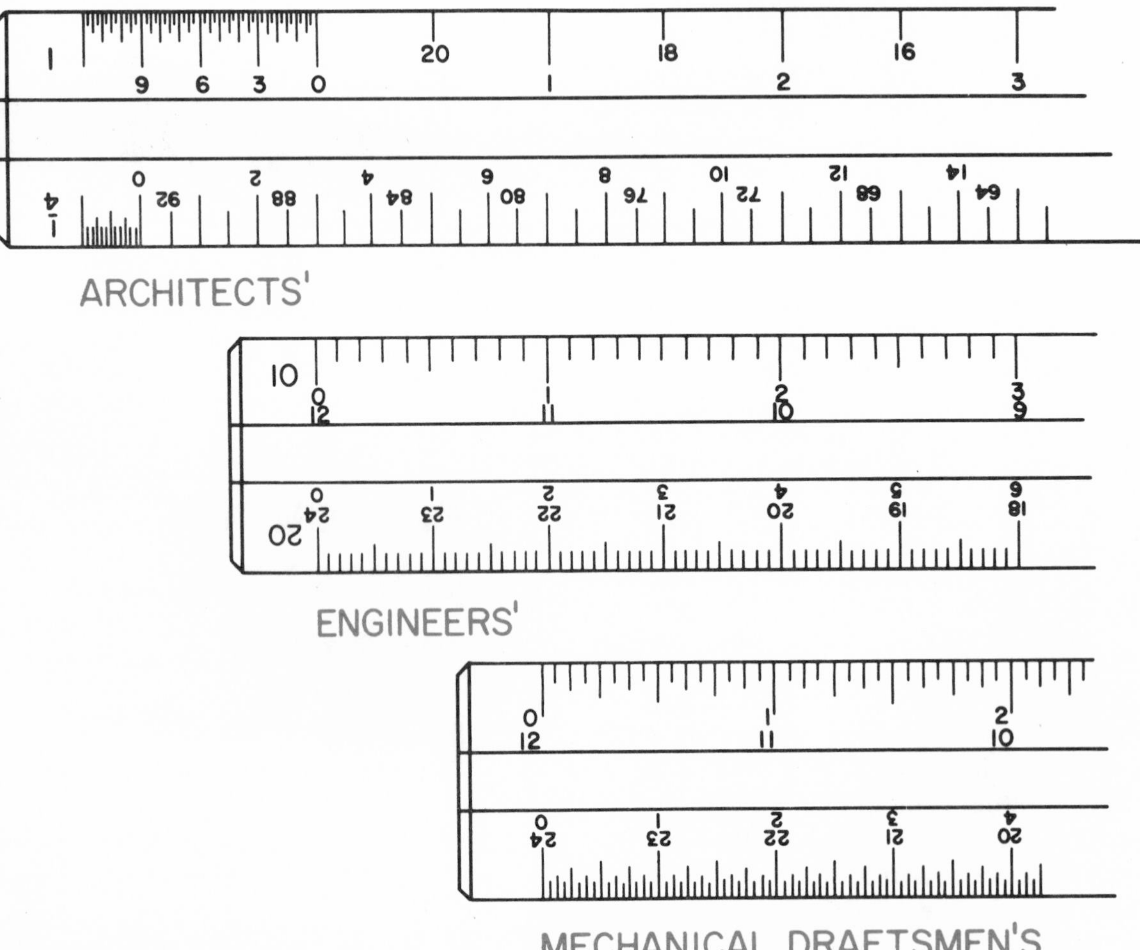

Fig. 3-24. Drafting scales.

SCALES

SCALES, Fig. 3-24, are in constant use on the drawing board because almost every line on a mechanical drawing must be of a measured length. Accurate drawings require accurate measurements.

Because of the diversity of the work that must be drawn, the scales used by the draftsman are made in many shapes, lengths and measurement graduations. They may be made of wood, plastic, or a combination of both materials. Graduations are printed on inexpensive scales, and are machine-engraved on the more costly ones.

For convenience, scales are classified according to their common uses:

ARCHITECTS' SCALE. Each division represents one foot and is divided into twelve parts, each part being equal to one inch.

ENGINEERS' SCALE. Used mostly where large reductions are required. It is divided into 10, 20, 30, 40, 50, 60, 80 and 100 units (each represents one foot) per inch.

MECHANICAL DRAFTSMEN'S SCALE. Most commonly divided into the following graduations: full size, and 3/4, 1/2, 1/4 and 1/8 in. to 1 foot.

TEST YOUR KNOWLEDGE - UNIT 3

1. The drawing board provides the ________ ________________________________.
2. A piece of heavy drawing paper or a special vinyl board cover is sometimes attached to the working surface of the board to:
   a. Provide a drawing surface.
   b. Protect its surface.
   c. Provides a surface to work out problems.
   d. All of the above.
   e. None of the above.
3. The T-square is used to ____________ ________________________________.
4. Vertical and inclined lines are drawn with __________ .
5. The compass is used in drafting to draw ________________________________.
6. The ____________ is used to repoint a pencil when it starts to show signs of dullness.
7. The erasing shield is used to __________ ________________________________.
8. The three methods used to attach drawing paper to the drawing board are ____________________. Which method causes the least damage to the board?
9. Angles can be measured and laid out on drawings by using a ________________.
10. The ________ or ________ pencil is recommended for layout work.
11. An ________ or ________ pencil is used to darken the lines and for lettering.
12. Scales are important to the draftsman because ________________________.
13. List the three classifications of scales:
    a. ________________________
    b. ________________________
    c. ________________________

Unit 4
DRAFTING TECHNIQUES

You were introduced to the ALPHABET OF LINES in the Unit on Sketching. The lines you sketched can be drawn more uniformly and accurately with drafting instruments. See Fig. 4-1.

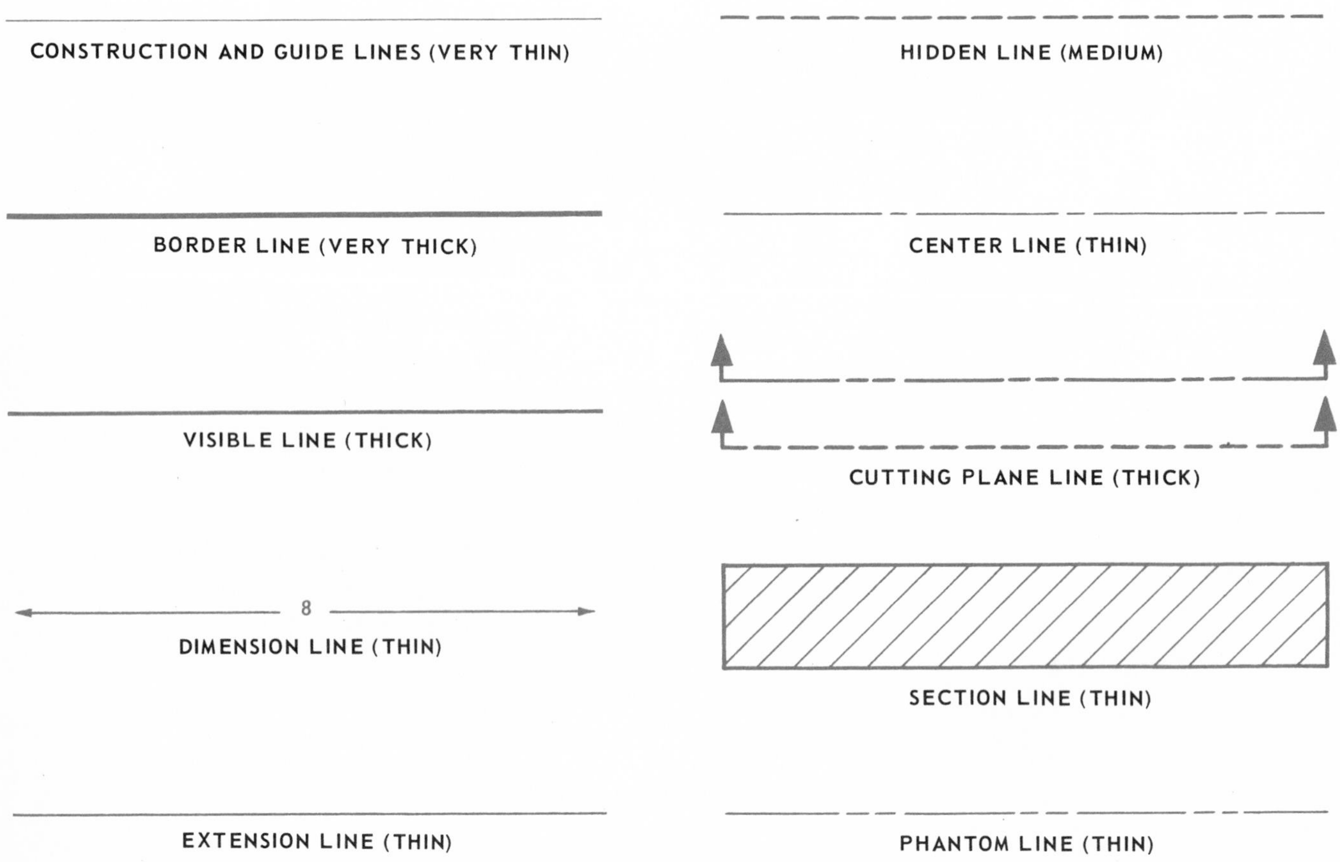

CONSTRUCTION AND GUIDE LINES (VERY THIN)

HIDDEN LINE (MEDIUM)

BORDER LINE (VERY THICK)

CENTER LINE (THIN)

VISIBLE LINE (THICK)

CUTTING PLANE LINE (THICK)

8

DIMENSION LINE (THIN)

SECTION LINE (THIN)

EXTENSION LINE (THIN)

PHANTOM LINE (THIN)

Fig. 4-1. The alphabet of lines.

To understand the LANGUAGE OF INDUSTRY, it is necessary that you know the characteristics of the various lines and the correct way to use them in a drawing, Fig. 4-2. In drafting room language the characteristics and uses of lines are known as LINE CONVENTIONS.

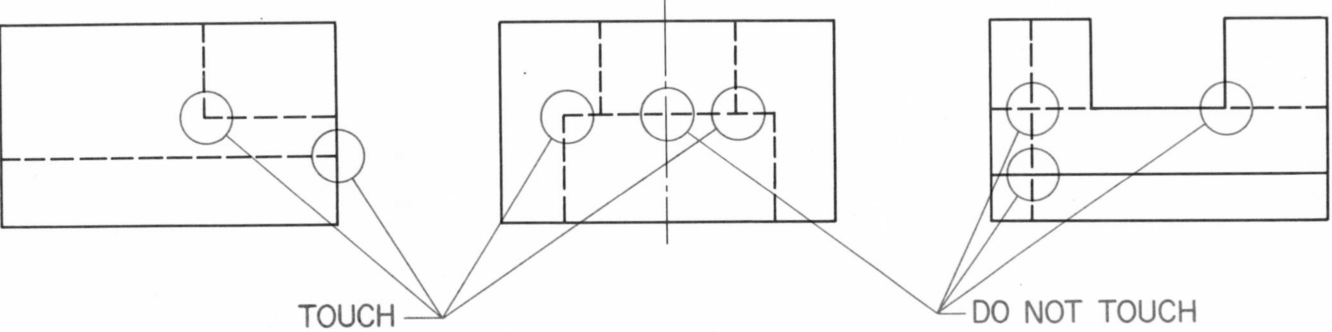

Fig. 4-2. Correct use of hidden lines.

Each type or kind of line has a specific meaning. It is most important that each line is drawn properly, is opaque and uniform in width throughout its entire length. See Fig. 4-3.

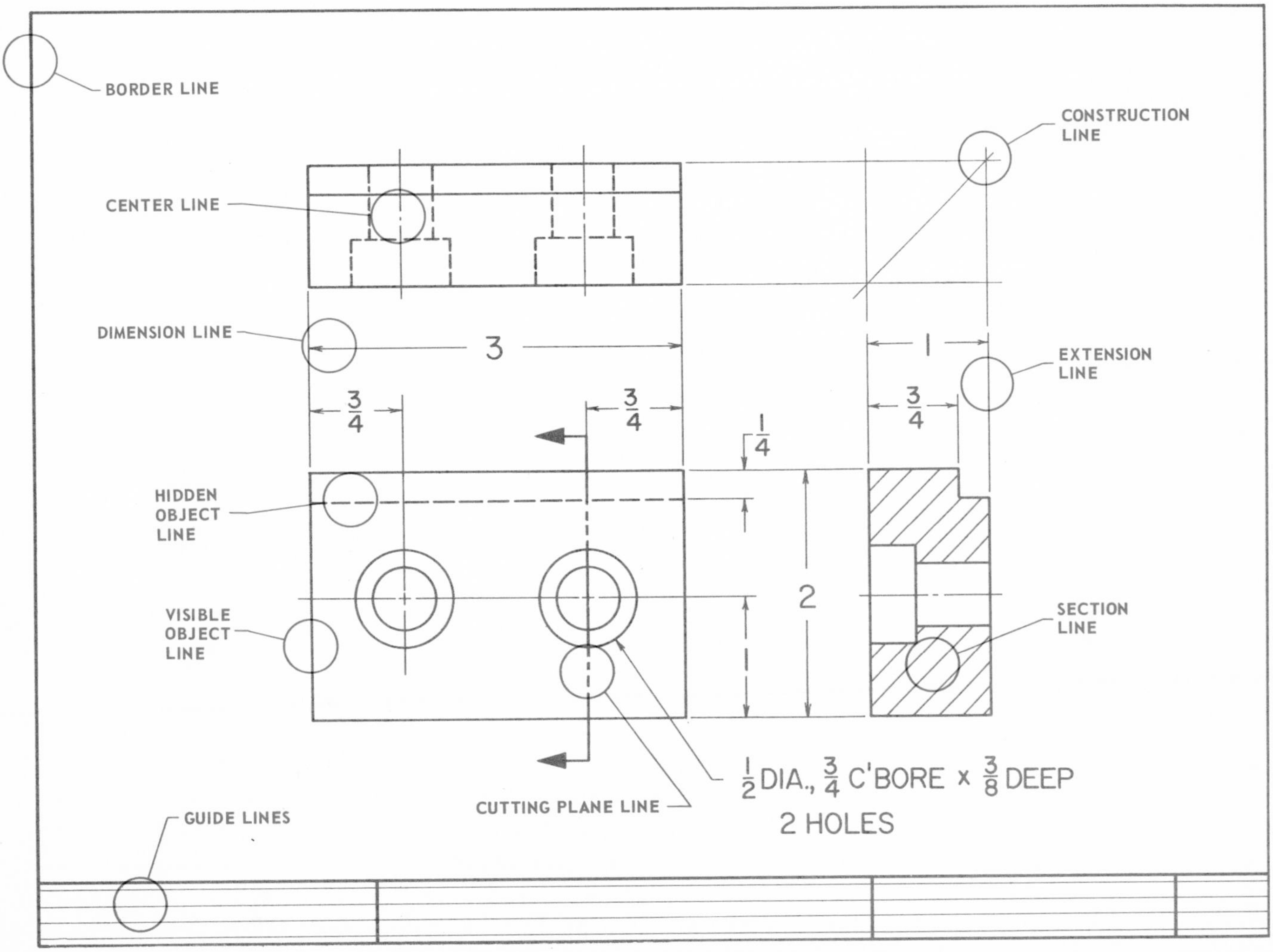

Fig. 4-3. How the various lines are used.

ALPHABET OF LINES

CONSTRUCTION AND GUIDE LINES (VERY THIN)

CONSTRUCTION LINES are drawn very lightly. They are used to block in drawings and as guide lines for lettering. They may be erased, if necessary, after they have served their purpose.

BORDER LINE (VERY THICK)

The BORDER LINE is the heaviest weight line used in drafting. It varies from 1/32 in. to 1/16 in. depending upon the size of the drawing sheet.

VISIBLE LINE (THICK)

The VISIBLE OBJECT LINE (also VISIBLE LINE) is used to outline the visible edges of the object being drawn. They should be drawn so that the views stand out clearly on the drawing. All of the visible object lines on the drawing should be the same weight.

8

DIMENSION LINE (THIN)

The DIMENSION LINE is usually capped at each end with arrowheads and is placed between two extension lines. With few exceptions it is broken with the dimension placed at midpoint between the arrowheads. The dimension line is a light line a bit heavier than the construction line. It is placed 1/4 to 1/2 in. away from the drawing.

EXTENSION LINE (THIN)

The EXTENSION LINE is the same weight as the dimension line. It extends the dimension beyond the outline of the view so that the dimension can be read easily. The line starts about 1/16 in. beyond the object and extends about 1/8 in. past the last dimension line.

HIDDEN LINE (THICK)

The HIDDEN OBJECT LINE (also HIDDEN LINE) is used to show the hidden features of the object. It is drawn the same weight as the visible object line and is composed of short lines approximately 1/8 in. long separated by spaces approximately 1/16 in. They may vary slightly according to the size of the drawing. Hidden object lines should always start and end with a dash in contact with the visible object line. See Fig. 4-2 for the correct uses of the hidden object line.

CENTER LINE (THIN)

The CENTER LINE is used to indicate the center of symmetrical objects. It is a fine dark line composed of alternate long (3/4 in.) and short (1/8 in.) dashes with 1/16 in. spaces between the dashes. The center line should extend uniformly only a short distance beyond the circle or view. They start and end with long dashes and should not cross at the spaces between the dashes.

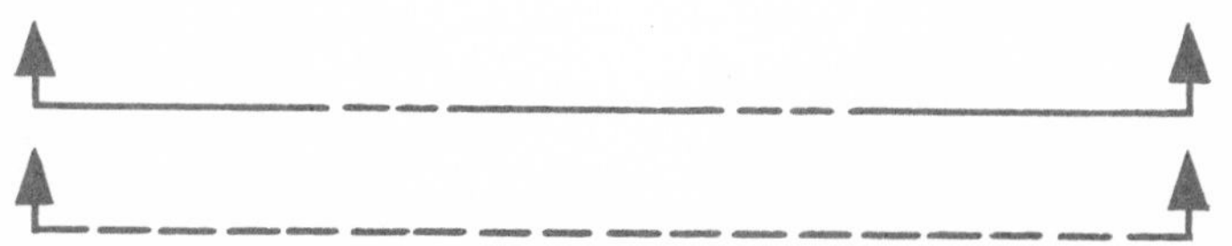

CUTTING PLANE LINE (THICK)

The CUTTING PLANE LINE is a heavy line. It is used to indicate where the sectional view will be taken. Two forms are recommended for general use. The first form is composed of a series of long (3/4 to 1 1/2 in.) and short (1/8 in. with 1/16 in. space) dashes. The second form is composed of equal dashes about 1/4 in. long with 1/16 in. spacing. The cutting plane line will be further explained in Unit 9 on Sectional Views.

SECTION LINE (THIN)

SECTIONAL LINES are used when drawing the inside features of the object. They indicate material cut by the cutting plane line, and also indicate the general classification of the material. The lines are fine dark lines.

PHANTOM LINE (THIN)

The PHANTOM LINE is used to indicate alternate positions of moving parts and for repeated detail like threads and springs. It is a thin dark line made of long dashes (3/4 to 1 1/2 in.) long, alternated with pairs of short dashes 1/8 in. in length, with 1/16 in. spaces.

HOW TO MEASURE

Almost every line on a mechanical drawing must be a measured line. If your drawings are to be made accurately, you must be able to make accurate measurements.

Measurements are made in the drafting room with SCALES, Fig. 4-4. The term SCALE is used to indicate both the device used to measure, and the size to which an object is to be drawn.

Scales have graduations on the edges which show lengths used to indicate larger units of measure (as 1/4 in. equals one foot).

While several different shapes of scales are available, Fig. 4-5, triangular shaped scales are most widely used in the school drafting room.

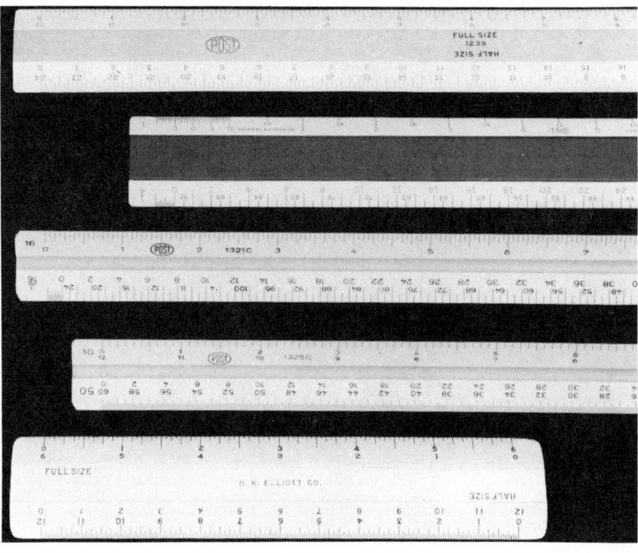

Fig. 4-4. Various types of drafting scales found in a drafting room.

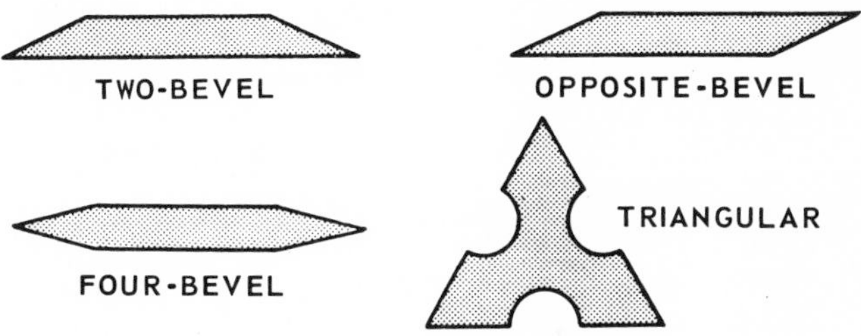

Fig. 4-5. Several different shapes of scales are available.

A SCALE CLIP, Fig. 4-6, provides a means to lift the triangular scale and keeps the desired scale edge in an upright position.

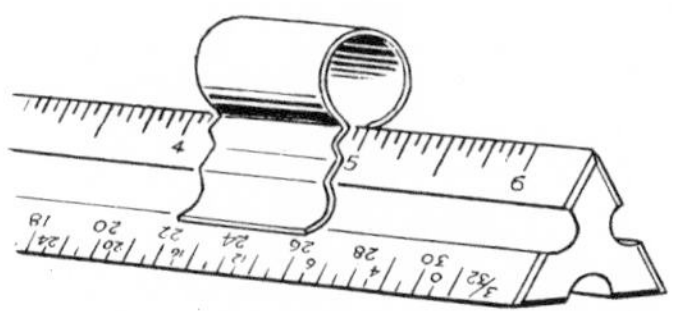

Fig. 4-6. The scale clip keeps the desired scale edge in an upright position.

There are three types of scales in common usage. They are classified as ARCHITECT'S, ENGINEER'S and MECHANICAL DRAFTSMEN'S SCALES, Fig. 4-7.

The triangular architect's scale has six faces, (left) Fig. 4-8. Each is graduated differently. You will find one face where the

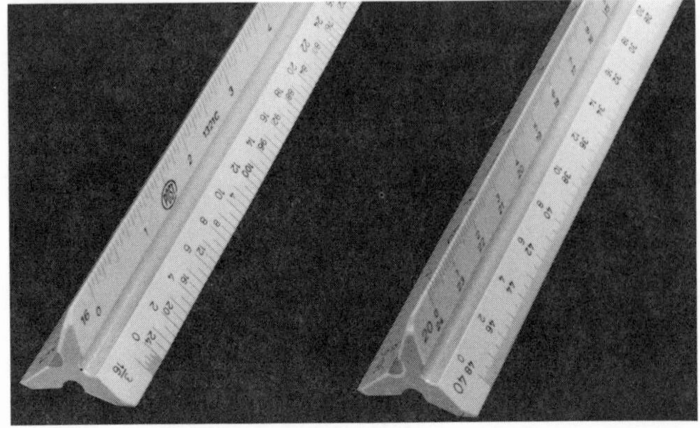

Fig. 4-7. Left. Three types of scales in common usage: Top–Architect's scale. Center–Engineer's scale. Bottom–Mechanical Draftsmen's scale. Fig. 4-8. Right. Triangular drafting scales. The architect's scale is on the left with the engineer's scale on the right.

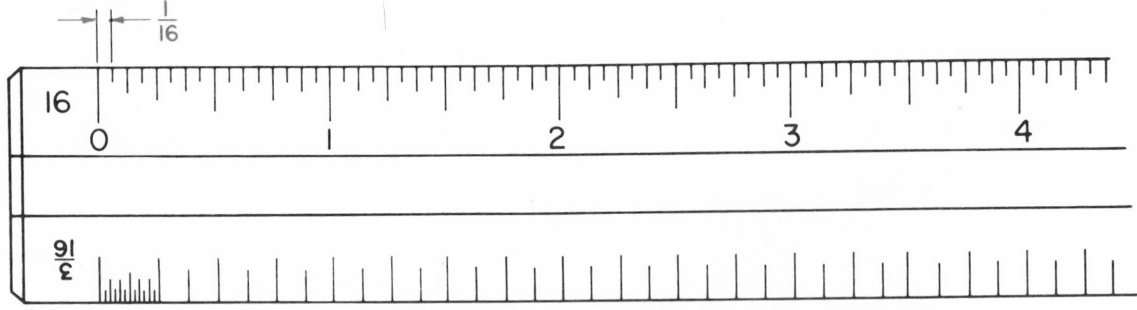

Fig. 4-9. One edge of the architect's scale is graduated in 1/16's.

inch divisions are each divided into sixteen (16) parts. That is, each division is equal to one-sixteenth (1/16) of an inch, Fig. 4-9.

To read the scale, imagine that the one-sixteenth (1/16) divisions are numbered as shown in Fig. 4-10. At first you may find it easier to count the spaces when you measure. However, after some practice this should not be necessary.

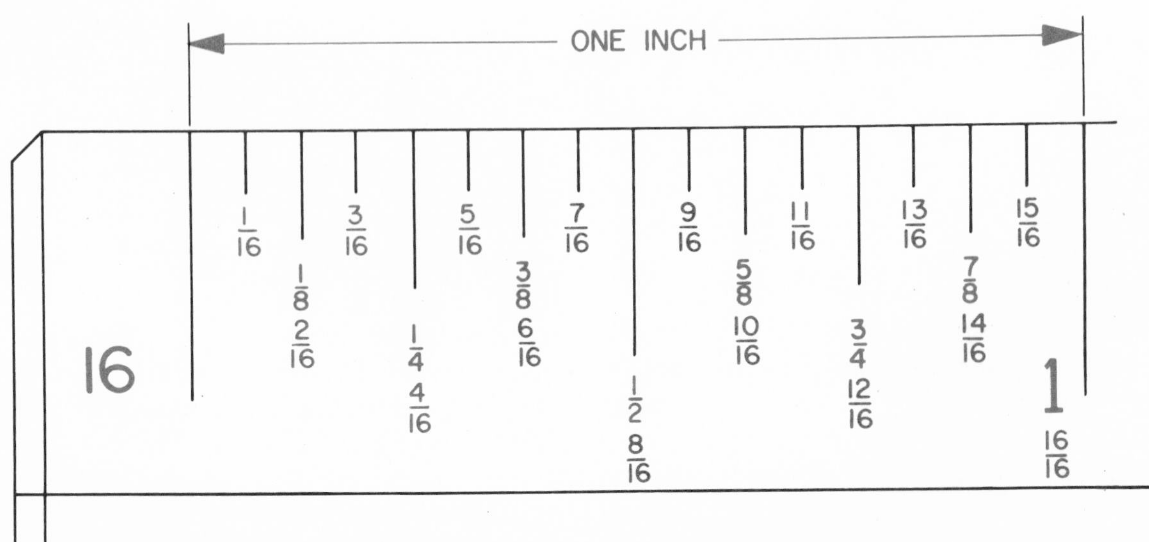

Fig. 4-10. The divisions numbered.

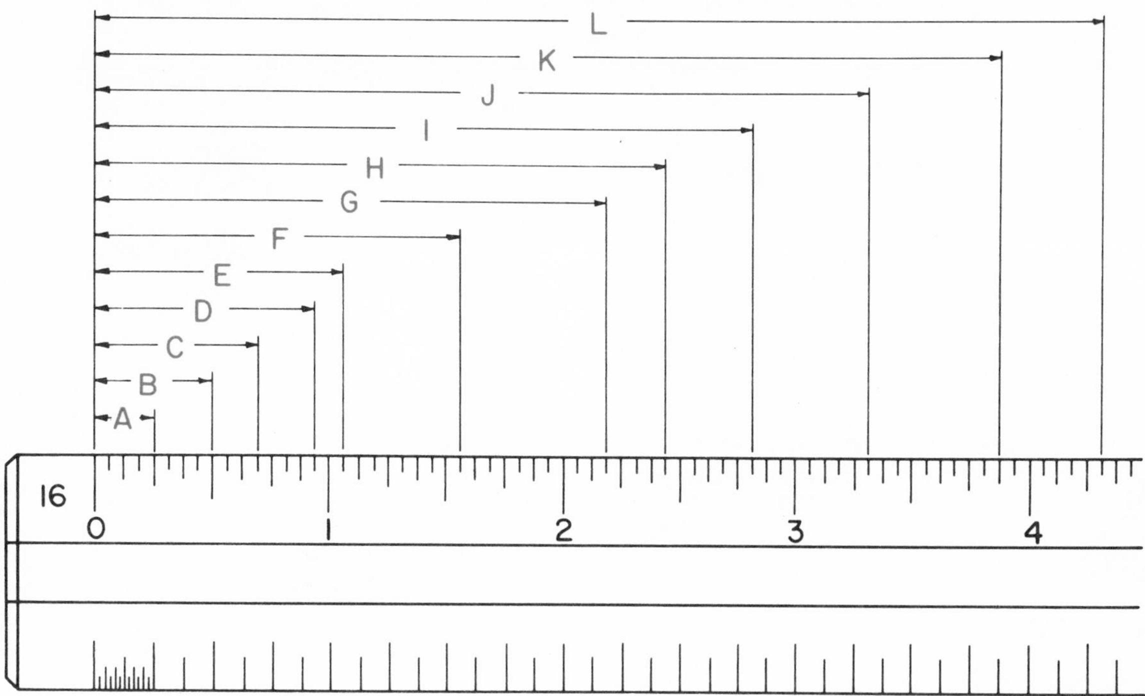

Fig. 4-11. How many can you answer correctly? On a piece of paper write the letters A to L. After each letter write the correct measurement. Ask your instructor to check your answer.

A section of the scale is shown in Fig. 4-11. How many of the measurements can you read correctly?

After you can read and make measurements accurately and quickly to one-sixteenth of an inch, note the remaining faces on the scale. Each is divided to represent a foot (12 in., 1'-0") of actual measurement reduced to a particular length.

There are two scales on each face and each is marked to show scale divisions of 3/32 and 3/16; 1/8 and 1/4; 1/2 and 1; 3/8 and 3/4; and 1 1/2 and 3. That is, the face marked with a 3 means that the foot (12 in.) has been reduced to 3 in. A drawing made using this scale would be one-quarter (1/4) actual size.

MAKING MEASUREMENTS

To make a measurement, observe the scale from directly above. Mark the desired measurement on the paper by using a light perpendicular line made with a sharp pencil, Fig. 4-12.

Keep the scale clean. Do not mark on it, or use it as a straightedge.

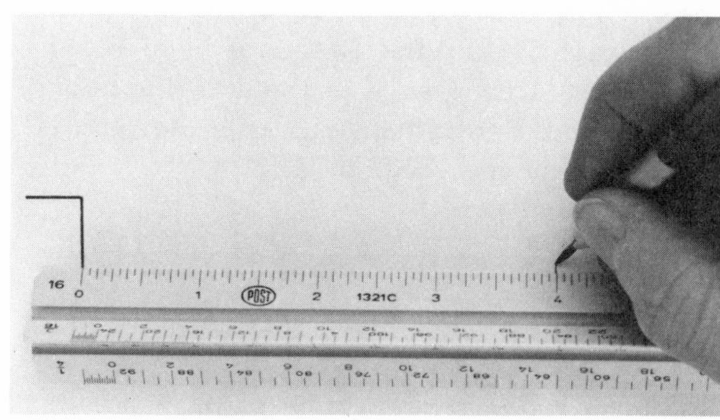

Fig. 4-12. When making a measurement, observe the scale directly from above.

HOW TO DRAW LINES WITH INSTRUMENTS

Care must be taken if lines are to be the same weight; that is, uniform in width and darkness.

When lines are drawn using instruments, hold the pencil perpendicular to the paper, and inclined at an angle of about 60 deg. in the direction the line is being drawn. To keep the lines uniform in weight, especially if a long line is being drawn, rotate the pencil as you draw. Rotating the pencil will also keep the point sharp.

Fig. 4-13. Use the T-square to draw horizontal lines. Note how the T-square head is held against the edge of the drawing board.

The T-square is used to make horizontal lines, Fig. 4-13. The lines are drawn from left to right. Hold the T-square head firmly against the LEFT edge of the drawing board (left-handed draftsmen will use the RIGHT edge of the board, Fig. 4-14).

Fig. 4-14. Left-handed draftsmen use the RIGHT edge of the drawing board.

Vertical lines may be drawn using a triangle and are drawn from the bottom to the top of the sheet, Fig. 4-15. The base of

Fig. 4-15. Vertical lines are drawn using the triangle. The lines are drawn from the bottom to the top of the sheet.

of the triangle must rest on the blade of the T-square. Use the left hand to hold the triangle in place.

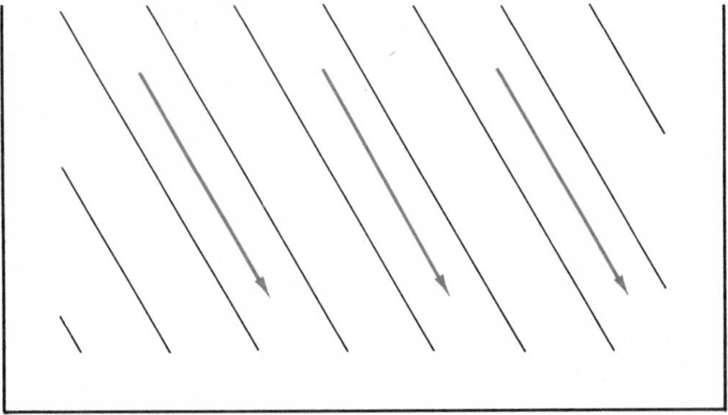

Fig. 4-16. Drawing lines that slant to the left.

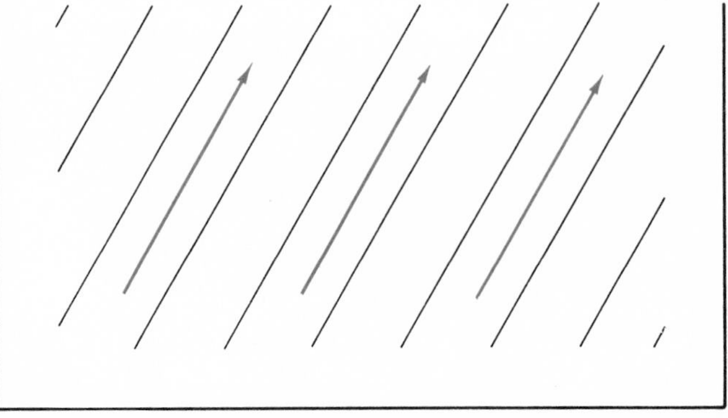

Fig. 4-17. Lines that incline to the right . . . from the bottom up.

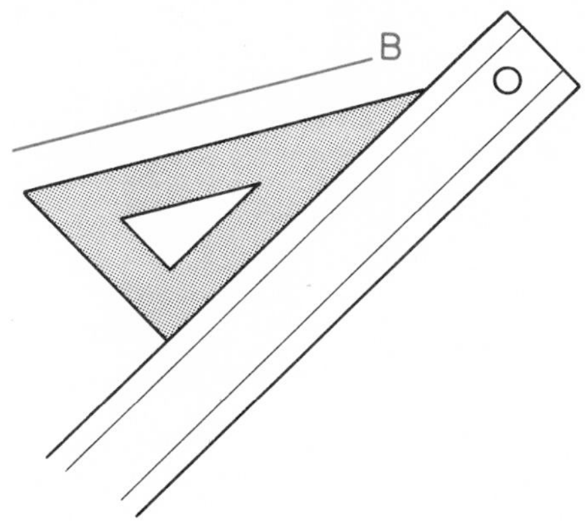

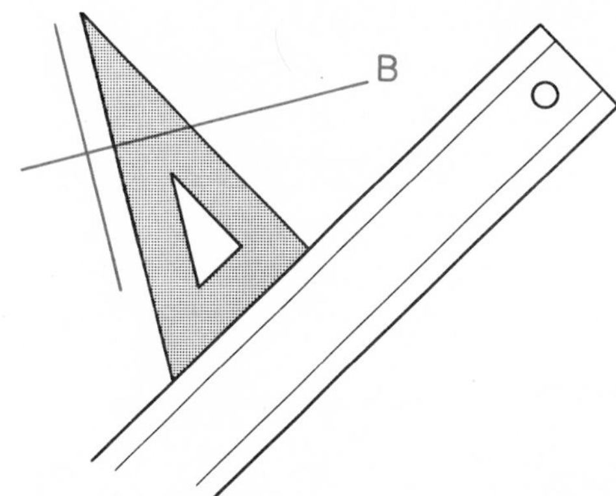

Fig. 4-18. How to draw a line perpendicular to a given line using the T-square and triangle.

The procedure to follow in drawing inclined lines; lines that are not vertical or horizontal, depends on the direction of the slope. Lines that incline to the left, Fig. 4–16, are drawn more easily from the top of the sheet down. Those that incline to the right, Fig. 4–17, should be drawn from the bottom of the sheet up.

HOW TO DRAW A LINE PERPENDICULAR TO A GIVEN LINE USING INSTRUMENTS

To draw a line perpendicular to a given line using instruments, place the hypotenuse (long edge) of any triangle parallel to the given line, Fig. 4–18. Support the angle on the T-square or another triangle. Rotate the first triangle around the 90 deg. corner to draw a line perpendicular to the given line.

Another technique used to draw a line perpendicular to a given line requires that you place either leg of the triangle parallel to the given line, Fig. 4–19. Support it on the T-square or another triangle. Slide the original triangle on the support and draw the line that will be perpendicular to the given line.

HOW TO ERASE

When drawing, every effort must be made to prevent mistakes. However, even the best draftsman must occasionally make changes or corrections on a drawing that will require erasing.

Some helpful suggestions to follow:

1. Keep your hands and instruments clean. This will help to keep "smudges"

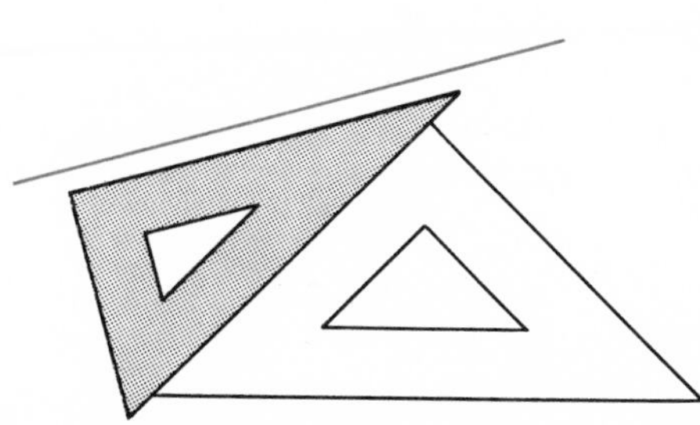

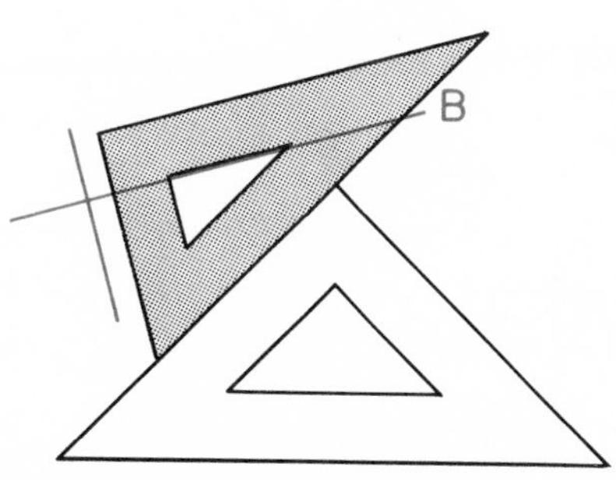

Fig. 4-19. How to draw a line perpendicular to a given line using two triangles.

Fig. 4-20. *Using the erasing shield.*

from forming.

2. Use an ERASING SHIELD, Fig. 4-20, whenever possible. Select an opening that will expose the area to be erased. The shield will protect the rest of the drawing while the erasure is made.

3. Clean the eraser crumbs from the board immediately after making an erasure. Remove them with a brush or clean cloth, not your hands.

4. Hold the paper with your free hand when erasing. This will keep the paper from wrinkling.

5. Erasing will remove the lead but it will not remove the pencil grooves from the paper. Avoid deep, wide grooves by first blocking in all views with light construction lines.

HOW TO USE A COMPASS

In general drafting work, the compass is used to draw circles and arcs. Care must be taken so that the line drawn with the compass is the same weight as the line produced with the pencil. To accomplish this, it is recommended that the compass lead should be several grades softer than that of the pencil.

Sharpen the lead and adjust the point as shown in Fig. 4-21. Do not forget to readjust the point after each sharpening.

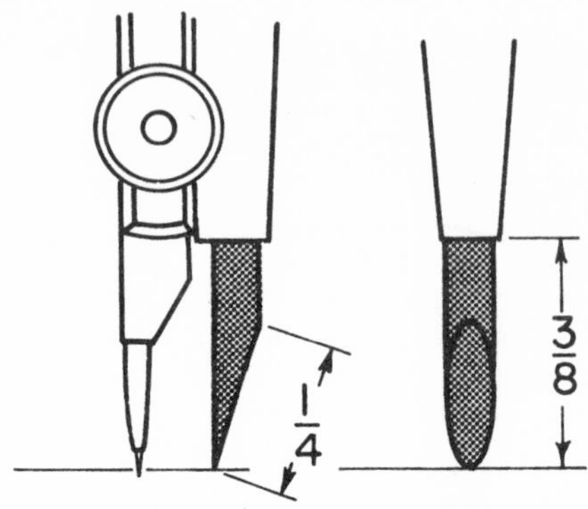

Fig. 4-21. *Sharpening and adjusting the compass lead. Do not forget to readjust the point after each sharpening.*

To set the compass, draw a line that is equal in length to the desired radius. Adjust the compass on this line, Fig. 4-22. Do not set the compass on the scale because the point will eventually ruin the division lines on the scale.

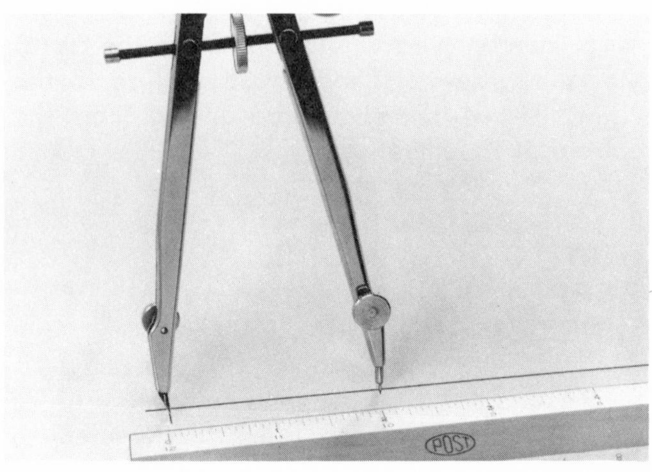

Fig. 4-22. *Adjust the compass to size on a measured line. Never set it on the scale.*

To draw the circle, rotate the compass in a clockwise direction, Fig. 4-23, with the tool inclined in the direction of rotation. Start and complete the circle on a center line. When drawing a series of concentric circles (circles with same center), draw the smallest circle first.

ATTACHING DRAWING SHEET TO BOARD

The drawing sheet can be attached to the board with drafting tape, thumbtacks or

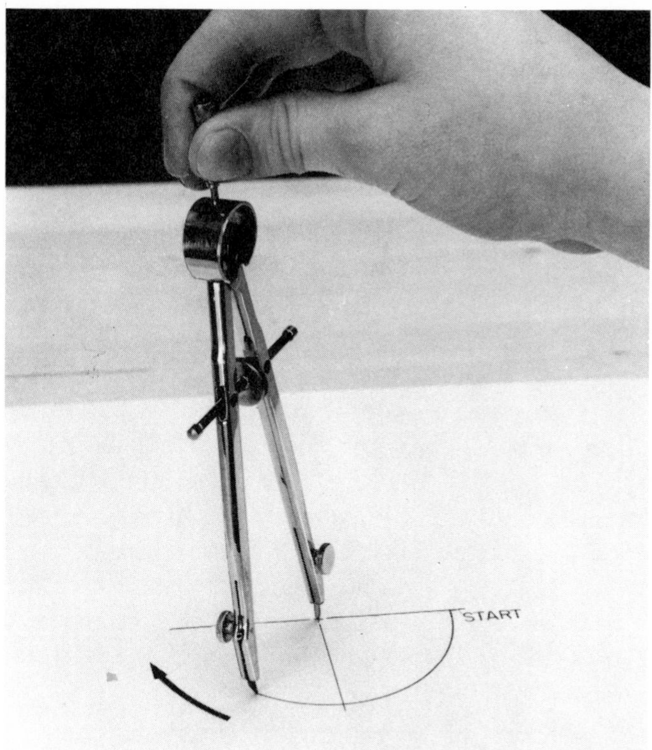

Fig. 4-23. Drawing a circle. Note that the compass is inclined in the direction of rotation.

staples. Tape is preferred because it does not damage the board.

Before attempting to attach the paper to the board, remove all eraser crumbs. To attach the paper, place the sheet on the board as shown in Fig. 4-24. Left-handed draftsmen should use the upper right-hand corner of the board.

Fig. 4-24. Locating the drawing sheet on the board.

Place the T-square on the board with the head firmly against the left edge. Slide it up until the top of the blade is in line with the top edge of the drawing sheet, Fig. 4-25. Position the sheet so the top edge is parallel with the T-square blade and fasten the sheet to the board. Larger sheets may also require fasteners on the bottom corners.

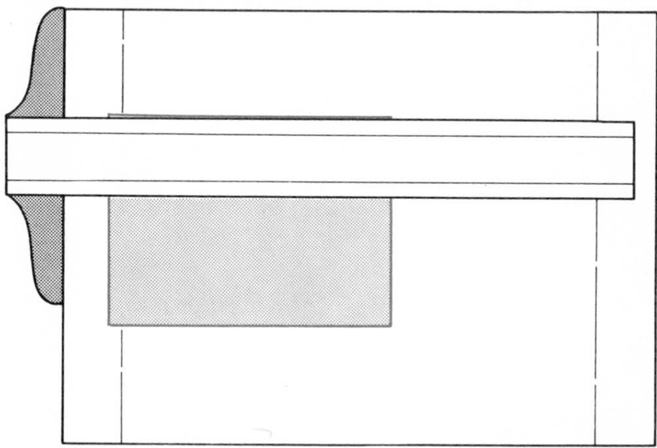

Fig. 4-25. Aligning the drawing sheet with the T-square.

This procedure is recommended when positioning and attaching A-size (8 1/2 x 11) drawing sheets:

1. Place the sheet on the board as shown in Fig. 4-24.
2. Place the head of the T-square firmly against the left edge of the board.
3. Align the drawing sheet by sliding the T-square blade until it contacts the bottom

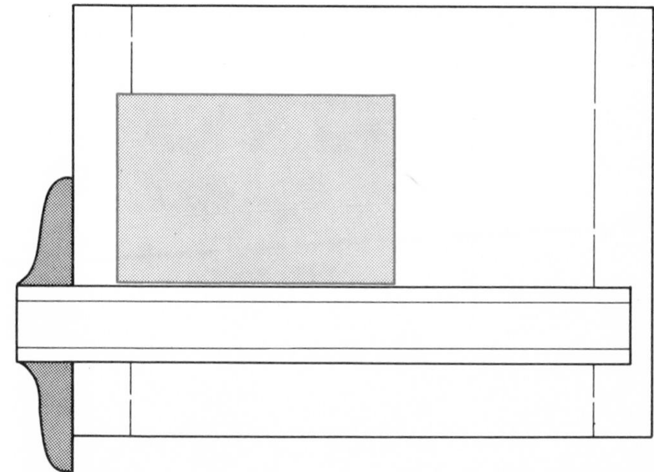

Fig. 4-26. The drawing sheet can also be aligned on the board by placing the bottom edge of the sheet on the edge of the T-square.

edge of the paper, Fig. 4-26. Align the sheet with this edge. Fasten the sheet to the board.

Lightweight paper, like tracing vellum, is slightly more difficult to attach to the board because it has a tendency to wrinkle. This is aligned on the drawing board with the T-square and the sheet is attached in the sequence shown in Fig. 4-27.

DRAFTING SHEET FORMAT

Most drafting rooms use a standard format in the layout of their drawing sheets. In general, the format consists of the border, title block and standard notes, Fig. 4-28.

The border is included to define the drawing area of the sheet. The title block and standard notes provide information that is necessary for the manufacture or assembly of the object described on the drawing sheet.

Most industrial firms use standard drawing sheet sizes. Drawings made on standard size sheets are easier to file and present less difficulty when prints are made from them.

With few exceptions, the drawings in this text should be drawn on 8 1/2 in. by 11 in., or 11 in. by 17 in. size sheets. The small sheet is known as an A-size sheet, and the large sheet as a B-size sheet.

Plan your work carefully. Do not use a large sheet if the smaller size sheet will do.

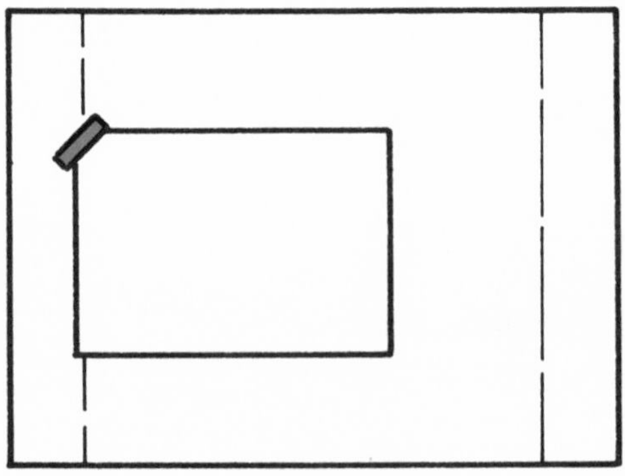

1. USE THE T-SQUARE TO LINE UP THE SHEET ON THE BOARD. FASTEN THE UPPER LEFT CORNER OF THE SHEET.

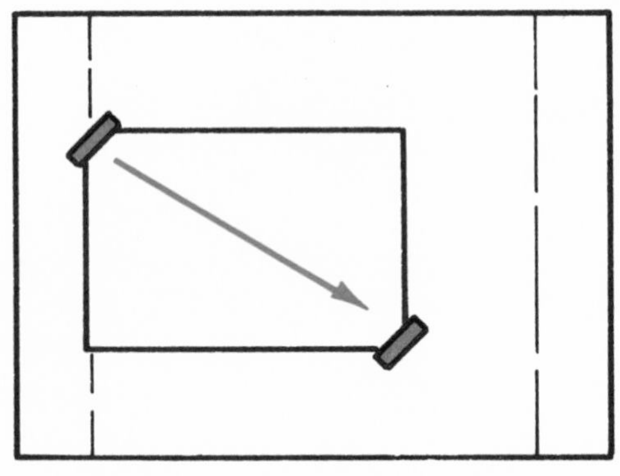

2. SMOOTH TO THE LOWER RIGHT CORNER. ATTACH SHEET.

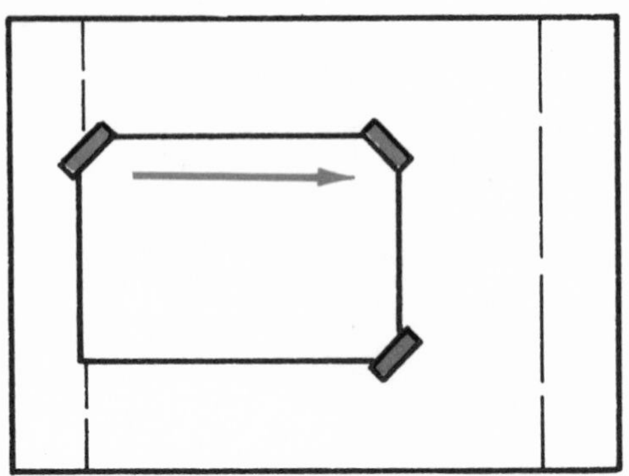

3. SMOOTH TO THE UPPER RIGHT CORNER. ATTACH SHEET.

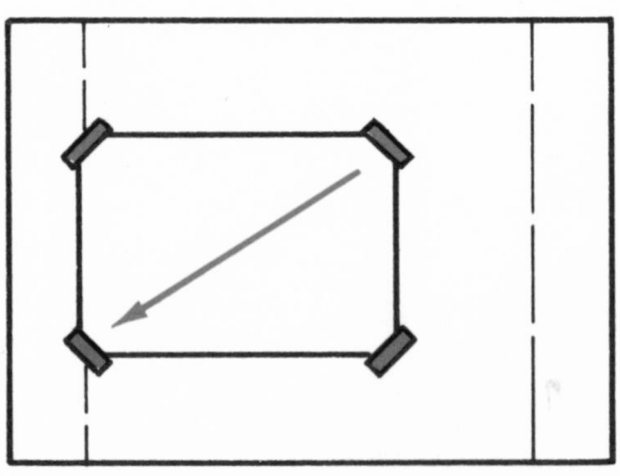

4. SMOOTH TO THE LOWER LEFT CORNER. FINISH ATTACHING THE SHEET.

Fig. 4-27. Sequence recommended for attaching lightweight papers (such as tracing vellums) to the board.

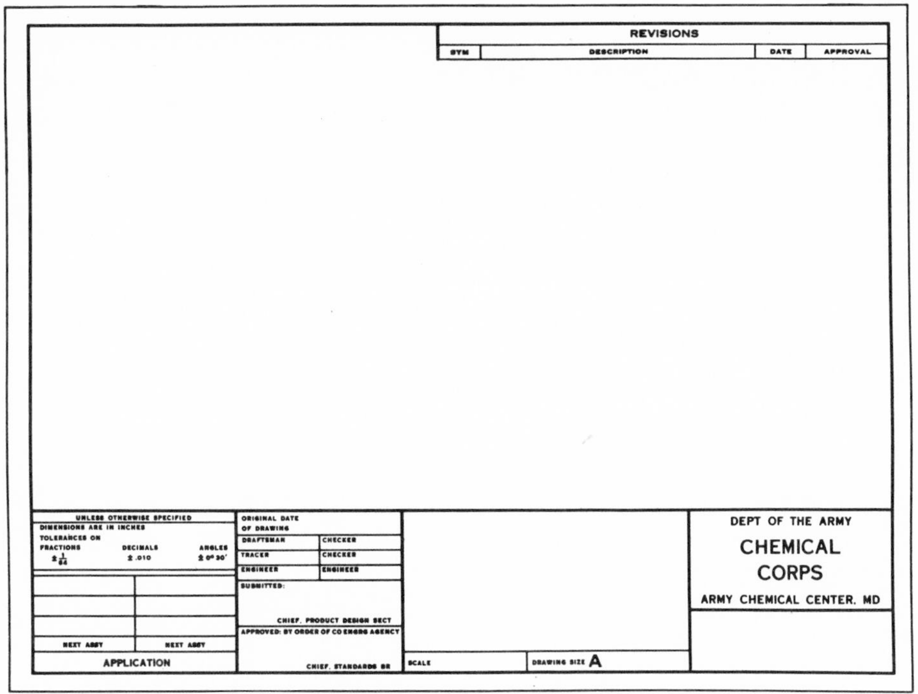

Fig. 4-28. Preprinted drawing sheets save a great deal of time for the draftsman.

RECOMMENDED DRAWING SHEET FORMAT

The following drawing sheet format is recommended for most of the problems presented in this text:

1. Put a 1/2 in. border on the sheet, Fig. 4-29. Use a short light pencil stroke, not a dot, as the guide for drawing the border line. The short light guide mark

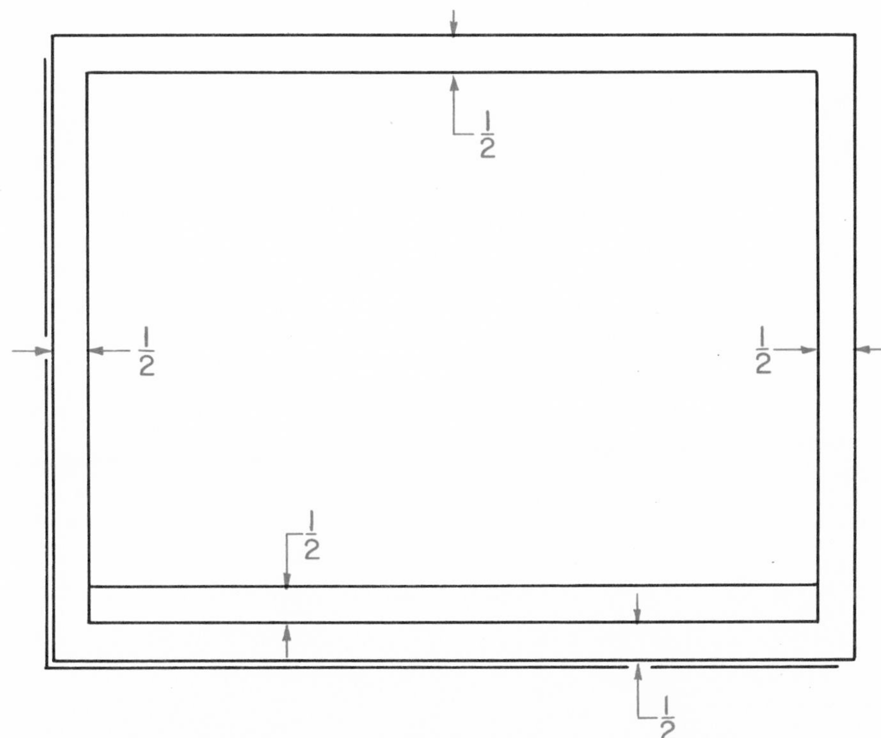

Fig. 4-29. How to start laying out the drawing sheet.

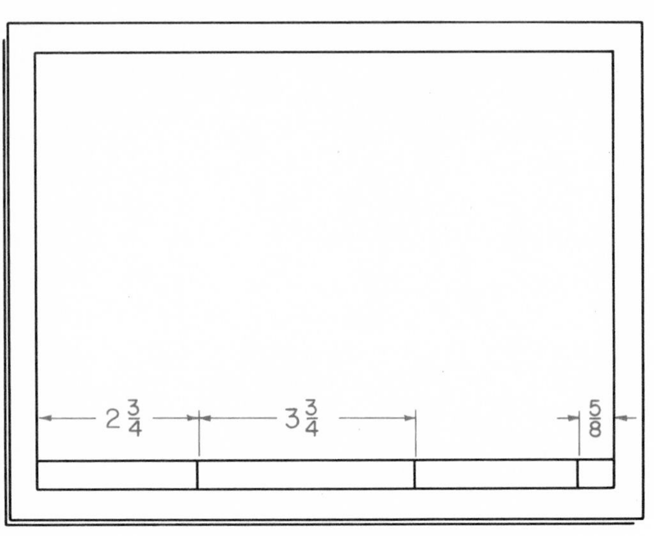

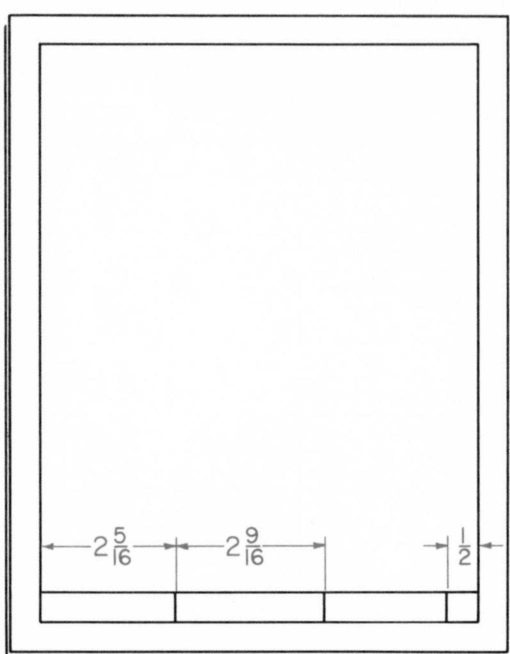

Fig. 4-30. Left. Horizontal drawing sheet format. Fig. 4-31. Right. Vertical drawing sheet format.

should be covered when the border is drawn.

2. Allow another 1/2 in. for the title block.

3. Divide the title block as shown in Figs. 4-30 and 4-31.

4. Guide lines for the title block are drawn, Fig. 4-32. The guide lines should be drawn VERY lightly.

5. Letter in the necessary information, Fig. 4-33.

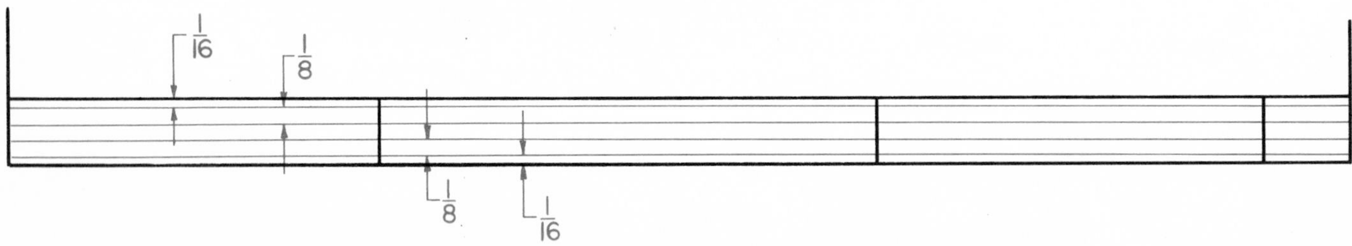

Fig. 4-32. Layout of guide lines for lettering.

DRAWING NUMBER

NAME OF YOUR SCHOOL	TITLE OF DRAWING	YOUR NAME SECTION DATE	5-3
SCHOOL LOCATION		SCALE CHECKED BY	

Fig. 4-33. Information suggested to be lettered on drawing sheets for material in this text.

TEST YOUR KNOWLEDGE - UNIT 4

1. Identify these lines:

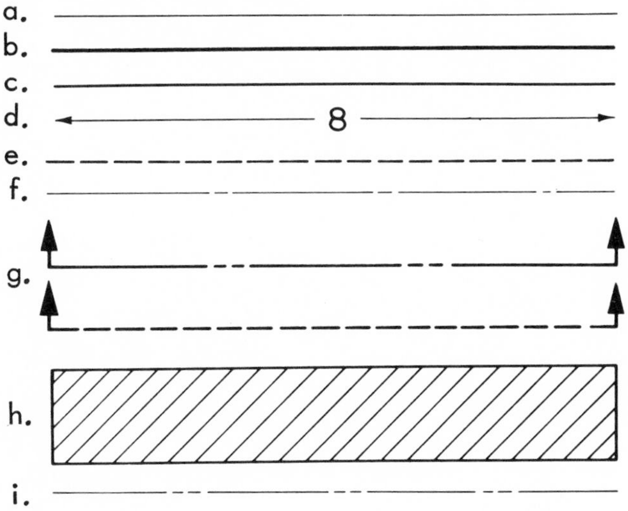

a. ______

b. ______

c. ______

d. ← 8 →

e. — — — — — —

f. ——— - ———

g.

h.

i. ——— -- ———

2. In drafting room language, the characteristics of the above lines and their correct use are known as ______.
3. Why is it important for a draftsman to be able to measure accurately?
4. In drafting, the term SCALE has two meanings. What are they?
5. Prepare sketches which show four different shapes of scales available.
6. List three types of scales in common use.
7. In drafting, horizontal lines are drawn using the ______.
8. Vertical lines are drawn using ______.
9. When erasing, the ______ is often used to protect surrounding areas.
10. Circles and arcs are drawn with a ______.
11. List three methods used to attach the drawing sheet to the board.
    a. ______.
    b. ______.
    c. ______.
12. The method of attaching drawing sheets used most is ______.

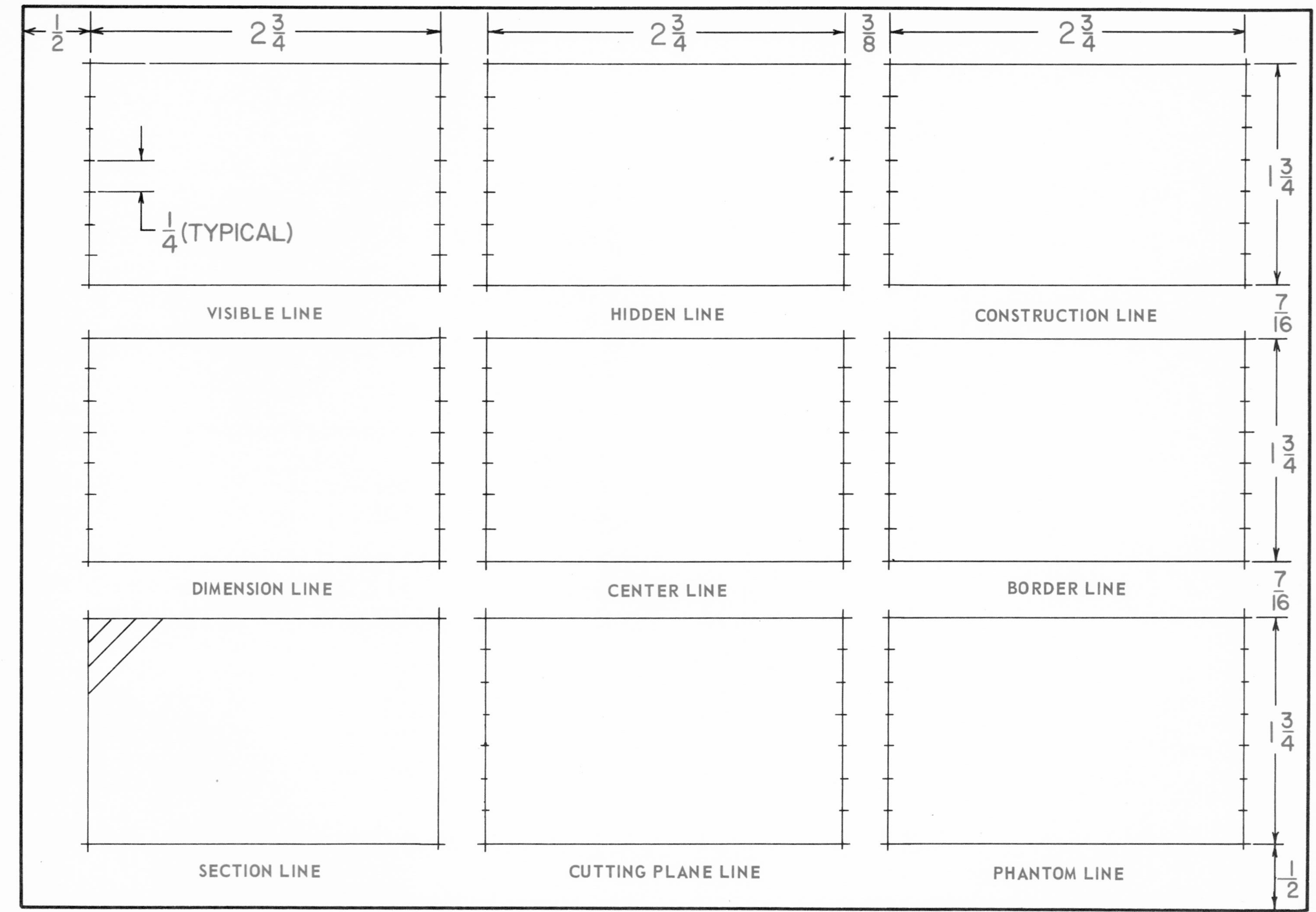

PROBLEM SHEET 4–1. ALPHABET OF LINES. *Duplicate this drawing on a separate sheet of paper. Construct the nine different lines called for.*

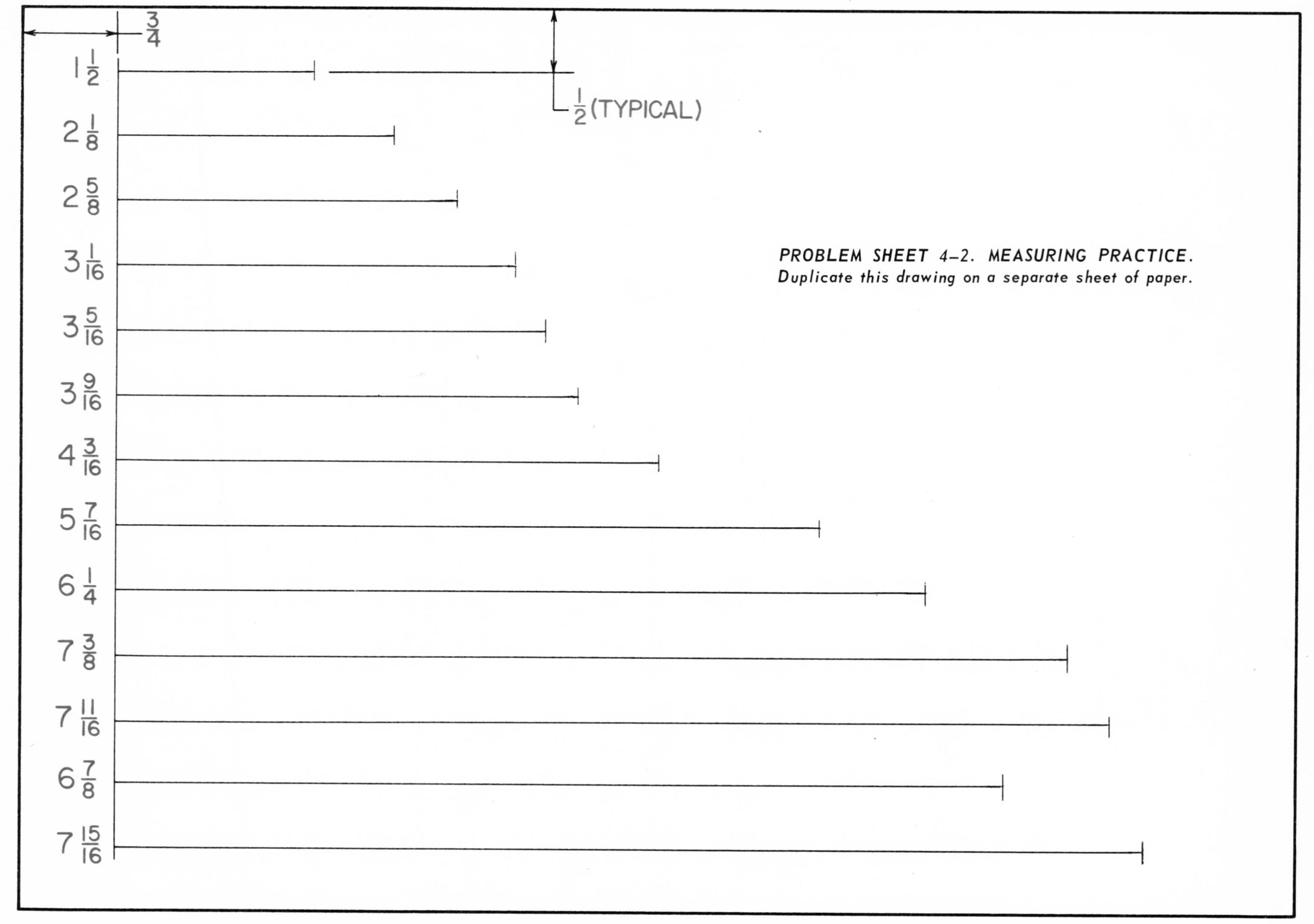

PROBLEM SHEET 4–2. MEASURING PRACTICE.
Duplicate this drawing on a separate sheet of paper.

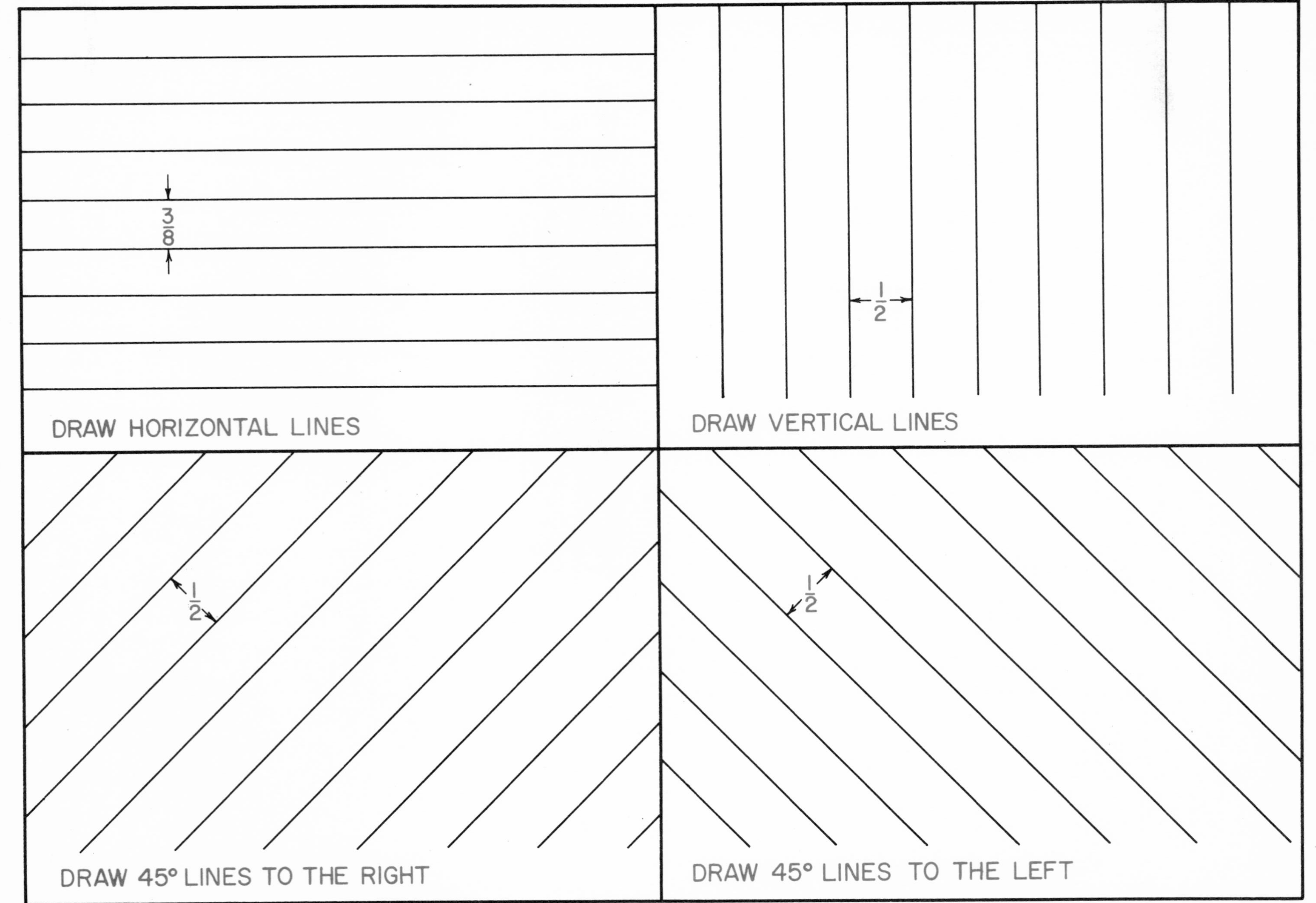

PROBLEM SHEET 4–3. INSTRUMENT PRACTICE.

$\frac{1}{2}$

DRAW 60° LINES TO THE RIGHT

$\frac{1}{2}$

DRAW 60° LINES TO THE LEFT

$\frac{1}{2}$

DRAW 30° LINES TO THE RIGHT

$\frac{1}{2}$

DRAW 30° LINES TO THE LEFT

PROBLEM SHEET 4–4. INSTRUMENT PRACTICE.

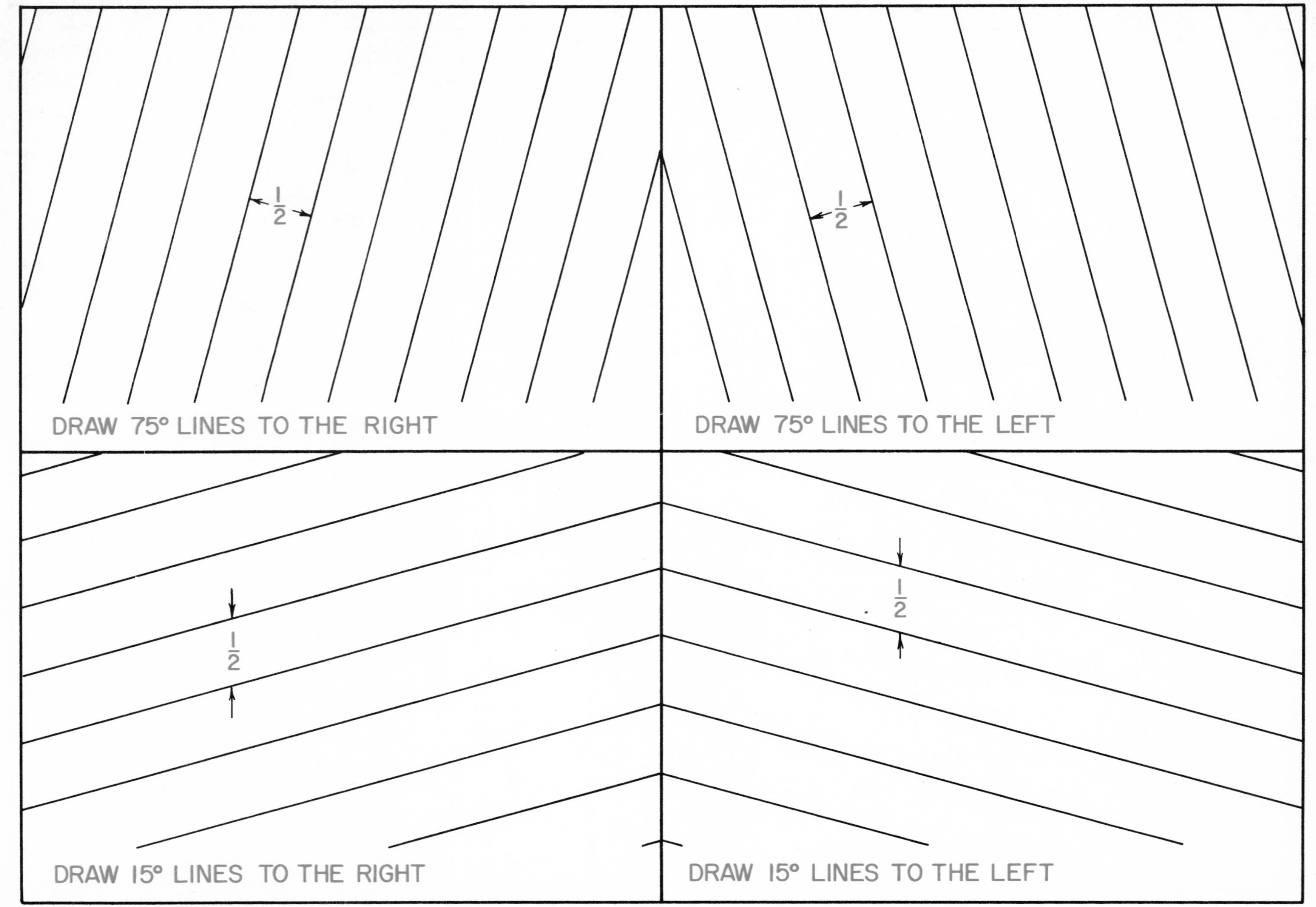

PROBLEM SHEET 4-5. INSTRUMENT PRACTICE.

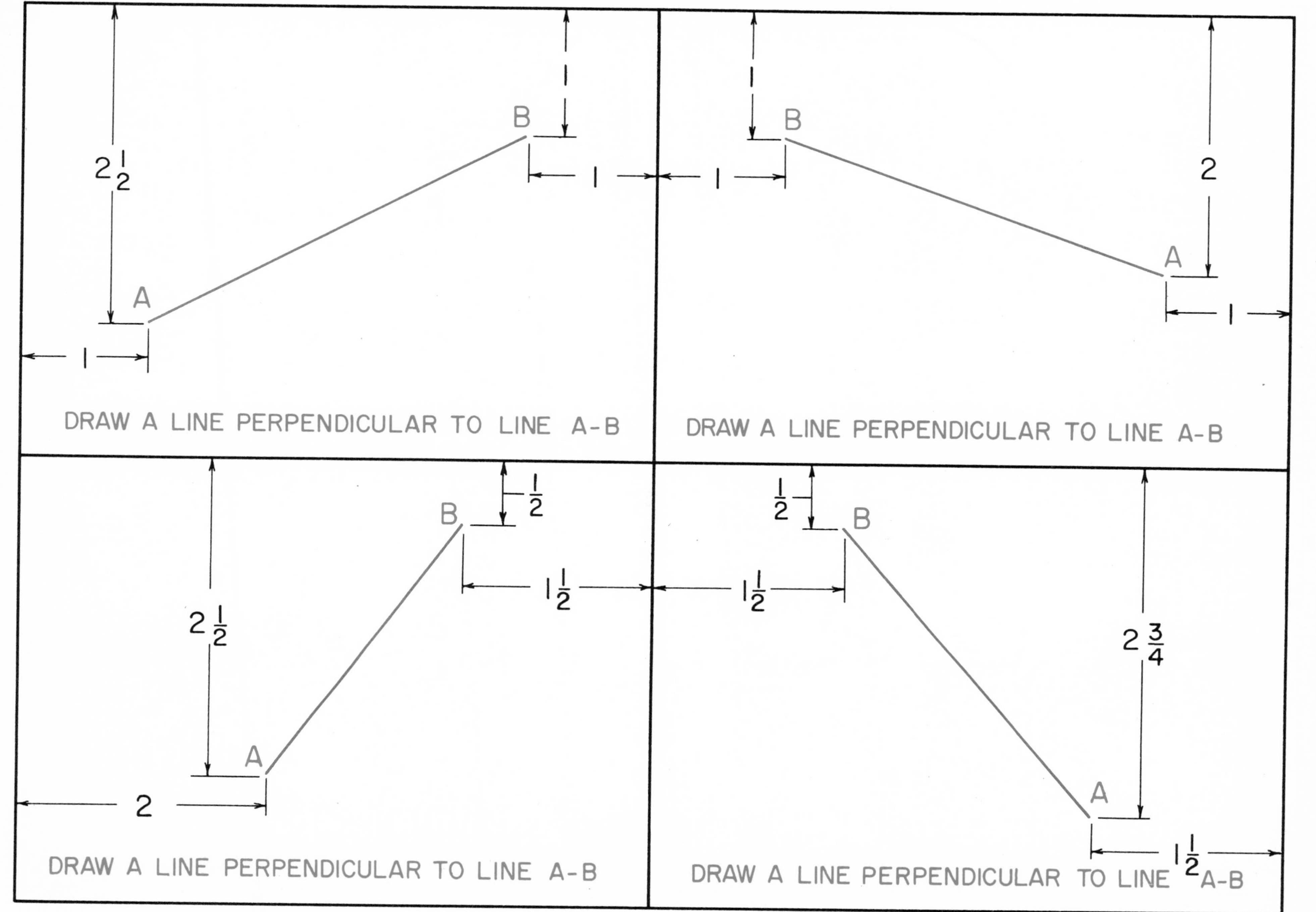

PROBLEM SHEET 4-6. INSTRUMENT PRACTICE.

Fig. 5-1. Wood project designs (novelty boxes) which involve several geometric figures.

Unit 5
GEOMETRICAL CONSTRUCTION

In your daily activities you have many encounters with geometry. The design of the airplane that flies overhead, and the automobile that passes on the street is based on geometrics. Buildings and bridges utilize squares, rectangles, triangles, circles and arcs in their construction. Every mechanical drawing and project is composed of one or more geometrial shapes. See Fig. 5-1.

It is important that you develop the ability to visualize and draw the basic geometric shapes presented in this Unit. This will aid you in solving drafting problems, and provide an opportunity for you to improve your skill in using drafting instruments.

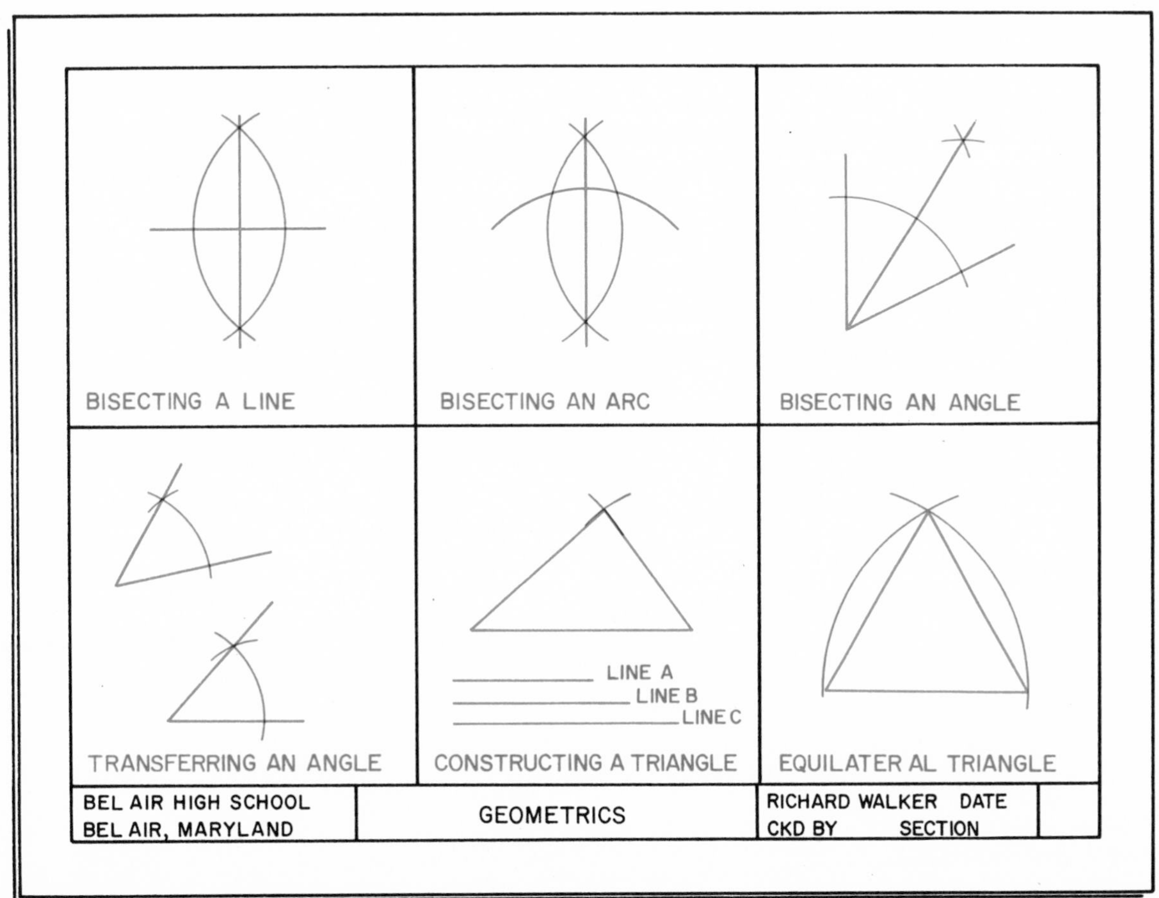

Fig. 5-2. In drawing geometric shapes, several figures may be placed on a single sheet.

DRAWING BASIC GEOMETRIC PROBLEMS

Since size does not enter into the solution of most of the geometric problems presented in this Unit, no dimensions are given. Most of the problems require little space in their solution; therefore, several problems may be included on a single drawing sheet. A suggested sheet layout is given in Fig. 5-2.

The drawing sheets can be made more interesting and attractive if colored pencils are used to draw the lines used to construct the geometric figures.

HOW TO BISECT OR FIND THE MIDDLE OF A LINE

(The bisecting line will be at right angles (90 deg.) to the given line.)

1. Let line A-B be the line to be bisected.
2. Set your compass to a distance larger than one half the length of the line to be bisected. Using this setting as the radius and the end of the line at A as the center point, draw arc C-D. Using the same compass setting but the end of the line at B as the center point, draw arc E-F.
3. Draw a line through the points where the arcs intersect. This line will be at right angles (90 deg. or perpendicular) to and bisect the original line A-B.

HOW TO BISECT AN ARC

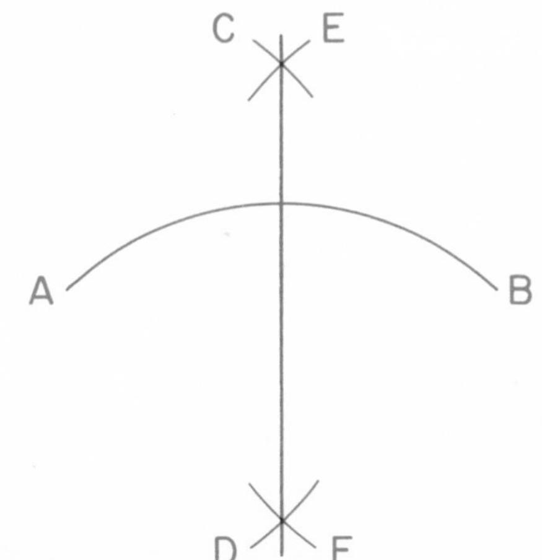

1. An arc or part of a circle is bisected by the same method described in HOW TO BISECT OR FIND THE MIDDLE OF A LINE.

HOW TO BISECT AN ANGLE

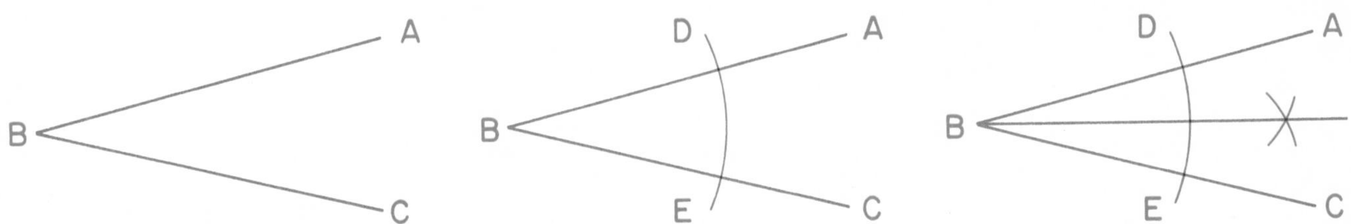

1. Let lines A-B and B-C be the angle to be bisected.
2. With B as the center, draw an arc intersecting the angle at D and E.
3. Using a compass setting greater than one half D-E as centers, draw intersecting arcs. A line through this intersection and B will bisect the angle.

HOW TO TRANSFER OR COPY AN ANGLE

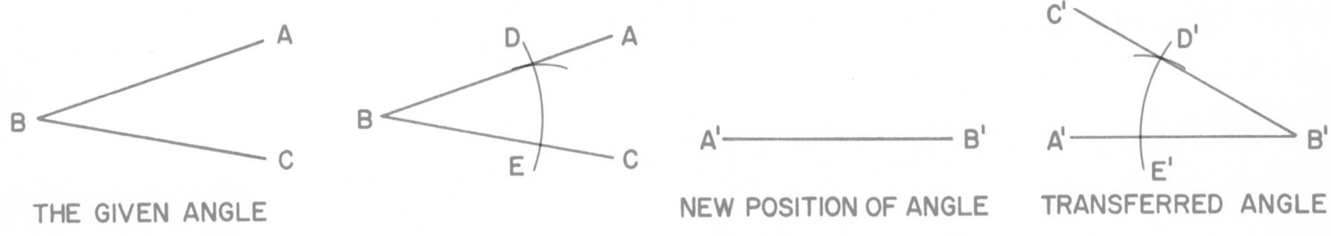

1. Let lines A-B and B-C be the angle to be transferred or copied.
2. Locate the new position of the angle and draw line A'-B'.
3. With B as the center point, draw an arc of any convenient radius on the given angle. This arc intersects the given angle at D and E.
4. Using the same radius and B' as the center draw arc D'-E'.
5. Set your compass equal to D-E. With point D' as a center and D-E as the radius, strike an arc which intersects the first arc at point E'.
6. Draw a line through the intersecting arcs to complete the transfer of the given angle.

HOW TO CONSTRUCT A TRIANGLE FROM GIVEN LINE LENGTHS

1. Let lines A, B and C be the sides of the required triangle.
2. Draw a line that is equal in length to line A.
3. Set your compass to a length equal to line B. Use one end of line A as a center and strike an arc. Reset your compass to a length equal to line C and with the other end of line A as a center strike another arc.
4. Connect the ends of line A to the points where the two arcs intersect.

HOW TO CONSTRUCT AN EQUILATERAL TRIANGLE

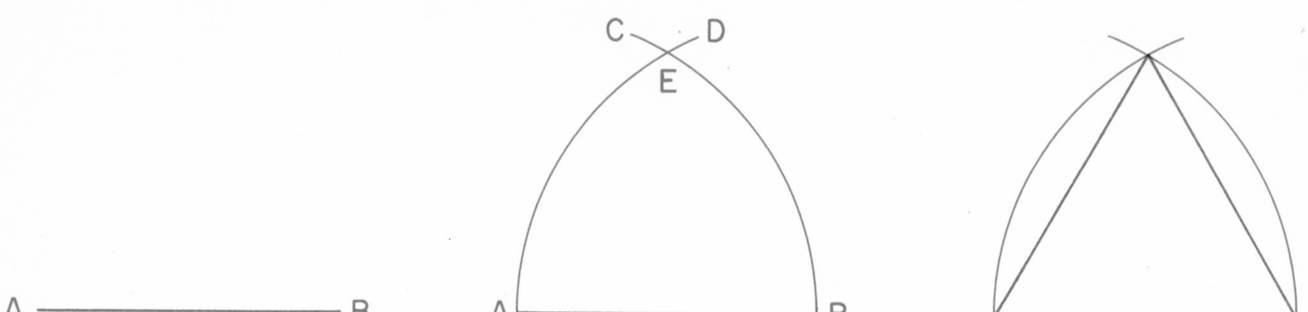

(A triangle having all sides equal in length.)

1. Let line A-B be the length of the sides of the triangle.
2. With A as the center and with the compass setting equal to the length of line A-B, strike the arc B-C. Using the same compass setting but with B as the center strike the arc A-D. These arcs intersect at E.
3. Complete the triangle by connecting A to E and B to E.

HOW TO DRAW A SQUARE WITH THE DIAGONAL GIVEN

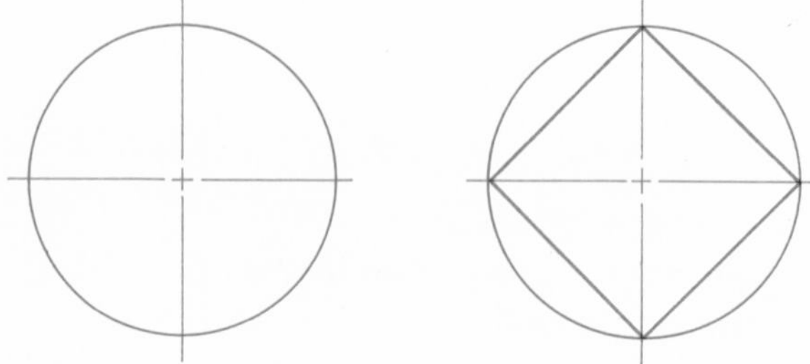

1. Draw a circle with a diameter equal to the length of the diagonal.
2. Connect the points where the center lines intersect the circle.

HOW TO DRAW A SQUARE WITH THE SIDE GIVEN

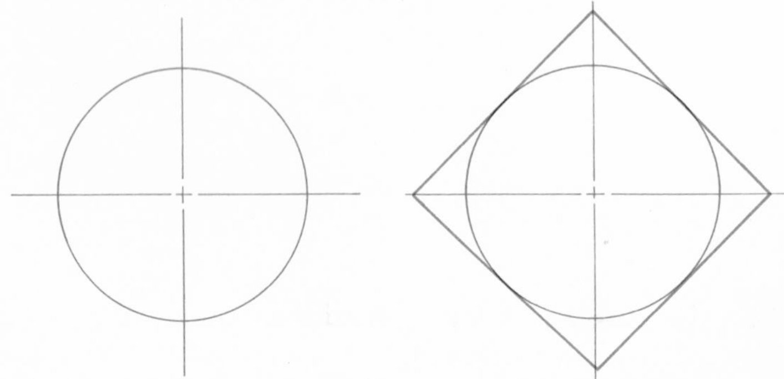

1. Draw a circle with a diameter equal to the length of the side.
2. Draw tangents at a 45 deg. to the center lines.

HOW TO CONSTRUCT A PENTAGON OR FIVE POINT STAR

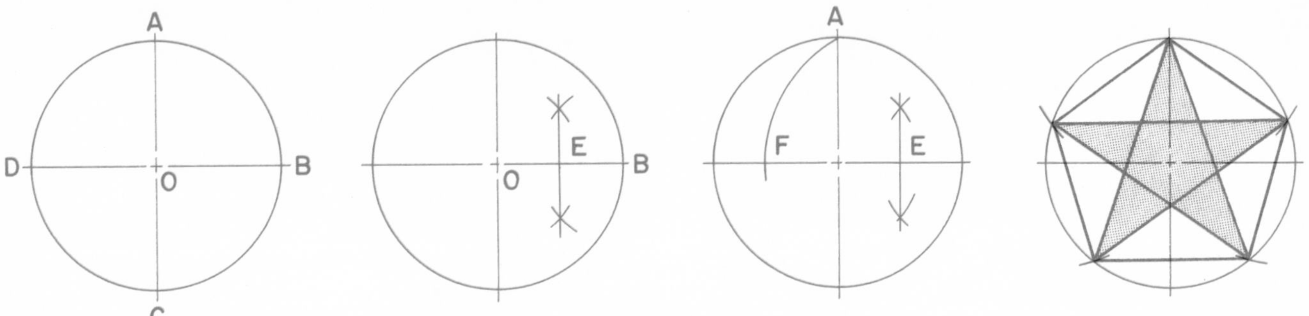

1. Draw a circle. Let A-C and B-D be the center lines and O be the point where the center lines intersect.
2. Bisect the line (radius) O-B. This will locate point E.
3. With E as the center and with the compass set to the radius E-A strike the arc A-F.
4. The distance A-F is one-fifth the circumference of the circle. Set your compass or dividers to this distance and circumscribe the circle starting at A. Connect these points as a pentagon or as a five point star.

HOW TO CONSTRUCT A HEXAGON (FIRST METHOD)

1. Draw a circle.
2. With a compass setting equal to the radius of the drawn circle start at point A and circumscribe the circle locating points B, C, D, E, F.
3. Connect the points with a straightedge to complete the hexagon.

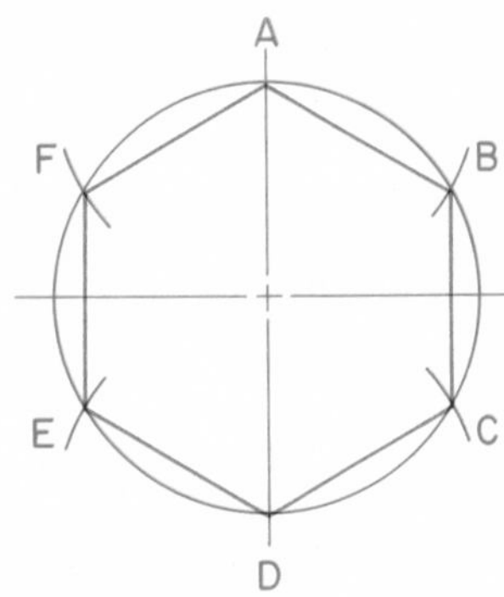

HOW TO CONSTRUCT A HEXAGON (SECOND METHOD)

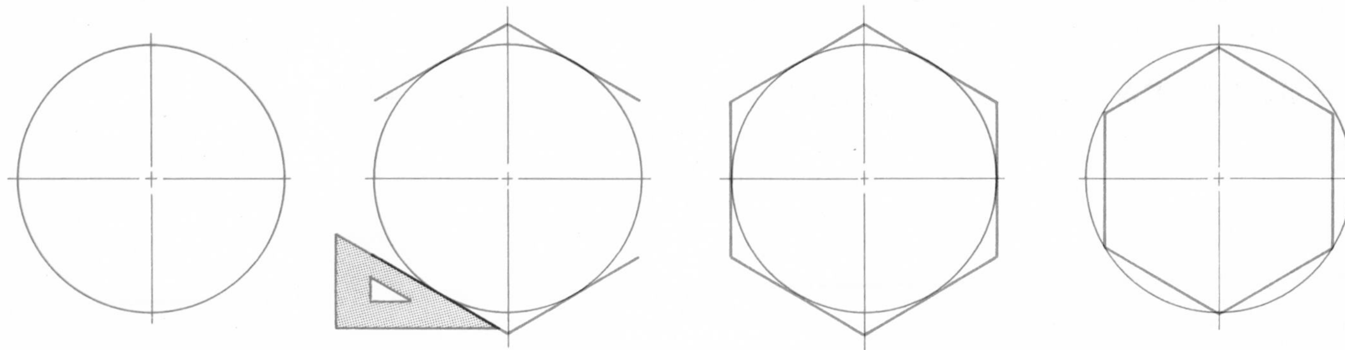

1. Draw a circle.
2. Use the 30-60 deg. triangle and draw construction lines at 60 deg. to the vertical center line and tangent to the circle.
3. Draw two vertical construction lines tangent to the circle. Fill in the construction lines with visible object lines to complete the hexagon.
4. The same basic technique can be used to draw (inscribe) a hexagon inside the circle.

HOW TO DRAW AN OCTAGON USING A CIRCLE

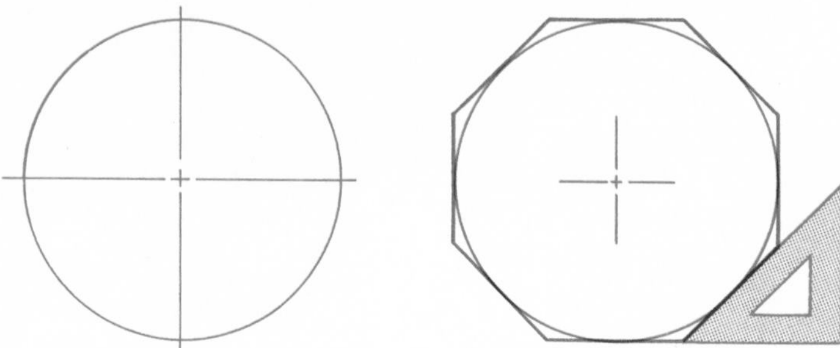

1. Draw a circle with a diameter equal to the distance across the flats of the desired octagon.
2. Draw the vertical and horizontal lines tangent to the circle. Use construction lines.
3. Complete the octagon by drawing the 45 deg. angle lines tangent to the circle. Fill in the construction lines with visible object lines.

HOW TO DRAW AN OCTAGON USING A SQUARE

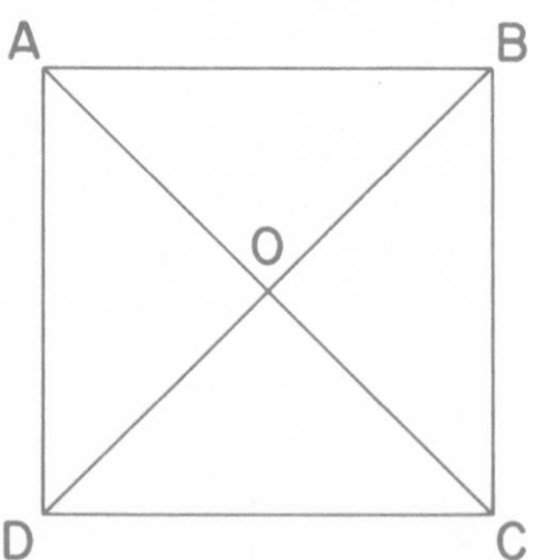

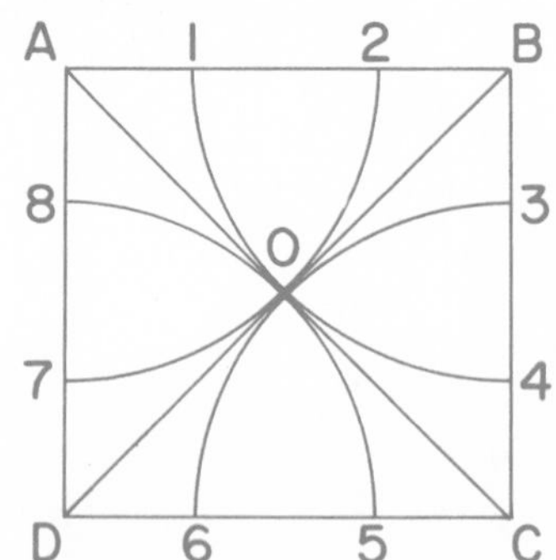

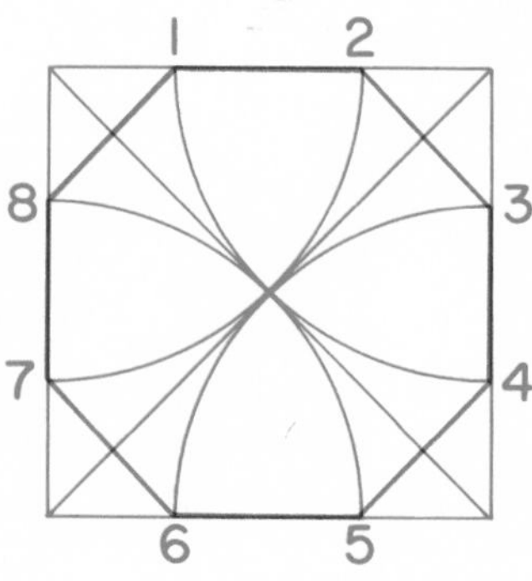

1. Draw a square with the sides equal in length to the distance across the flats of the required octagon. Draw the diagonals A-C and B-D. The diagonals intersect at O.
2. Set your compass to radius A-O and with corners A, B, C and D as centers, draw arcs that intersect the square at points 1, 2, 3, 4, 5, 6, 7 and 8.
3. Complete the octagon by connecting point 1 to 2, 2 to 3, 3 to 4, 4 to 5, 5 to 6, 6 to 7, 7 to 8 and 8 to 1 with object lines.

HOW TO DRAW AN ARC TANGENT TO TWO LINES AT A RIGHT ANGLE

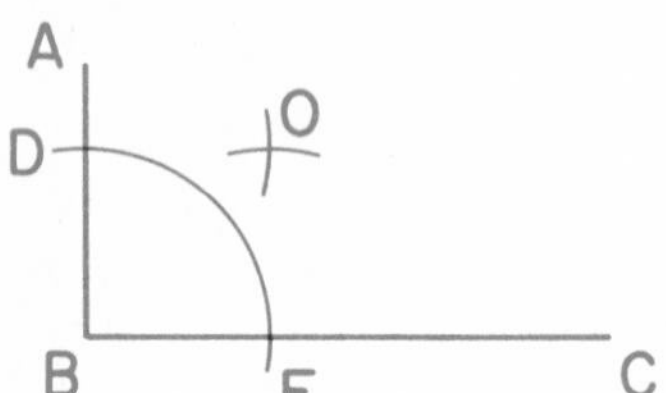

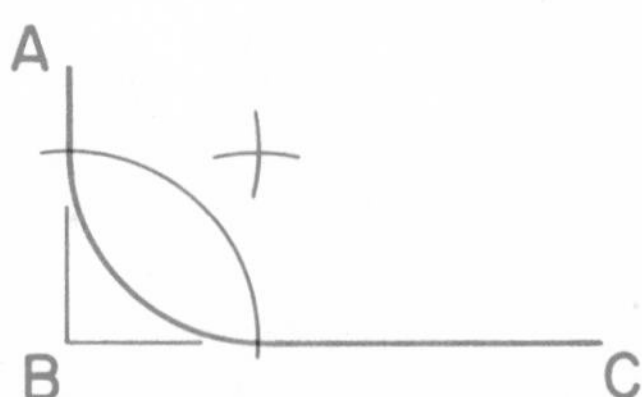

1. Let A-B and B-C be the lines that form the right angle.
2. Set your compass to the radius of the required arc and using B as the center strike arc D-E. With D and E as centers and with the same compass setting draw the arcs that intersect at O.
3. With O as the center and with the compass at the same setting draw the required arc. It will be tangent to lines A-B and B-C at points D and E.

HOW TO DRAW AN ARC TANGENT TO TWO STRAIGHT LINES

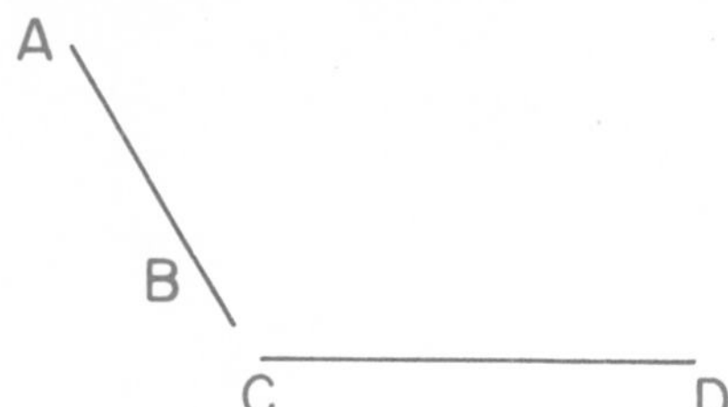

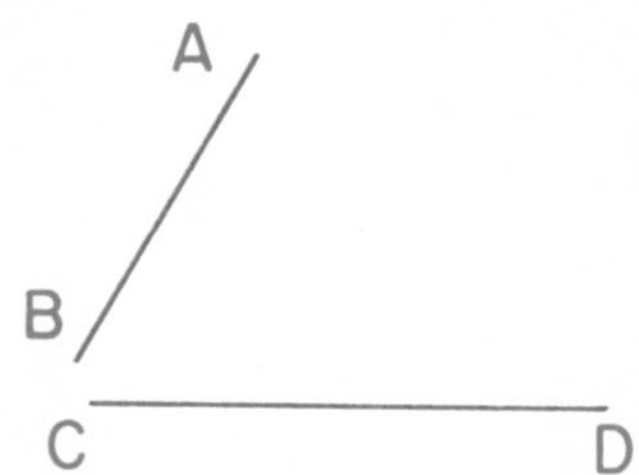

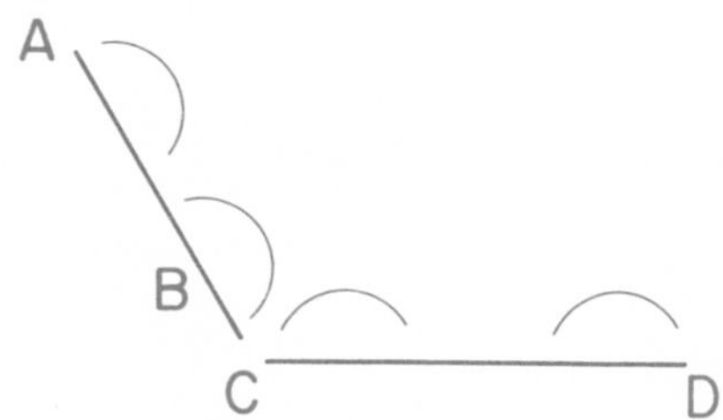

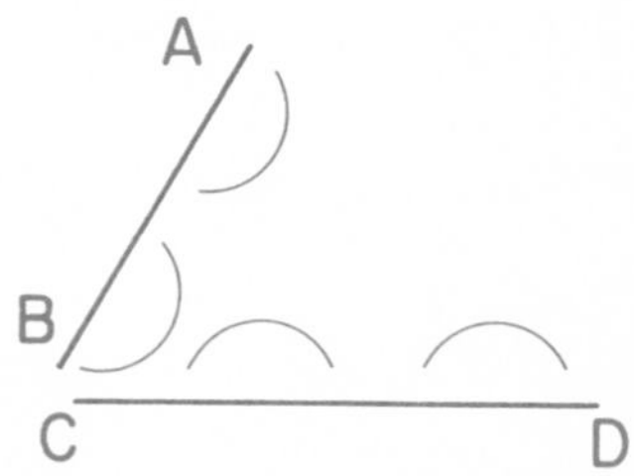

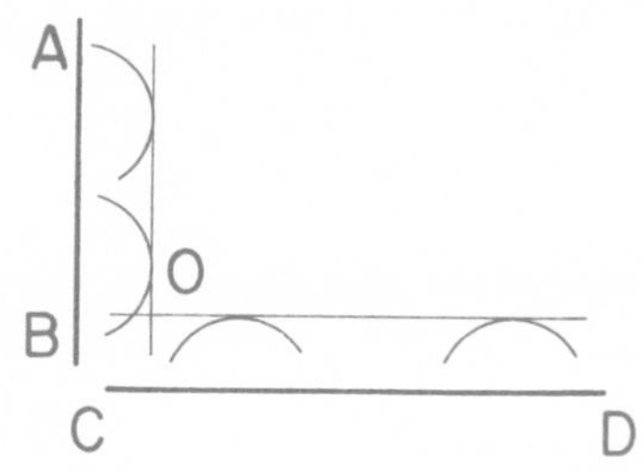

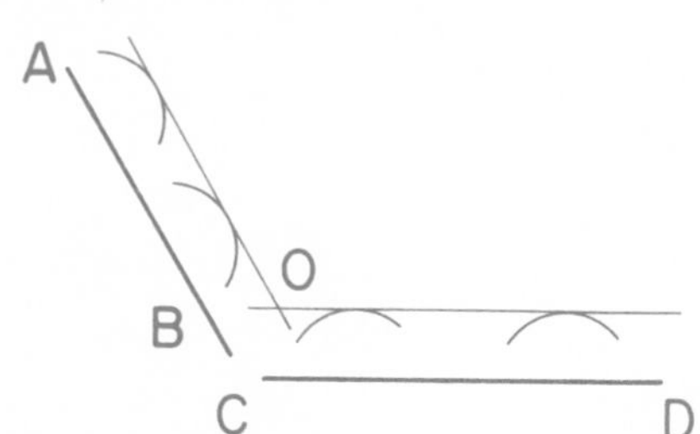

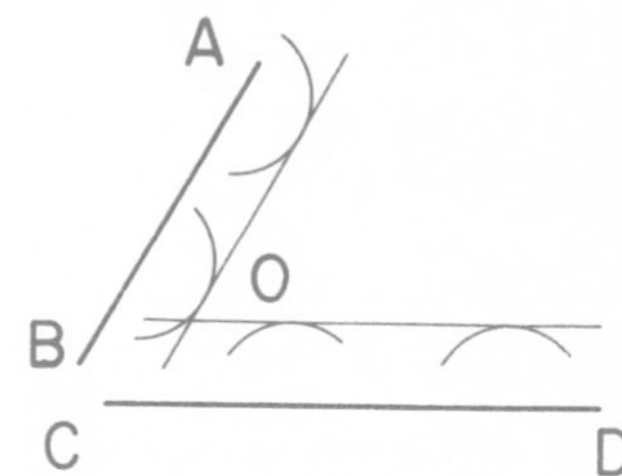

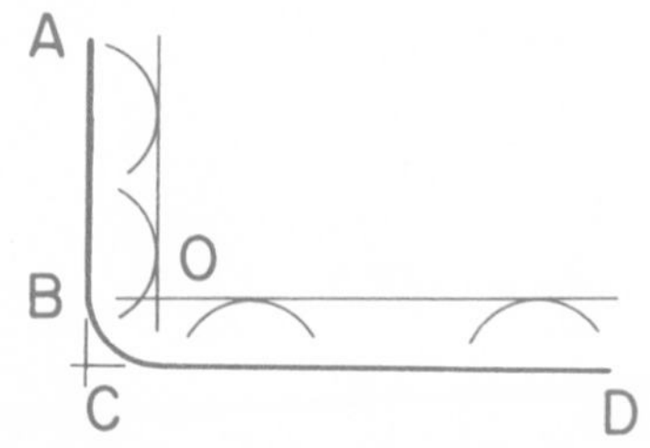

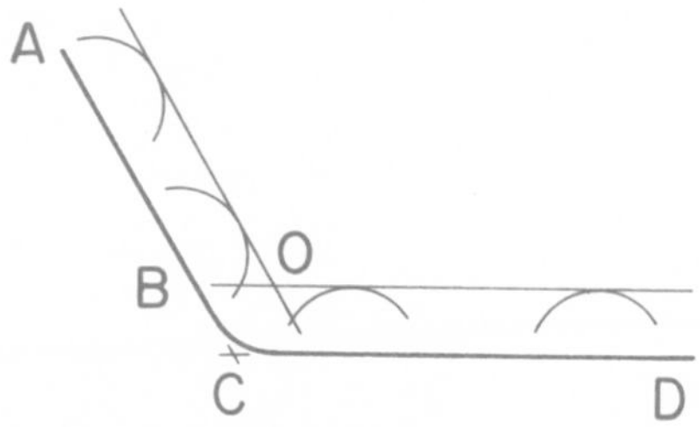

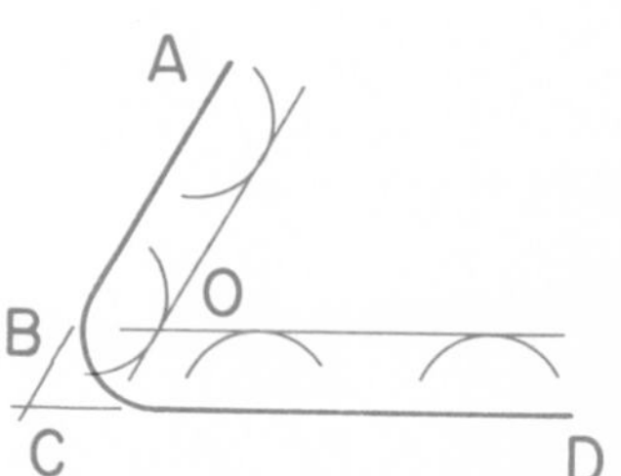

1. Let lines A-B and C-D be the two straight lines.
2. Set your compass to the radius of the arc to be drawn tangent to the two straight lines and with points near the ends of the lines A-B and C-D as centers strike two arcs on each line.
3. Draw straight construction lines tangent to the arcs.
4. The point where the two lines intersect (0) is the center for drawing the required arcs.

HOW TO DRAW AN ARC TANGENT WITH A STRAIGHT LINE AND A GIVEN ARC

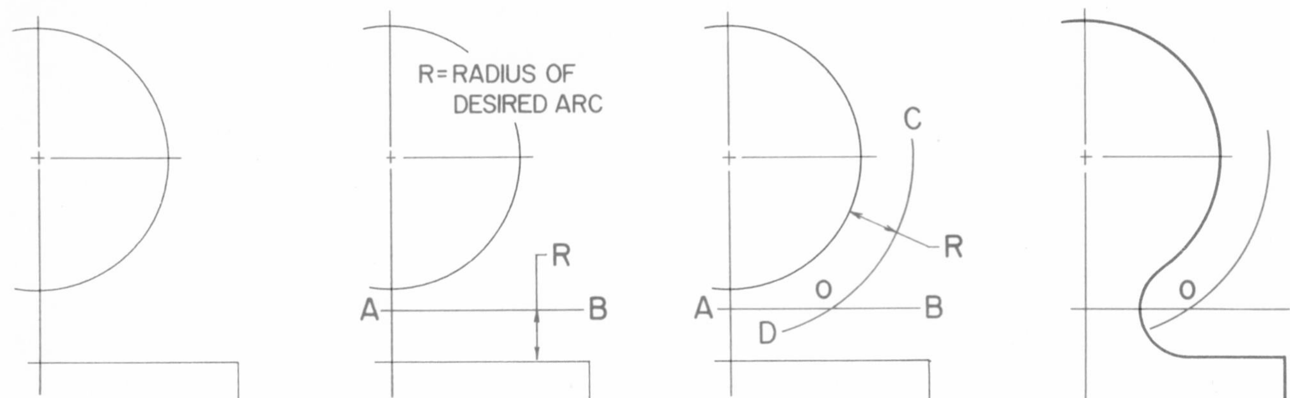

1. Draw the given arc and the straight line in proper relation to one another. Use construction lines.
2. Draw line A-B a distance equal to the radius of the desired arc from and parallel to the straight line.
3. Draw arc C-D by setting your compass to a radius equal to the radius of the given arc plus the radius of the desired arc. This arc intersects with line A-B at point O.
4. Using point O as the center and with the compass set to the radius of the desired arc draw the required arc. It will be tangent to the given arc and to the straight line. Fill in construction lines with object lines.

HOW TO DRAW TANGENT ARCS

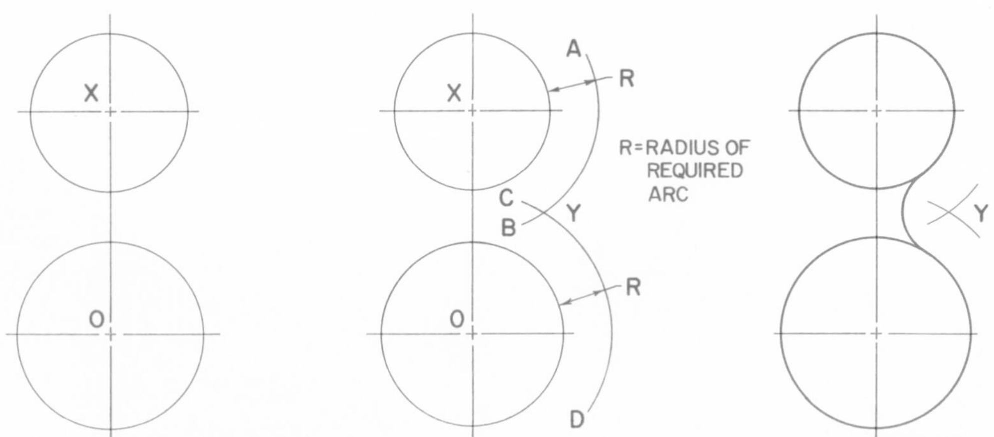

1. Draw the two arcs in proper relation to each other that require the tangent arc to join them. Let O and X be the centers of these arcs.
2. Set the compass to a distance equal to the radius of the given circle plus the radius of the required arc. Using X as the center draw arc A-B. Reset the compass to a distance equal to the radius of the second given circle plus the radius of the required arc. Using O as the center draw arc C-D. These arcs intersect at Y.
3. Set the compass to the radius of the required arc. With Y as the center draw the desired arc. This arc will be tangent to the given arcs.

HOW TO DIVIDE A LINE INTO A GIVEN NUMBER OF EQUAL DIVISIONS

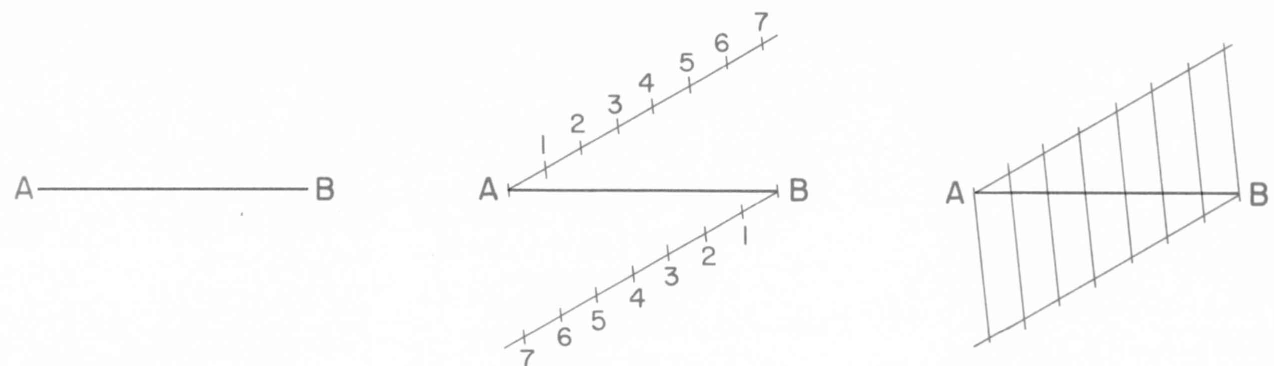

1. PROBLEM: Divide line A–B into seven equal divisions.
2. Project construction lines from A and B that are parallel to one another.
3. Open your compass or dividers to a suitable setting (approximately 1/4 in.) and starting at A step off seven spaces on the projected line. Starting at B step off seven spaces on the other projected line.
4. Connect the points with construction lines as shown. The points of division are where these lines pass through the given line A–B.

HOW TO DRAW AN ELLIPSE USING CONCENTRIC CIRCLES

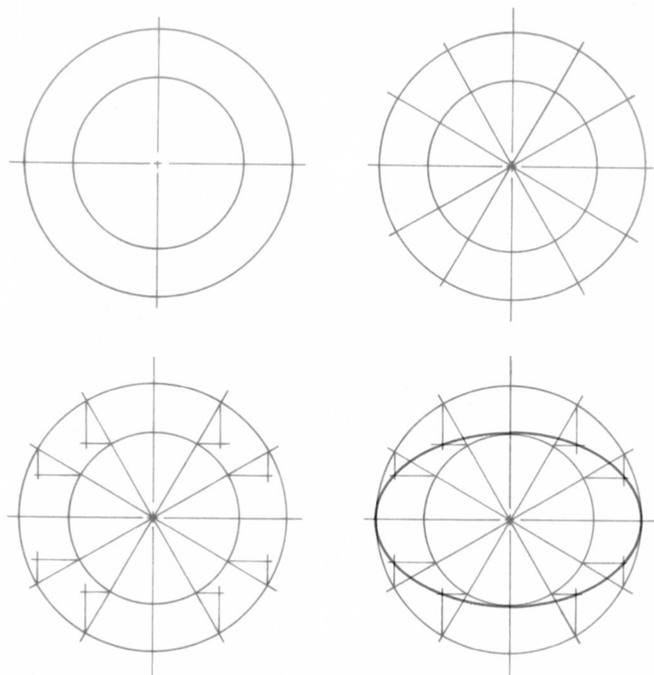

1. Draw two concentric circles. The diameter of the large circle is equal to the length of the large axis of the desired ellipse. The diameter of the small circle is equal to the length of the small axis of the desired ellipse.
2. Divide the two circles into twelve equal parts. Use your 30–60 deg. triangle.
3. Draw horizontal lines from the points where the dividing lines intersect the small circle. Vertical lines are drawn from the points where the dividing lines intersect the large circle.
4. Connect the points where the vertical and horizontal lines intersect with a French curve.

HOW TO DRAW AN ELLIPSE USING THE PARALLELOGRAM METHOD

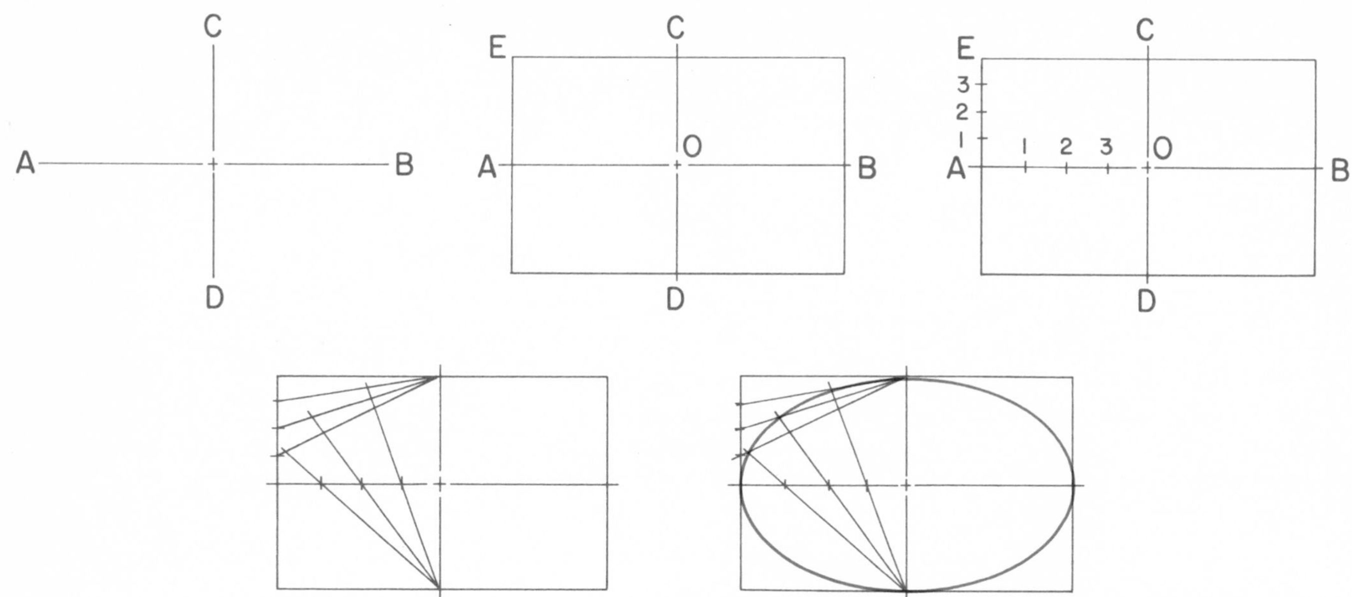

This method is satisfactory for drawing large ellipses.

1. Let line A-B be the major axis and line C-D be the minor axis of the required ellipse.
2. Construct a rectangle with sides equal in length and parallel to the axes.
3. Divide A-0 and A-E into the same number of equal parts.
4. From C draw a line to point 1 on line A-E. Draw a line from D through point 1 on line A-C. The point of intersection of these two lines will establish the first point of the ellipse. The remaining points in this section are completed as are similar points in the other three sections (quadrants).
5. Connect the points with a French curve to complete the ellipse.

OUTSIDE ACTIVITIES

1. Prepare a bulletin board display that illustrates the use of geometric shapes in buildings and bridges.
2. Develop a list of everyday items that make use of geometric shapes. For example: Nut and bolt heads are round, square and hexagonal.
3. Secure a photo or a drawing of a modern airplane. Place a sheet of tracing vellum over it and sketch in the various geometric shapes used in its design.
4. Do the same using a photo or drawing of a late model automobile.

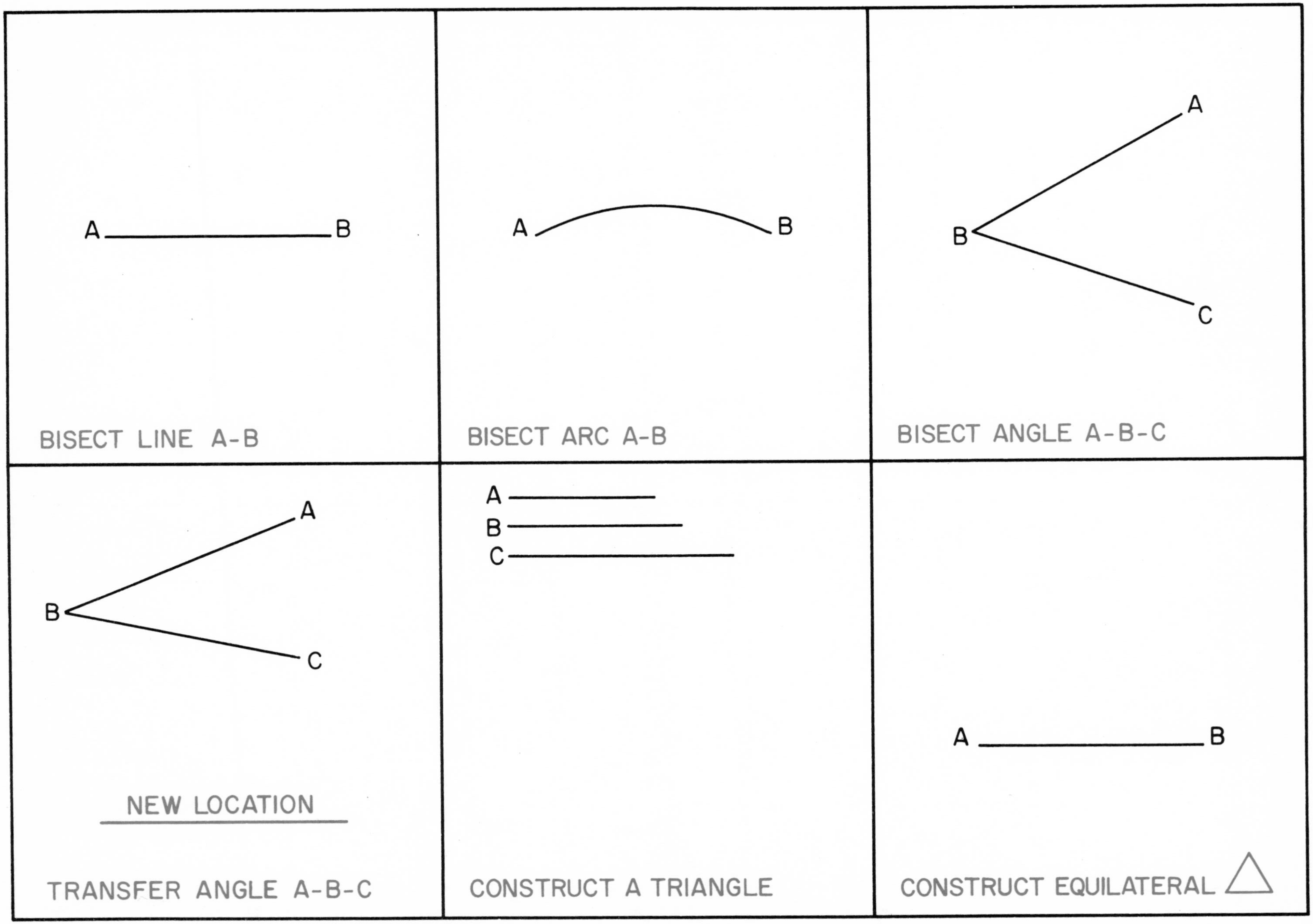

A B

BISECT LINE A-B

A B

BISECT ARC A-B

A B C

BISECT ANGLE A-B-C

A B C

NEW LOCATION

TRANSFER ANGLE A-B-C

A
B
C

CONSTRUCT A TRIANGLE

A B

CONSTRUCT EQUILATERAL △

PROBLEM SHEET 5–1. GEOMETRICS.

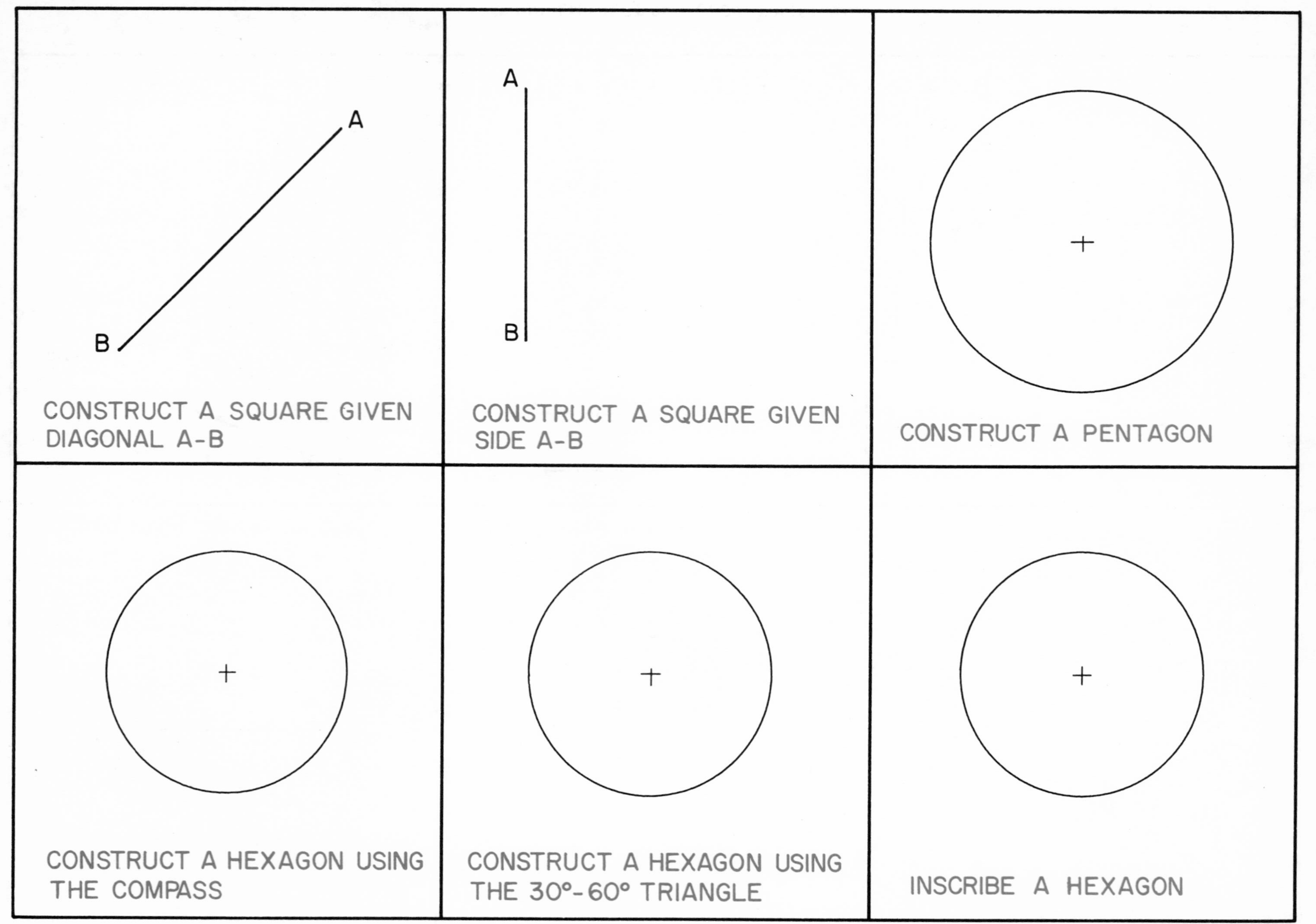

PROBLEM SHEET 5-2. GEOMETRICS.

CONSTRUCT AN OCTAGON

CONSTRUCT AN OCTAGON

A B

DIVIDE LINE A-B INTO NINE (9) EQUAL PARTS

A B C

DRAW AN ARC TANGENT TO LINES A-B AND B-C

A B C

DRAW AN ARC TANGENT TO LINES A-B AND B-C

A B C

DRAW AN ARC TANGENT TO LINES A-B AND B-C

PROBLEM SHEET 5–3. GEOMETRICS.

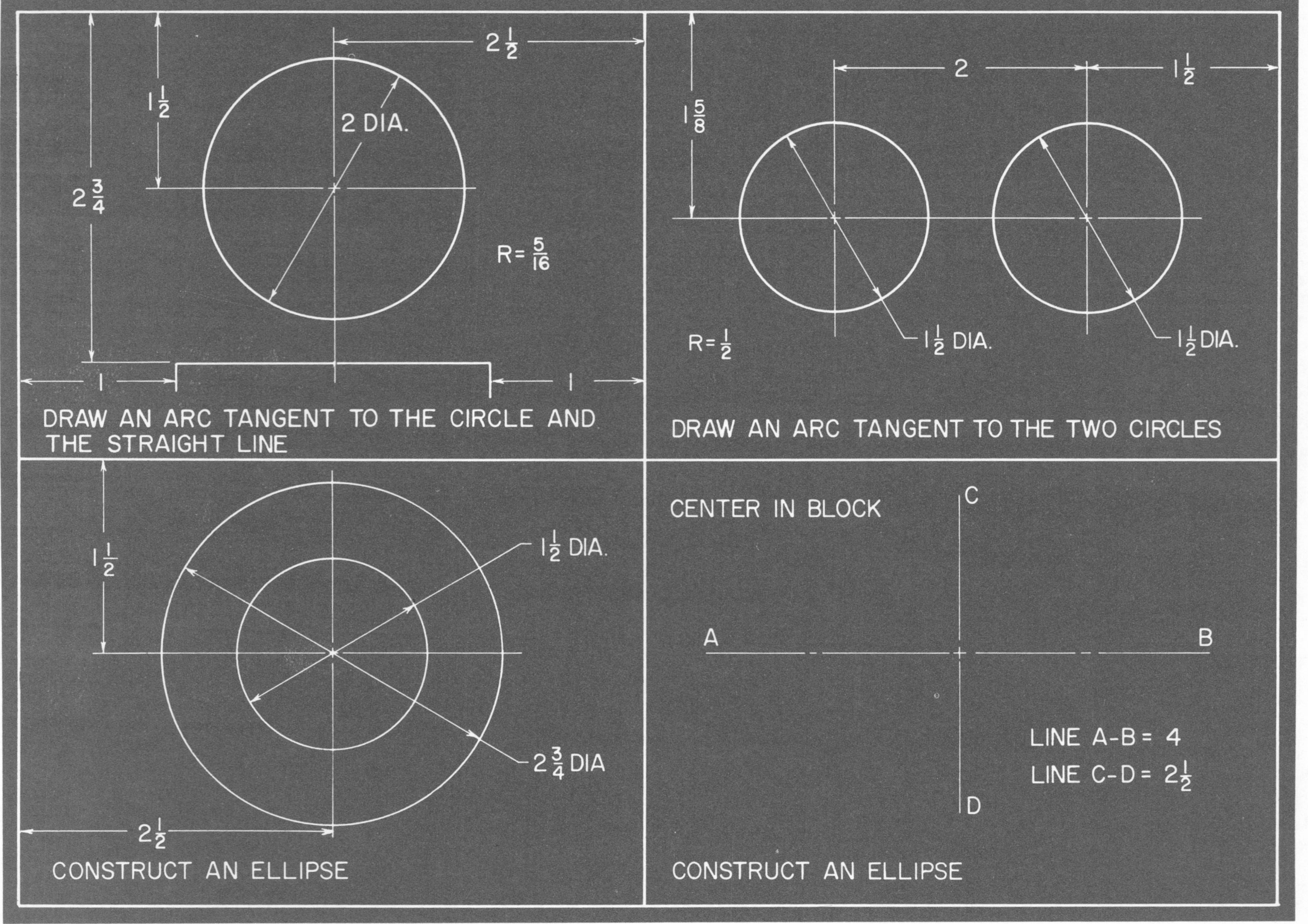

PROBLEM SHEET 5–4. GEOMETRICS.

AIRCRAFT INSIGNIA OF THE WORLD

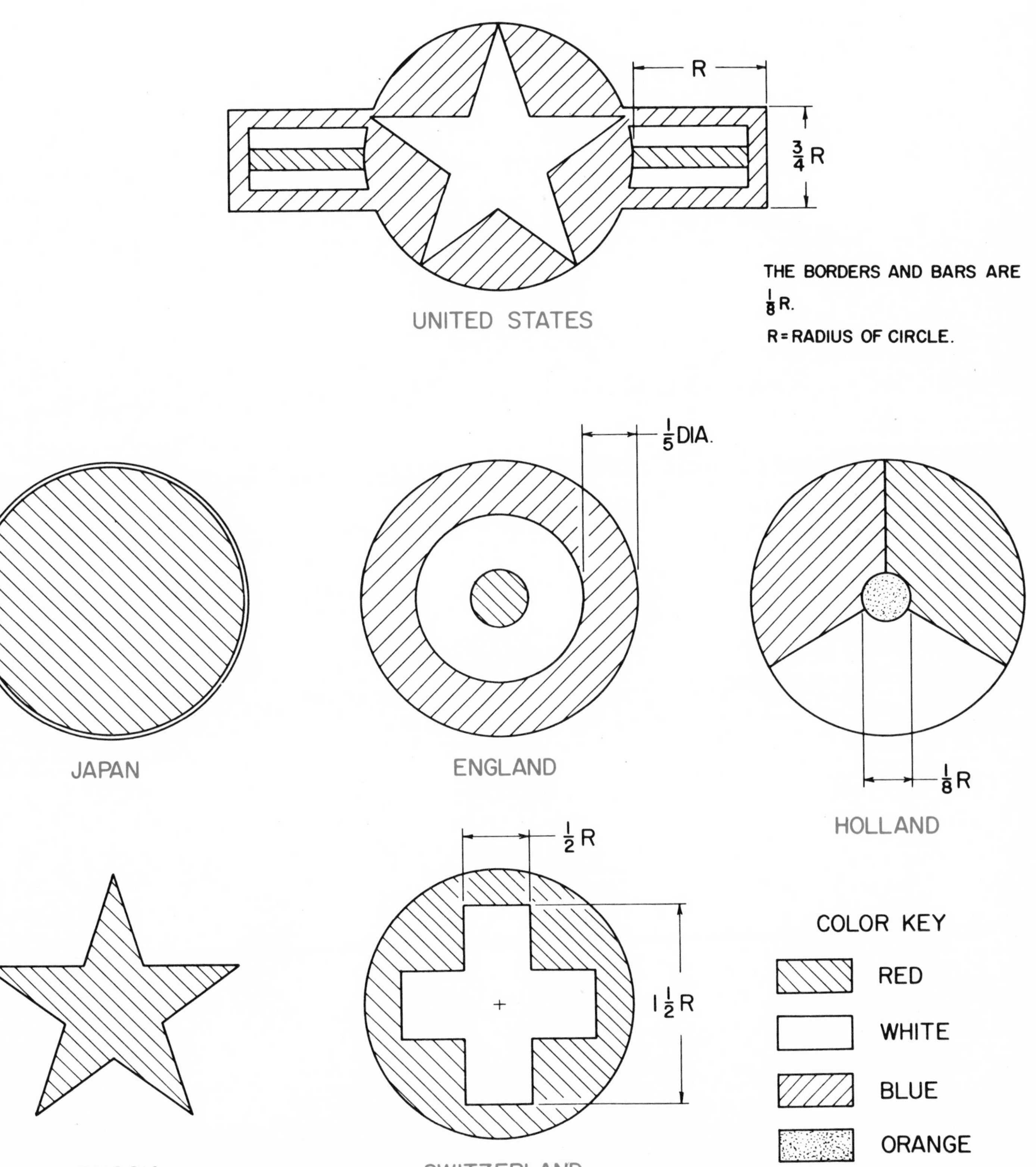

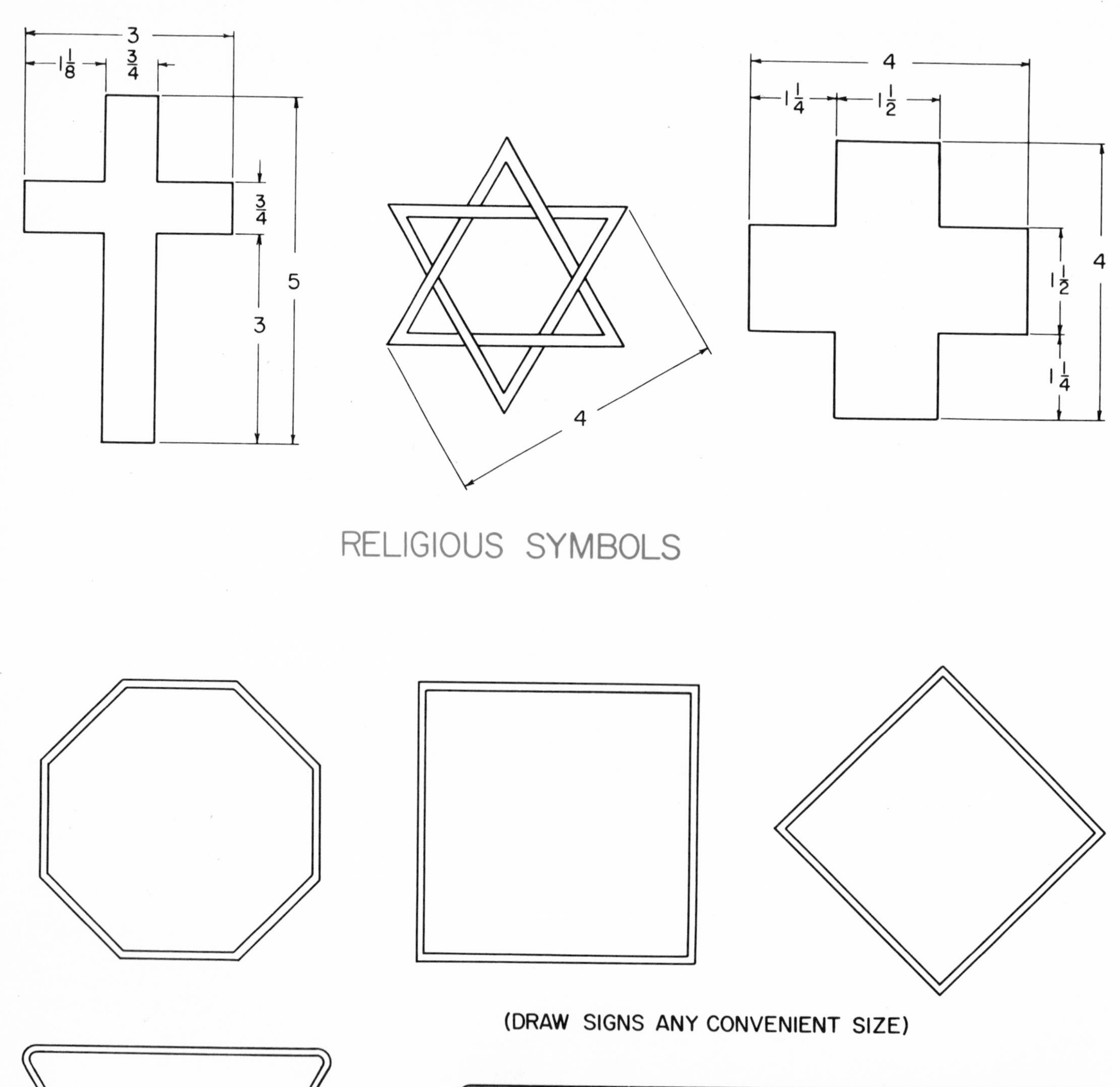

RELIGIOUS SYMBOLS

(DRAW SIGNS ANY CONVENIENT SIZE)

TRAFFIC WARNING SIGNS

HOW IS EACH SIGN USED?

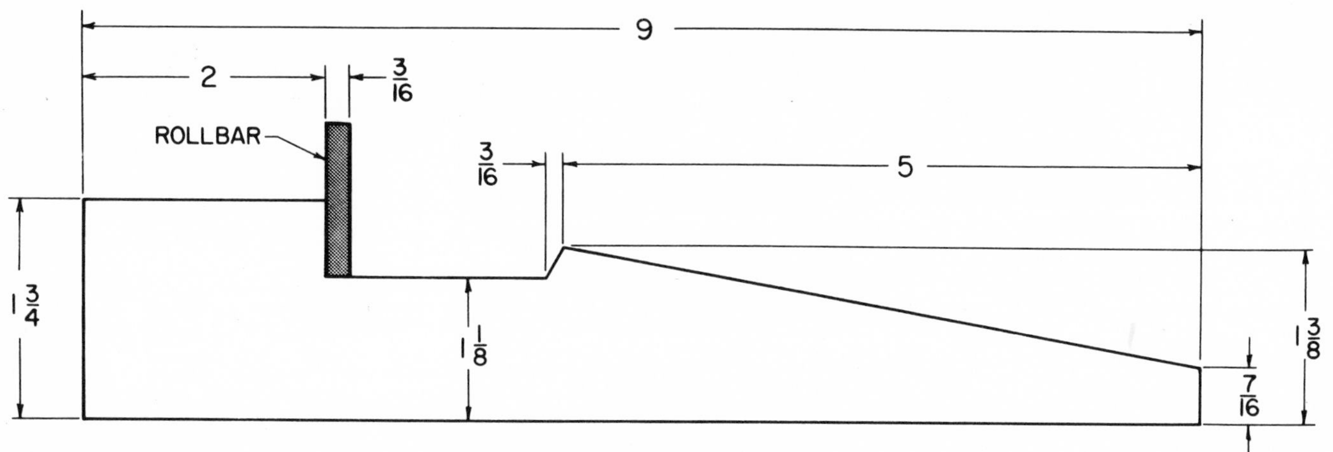

DESIGN PROBLEM 1

DEVELOP DISTINCTIVE RACING STRIPES FOR THIS FORMULA "V" BODY

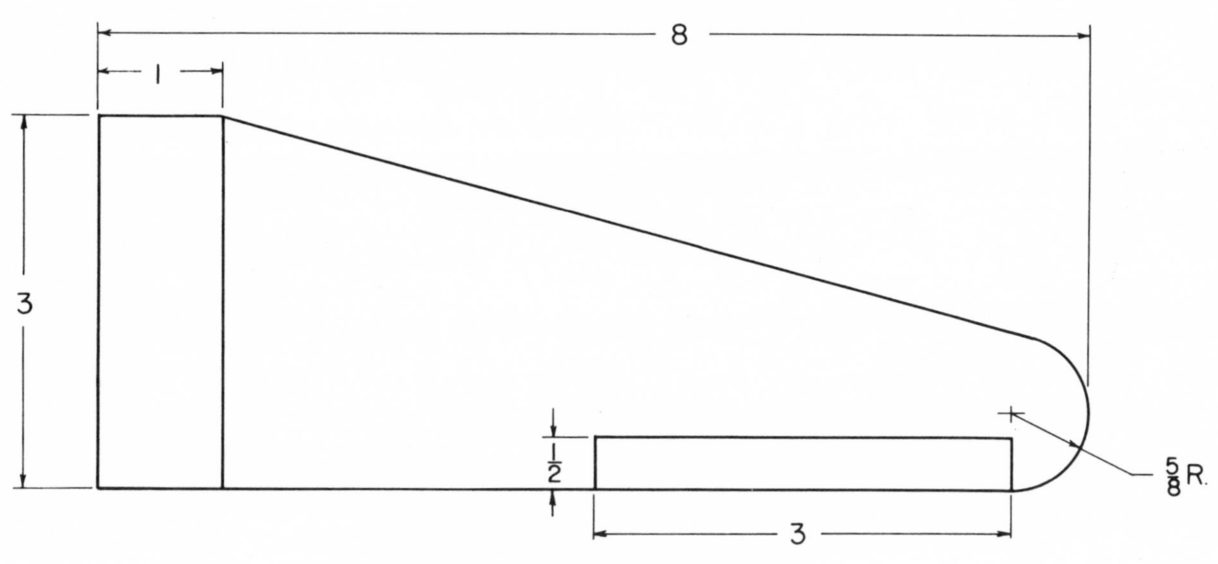

DESIGN PROBLEM 2

DESIGN A UNIQUE COLOR SCHEME FOR THE AMERICAN AIRPLANE TO BE USED IN THE WORLD AEROBATIC CHAMPIONSHIP. USE GEOMETRIC FIGURES.

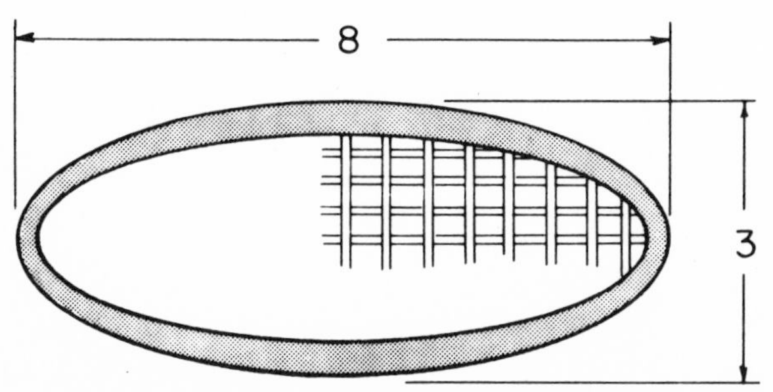

DESIGN PROBLEM 3

DESIGN A GRILL FOR THIS SPORT CAR FRONT.

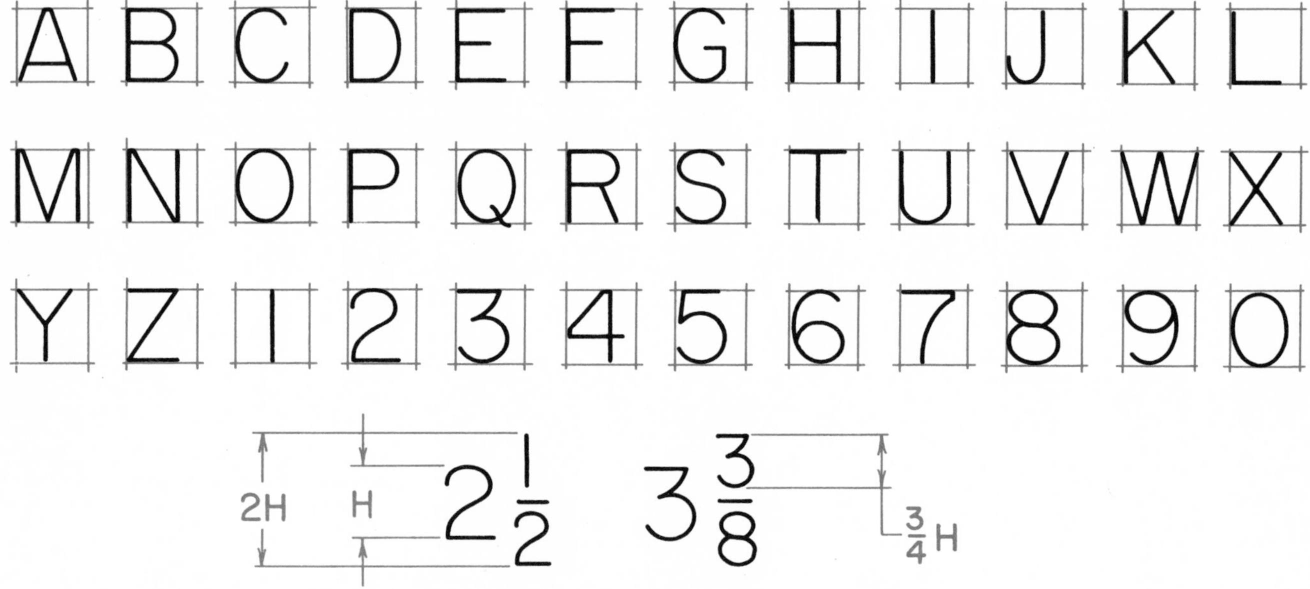

VERTICAL SINGLE STROKE GOTHIC ALPHABET

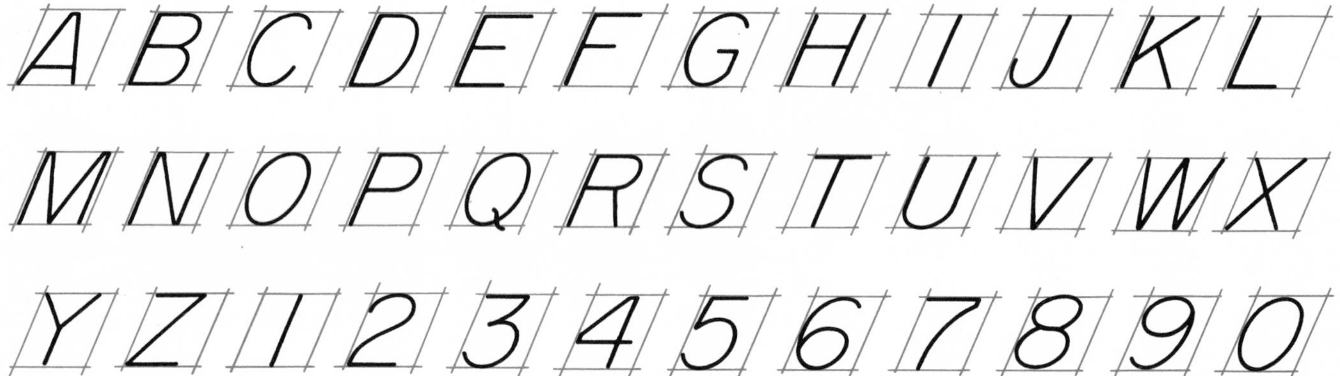

INCLINED SINGLE STROKE GOTHIC ALPHABET

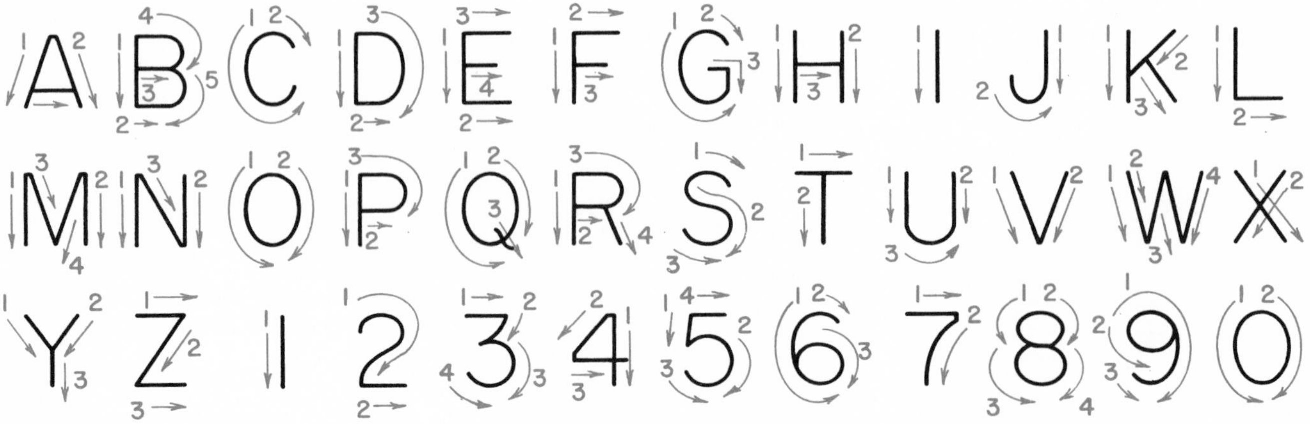

RECOMMENDED SEQUENCE FOR MAKING SINGLE STROKE GOTHIC ALPHABET

Fig. 6-1. The Single Stroke Gothic Alphabet.

Unit 6
LETTERING

LETTERING is used on drawings to give dimensions and other pertinent information needed to fully describe the item. The lettering must be neat and legible if it is to be easily read and understood.

A drawing will be improved by good lettering. A good drawing will look sloppy and unprofessional if the lettering is poorly done.

SINGLE STROKE GOTHIC ALPHABET

The American National Standards Institute (originally ASA) recommends that SINGLE STROKE GOTHIC ALPHABET, Fig. 6-1, be the accepted lettering standard because it can be drawn rapidly and is highly legible. It is called single stroke lettering not because each letter is made with a single stroke of the pencil (most letters require several strokes to complete), but because each line is only as wide as the point of the pencil or pen.

Single stroke lettering may be vertical or inclined, as there is no definite rule stating that it should be one or the other. However, mixing them on a drawing should be avoided, Fig. 6-2.

ONLY ONE FORM OF LETTERING SHOULD APPEAR ON A DRAWING.

AVOID COMbINING SEVERAL fORMS Of LETTERING.

Fig. 6-2. Avoid mixing several forms of lettering on a drawing.

YOU can do first class lettering if you learn the basic shapes of the letters, the proper stroke sequence for making them, and the recommended spacing between letters and words. You must also practice regularly.

LETTERING WITH A PENCIL

A H or 2H pencil is used by most draftsmen for lettering. Sharpen the pencil to a sharp conical point, and rotate it as you letter, Fig. 6-3, to keep the point sharp and the letters uniform in weight and line width. Resharpen the pencil when the lines become wide and "fuzzy."

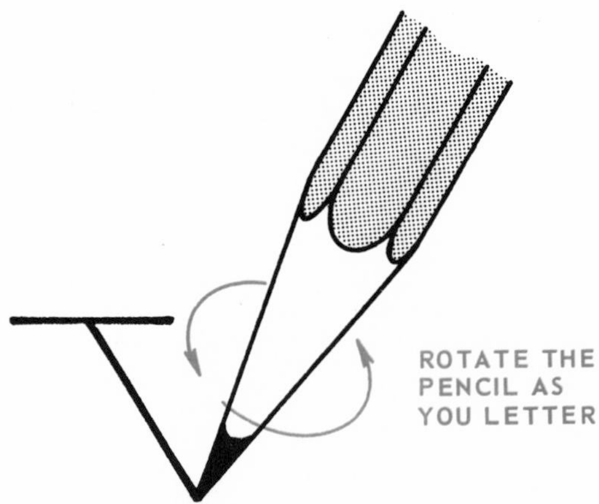

Fig. 6-3. Rotate the pencil point as you letter to keep the point sharp and the lettering uniform in weight.

GUIDE LINES

Good lettering requires the use of GUIDE LINES, Fig. 6-4. Guide lines are very fine lines made with a "needle sharp" 4H or 6H pencil. Guide lines should be drawn so lightly they will not show up on a print made from the drawing. VERTICAL GUIDE LINES, Fig. 6-5, may be used to assure that the letters will be vertical. Use INCLINED GUIDE LINES, Fig. 6-6, drawn at 67 1/2 deg. to the horizontal line, where inclined lettering is to be used.

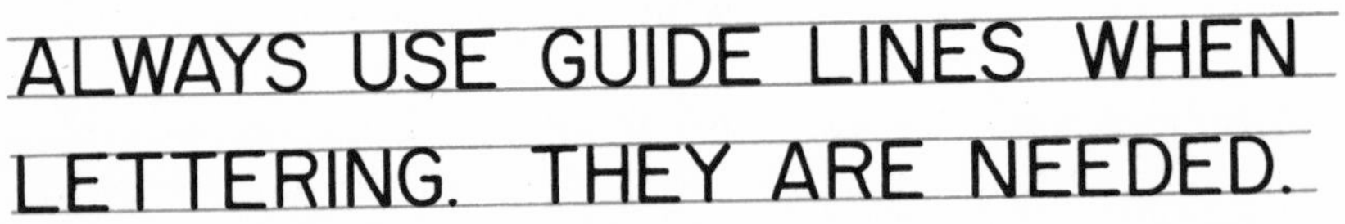

ALWAYS USE GUIDE LINES WHEN
LETTERING. THEY ARE NEEDED.

Fig. 6-4. Guide lines must be used when lettering to keep the letters uniform in height.

Fig. 6-5. Vertical guide lines.

INCLINED GUIDE LINES HELP KEEP
INCLINED LETTERING UNIFORM.

Fig. 6-6. Inclined guide lines.

SPACING

In lettering, proper spacing of the letters is important. There is no hard and fast rule which indicates how far apart the letters should be spaced. The letters should be placed so spaces between the letters appear to be about the same. Adjacent letters with straight lines require more space than curved letters. Letter spacing is judged by eye rather than by measuring, Fig. 6-7.

DEVALUATION

SPACED BY MEASURING

DEVALUATION

SPACED VISUALLY

Fig. 6-7. Letter spacing is judged by eye rather than by measuring.

Spacing between words and between sentences is another matter. The spacing between words and sentences should be equal to the height of the letters, Fig. 6-8.

WORDS AND LETTERS MUST BE CLEARLY SEPARATED. SPACING BETWEEN WORDS AND SENTENCES IS EQUAL TO THE HEIGHT OF THE LETTER USED.

Fig. 6-8. Spacing between words and sentences.

Spacing between lines of lettering should be equal to, or slightly less than, the height of the letters, Fig. 6-9.

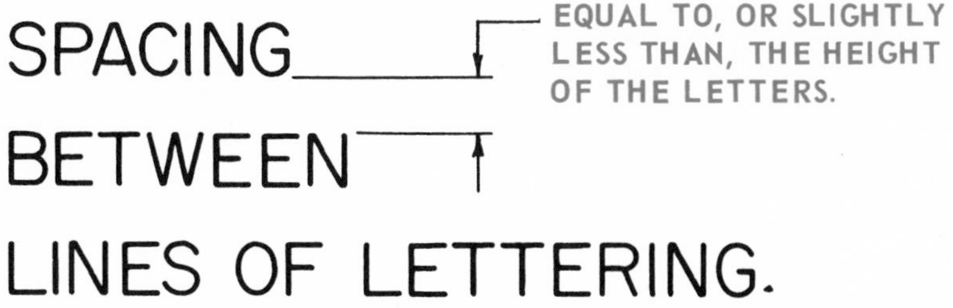

Fig. 6-9. Spacing between lines of lettering.

LETTER HEIGHT

On most drawings, letters 1/8 in. high will be satisfactory. Titles are 3/16 to 1/4 in. high. To make the information easier to read, these sizes may be increased on large drawings.

LETTERING AIDS AND DEVICES

The BRADDOCK-ROWE LETTERING TRIANGLE, Fig. 6-10, and the AMES LETTERING INSTRUMENT, Fig. 6-11, are devices that may be used as aids for drawing guide lines. The

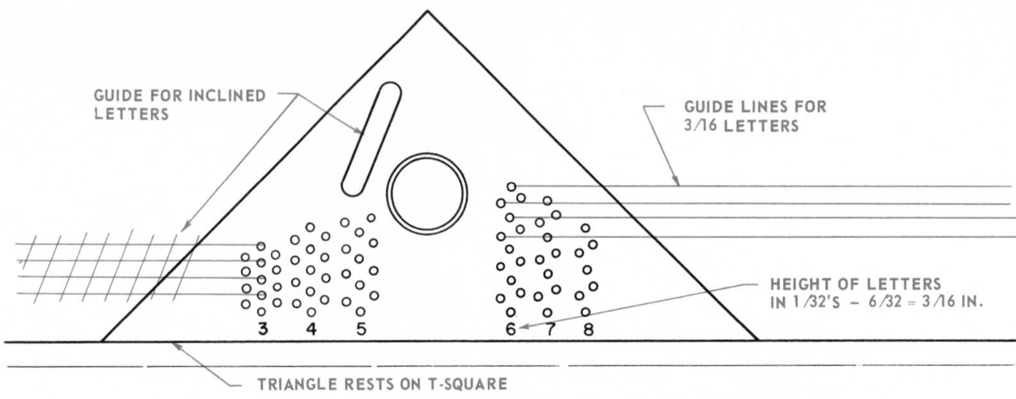

Fig. 6-10. The Braddock-Rowe lettering triangle.

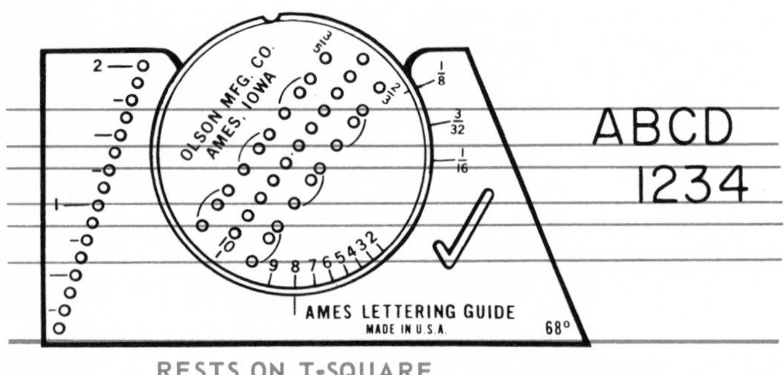

Fig. 6-11. The Ames lettering instrument.

numbers engraved below each series of holes in the triangle indicate the height of the letters in thirty-seconds. The series marked 3 indicates that the guide lines will be 3/32 in. apart; 4 indicates that they will be 4/32 or 1/8 in. apart. The numbers on the disk of the Ames lettering instrument are rotated until they are even with a line engraved on the base of the tool. The numbers also indicate the spacing of the guide lines in thirty-seconds.

Fig. 6-12. A special typewriter may be used to put information on drawings. (Mechanical Enterprises, Inc.)

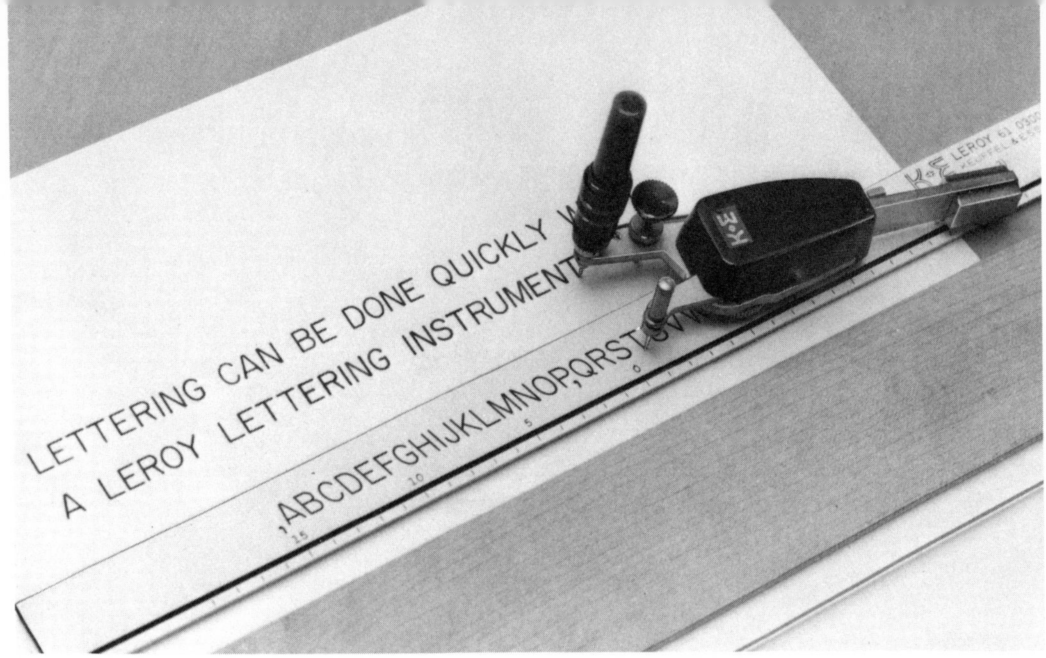

Fig. 6-13. LEROY lettering equipment produces lettering mechanically. Many styles of lettering are available.

MECHANICAL LETTERING DEVICES

Hand lettering is expensive because it requires so much time. For this reason, modern industry utilizes many mechanical devices to save time, and to improve the legibility of the letters.

Many drafting and engineering offices are making use of typewriters, Fig. 6-12, to put information on drawings. This releases highly skilled draftsmen for other work.

With the advent of microfilm, it became necessary for lettering to be highly legible (the drawings are reduced and enlarged photographically). Good lettering can be done easily and rapidly with mechanical lettering devices, such as shown in Fig. 6-13. This type equipment is available for a great variety of lettering styles, alphabets and symbols in a wide range of sizes and thicknesses.

Lettering equipment as shown in Fig. 6-14, is used for making signs and posters.

Fig. 6-14. The Wrico Sign-Maker utilizes a plastic stencil to make characters. This model is used for making large lettering for signs and posters. (Wood-Regen Instrument Co., Inc.)

TEST YOUR KNOWLEDGE - UNIT 6

1. Why is lettering needed on drawings?
2. The single stroke gothic letter is recommended because ______________________________.
3. The single stroke lettering used on a drawing may be ______________ or ______________ because there is no definite rule stating that it has to be one or the other.
4. The ________ or ________ pencil is usually used for lettering.
5. Why are guide lines used when lettering?
6. Name three types of lettering aids and mechanical lettering devices.
7. Why does industry use mechanical lettering devices?

OUTSIDE ACTIVITIES

1. Using signs and posters from your school's bulletin board, make a display of different lettering examples. Point out good spacing and poor spacing. Check to see if lettering styles are mixed in each example.
2. Research the variety of lettering styles, alphabets, and symbols available with mechanical lettering devices.
3. Letter the sentence "Good lettering technique requires practice and concentration." by hand, using Vertical Single Stroke Gothic. Letter the same sentence using a mechanical lettering device such as the LEROY lettering equipment. Time yourself and report to the class which is faster and which is easier to read.

YOUR NAME

YOUR SCHOOL

A GOOD DRAFTSMAN LETTERS

NEATLY AND RAPIDLY.

$\frac{1}{2}$ $\frac{1}{2}$ $\frac{3}{4}$ $\frac{3}{16}$ $\frac{1}{2}$

BEL AIR HIGH SCHOOL BEL AIR, MARYLAND	LETTERING PRACTICE-1	RICHARD WALKER SECTION CHECKED BY: DATE	6-1

PROBLEM SHEET 6–1. LETTERING PRACTICE.

1 2 3 4 5 6 7 8 9 0 $\frac{1}{2}$ $\frac{1}{16}$ $\frac{3}{8}$ $\frac{5}{32}$

THE QUICK RED FOX JUMPED
OVER THE LAZY BROWN DOG.

	LETTERING PRACTICE-2		6-2

PROBLEM SHEET 6–2. LETTERING PRACTICE.

"ONE SMALL STEP FOR A MAN,
ONE GIANT LEAP FOR MANKIND."

"HERE MEN FROM THE PLANET
EARTH FIRST SET FOOT UPON
THE MOON JULY 1969, A.D. WE
CAME IN PEACE FOR ALL MAN-
KIND."

	LETTERING PRACTICE-3		6-3

PROBLEM SHEET 6–3. LETTERING PRACTICE.

1 2 3 4 5 6 7 8 9 0 $\frac{1}{2}$ $\frac{3}{4}$ $\frac{5}{8}$ $\frac{7}{9}$

PACK EACH BOX WITH SEVEN DOZEN GIANT JUGS.

	LETTERING PRACTICE-4		6-4

PROBLEM SHEET *6–4.* ***LETTERING PRACTICE.***

SELECT A FAVORITE SAYING
OR QUOTATION AND LETTER
IT IN $\frac{1}{8}$, $\frac{3}{16}$ AND $\frac{1}{4}$ VERTICAL
OR INCLINED LETTERS.

	LETTERING PRACTICE-5		6-5

PROBLEM SHEET 6–5. LETTERING PRACTICE.

Fig. 7-1. Most of the drawings used by industry are of the multiview type. This huge truck was designed to haul salt in Mexico. Several thousand multiview drawings were used in its manufacture. (KW-Dart Truck Co.)

Unit 7
MULTIVIEW DRAWINGS

When a drawing is made with the aid of instruments, it is called a MECHANICAL DRAWING. Straight lines are drawn using a T-square and triangle or drafting machine, while circles, arcs and curves are drawn with a compass and French curve.

Almost all drawings used by industry are made using instruments and are in the form of MULTIVIEW DRAWINGS, Fig. 7-1. That is, more than one view is required to give a shape description of the object being drawn. In developing the needed views, the object is normally viewed from six directions, Fig. 7-2.

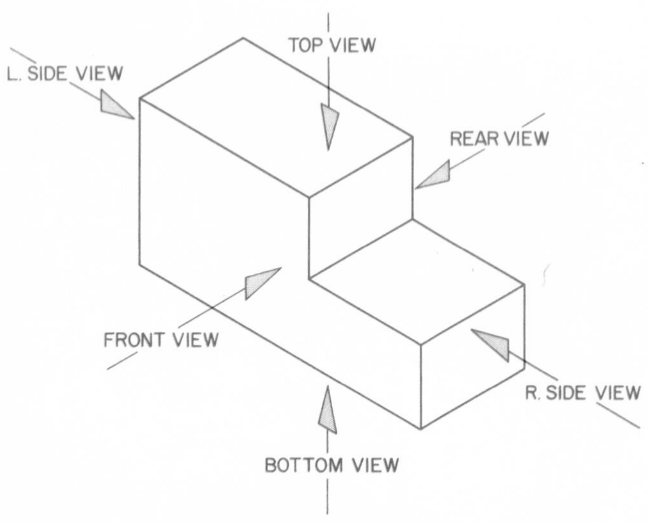

Fig. 7-2. An object is normally viewed from six different directions.

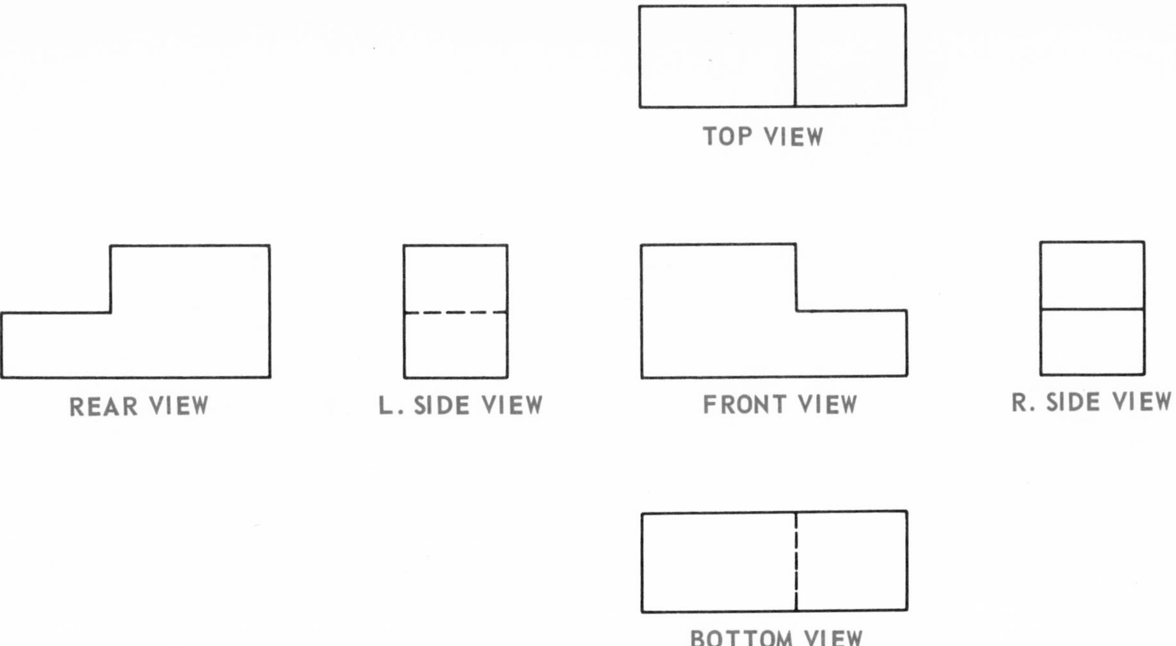

Fig. 7-3. The six directions of sight give these views.

The various directions of sight will give us the FRONT, TOP, RIGHT SIDE, LEFT SIDE, REAR and BOTTOM VIEWS, Fig. 7-3. To obtain the views, let's think of the object as being enclosed in a hinged glass box, Fig. 7-4.

Fig. 7-4. The object is enclosed in a hinged glass box.

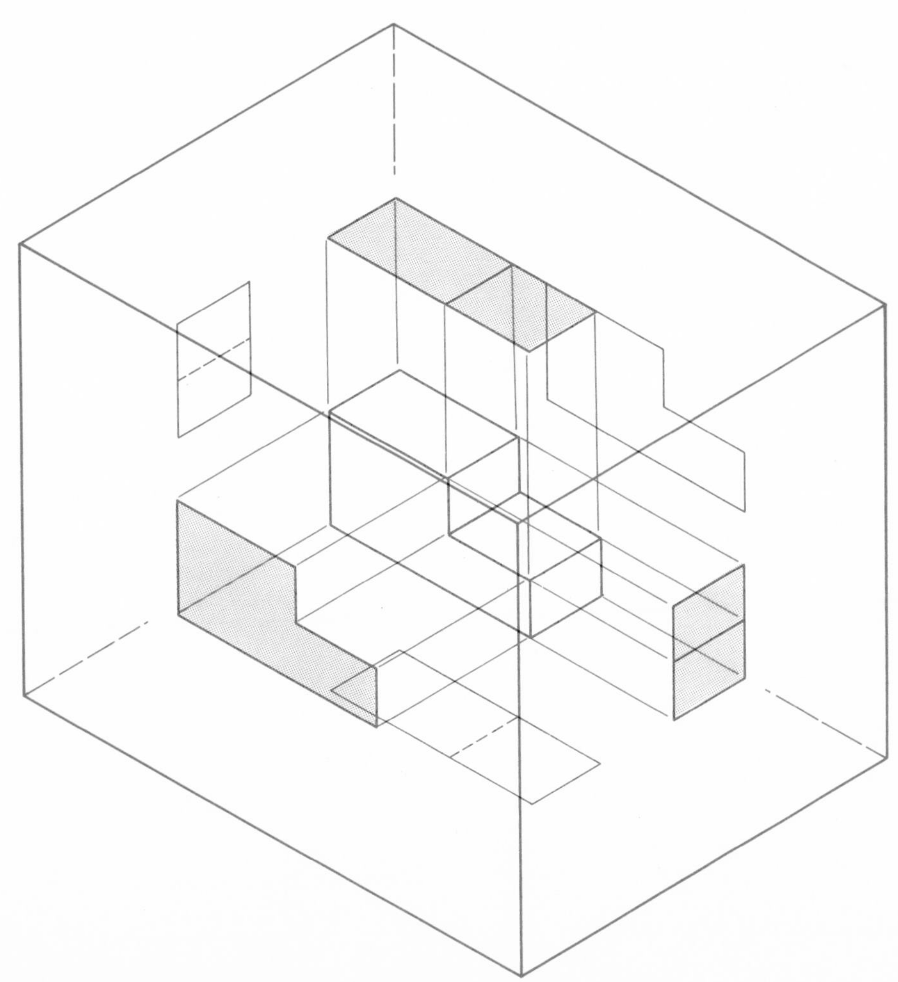

Now imagine that the views are projected to the six sides of the box - the top view of the object is seen on the top of the box, the front view on the front of the box, and so on for the remaining views.

This method of developing views is called ORTHOGRAPHIC PROJECTION and permits a three-dimensional object to be drawn on a flat sheet of paper having only two dimensions. This technique forms the starting point of engineering drawing. With orthographic projection, in the United States at least, the top view is always drawn directly above the front view, the right side view is drawn to the right of the front view and in line with it, and the remaining views are in the positions shown when the glass box is opened, Fig. 7-5.

Fig. 7-5. The glass projection box opened out.

SELECTING VIEWS TO BE USED

As you can see, we can draw at least six views of the object. This does not mean that all of these views must be used, or are needed. Only those views needed to give a shape description of the object should be drawn. Any view that repeats the same shape description as another view need not be used, Fig. 7-6.

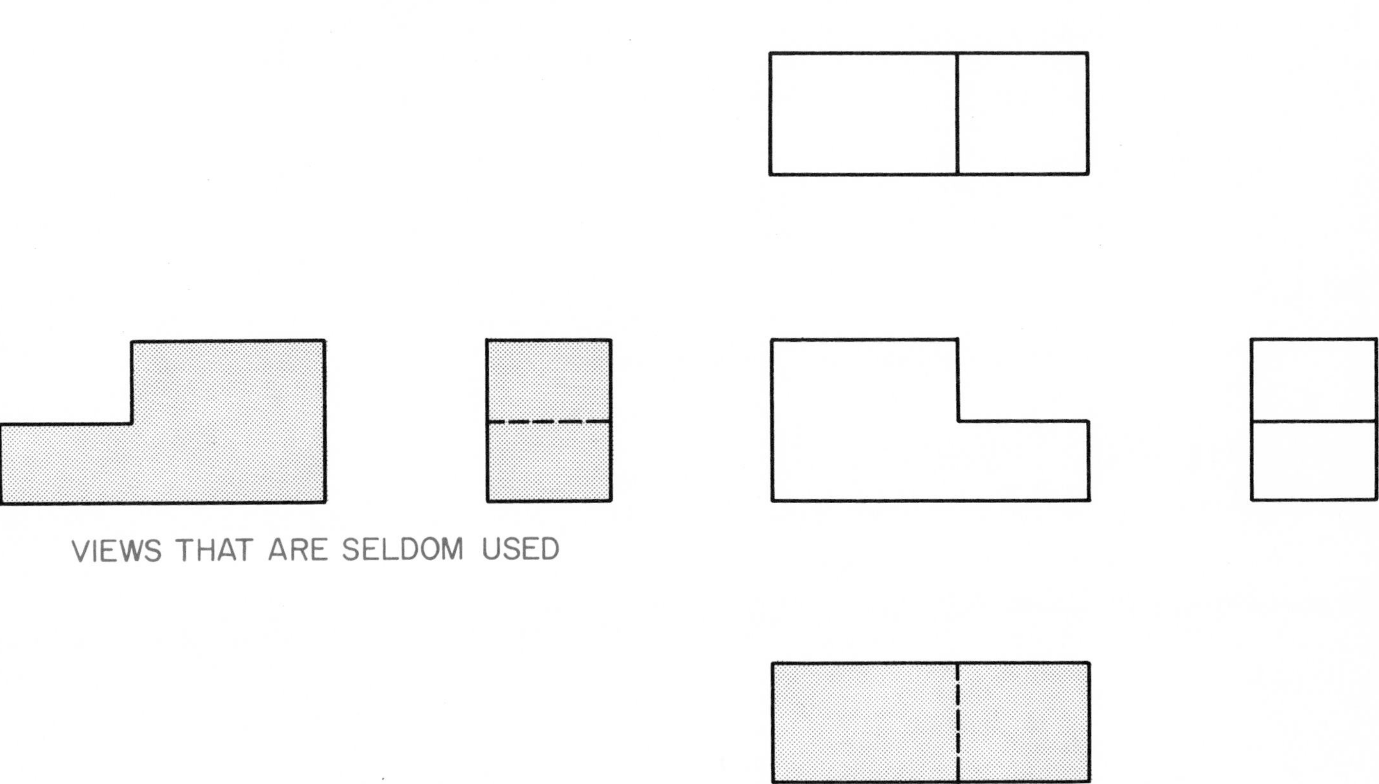

Fig. 7-6. Not all views are needed. Eliminate any view that repeats the same shape description as another view.

In most instances, two or three views are sufficient to show the shape of an object.

Views showing a large number of hidden lines should be used only if absolutely necessary since too many hidden lines tend to make the drawing confusing. Use another view or a sectional view.

TRANSFERRING POINTS

Each view will show a minimum of two dimensions. Any two views of an object will have at least one dimension in common. Time can be saved if a dimension from one view is projected to the other view instead of measuring, Fig. 7-7. Construction lines are used when transferring points.

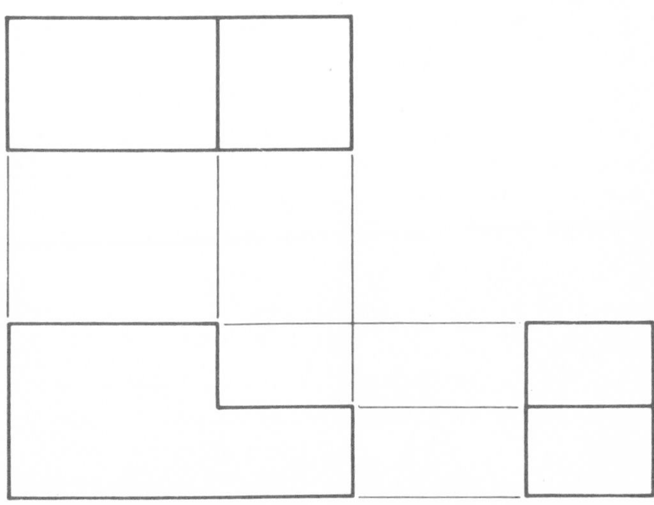

Fig. 7-7. Transferring points from view to view.

Additional time can be saved in transferring the depth of the top view to the side view. Two methods are shown in Fig. 7-8.

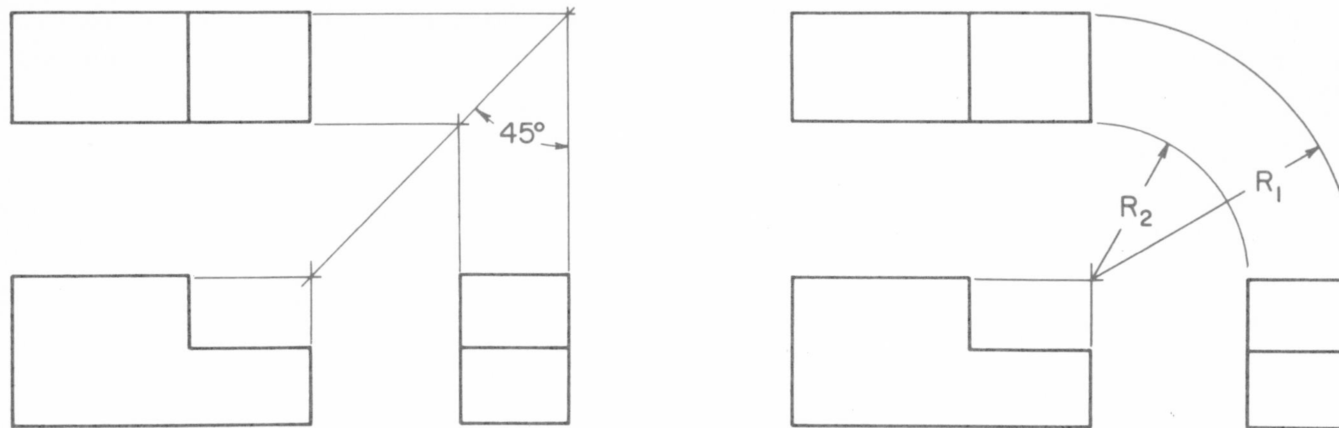

Fig. 7-8. Two accepted methods used to transfer the depth of the top view to the side view.

HOW TO CENTER DRAWING ON SHEET

A drawing looks more professional if the views are evenly spaced and centered on the drawing sheet. Centering the views on the sheet is not difficult.

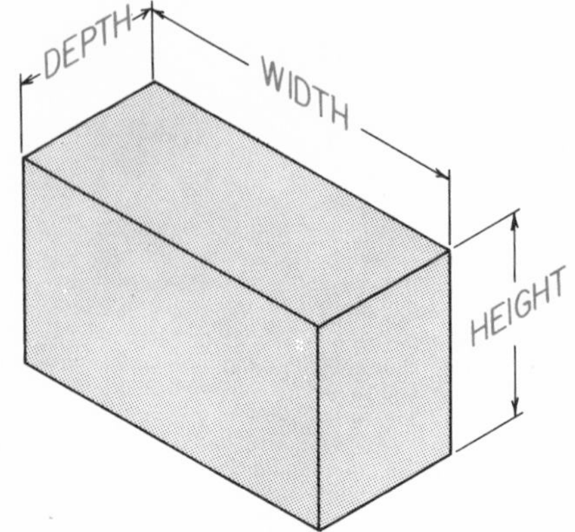

Fig. 7-9. How an object is described.

The following procedure is recommended:

1. Examine the object that is to be drawn. Observe its dimensions - width, depth and height, Fig. 7-9. Determine the position in which the object is to be drawn.

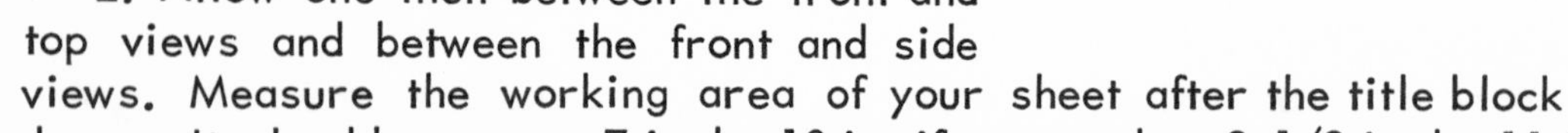

2. Allow one inch between the front and top views and between the front and side views. Measure the working area of your sheet after the title block and border have been drawn. It should measure 7 in. by 10 in. if you used an 8 1/2 in. by 11 in. drawing sheet.

3. To locate the front view, add the width of the front view to the depth of the right side view, plus the one inch space between the views. Subtract this total from the horizontal width of the working surface. Divide this answer by two. This will be the starting point for laying out the sheet horizontally. Measure in the resulting answer from the left border line and draw a vertical construction line. With this construction line as the reference point, measure over a distance equal to the width of the front view and draw another vertical construction line through this point, Fig. 7-10.

4. The same procedure is used to center the views vertically. However, the height of the front view and the depth of the top view are used. A one inch space will separate the views. Add these distances, and subtract the answer from the vertical working distance of the paper. Divide the resulting answer by two. This distance is measured vertically from the top of the title block. From this starting point measure up the height of the front view, the one inch that separates the views and the depth of the top view. Draw construction lines through these points, Fig. 7-11.

5. Use either the 45 deg. angle method or the radius method, to transfer the depth of the

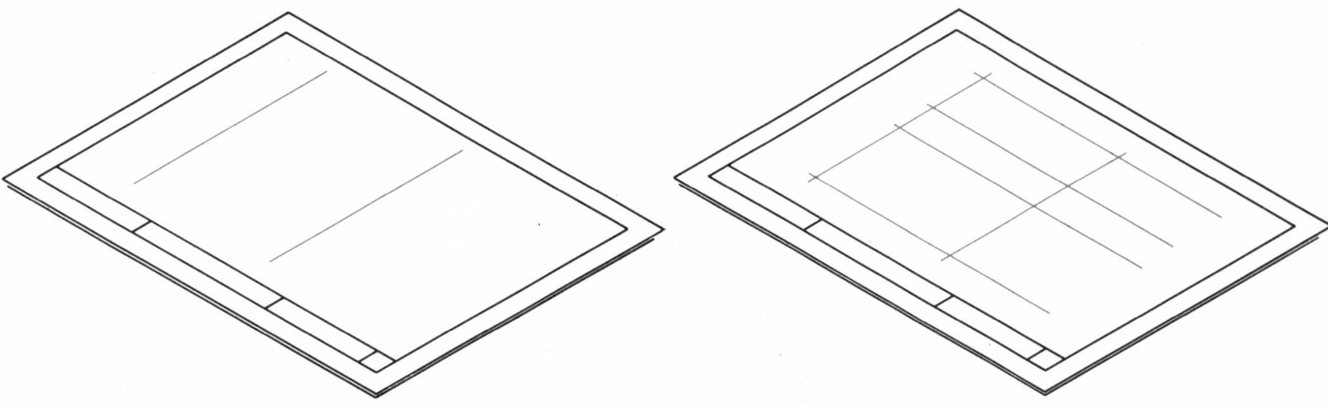

Fig. 7-10. Left. The first step in locating the front and top views on the drawing sheet. Fig. 7-11. Right. The front and top views are blocked in.

top view to the right side view of the object, Fig. 7-12.

6. Draw in the right side view. Use construction lines.

7. Complete the drawing by going over the proper construction lines. Use the correct weight and type of line (object line, hidden object line, center lines, etc.), Fig. 7-13.

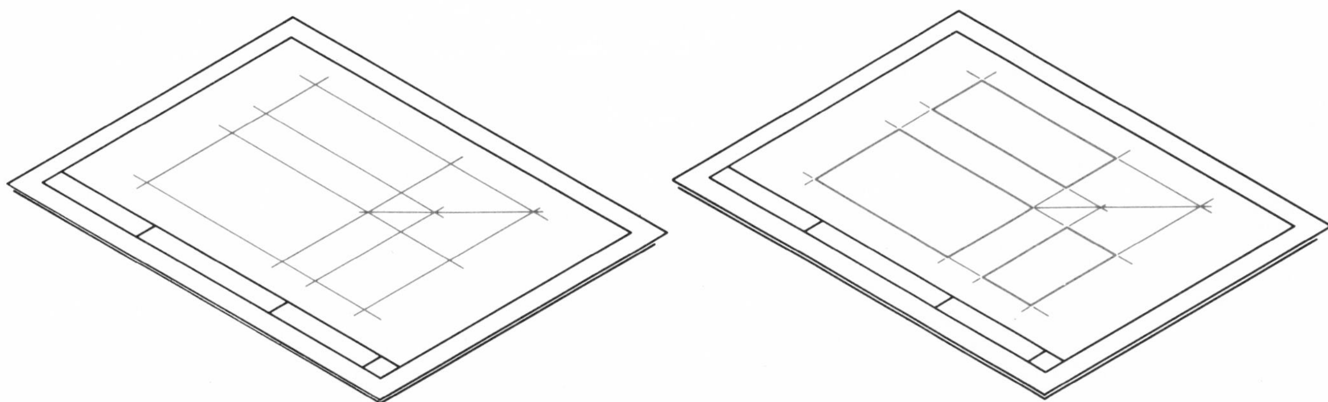

Fig. 7-12. Left. Projecting blocking in the right side view. Fig. 7-13. Right. Draw in the object lines to complete the drawing. Construction lines may be erased.

TEST YOUR KNOWLEDGE - UNIT 7

1. A drawing is said to be a MECHANICAL DRAWING when it was drawn using ____________.
2. A drawing that uses two or more views to describe an object is known as a ______________.
3. List the six views normally seen when making the drawing listed above.
   a. ____________________.
   b. ____________________.
   c. ____________________.
   d. ____________________.
   e. ____________________.
   f. ____________________.
4. The method used to develop these six views is called ____________________.
5. What views of the six views are normally used to describe an object?

OUTSIDE ACTIVITIES

1. Collect props for the class to draw using instruments. One prop should require only a two-view drawing; another prop should require a three-view drawing. Find other props which require more than three views to give a complete shape description.
2. Build a hinged box out of clear plastic which can be used to demonstrate the unfolding of an object into its multiview parts; the front, top, bottom, and sides. Place a prop inside the box, trace the profile of the object on the side of the plastic box with chalk, then unfold the box.
3. Make a large poster for your drafting room showing the step-by-step procedure to follow in centering a drawing on a sheet.

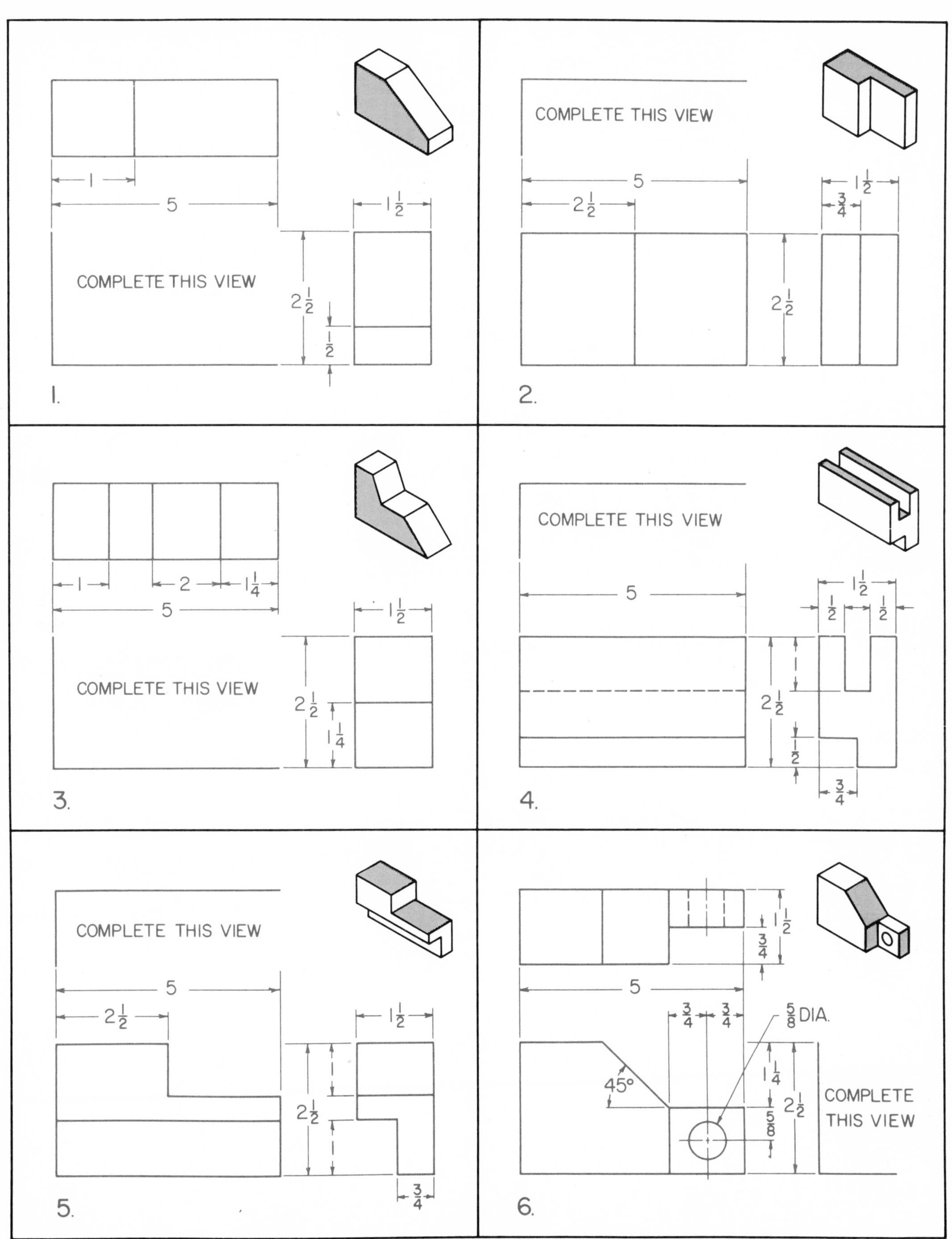

PROBLEM SHEET 7–1. MULTIVIEW DRAWINGS. *Draw each problem on a separate sheet and complete as indicated.*

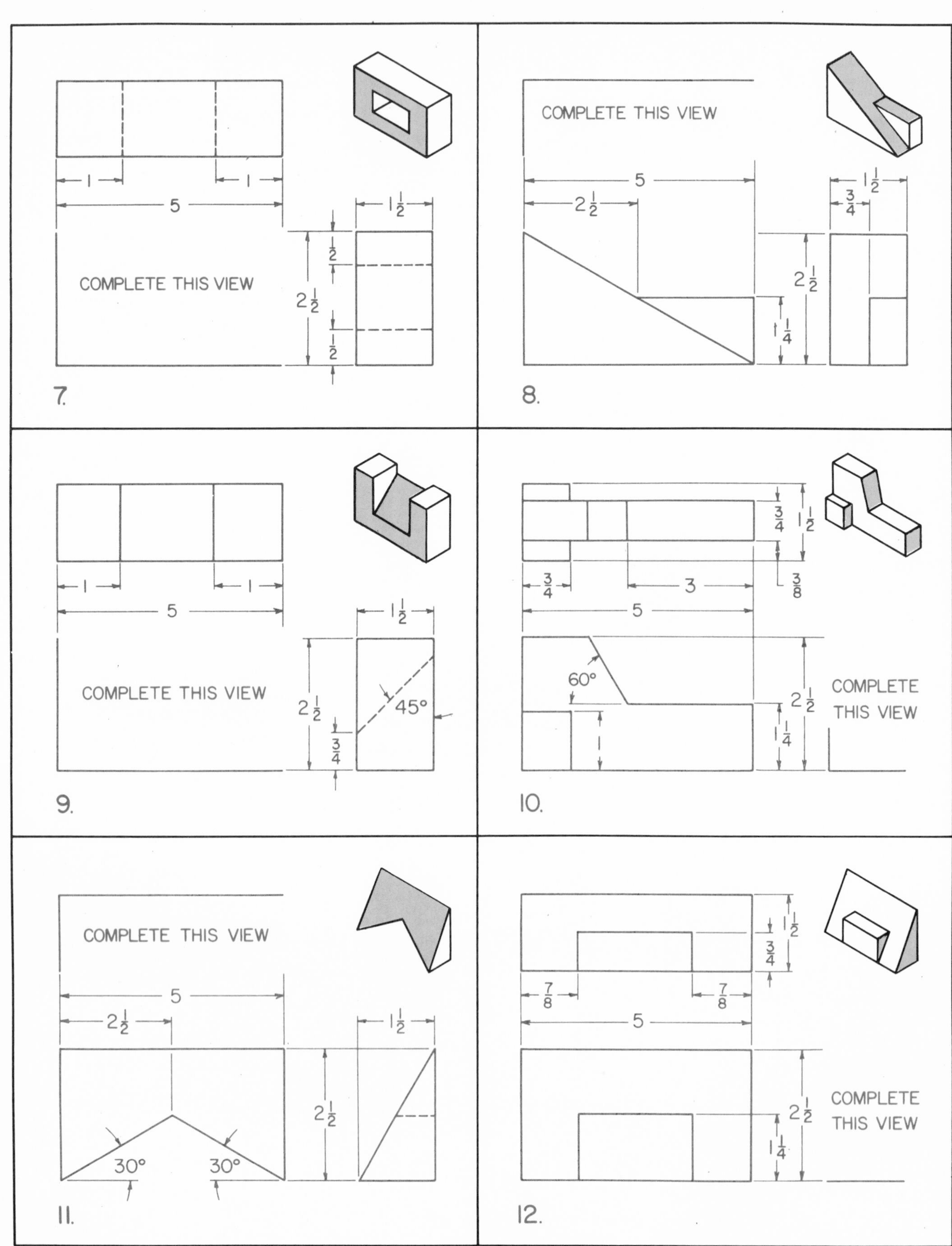

PROBLEM SHEET 7–2. MULTIVIEW DRAWINGS. *Draw each problem on a separate sheet and complete as indicated.*

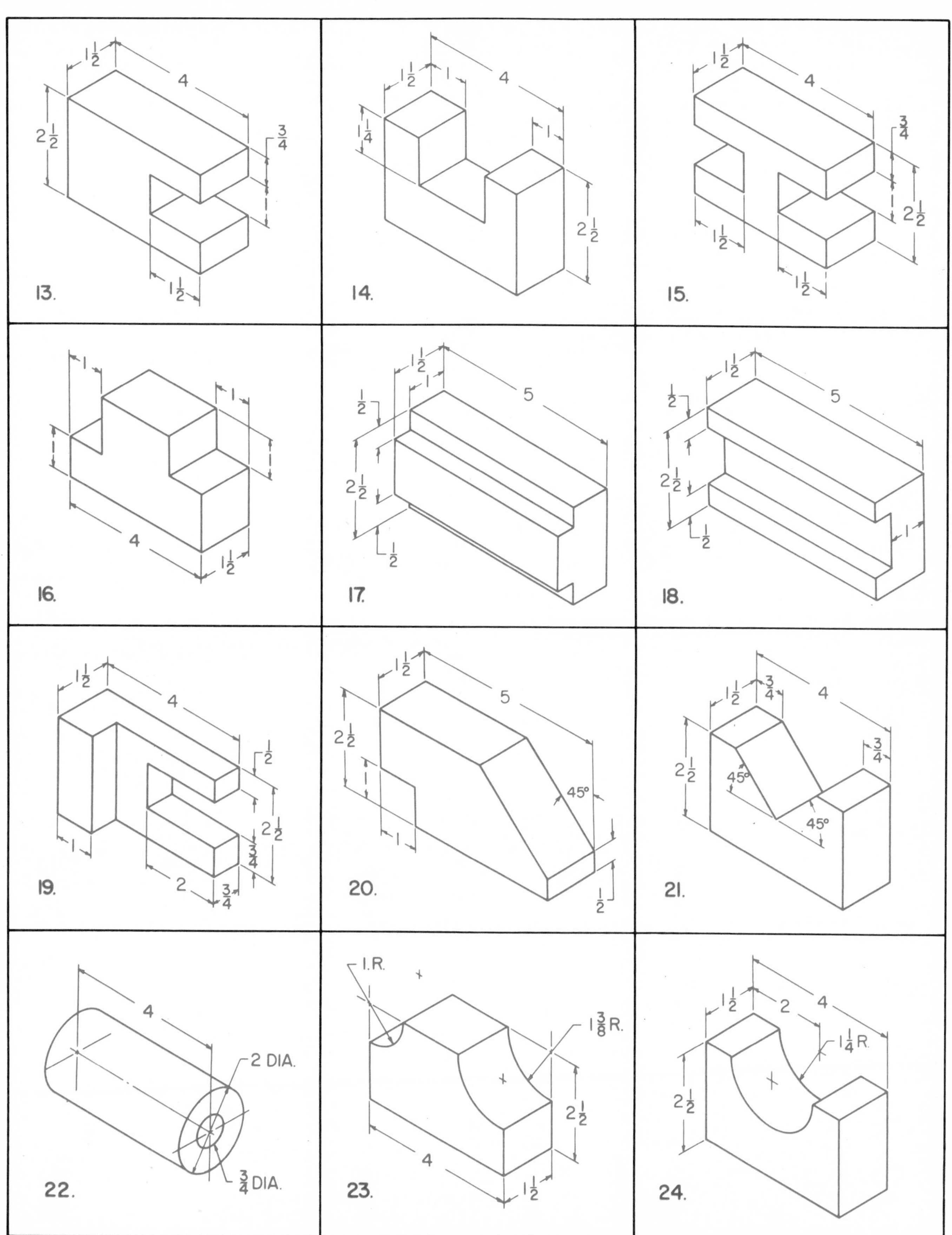

PROBLEM SHEET 7–3. MULTIVIEW DRAWINGS. *Draw each problem on a separate sheet. Draw as many views as necessary to fully describe each problem.*

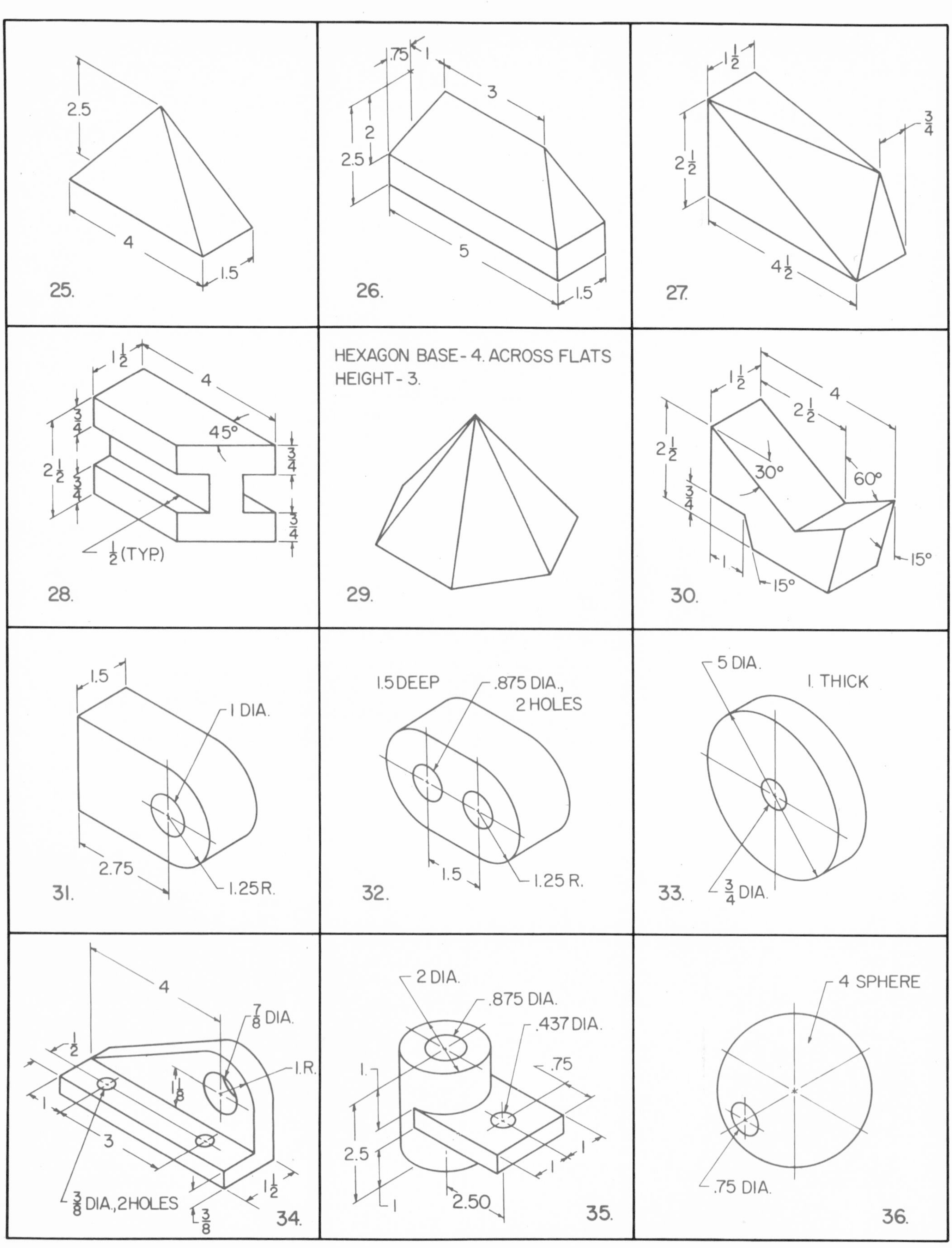

PROBLEM SHEET 7–4. MULTIVIEW DRAWINGS. *Draw each problem on a separate sheet. Draw as many views as necessary to fully describe each problem.*

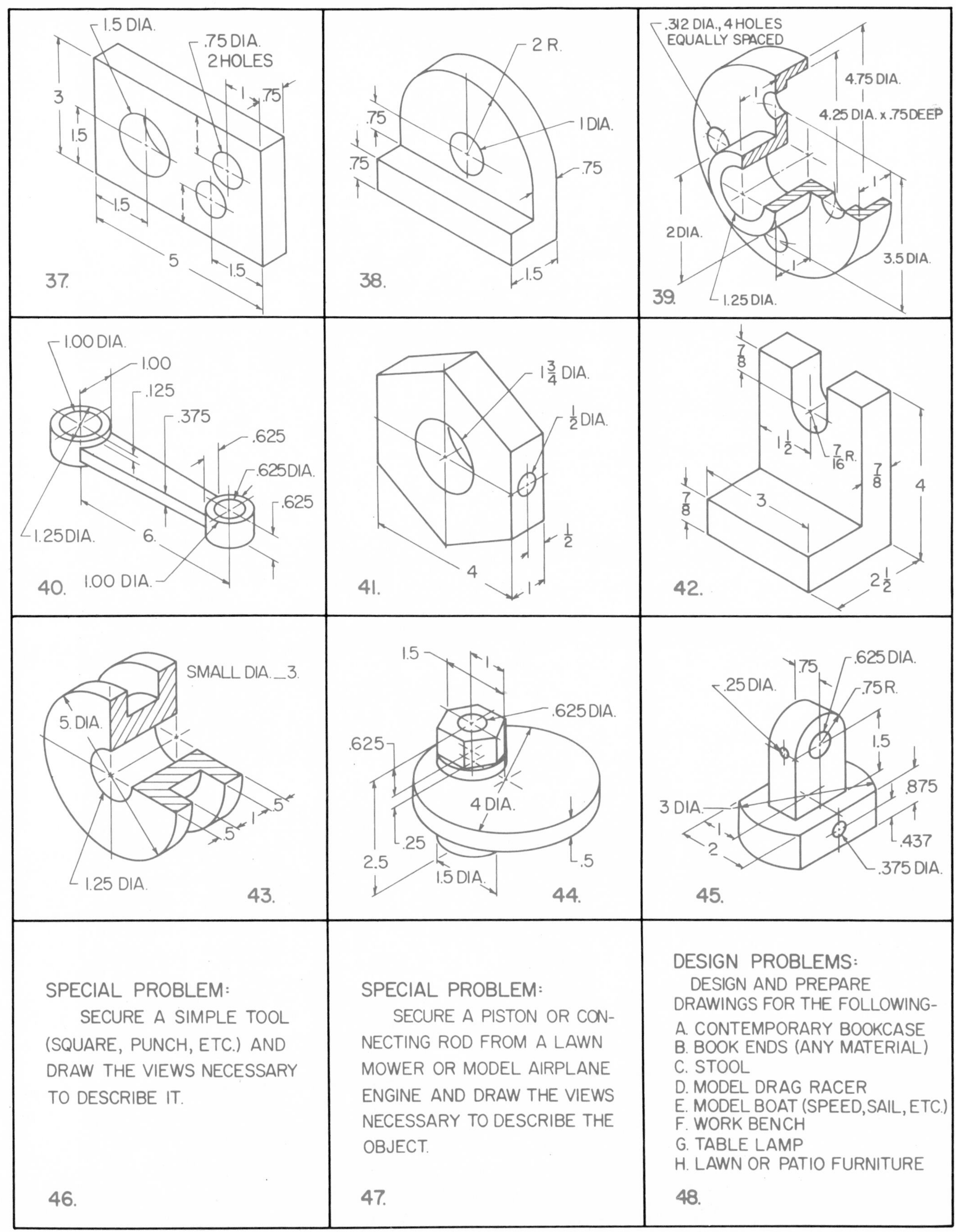

PROBLEM SHEET 7–5. MULTIVIEW DRAWINGS. *Draw each problem on a separate sheet. Draw as many views as necessary to fully describe each problem.*

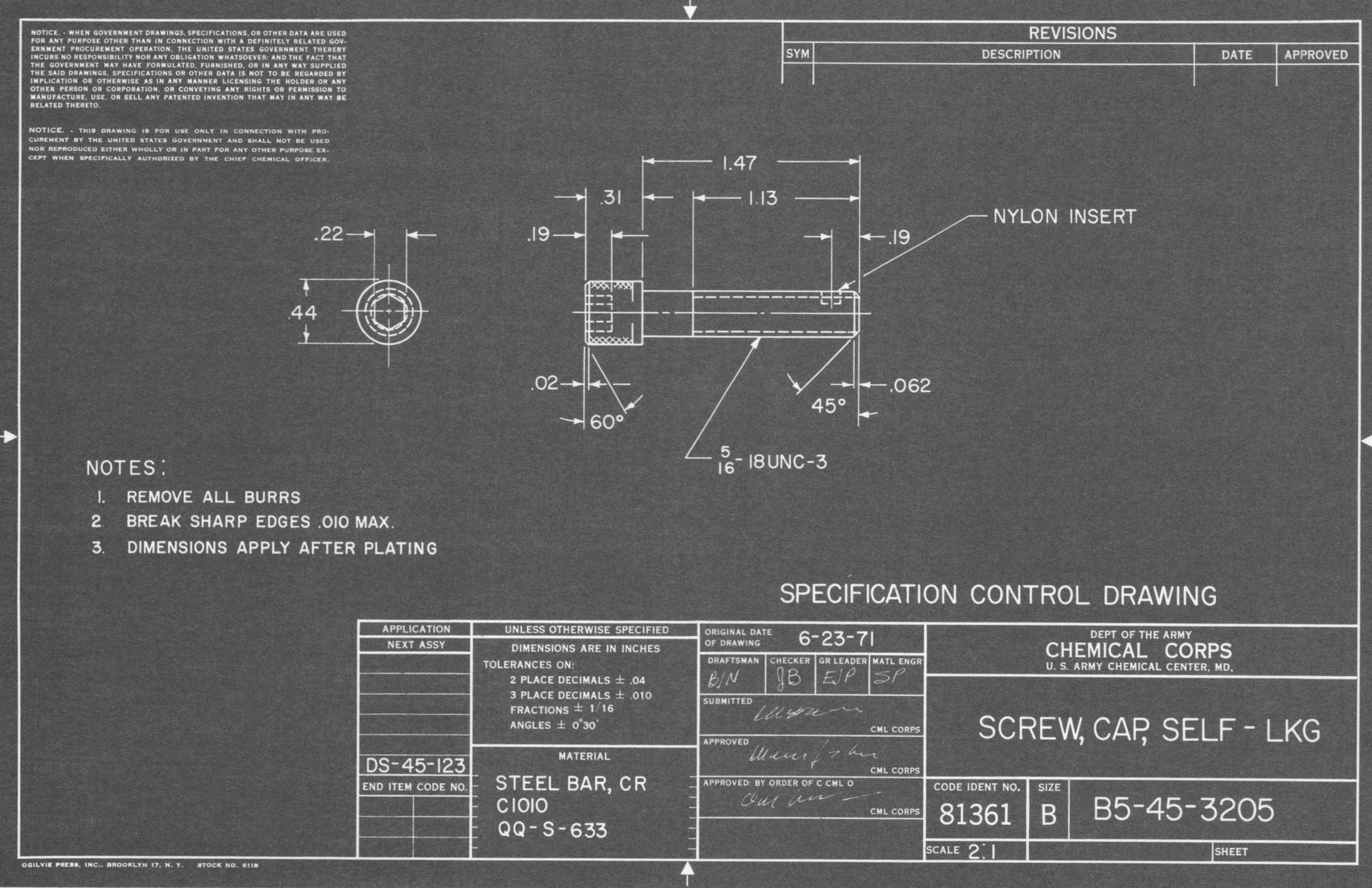

Fig. 8-1. Industry drawing showing dimensions and shop notes.

Unit 8
DIMENSIONING AND SHOP NOTES

If an object is to be manufactured according to the designer's specifications, the craftsman usually needs more information than that furnished by a scale drawing of its shape. DIMENSIONS and SHOP NOTES are needed, Fig. 8-1.

DIMENSIONS define the sizes of the geometrical features of an object. They are presented in appropriate units of measure--inches, millimeters, etc.

SHOP NOTES provide additional information along with size.

READING DIRECTION FOR DIMENSIONS

Dimensions are placed on the drawing in either an UNIDIRECTIONAL or an ALIGNED manner, Fig. 8-2. Unidirectional dimensioning is preferred.

Unidirectional dimensions are placed to be read from the bottom of the drawing.

Aligned dimensions are placed parallel to the dimension line. The numerals are read from the bottom or right side of the drawing.

Regardless of the method used, dimensions shown with leaders and notes are lettered parallel to the bottom of the drawing.

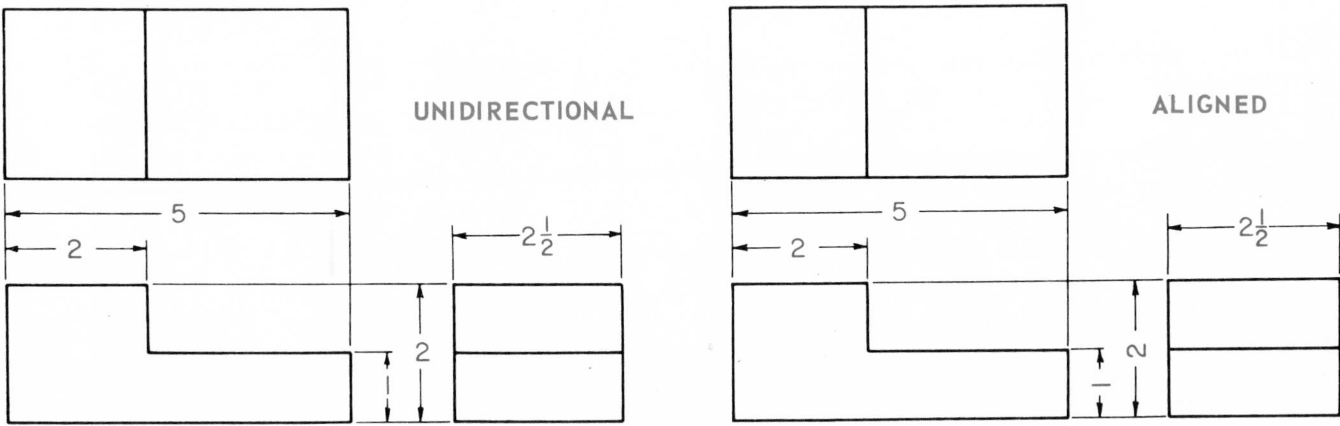

*Fig. 8-2. The two accepted methods of dimensioning drawings. The **UNIDIRECTIONAL** method is preferred.*

DIMENSIONING A DRAWING

From your study of the ALPHABET OF LINES, you will remember that special lines are used for dimensioning, Fig. 8-3. The dimension line is a fine solid line used to indicate distance and location. It should be fine enough to contrast with the object lines, and is broken near the center for the insertion of the dimension. The line is capped with arrowheads.

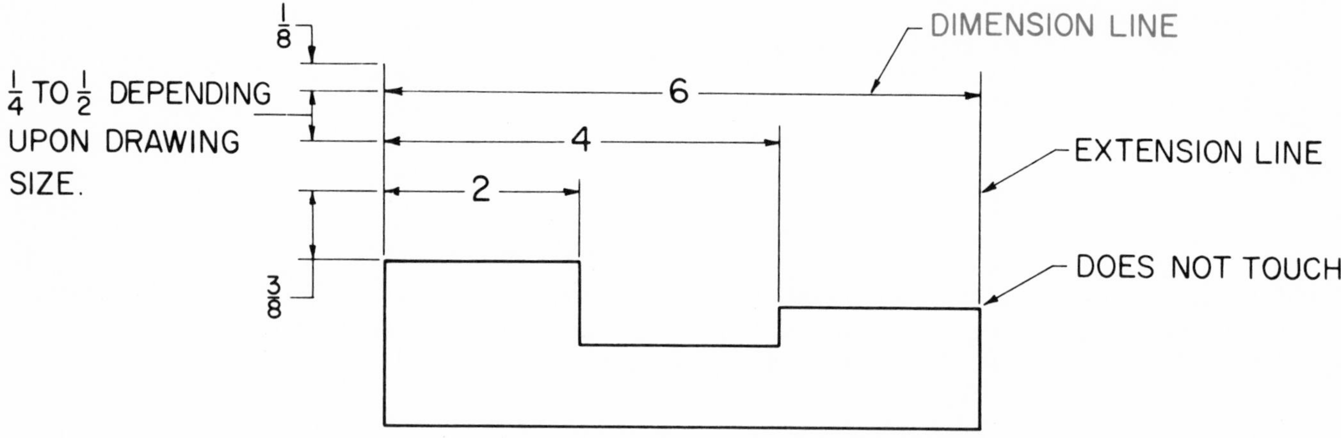

Fig. 8-3. The lines used for dimensioning. Note how they contrast with the object line.

The line that extends out from the drawing is called an extension line and projects 1/8 in. beyond the last dimension line. It should not touch the drawing. The smaller or detail dimensions are nearest the view, while the larger or overall dimensions are farthest from the view.

Arrowheads are drawn freehand and should be carefully made, Fig. 8-4. The solid arrowhead is generally preferred. It is made narrower and slightly longer than the open arrowhead. For most applications, arrowheads 1/8 in. long are satisfactory.

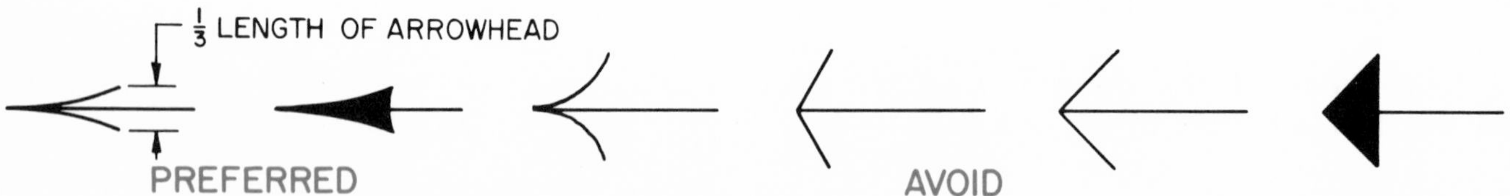

Fig. 8-4. Arrowheads are drawn freehand. The solid arrowhead is generally preferred.

GENERAL RULES FOR DIMENSIONING

Dimensions, if they are to be easily understood, should conform to the following general rules:

1. Place dimensions on the views that show the true shape of the object, Fig. 8-5.
2. Unless absolutely necessary, dimensions should not be placed within the views.
3. If possible, dimensions should be grouped together rather than scattered about the drawing, Fig. 8-6.
4. Dimensions must be complete so that no scaling of the drawing is required. Also, it

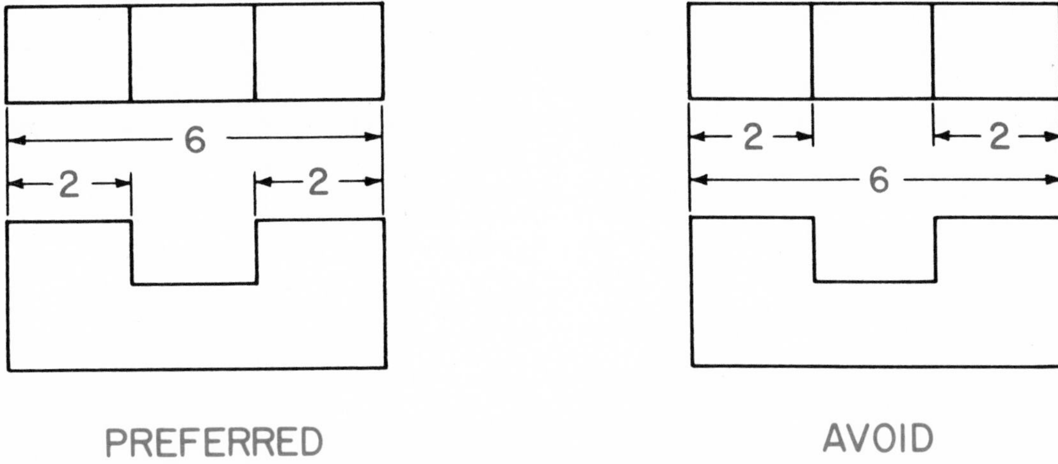

Fig. 8-5. Keep the dimensions on the view that shows the true shape of the object.

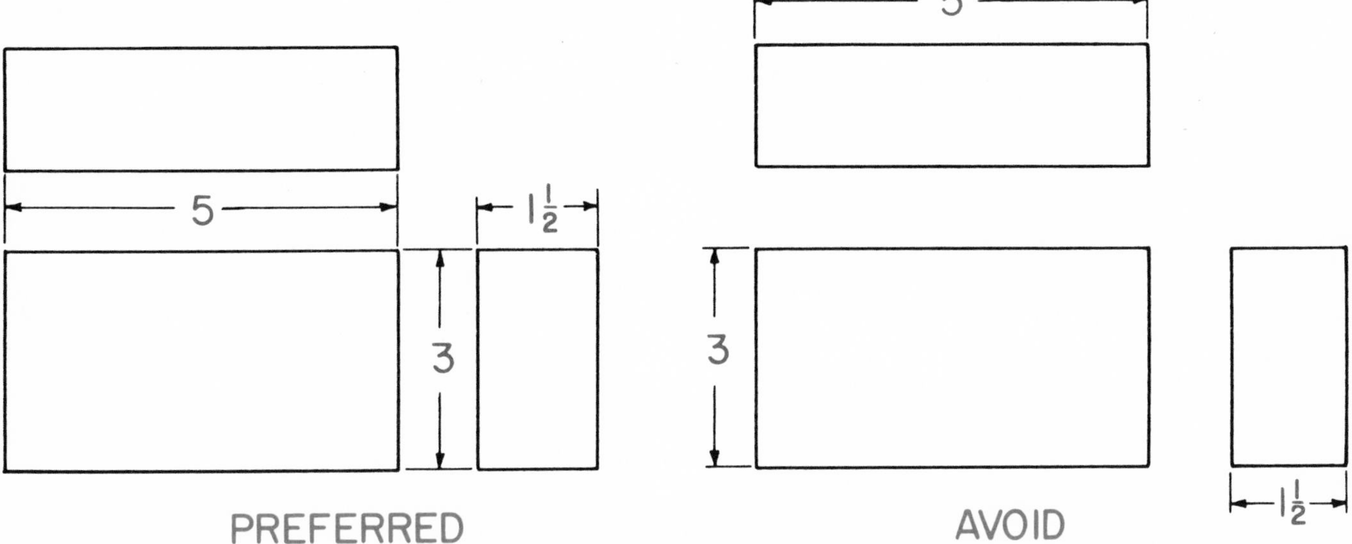

Fig. 8-6. Keep dimensions grouped for easier understanding of the drawing.

should be possible that sizes and shapes may be determined without assuming any measurements.

5. Draw dimension lines parallel to the direction of measurement. If there are several parallel dimension lines, the numerals should be staggered to make them easier to read, Fig. 8–7.
6. Dimensions should not be duplicated unless they are absolutely necessary to the understanding of the drawing. Omit

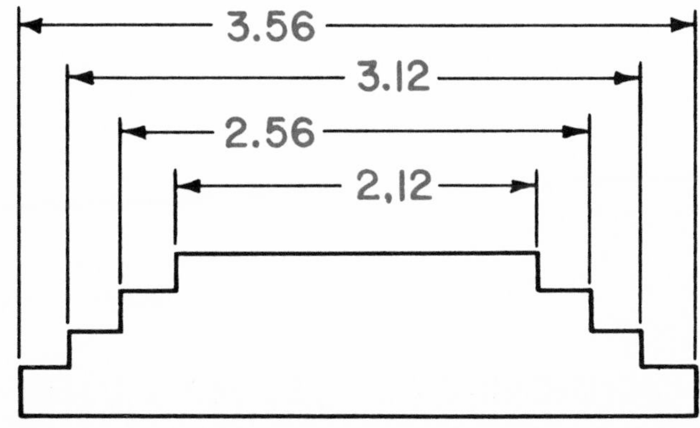

Fig. 8-7. When there are several parallel dimension lines the numerals should be staggered to make them easier to read.

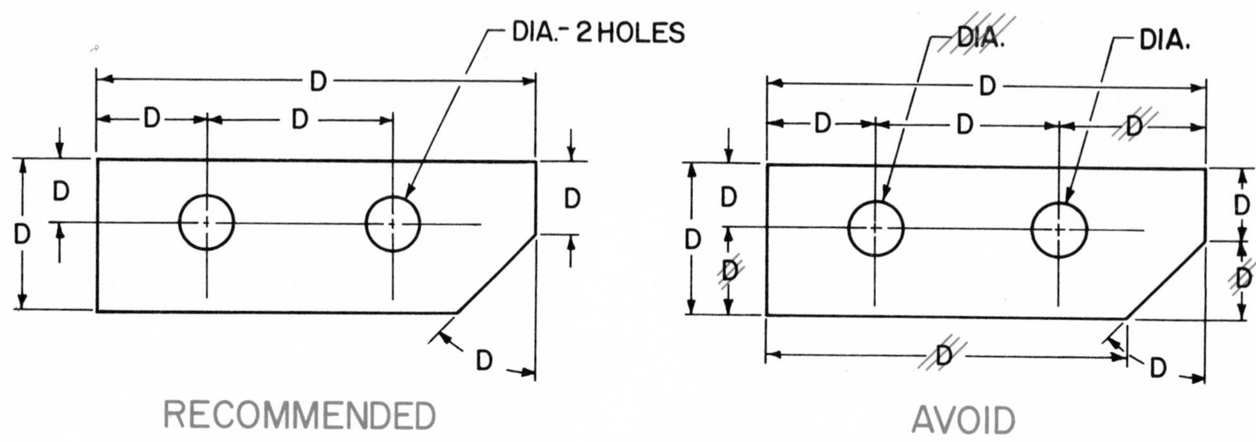

Fig. 8-8. Avoid duplicating dimensions unless they are necessary to the understanding of the drawing.

unnecessary dimensions, Fig. 8-8.

7. Plan your work carefully so that dimension lines do not cross extension lines, Fig. 8-9.
8. When all dimensions on a drawing are in inches, the inch symbol (") need not be used.
9. Numerals and fractions must be drawn in proper relation to one another, Fig. 8-10.

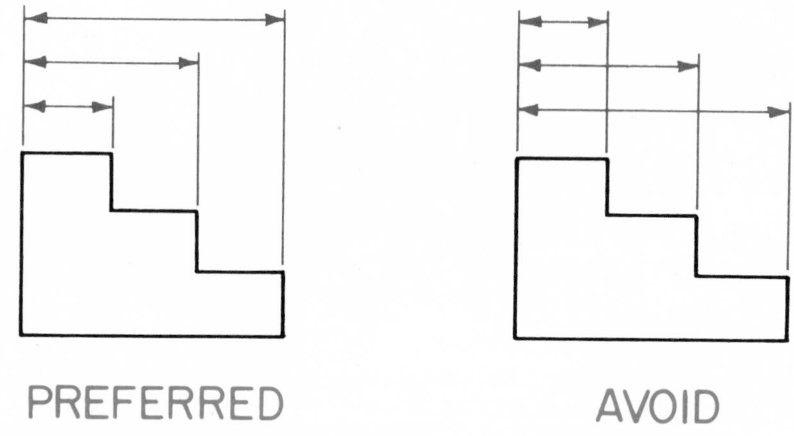

Fig. 8-9. Line crossing can be kept to a minimum if the shortest dimension lines are placed next to the object outline.

Fig. 8-10. Numerals and fractions must be drawn in proper relation to one another.

DIMENSIONING CIRCLES, HOLES AND ARCS

Circles and round holes are dimensioned as shown in Fig. 8-11. The size represents the diameter. If the diameters of several concentric circles must be dimensioned on a drawing, it may be more convenient to show them on the front view, Fig. 8-12.

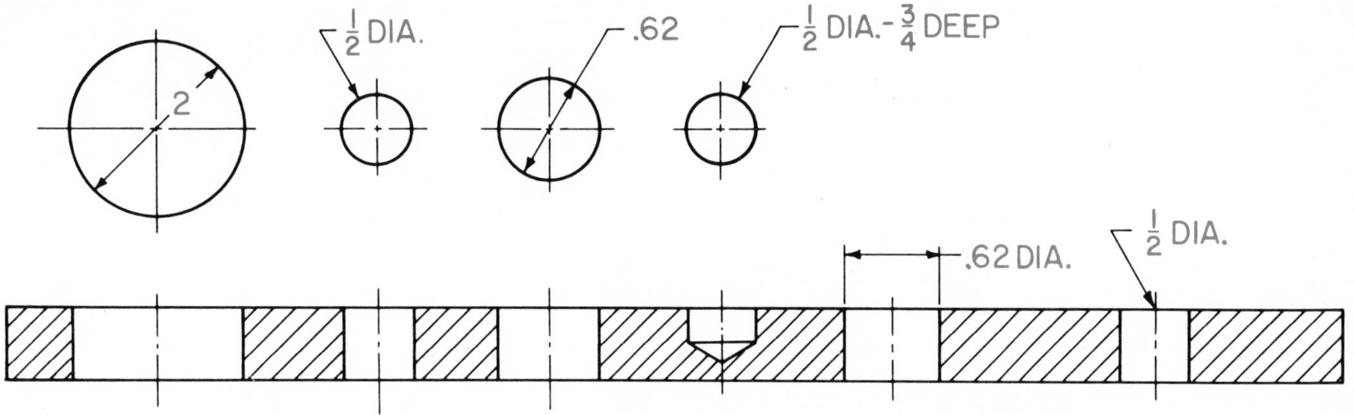

Fig. 8-11. Circles and holes are dimensioned by giving the diameter.

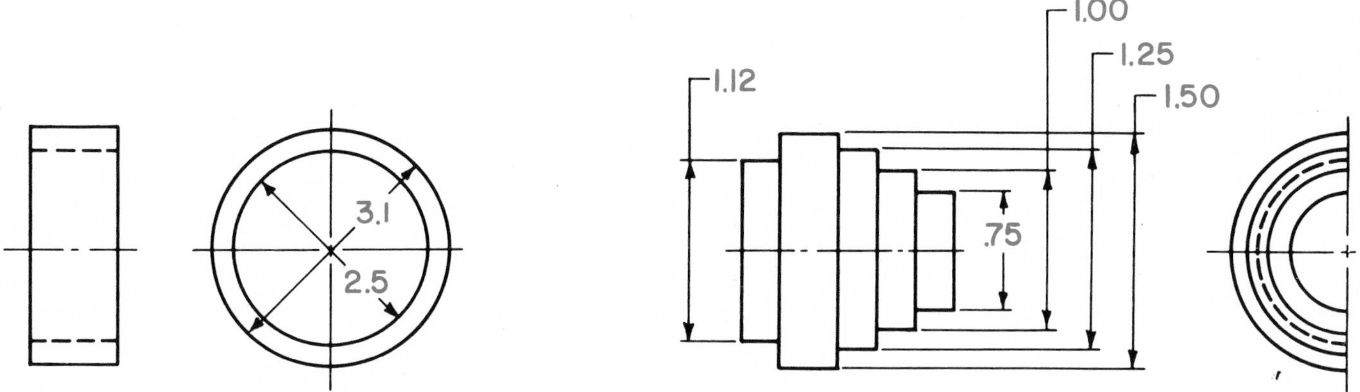

Fig. 8-12. Recommended ways to dimension concentric circles. Note that when the method to the right is used only half a right side view is needed.

The correct way to use a leader to indicate a diameter is shown in Fig. 8-13. The leader ALWAYS points to the center of the diameter.

Fig. 8-13. A leader is used to direct attention to a note or to indicate sizes of arcs and circles. When used with arcs and circles the leader should radiate from their centers.

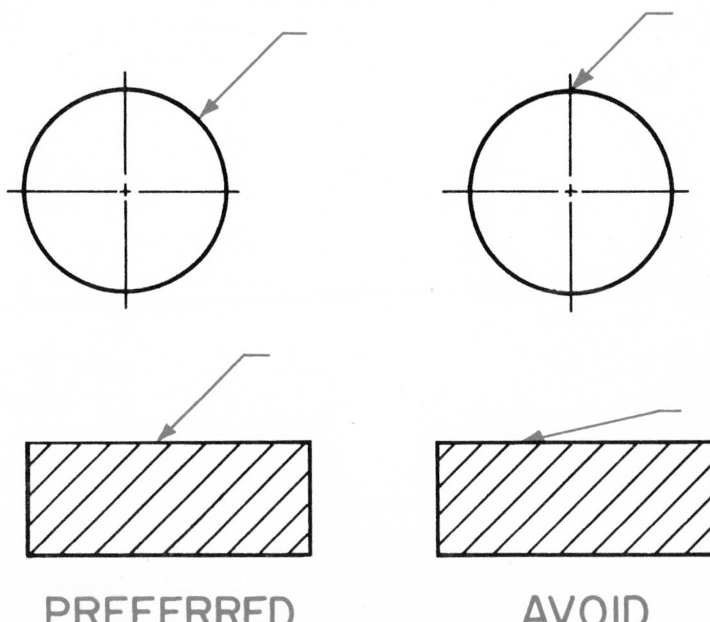

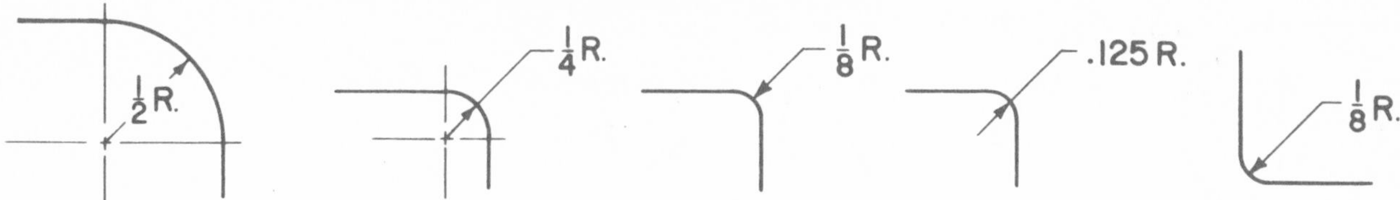

Fig. 8-14. Recommended ways of dimensioning arcs.

Arcs are dimensioned by giving the radius, Fig. 8-14. Round holes and cylindrical parts are dimensioned from centers, never from the edges, Fig. 8-15.

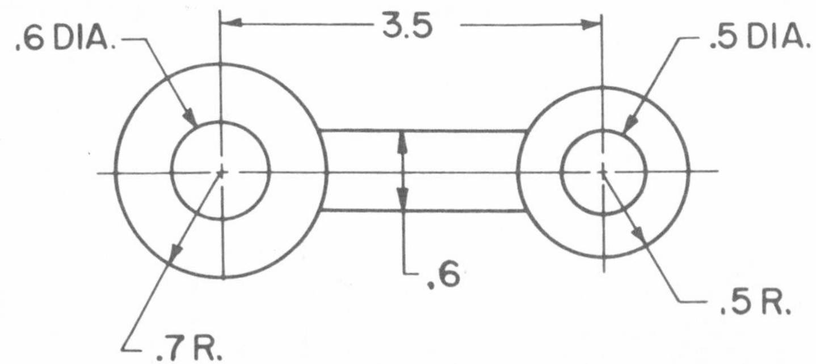

Fig. 8-15. Where feasible, round holes and cylindrical parts are dimensioned from centers.

When it is necessary to dimension a series of holes around a circle, use a note to designate the number of holes, their size and the diameter of the circle, Fig. 8-16.

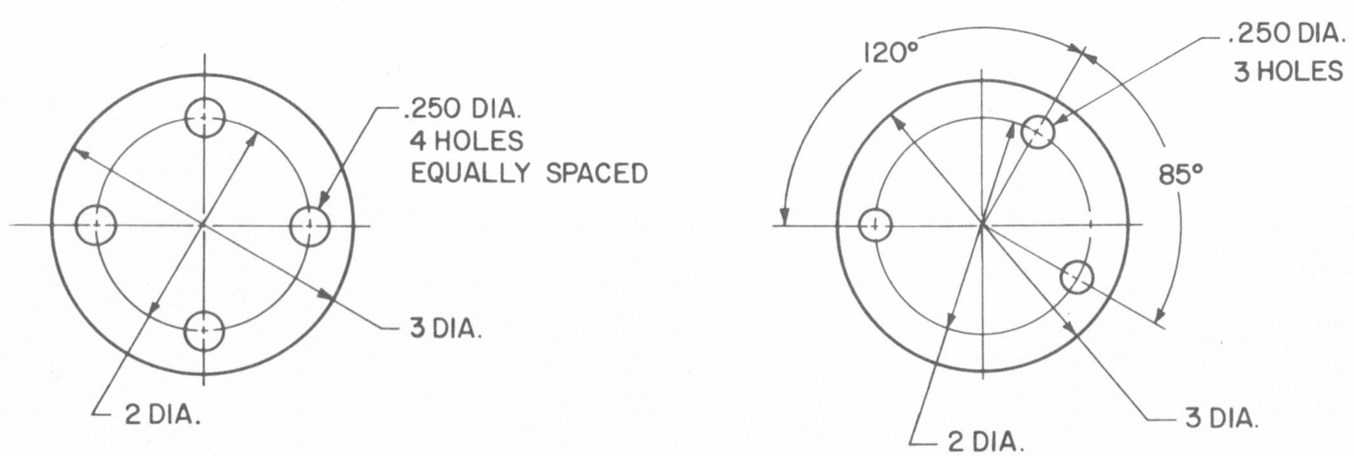

Fig. 8-16. Left. The correct way to dimension equally spaced holes. Right. Holes that are not equally spaced are dimensioned this way.

DIMENSIONING ANGLES

Angular dimensions are expressed in degrees, minutes and seconds. Angles are dimensioned as shown in Fig. 8-17.

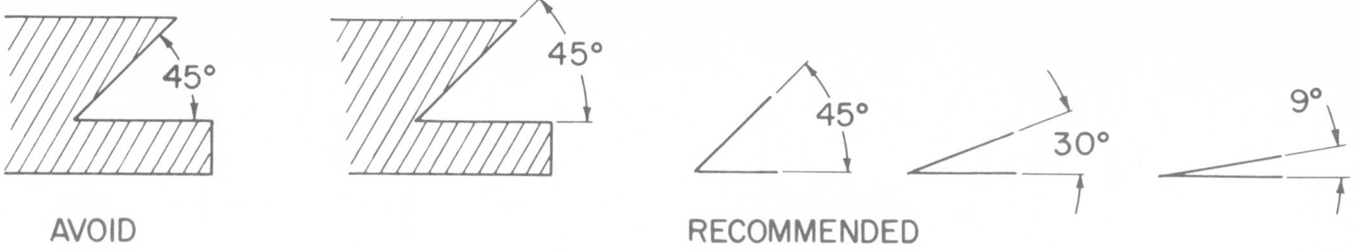

Fig. 8-17. Dimensioning angles.

DIMENSIONING SMALL PORTIONS OF AN OBJECT

When the space between extension lines is too small to place both the numerals and the arrowheads, dimensions are indicated as shown in Fig. 8-18.

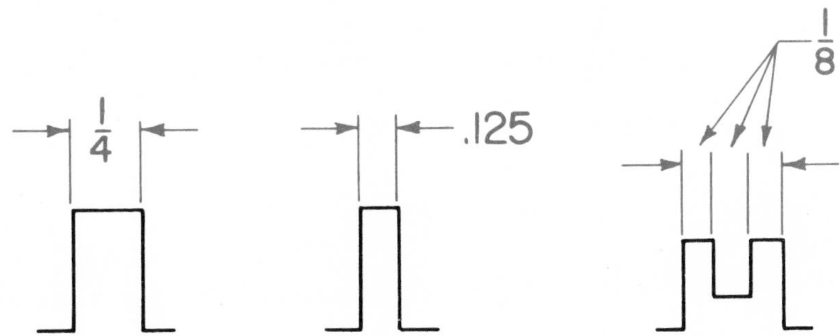

Fig. 8-18. When the space between extension lines is too small to place both arrowheads and numerals, dimensions may be indicated as shown.

TEST YOUR KNOWLEDGE - UNIT 8

1. Dimensions and shop notes are needed if ____________________.
2. Dimensions define ____________________.
3. Shop notes provide ____________________.
4. Unidirectional dimensions are read from ____________________.
5. Aligned dimensions are read from ____________________.
6. The dimension line is a:
   a. Heavy solid line.
   b. Fine solid line.
   c. Fine dotted line.
   d. None of the above.
7. The dimension line is capped with ____________________.
8. Dimensions are placed on the views that show ____________________.
9. Dimensions should be __________ rather than __________ about the drawing.
10. Dimensions must be complete so that:
    a. No scaling of the drawing is necessary.
    b. Sizes and shapes can be determined without assuming any measurements.
    c. Both of the above.
    d. None of the above.
11. When all dimensions are in inches the ____________________ need not be used.
12. When a leader is used to indicate a diameter it always points to the __________ of the diameter.

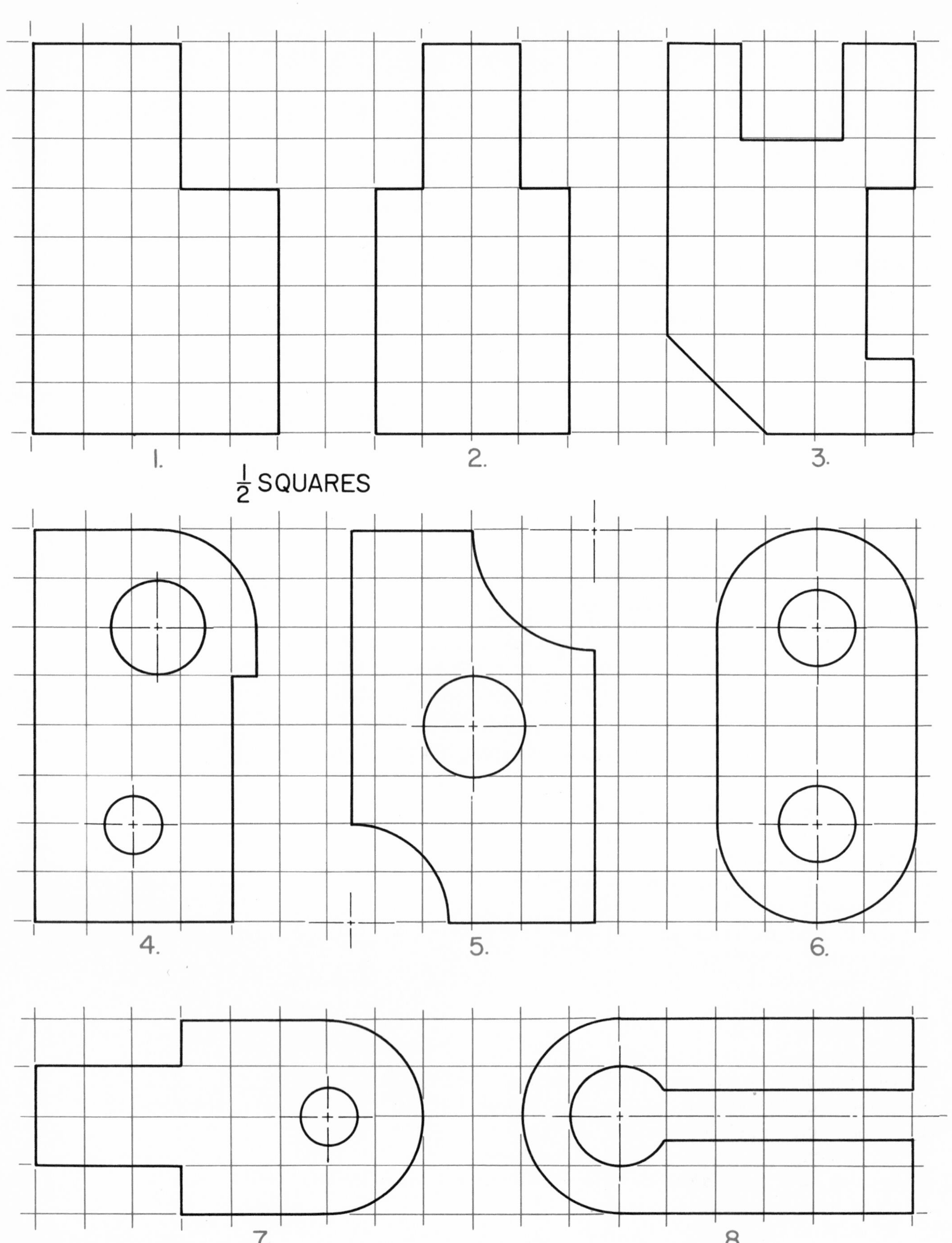

PROBLEMS 8–1 to 8–8. ***DIMENSIONING PROBLEMS.*** *Redraw and dimension. Two of these problems should be drawn on a sheet.*

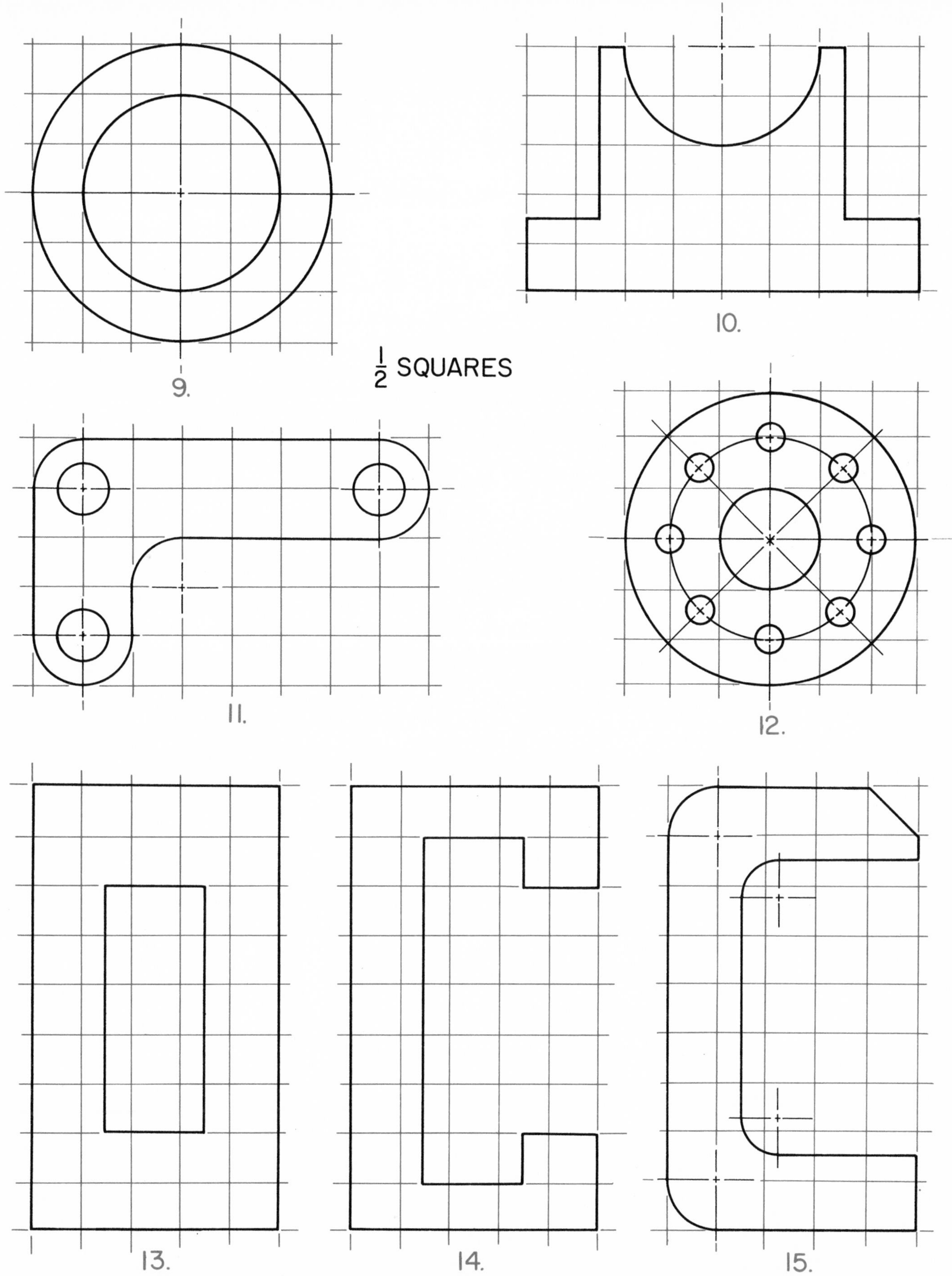

PROBLEMS 8–9 to 8–15. ***DIMENSIONING PROBLEMS.*** *Redraw and dimension problems.*

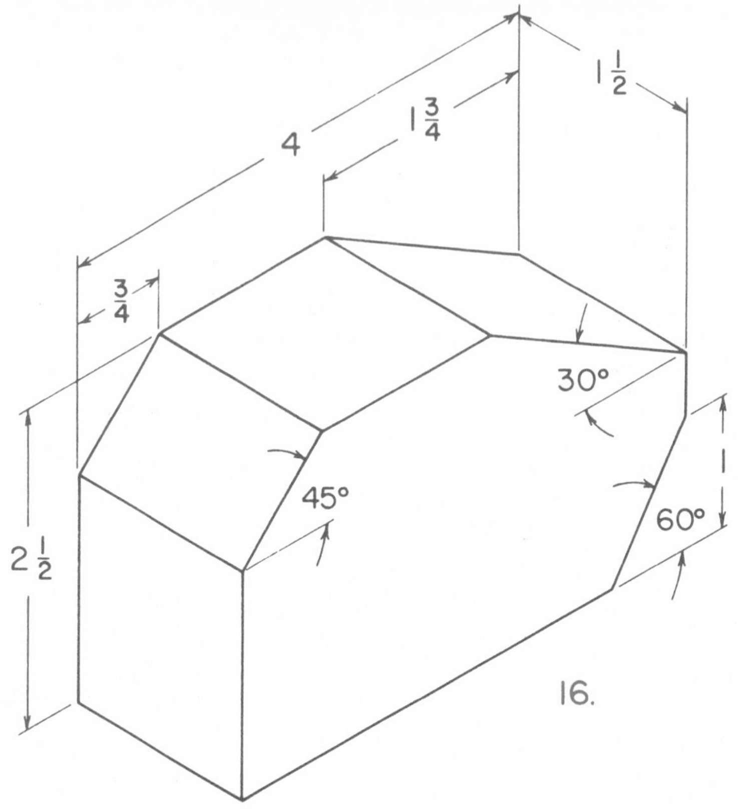

PROBLEM 8–16. *Above.* **SHIM.** **PROBLEM** 8–17. *Below.* **ALIGNMENT PLATE.** *Prepare the necessary views and correctly dimension them.*

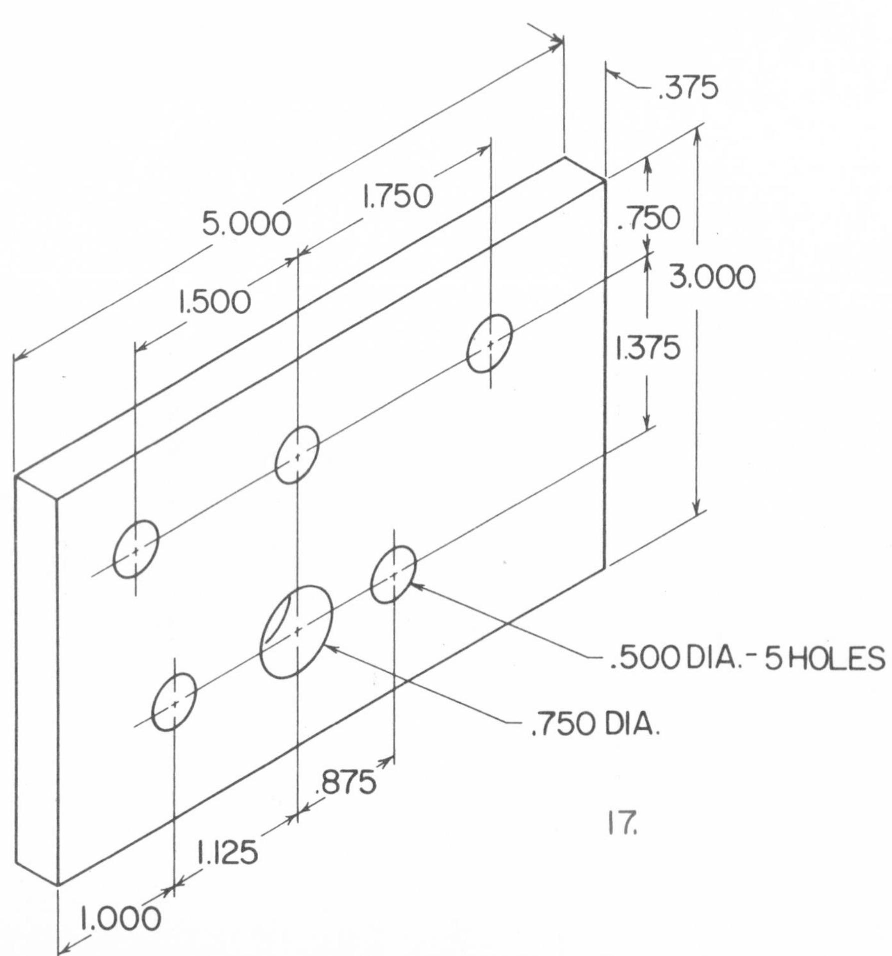

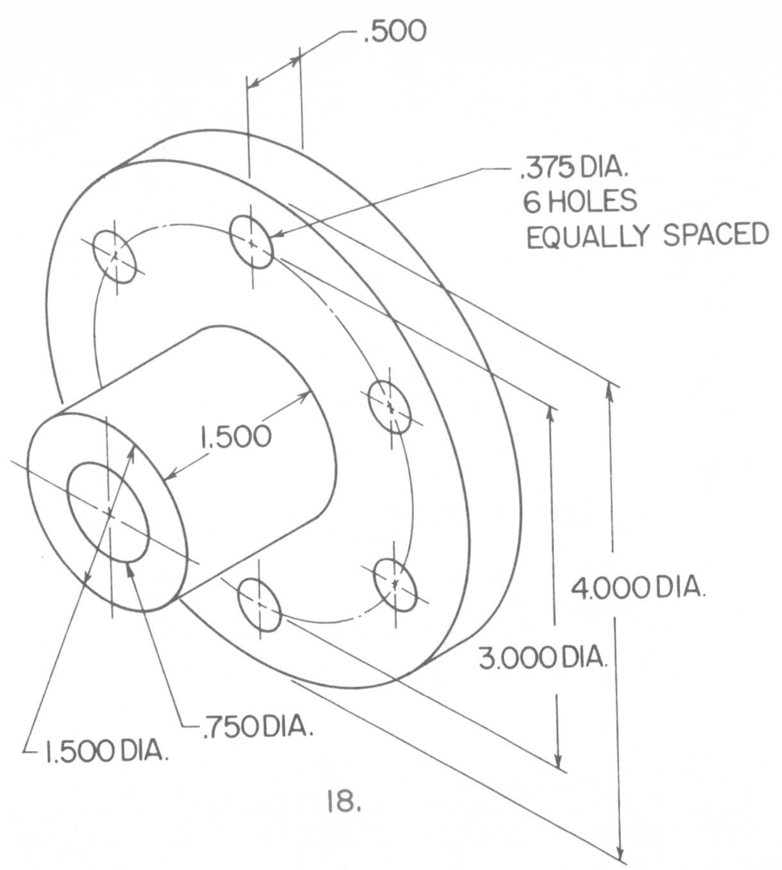

PROBLEM 8–18. Above. ***FACE PLATE.*** *Prepare a two view drawing and dimension correctly.* ***PROBLEM 8–19.*** *Below.* ***COVER PLATE.*** *Prepare the necessary views and dimension correctly.*

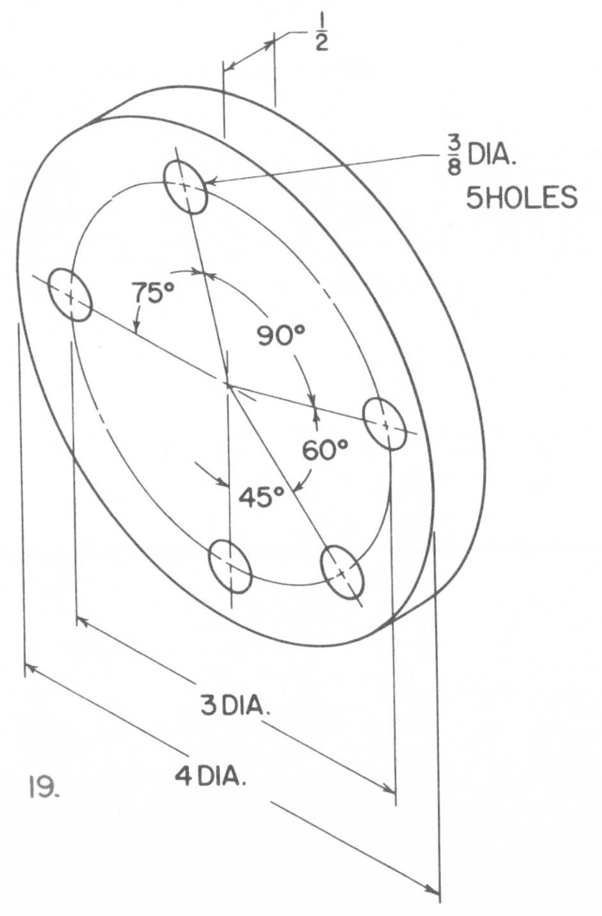

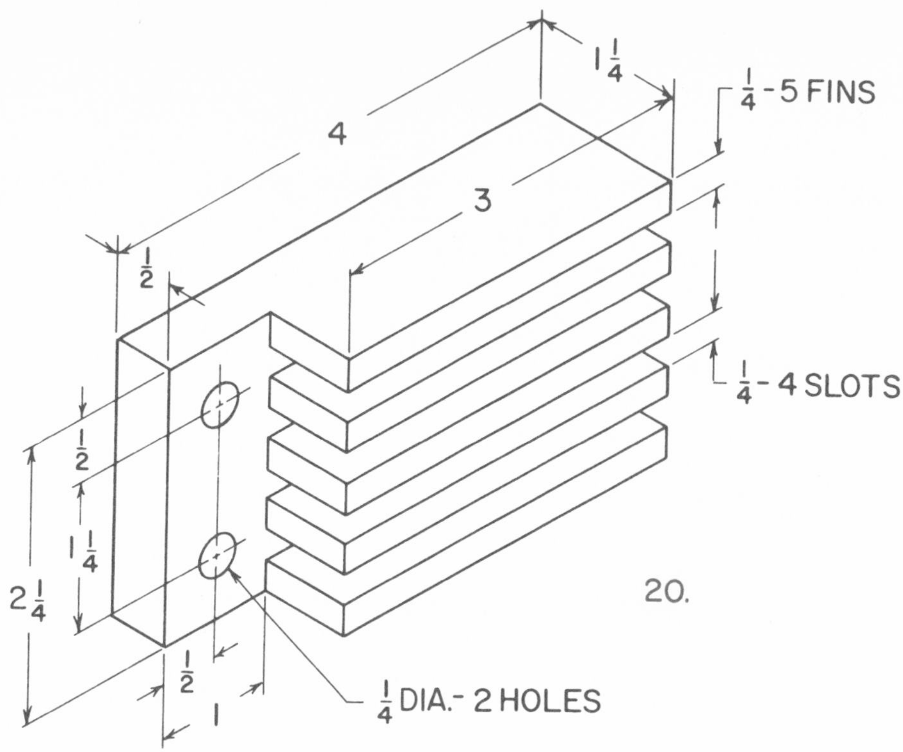

***PROBLEM** 8–20. Above. **HEAT SINK**. **PROBLEM** 8–21. Below. **STEP BLOCK**. Prepare the necessary views and dimension correctly.*

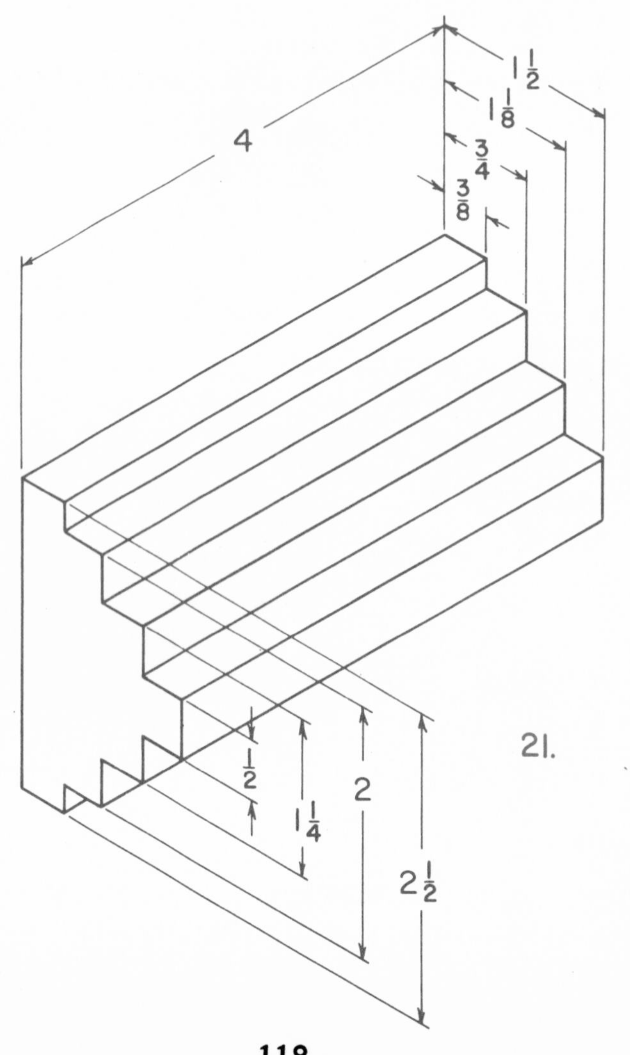

Unit 9
SECTIONAL VIEWS

Details on objects that are simple in design can be shown by using multiview projection. When an object has some of its design features hidden from view, as in Fig. 9-1, it is not easy to show the shape of the interior structure without using a "jumble" of hidden lines. SECTIONAL VIEWS are employed to make drawings of this type less confusing and easier to understand.

SECTIONAL VIEWS show how an object would look if a cut were made through it

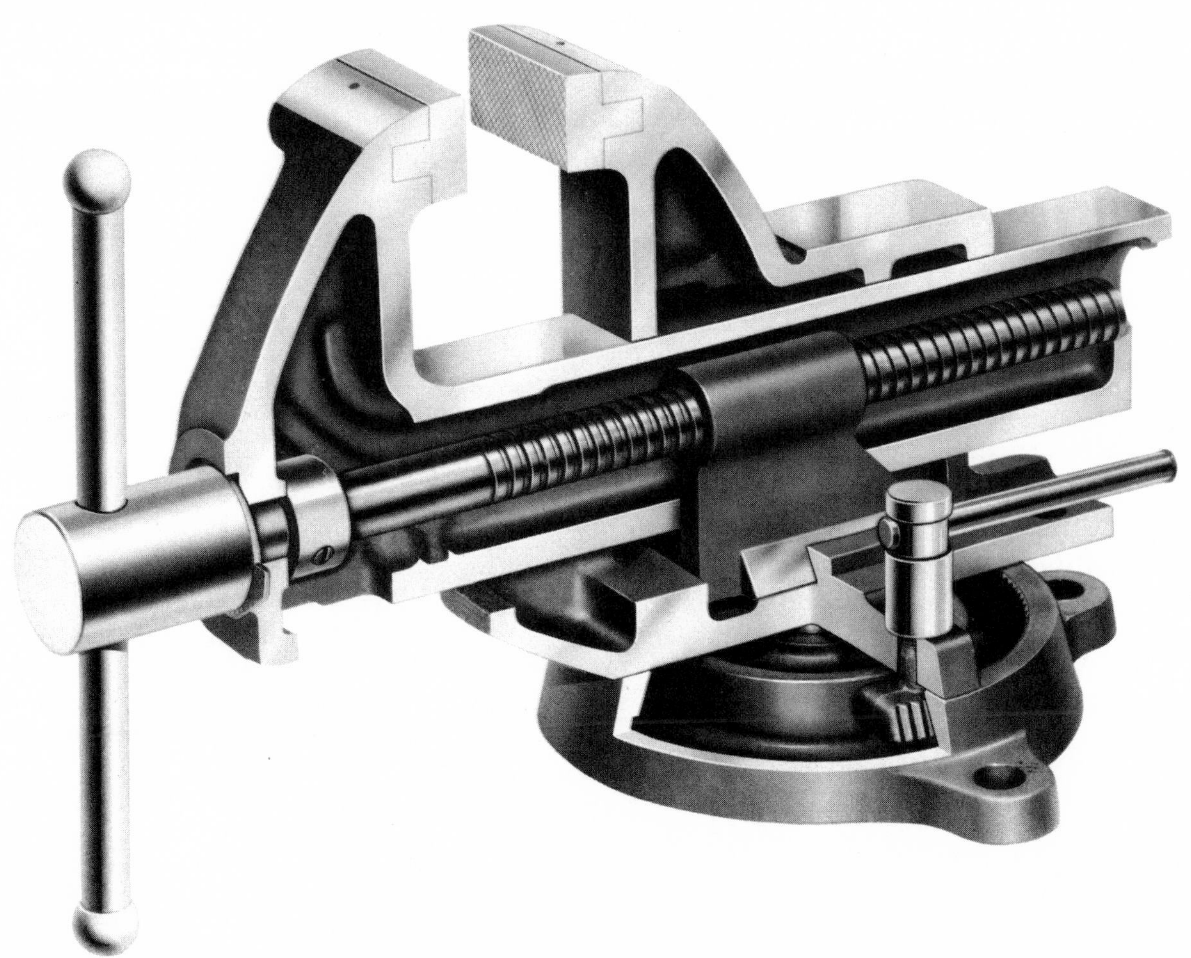

Fig. 9-1. Sectional view photo which shows the interior structure of a vise. (Columbian Vise and Mfg. Co.)

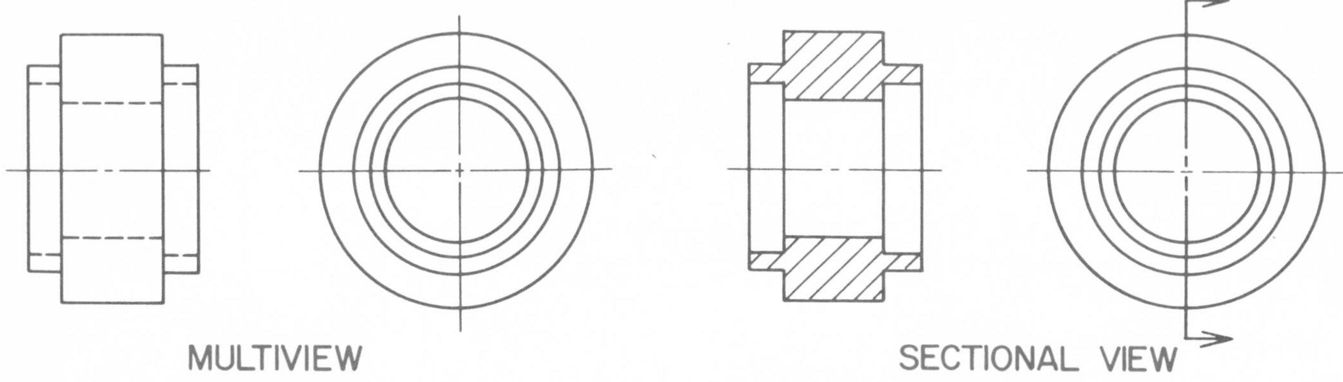

Fig. 9-2. Conventional multiview drawing and a drawing in section.

perpendicular to the direction of sight, Fig. 9-2. Sectional view drawings are necessary for a clear understanding of the shape of complicated parts.

CUTTING PLANE LINE

The function of a CUTTING PLANE LINE is to indicate the location of an imaginary cut made through the object to reveal its interior characteristics, Fig. 9-3. The extended lines are capped with arrowheads that indicate the direction of sight to view the section. Letters A-A, B-B, etc., are used to identify the section if it is moved to another position on the sheet, or if several sections are incorporated on a single drawing.

Fig. 9-3. Cutting plane lines. Either type may be used.

SECTION LINES

General purpose section lining (symbol for cast iron) is frequently used on drawings where the material specifications ("specs") are lettered elsewhere on the drawing.

Spacing of general purpose section lining is by eye and usually at 45 deg., Fig. 9-4.

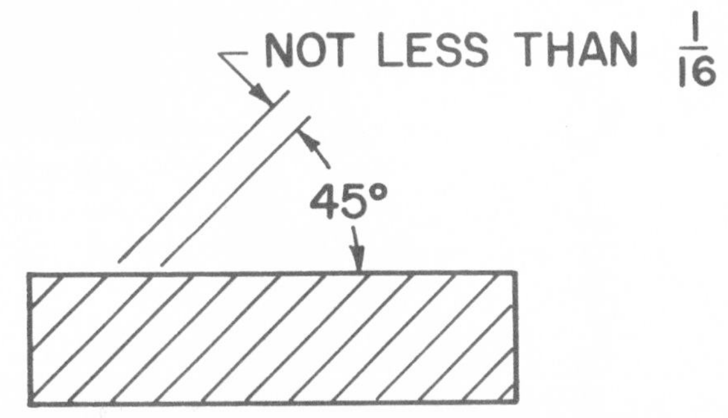

Fig. 9-4. General purpose section lining is spaced by eye and is usually drawn at 45 deg.

Line spacing is somewhat dependent on the size of the drawing or the area to be sec-

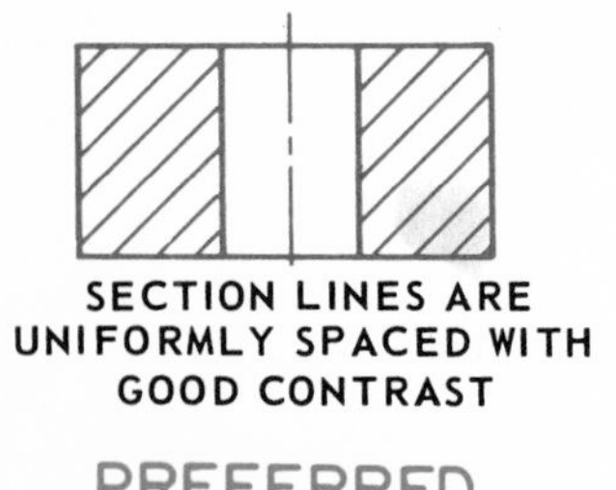

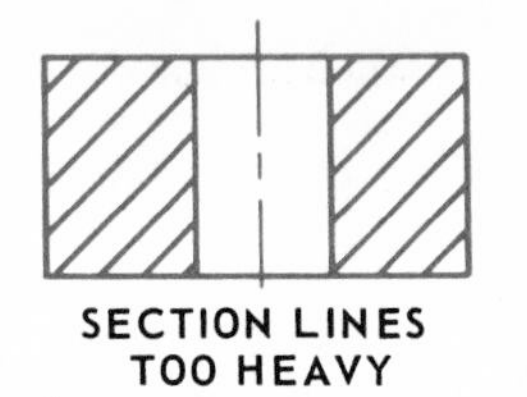

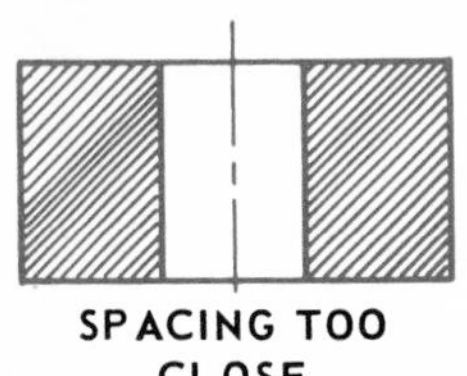

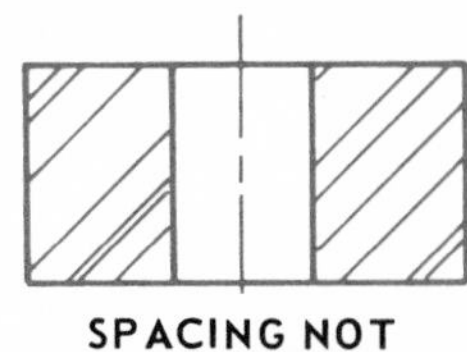

Fig. 9-5. Section line spacing.

tioned, and is usually about 1/16 in. for small areas, and up to 1/8 in. or more for large sections.

Do not use lines that are too thick or spaced too closely, Fig. 9-5. Also avoid lines that are not uniformly spaced or are drawn in different directions.

Where section lines will be parallel, or nearly parallel with the outline of the object, they should be drawn at some other angle, Fig. 9-6.

Fig. 9-6. The outline shape of the section may require the section lining to be drawn at other than 45 deg.

Should two or more pieces be shown in section, the section lines should be drawn in opposite directions or angles, to provide contrast, Fig. 9-7.

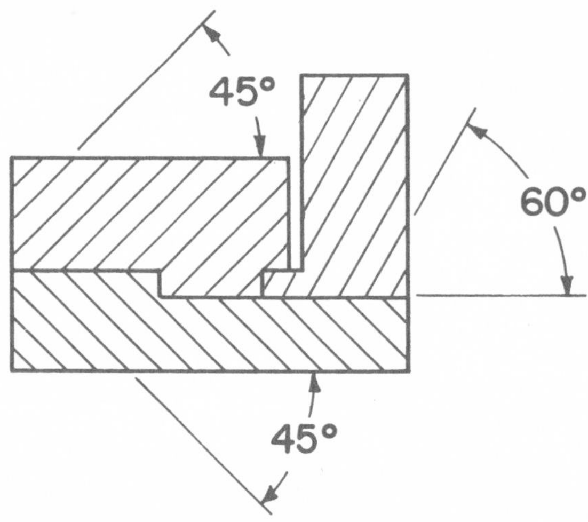

Fig. 9-7. Drawing section lines when several parts in the same sectional view are adjacent.

FULL SECTION

The FULL SECTION, Fig. 9-8, is developed by imagining that the cut has been made through the entire object. The part of the object between your eye and the cut is removed to reveal the interior features of the object. The resulting features are drawn as part of one of the regular multiview projections. Hidden lines behind the cutting plane are omitted unless they are necessary for a better understanding of the view or for dimensioning purposes.

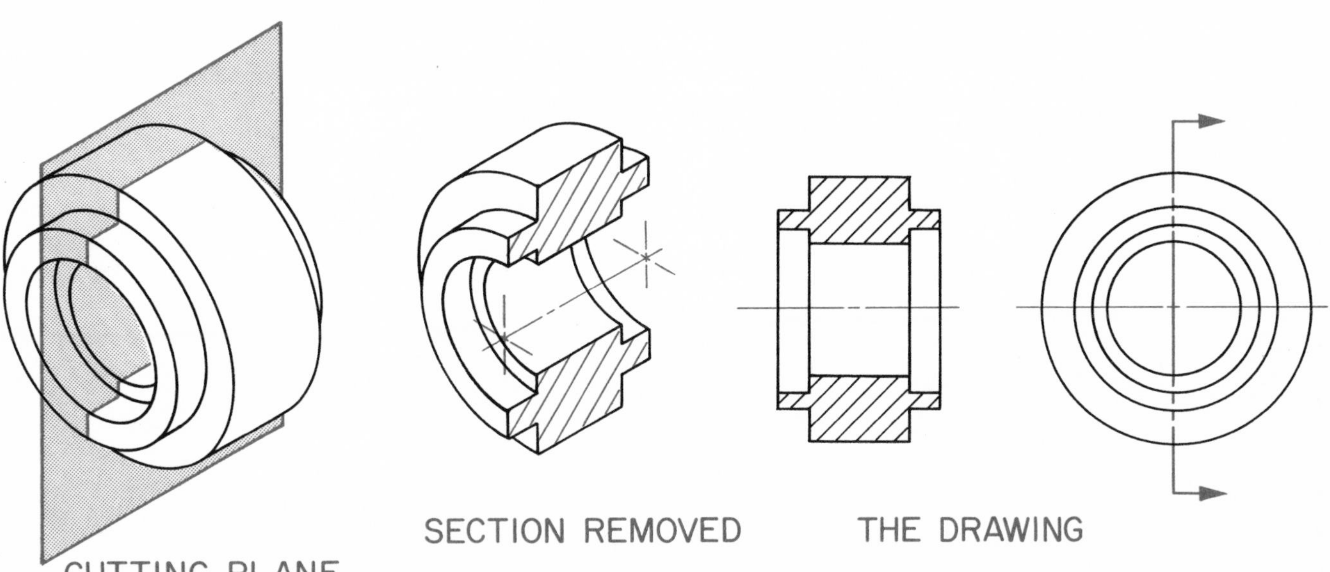

Fig. 9-8. The full section.

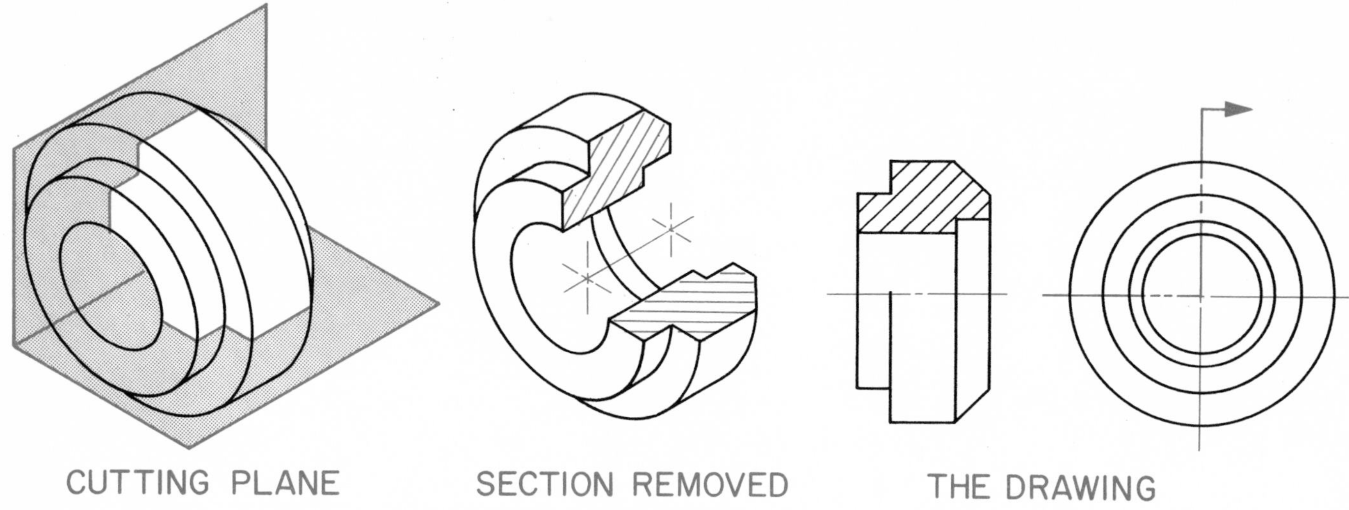

Fig. 9-9. The half section.

HALF SECTION

The shape of one-half of the interior features and one-half of the exterior features of an object are shown in the HALF SECTION, Fig. 9-9. Half sections are best suited for symmetrical objects. Cutting plane lines are passed through the piece at right angles to each other. One-quarter of the object is considered removed to show a half section of the interior structure. Unless needed for clarity, hidden lines are eliminated from half sections.

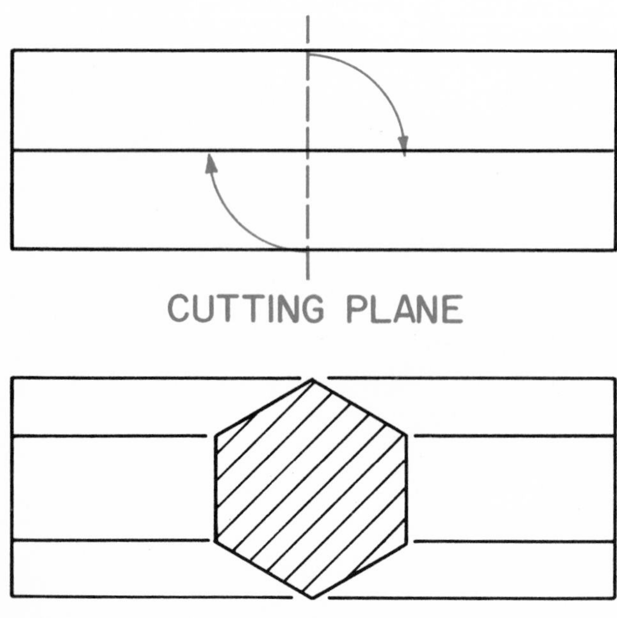

Fig. 9-10. A revolved section.

REVOLVED SECTION

REVOLVED SECTIONS are primarily utilized to show the shape of such things as spokes, ribs and stock metal shapes, Fig. 9-10. A revolved section is a drawing within a drawing. To prepare a revolved section, imagine that a section of the part to be shown is cut out and revolved 90 deg. in the same view.

Do not draw the object lines of a normal view through the revolved section.

ALIGNED SECTION

It is not considered good practice to make a true full section of a symmetrical

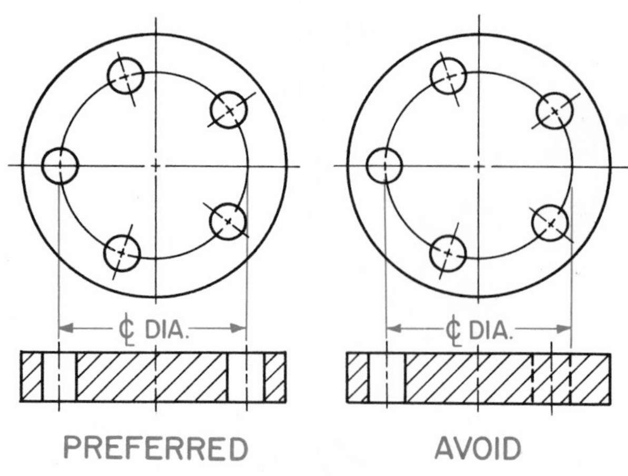

Fig. 9-11. An aligned section.

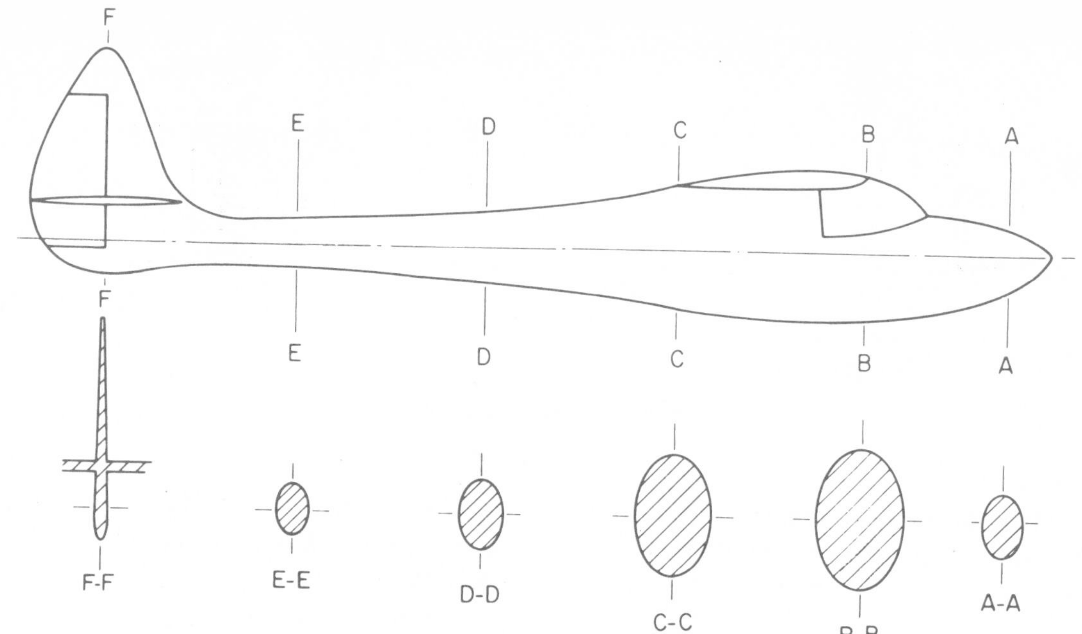

Fig. 9-12. A drawing showing the use of removed sections.

object that has an odd number of holes, webs or ribs. An ALIGNED SECTION is needed. With it, two of the divisions are depicted on the view, one of them being revolved into the plane of the other, Fig. 9-11.

REMOVED SECTION

There are times when it is not possible to draw a needed sectional view on one of the regular views. When this situation occurs, a REMOVED SECTION is used, Fig. 9-12. The section (or sections) are placed elsewhere on the sheet.

Use the removed section when the section must be enlarged for better understanding of the drawing.

OFFSET SECTION

While the cutting plane is ordinarily taken straight through the object, it may be necessary to show features not located on the cutting plane. When this occurs, the cutting plane may be stepped or offset to pass through these features, Fig. 9-13. This type of section is called an OFFSET SECTION.

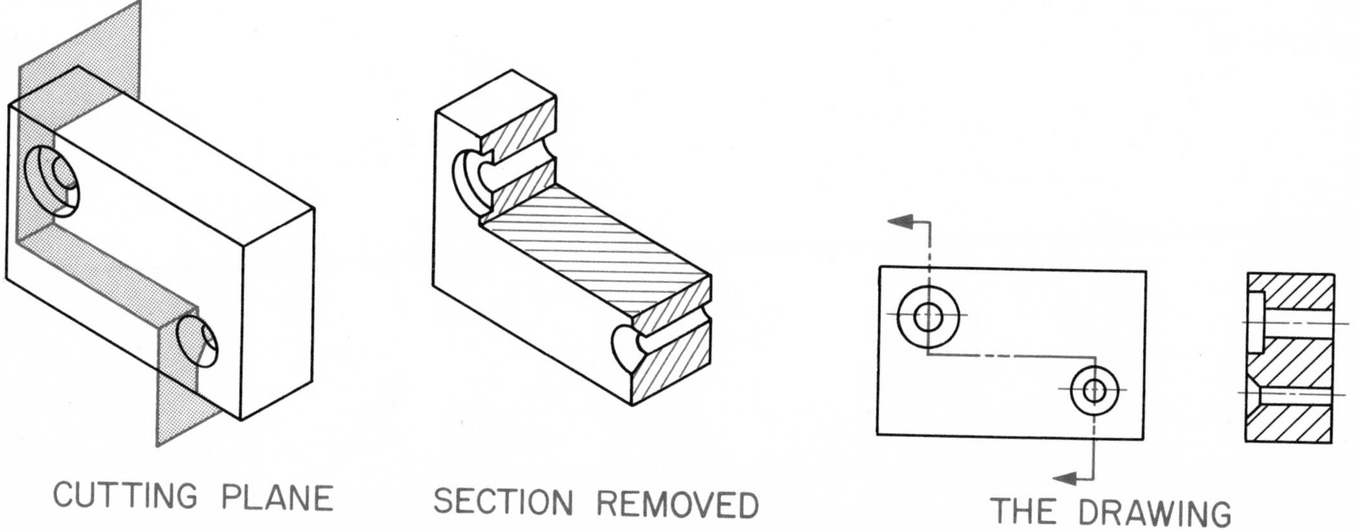

Fig. 9-13. An offset section.

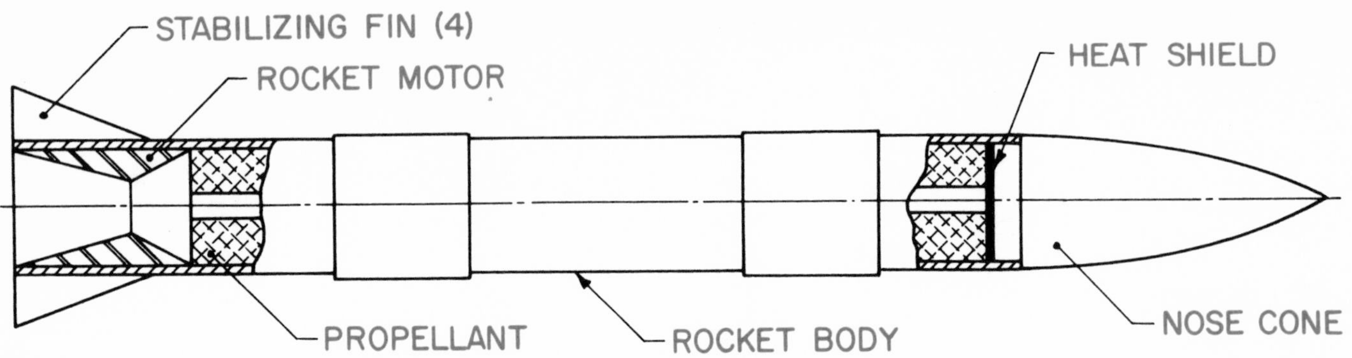

Fig. 9-14. Broken out section.

The offsets are not shown in the sectional view. If necessary, use reference letters A-A, B-B, etc., at the ends of the cutting plane line.

BROKEN OUT SECTION

The BROKEN OUT SECTION is employed when a small portion of a sectional view will provide the required information. Break lines define the section and are shown on one of the regular views, Fig. 9-14.

CONVENTIONAL BREAKS

When an object with a small cross section but of some length must be drawn, the draftsman is faced with several problems. If he draws it full size, it may be too long to fit on a standard size drawing sheet. If he uses a scale small enough to fit the part on the sheet, the details may be too small to give the required information or to be dimensioned.

In such cases, the object can be fitted to the sheet by reducing its length by means of the CONVENTIONAL BREAK, Fig. 9-15. This method permits a portion of the object to be deleted on the drawing.

The conventional break can only be used when the cross section of the part is uniform its entire length.

SYMBOLS TO REPRESENT MATERIALS

Fig. 9-16 shows symbols which may be used on section views to indicate a number of different materials. The lines should be drawn dark and thin to contrast with the heavier object lines.

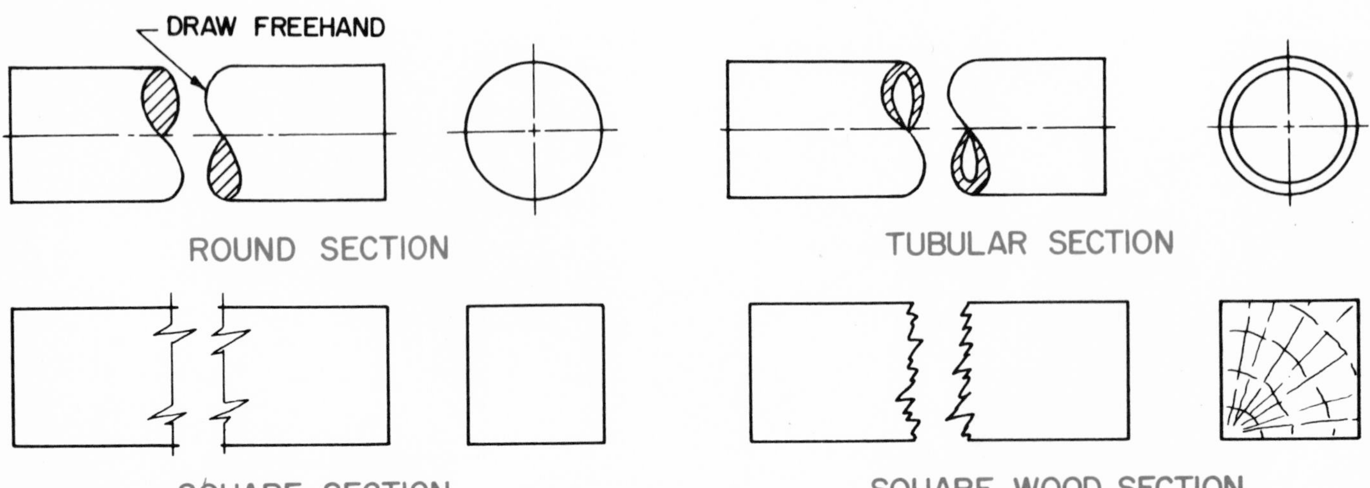

Fig. 9-15. Conventional breaks. By using breaks, an object with a small cross section but of some length may be drawn full size on a standard size drawing sheet.

Sectional Views

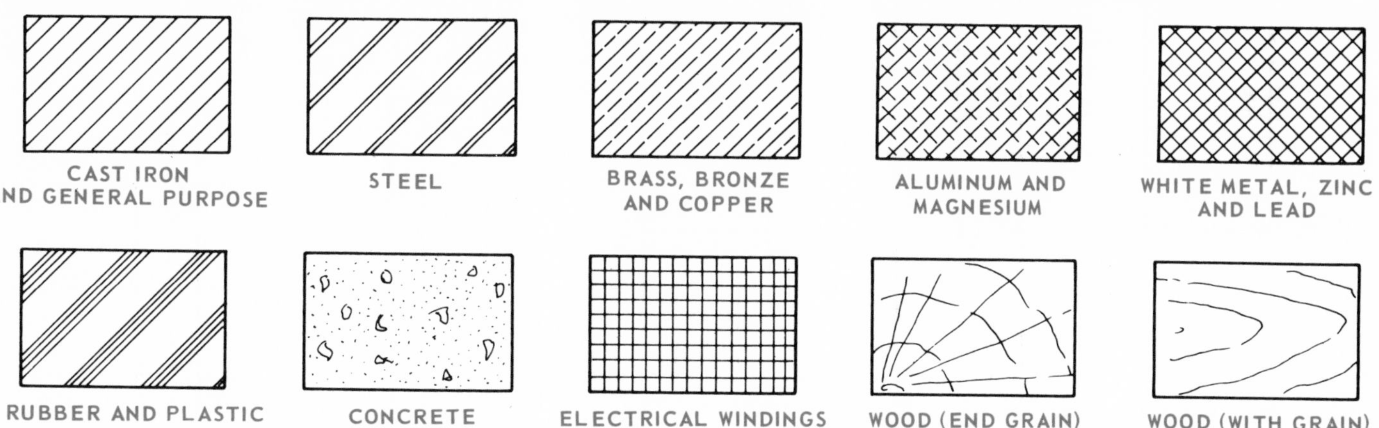

Fig. 9-16. Standard code symbols to use for various materials shown in section.

TEST YOUR KNOWLEDGE - UNIT 9

1. What are sectional views?
2. When are section views used?
3. The ______ _______ _______ indicates location of imaginary cut made through object to show its interior details.
4. Sections are identified by the use of ________.
5. The _________ section is used when the cut has been made through the entire object.
6. The ___________ section shows half of the interior and half of the exterior of the object.
7. _________ sections are primarily used to show the shape of such things as spokes, ribs and stock metal shapes.
8. The broken out section is used when ______________________________.
9. Objects with small cross sections but of considerable length can be fitted on a standard size sheet by making use of the ______________.
10. _________ lines may be used to indicate the material that has been cut.

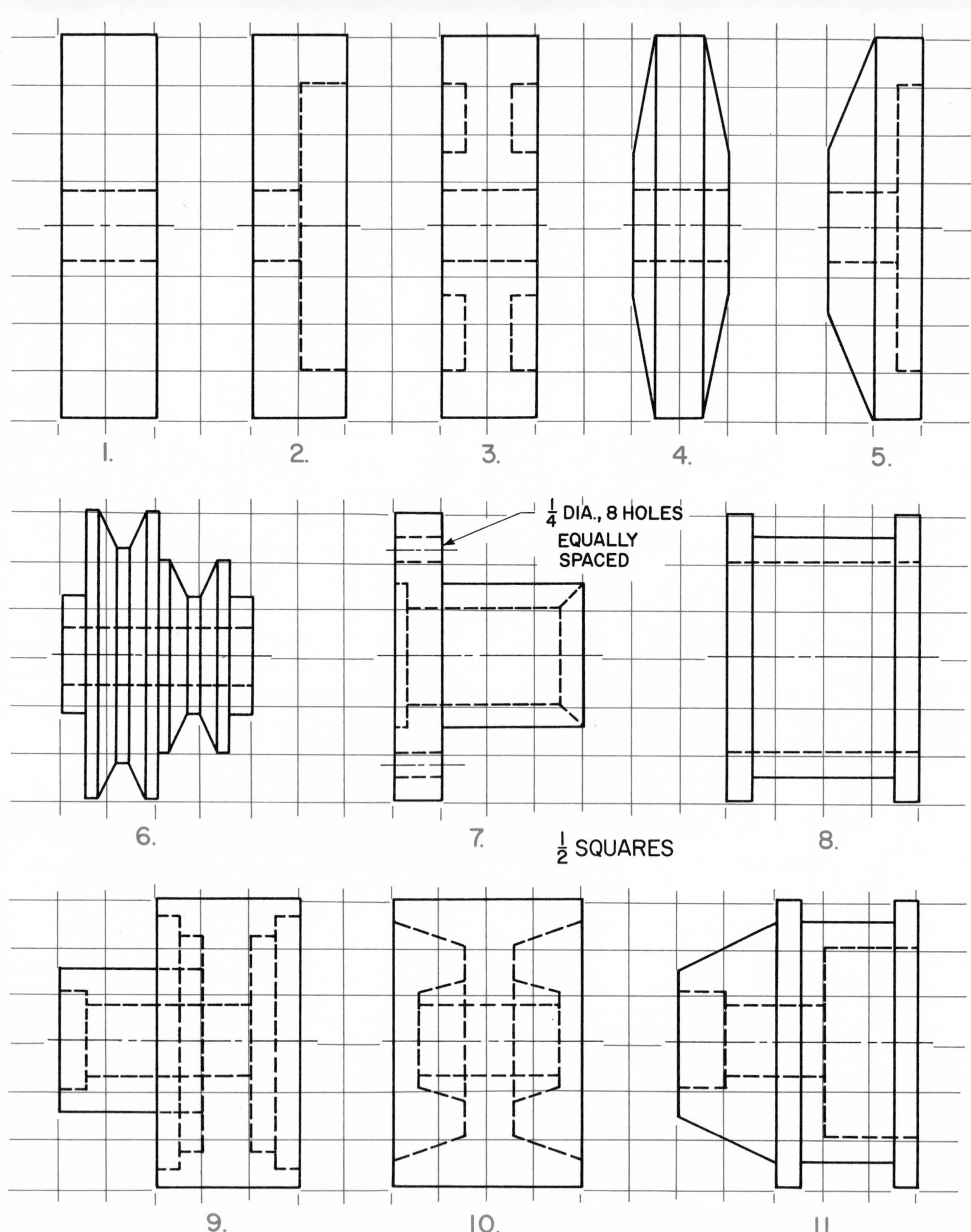

PROBLEMS 9–1 to 9–5. GRINDING WHEELS. Draw the views necessary to show the shape of each wheel. Make one view a full section or a half section as directed by your instructor. PROBLEM 9–6. PULLEY. Draw the views necessary to show the shape of this pulley. Draw one view as a half section. PROBLEM 9–7. COUPLING. Draw the views necessary to show the shape of this coupling. Draw one view as a full section. PROBLEM 9–8. BUSHING. Draw the views necessary to show the shape of this bushing. Draw one view as a half section. PROBLEM 9–9. SPACER. Draw the views necessary to show the shape of this spacer. Draw one view as a half section. PROBLEM 9–10. FLAT BELT PULLEY. Draw the views necessary to show the shape of this pulley. Draw one view as a full section. PROBLEM 9–11. ADAPTER BEARING. Draw the views necessary to show the shape of this bearing. Draw one view as a half section.

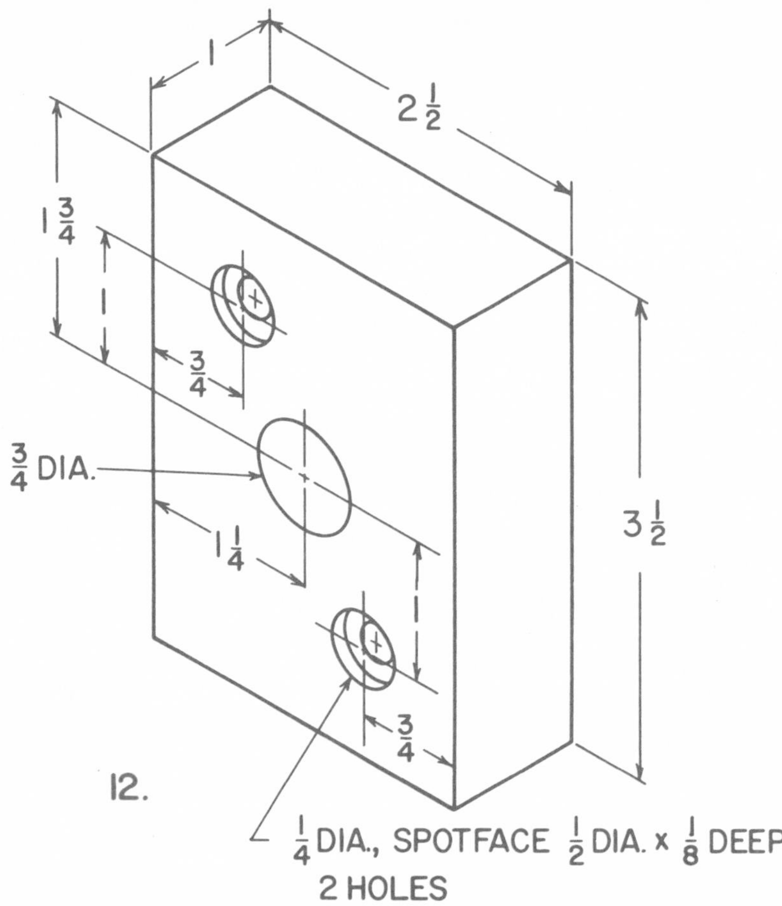

PROBLEM 9–12. Above. SPACER. Draw the views necessary to show the shape of the spacer. Draw one view as an offset section through the three holes. PROBLEM 9–13. Below. SPECIAL BUSHING. Draw the views necessary to show the shape of the bushing. Draw one view as a full section along the horizontal plane.

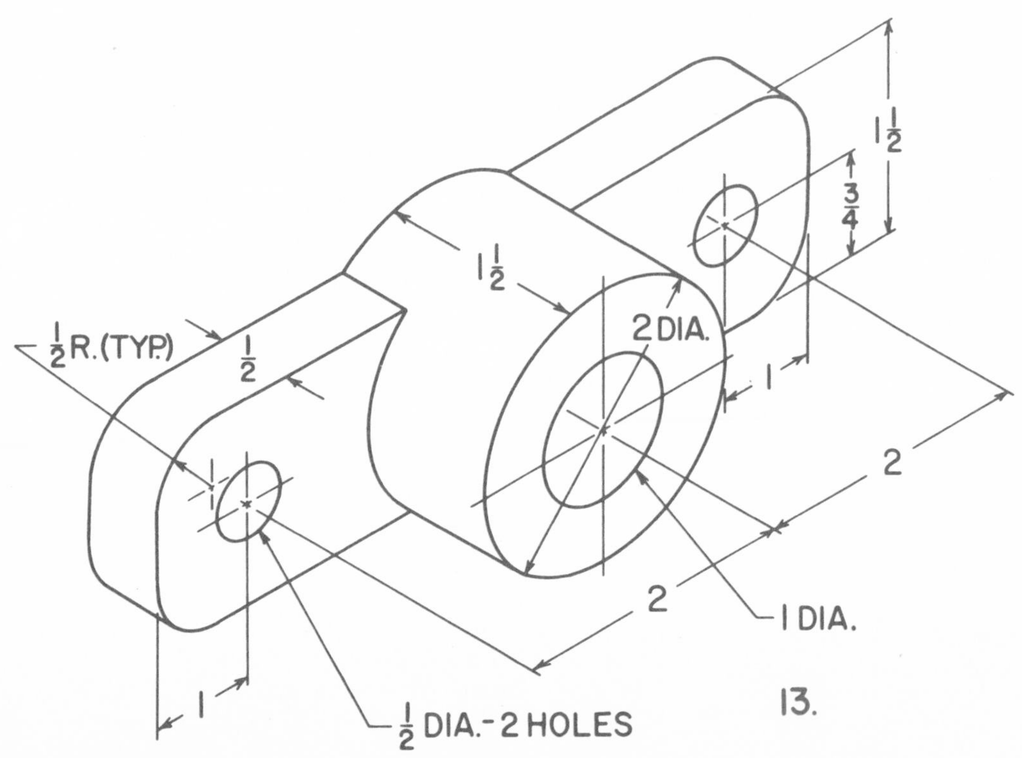

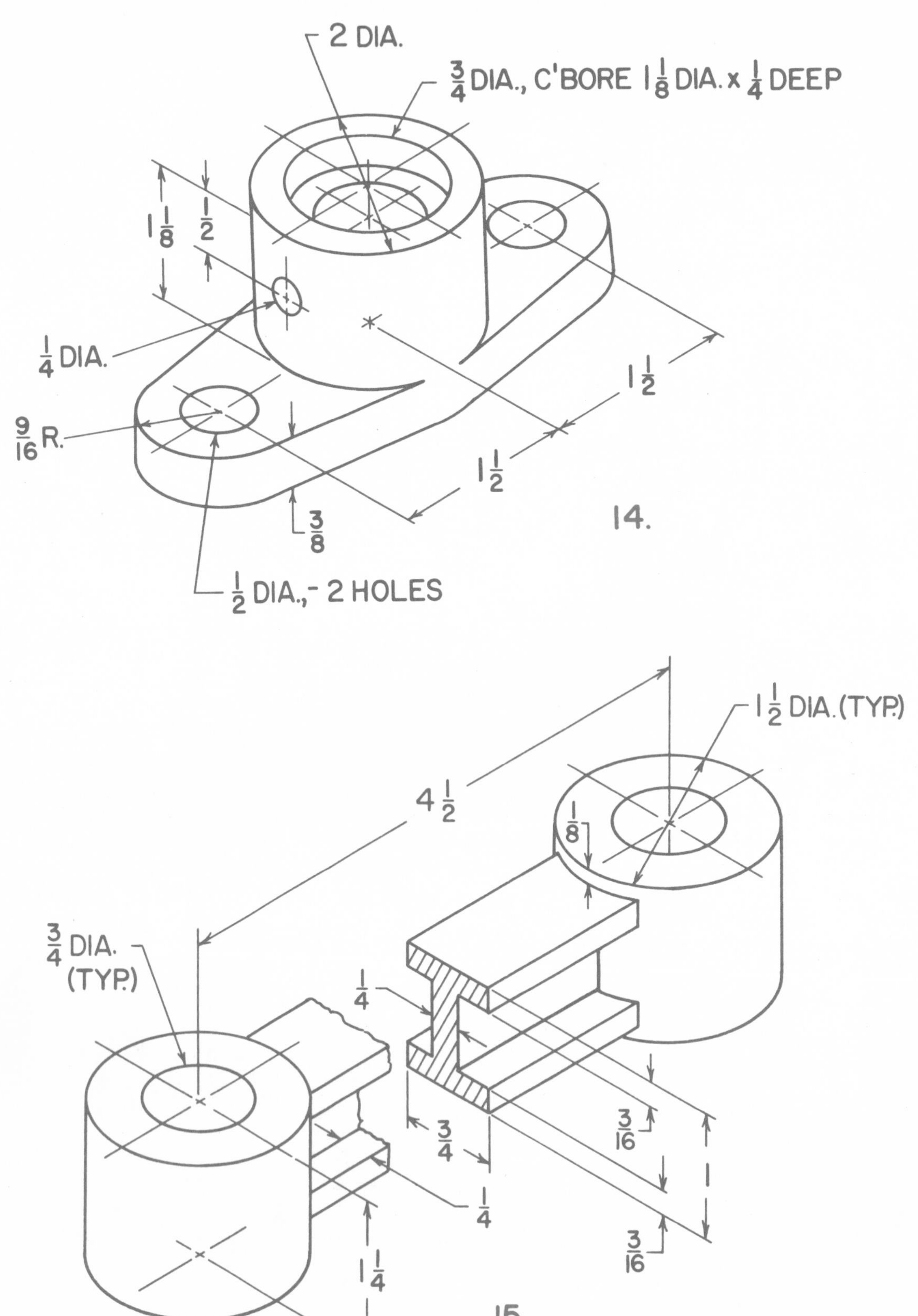

PROBLEM 9–14. BRACKET. Draw the views necessary to show the shape of the bracket. Include a broken out section through the 1/4 in. diameter hole on one view. PROBLEM 9–15. CONNECTING ROD. Draw the views necessary to show the shape of the rod. The cross section of the rod may be shown as a revolved section or as a removed section.

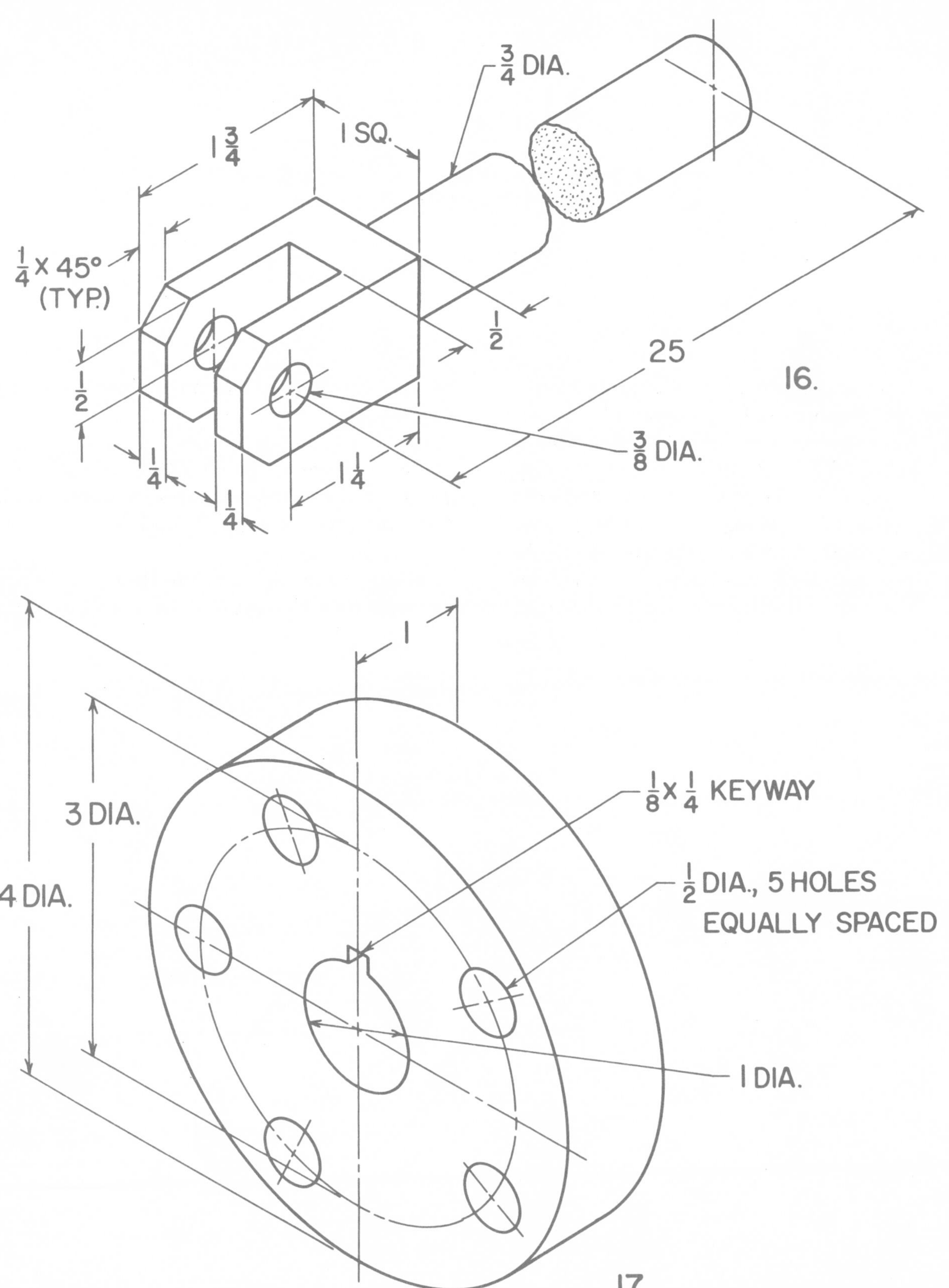

PROBLEM 9–16. TORQUE ROD. Draw the views necessary to show the shape of the rod. Use the conventional break to fit the object on the drawing sheet. PROBLEM 9–17. ADAPTER PLATE. Draw the views necessary to show the shape of the plate. Draw one view as a full section.

Unit 10
AUXILIARY VIEWS

The true shape and size of objects having angular or slanted surfaces cannot be shown using the regular (top, front, side) views. The TRUNCATED BLOCK shown in Fig. 10-1, is such an object (truncated means that the object has been cut off at an angle). The true length of the cut is shown on the front view, but this view does not show its width. The true width of the cut is shown on the top and side views but neither shows the true length.

An additional or AUXILIARY VIEW is needed to show the true length and true width of the angular surface, Fig. 10-2.

When drawing an auxiliary view, remember that the view is ALWAYS projected

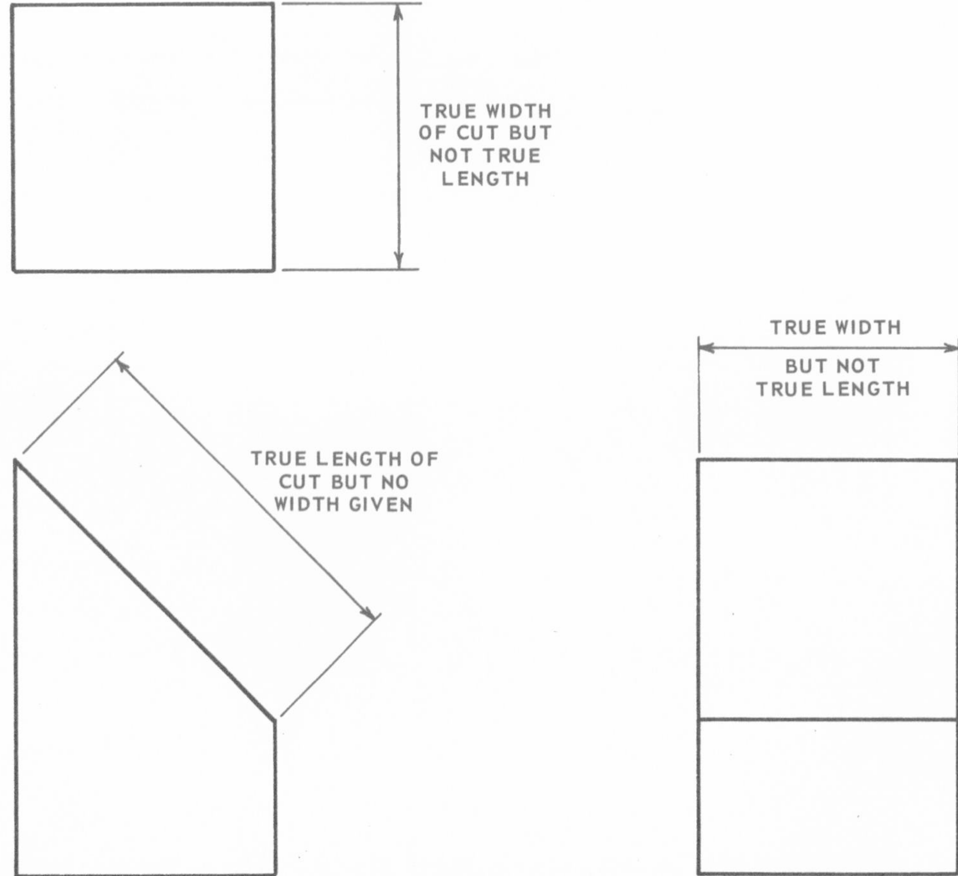

Fig. 10-1. Why AUXILIARY VIEWS are necessary. The single views do not show the true shape (length and width) of the cut off portion of the object.

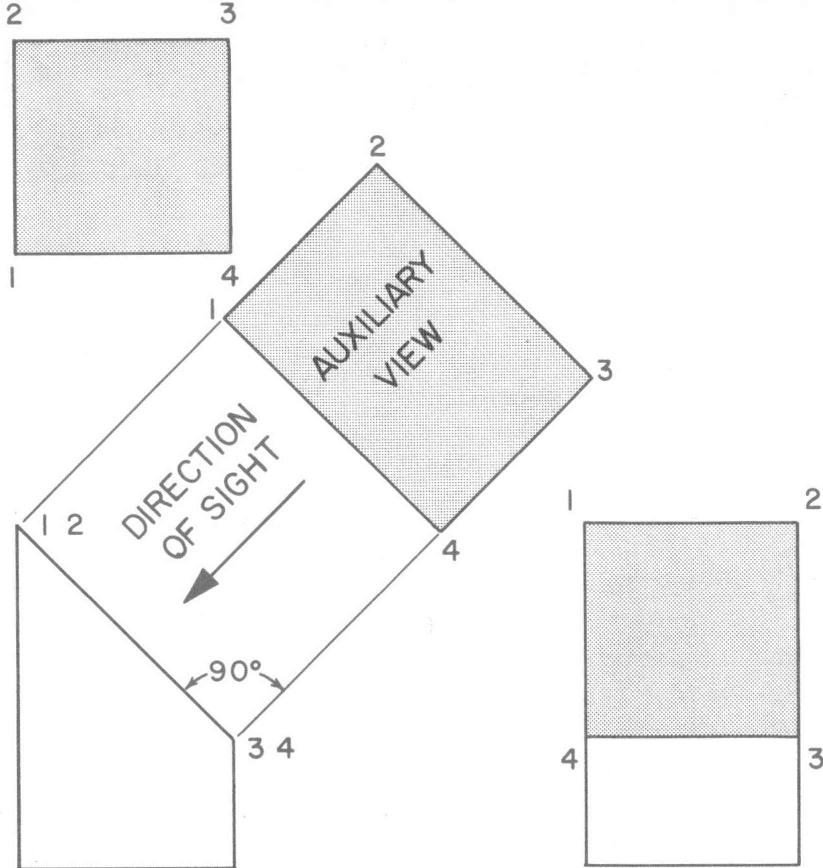

Fig. 10-2. The ***AUXILIARY VIEW*** *shows the true shape of the angular surface.*

from the regular view on which the inclined surface appears as a line. Also, the construction lines projecting from the inclined surface are ALWAYS at right angles to the cut.

When drawing auxiliaries, the usual practice is to show only the inclined portion of the view. It is seldom necessary to draw a full projection of the object, Fig. 10-3. It is often possible to eliminate one of the con-

Fig. 10-3. It is not necessary to draw a full projection of the object.

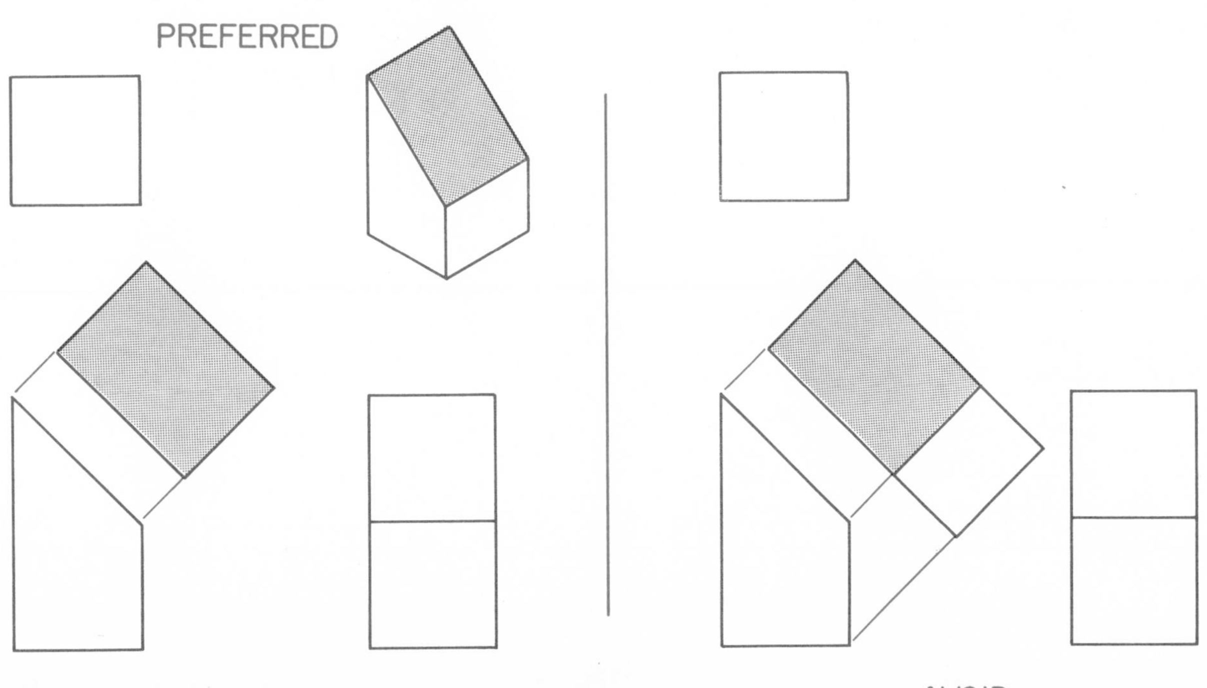

ventional views when using an auxiliary view. See Fig. 10-4.

An auxiliary view can be projected from any view that shows the inclined surface as a line. It would be called a FRONT AUXILIARY if projected from the front view, a TOP AUXILIARY if projected from the top view, and a RIGHT SIDE AUXILIARY if projected from the right side.

Considerable time can be saved when drawing auxiliary views of symmetrical objects by drawing only half of the view.

Auxiliary views that include rounded surfaces or circular openings may cause minor problems. Fig. 10-5 shows how such a rounded surface would be drawn. Proceed as follows:

1. Draw the needed views. Divide the circular view into 12 equal parts. Project the divisions to the other view.
2. At any convenient distance from the inclined face, draw center line A'-A' for the auxiliary view. This center line is parallel to the inclined face.
3. Project the necessary points from the inclined surface through center line A'-A'. These lines are at right angles (90°) to the inclined face.
4. Using dividers or a compass, transfer measurements A-a, A-b, and A-c to the appropriate lines to the right and left of center line A'-A' (to get points A'-a, A'-b, and A'-c).
5. Complete the auxiliary view by connecting the points with a French curve.

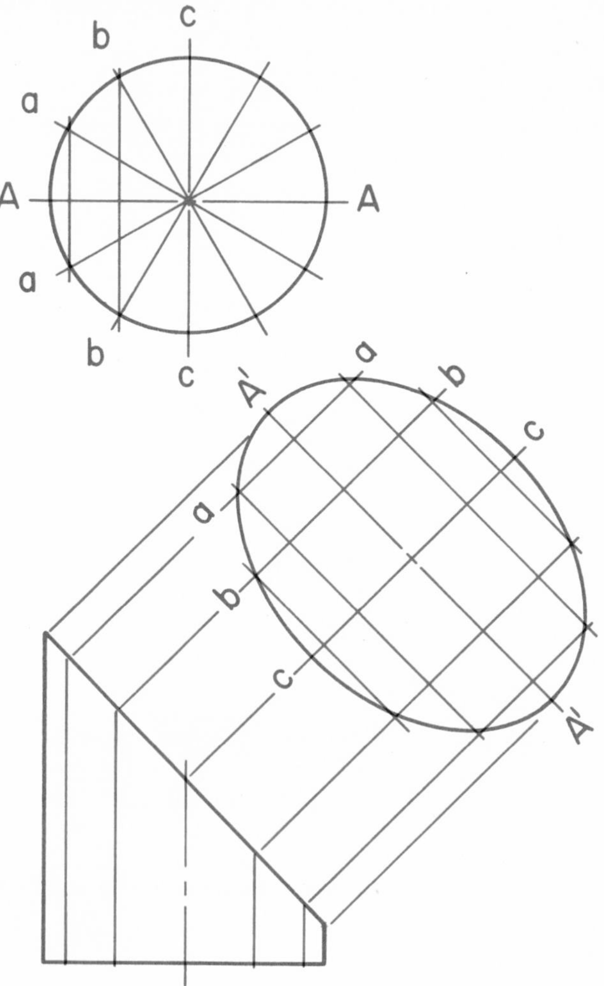

Fig. 10-5. Drawing an auxiliary view of a circular object.

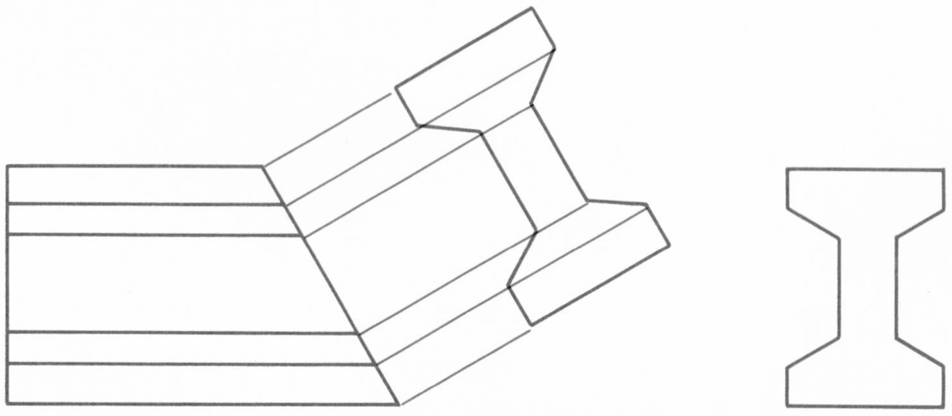

Fig. 10-4. One view may often be eliminated when using an auxiliary view.

TEST YOUR KNOWLEDGE - UNIT 10

1. Why are auxiliary views needed?
2. Make a sketch of an object that would require an auxiliary view.
3. The auxiliary view is always projected from the view that shows the inclined surface as a ____________________. The construction lines projecting from the inclined surface are always at ________ to the cut.
4. The auxiliary view when projected from the front view is called a ______________.
5. When drawing an auxiliary view of a symmetrical object, much time can be saved by drawing ______________________.

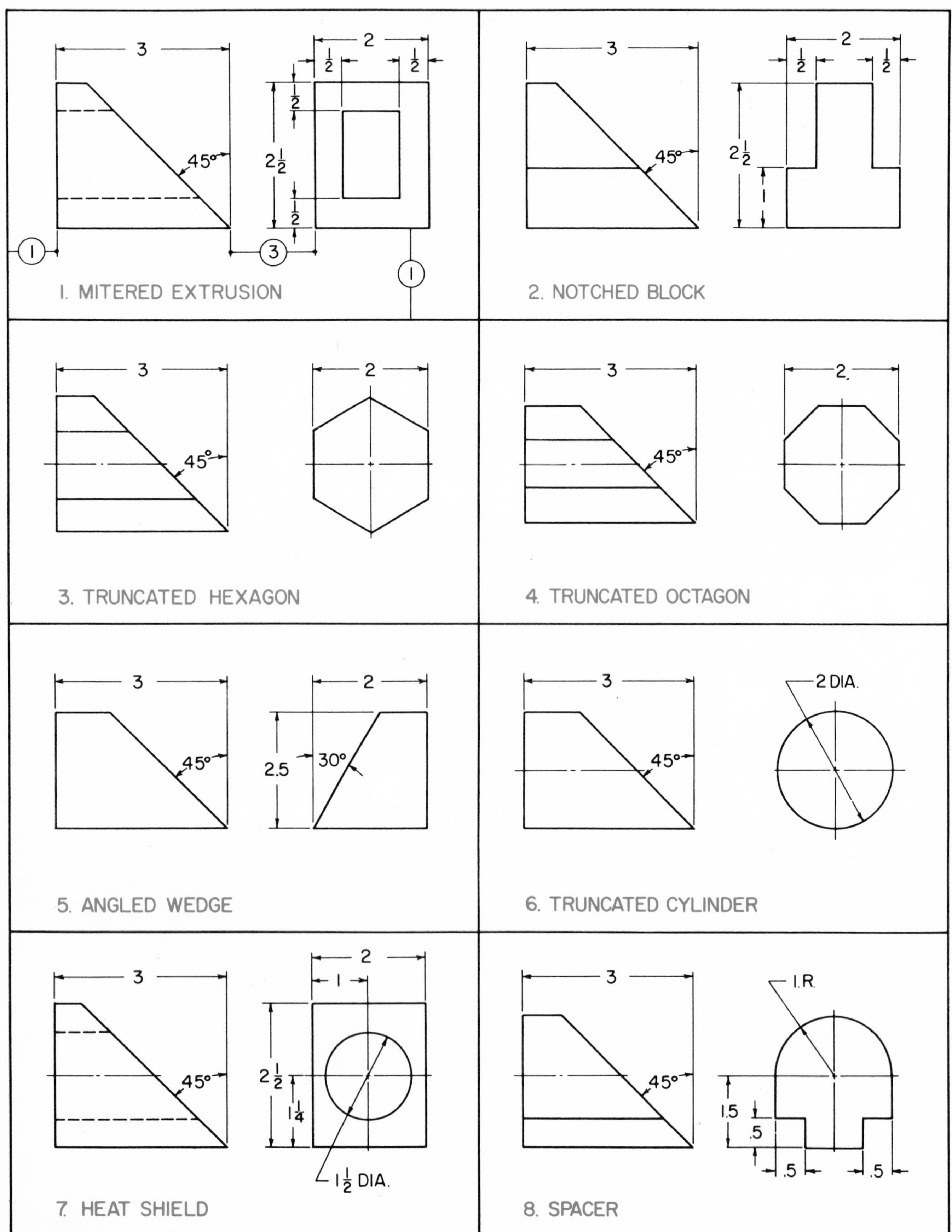

PROBLEMS 10–1 to 10–8. Space drawings according to the circled dimensions. The top views may be eliminated.

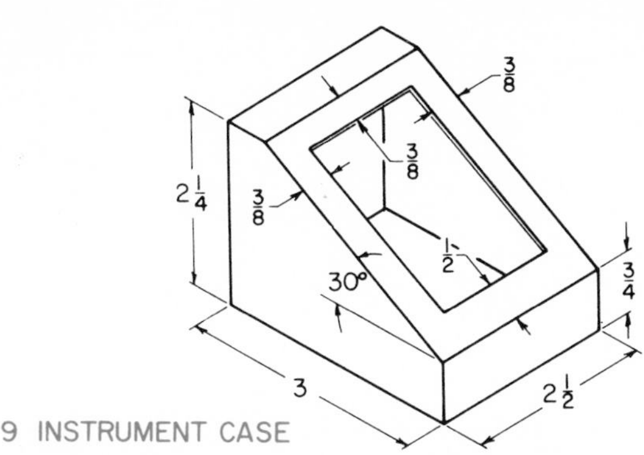

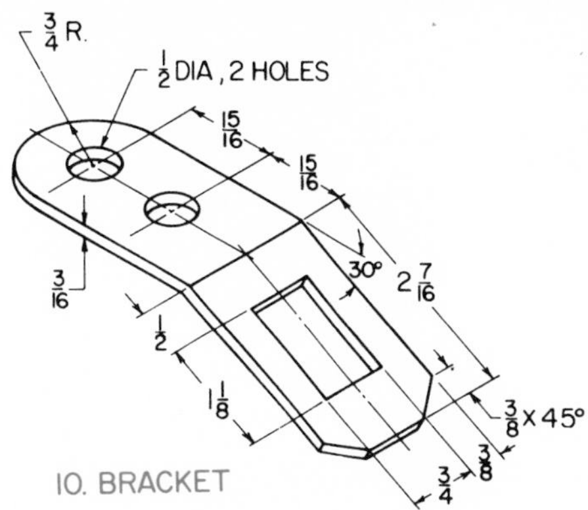

PROBLEM 10–9. Left. INSTRUMENT CASE. *1–Use a vertical drawing sheet format. 2–Allow 4 in. between front and top views. 3–Locate the auxiliary view 1 1/2 in. from the front view. 4–The front view is 1 in. from the left border.* ***PROBLEM 10–10. Right. BRACKET.*** *1–Use a horizontal drawing sheet format. 2–Allow 3 in. between the front and top views. 3–Locate the auxiliary view 1 in. from the front view. 4–The front view is 2 3/4 in. from the left border.*

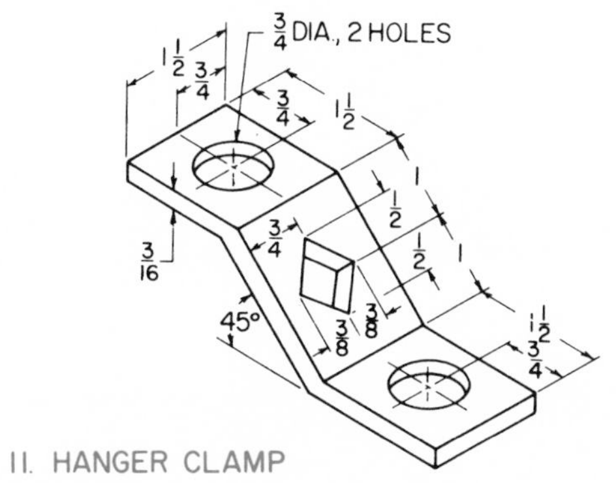

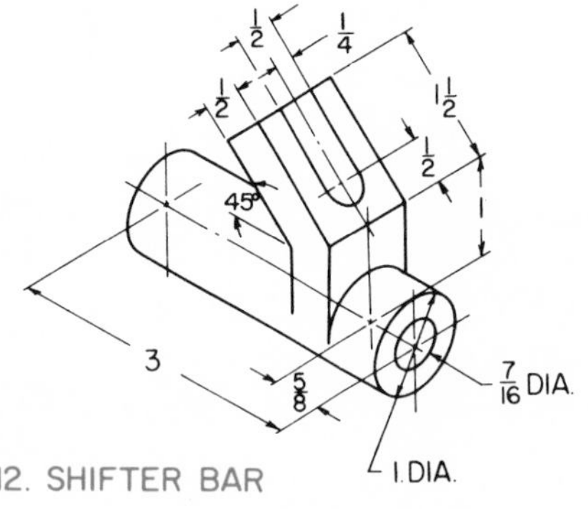

PROBLEM 10–11. Left. HANGER CLAMP. *1–Use a horizontal drawing sheet format. 2–Allow 3 in. between front and top views. 3–Locate the auxiliary view 1 1/2 in. from the front view. 4–Allow 2 1/2 in. between the front and right side view. 5–The front view is 3/4 in. from the left border.* ***PROBLEM 10–12. Right. SHIFTER BAR.*** *1–Use a horizontal drawing sheet format. 2–Allow 2 in. between the front and top views. 3–Locate the auxiliary view 1 in. from the front view. 4–Allow 2 in. between the front and right side view. 5–The front view is 2 in. from the left border.*

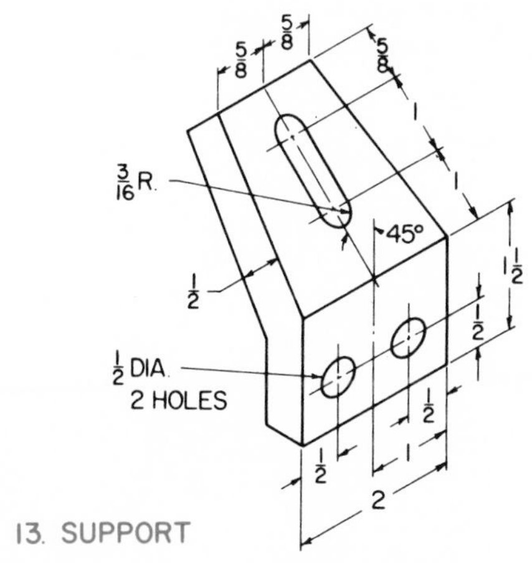

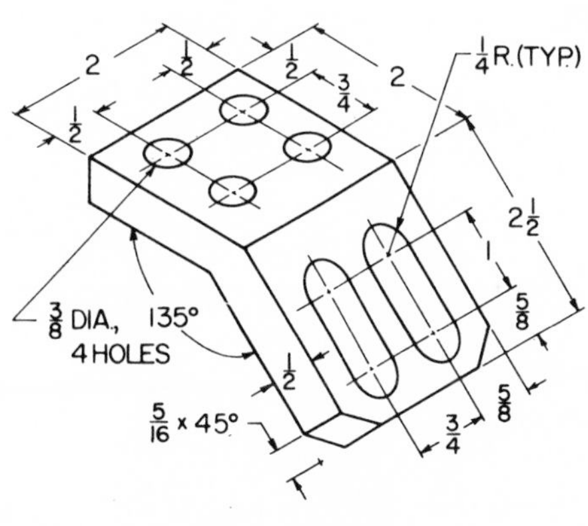

PROBLEM 10–13. Left. SUPPORT. *1–Use a vertical sheet format. 2–Allow 2 1/2 in. between front and top views. 3–Locate the auxiliary view 1 1/2 in. from the front view. 4–The front view is 1 1/4 in. from the left border.* ***PROBLEM 10–14. Right. ADJUSTABLE BRACKET.*** *1–Use a horizontal sheet format. 2–Allow 2 1/8 in. between the front and top views. 3–Locate the auxiliary view 1 in. from the front view. 4–Allow 2 3/4 in. between the front and right side views. 5–The front view is 3/4 in. from the left border.*

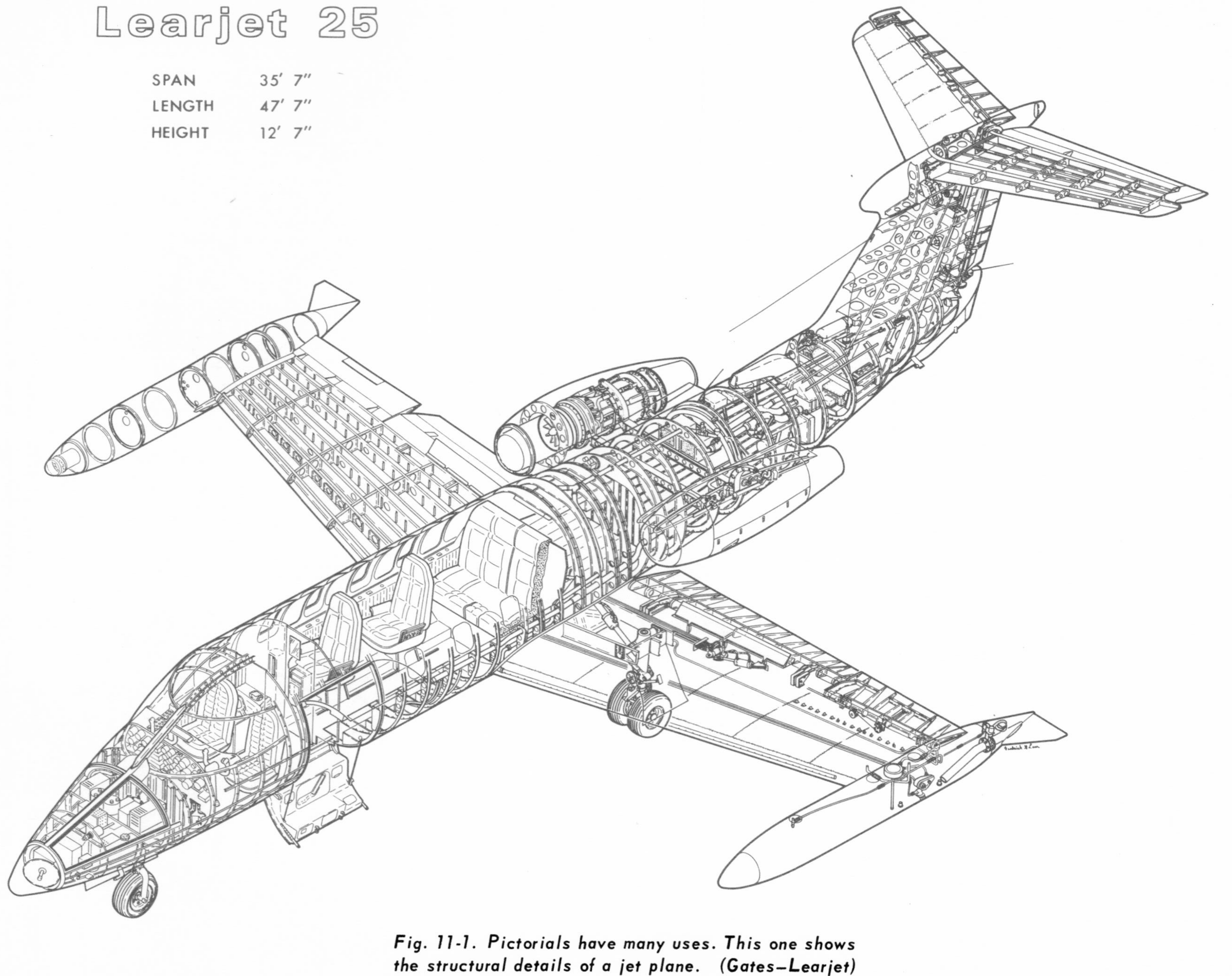

Fig. 11-1. Pictorials have many uses. This one shows the structural details of a jet plane. (Gates–Learjet)

Unit 11
PICTORIALS

PICTORIAL DRAWINGS

A PICTORIAL DRAWING shows a likeness (shape) of an object as viewed by the eye. The pictorial drawing of the jet plane, Fig. 11-1, shows many of the structural details of the craft.

If you have worked with radio or electronic kits, you are familiar with pictorial drawings which show the builder what needs to be done and how to do it to complete his project, Fig. 11-2.

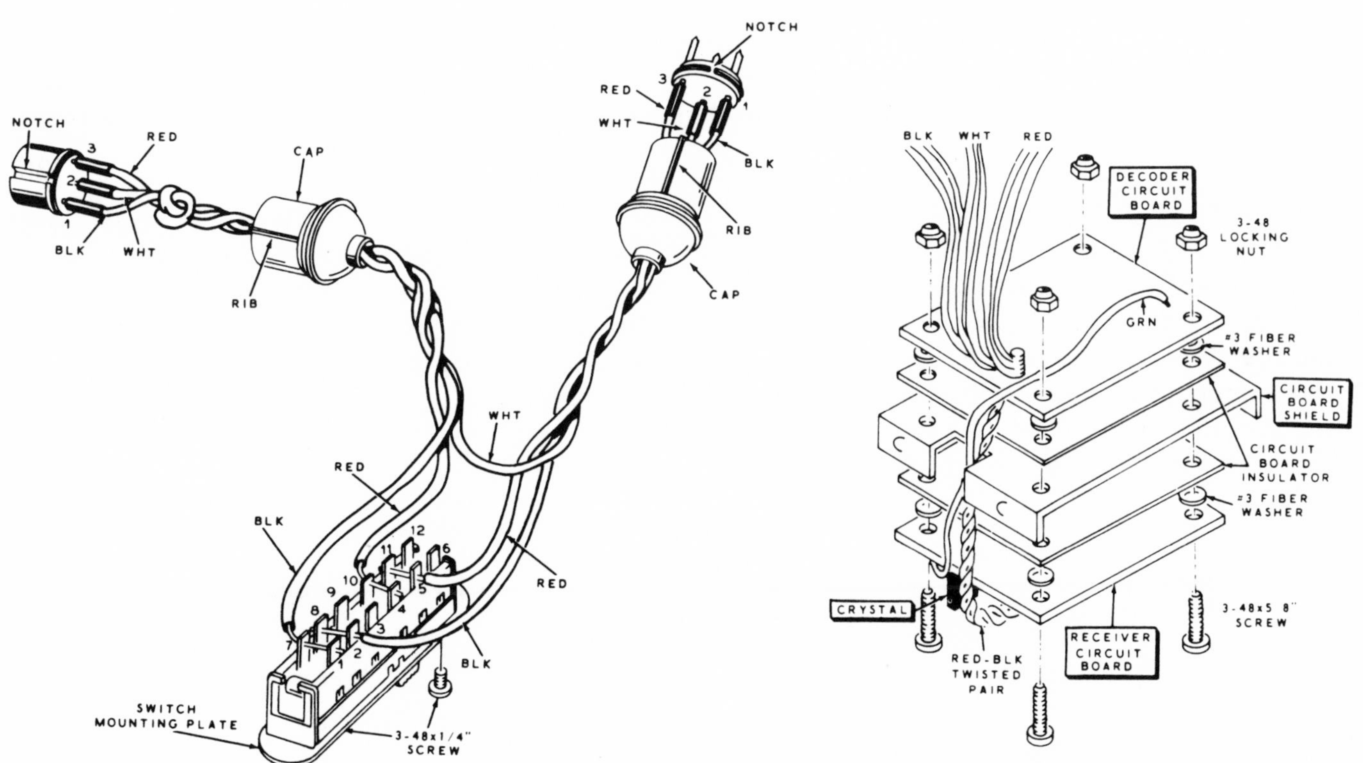

Fig. 11-2. Pictorials are used to give instructions in many hobby areas. These are from a radio construction manual. (Heath Co.)

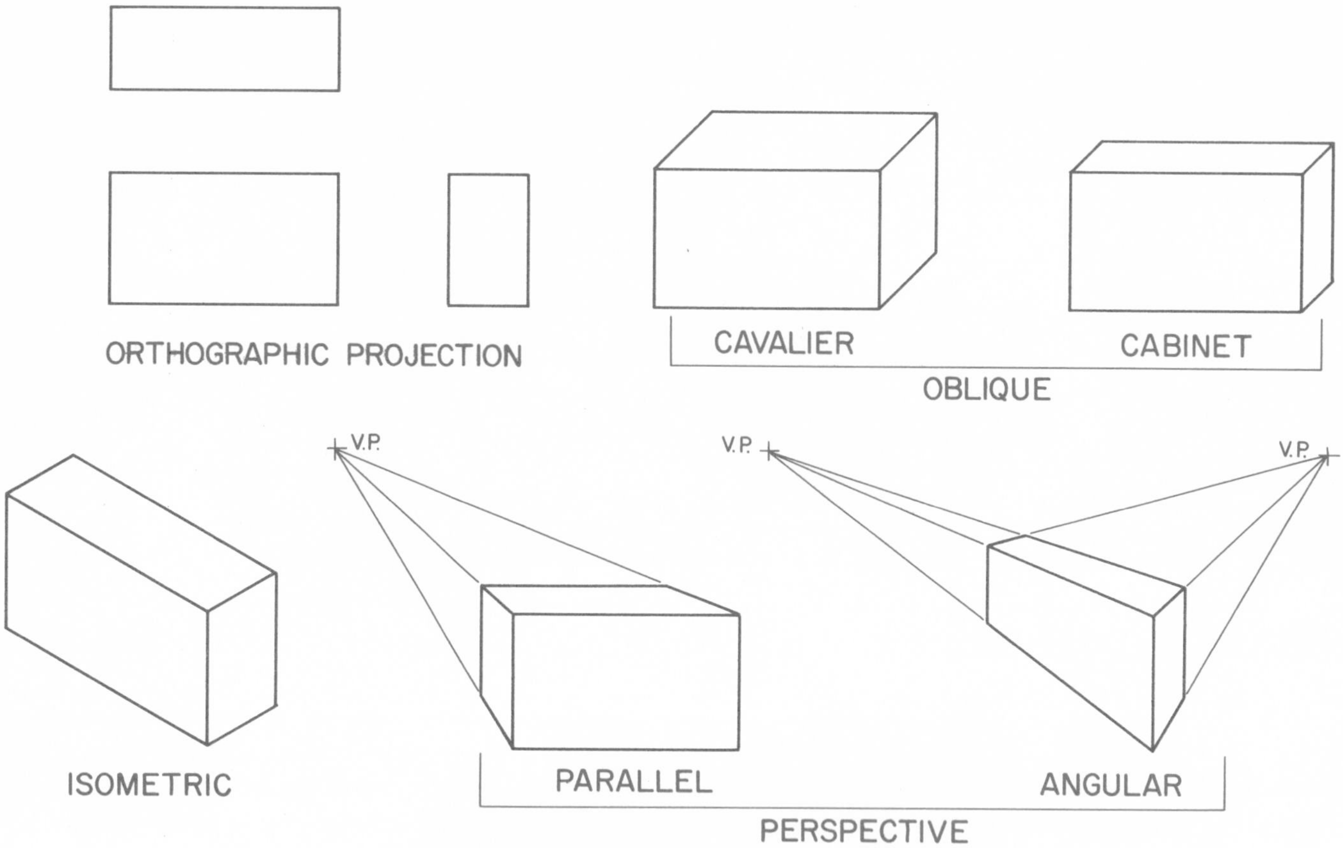

ORTHOGRAPHIC PROJECTION

CAVALIER CABINET

OBLIQUE

V.P. V.P. V.P.

ISOMETRIC

PARALLEL ANGULAR

PERSPECTIVE

Fig. 11-3. Types of pictorial drawings.

Several types of pictorial drawings with which you should become familiar are shown in Fig. 11-3.

ISOMETRIC DRAWINGS

An ISOMETRIC DRAWING shows an object as it is. All lines which show the width and depth are drawn full length (or in the same proportionate length). Edges which are upright are shown by vertical lines. The object is assumed to be in position with its corners toward you, and its horizontal edges sloping away at angles of 30 deg. to the right and to the left.

An ISOMETRIC DRAWING is made entirely with instruments, using three base lines, Fig. 11-4. One of the lines is vertical, the other two are drawn at an angle of 30 deg. to the horizontal. The base lines may be reversed if more information can be shown with the object in the reversed position, Fig. 11-5.

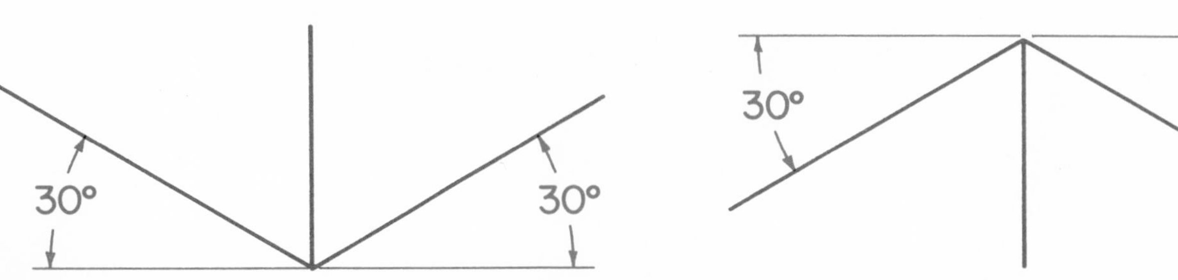

Fig. 11-4. Lines required to make isometric drawings.

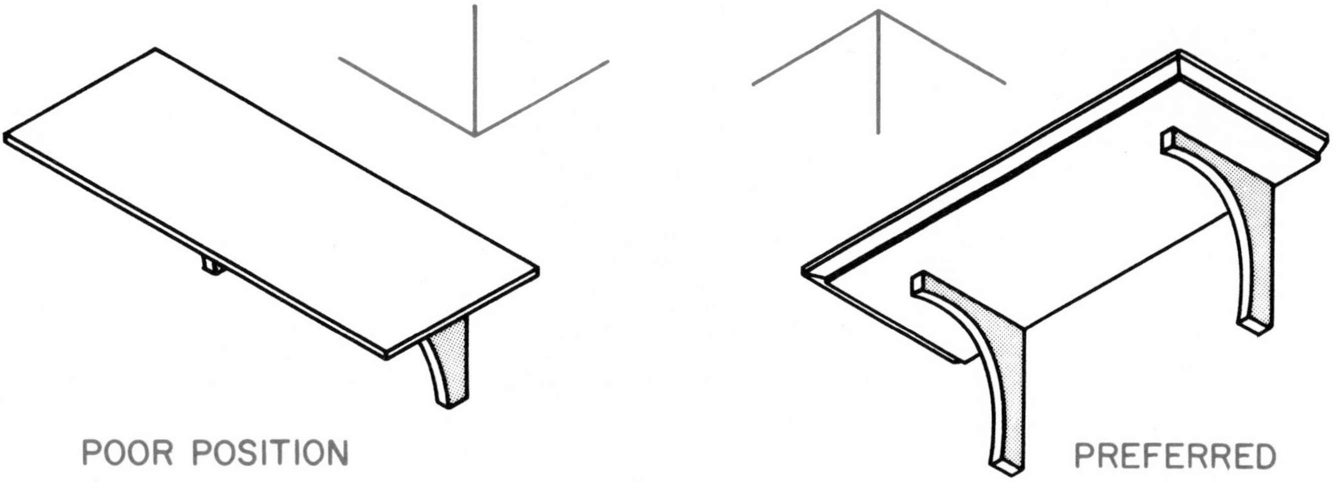

Fig. 11-5. Reverse the base lines if more information can be shown.

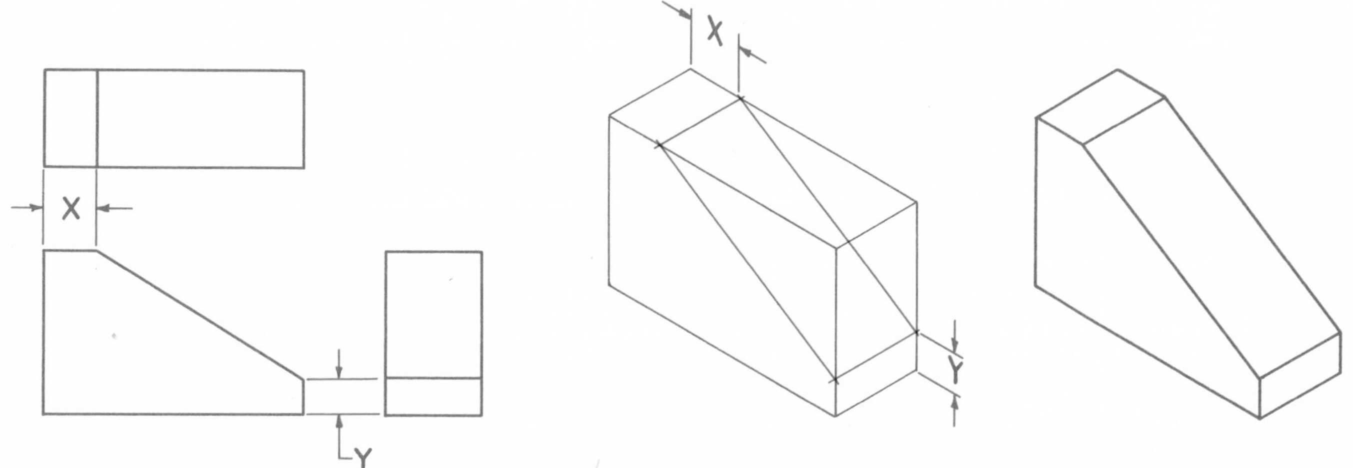

Fig. 11-6. Drawing non-isometric lines.

Lines not parallel to the three base lines are called NON-ISOMETRIC LINES. These may be drawn by transferring reference points from a multiview drawing to the isometric drawing, Fig. 11-6.

Isometric circles, Fig. 11-7, (circles which are not true ellipses) may be drawn

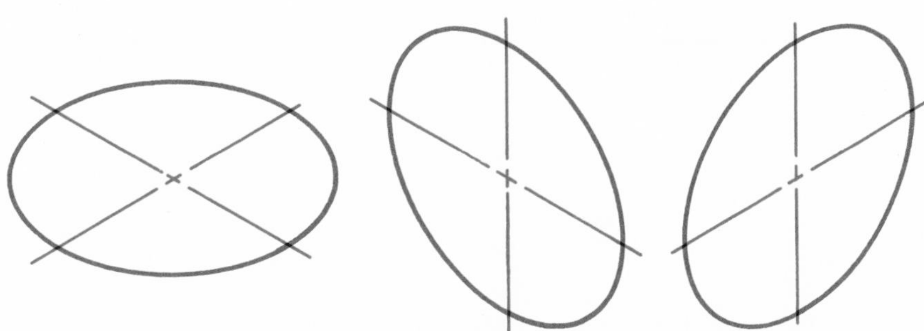

Fig. 11-7. Isometric circles.

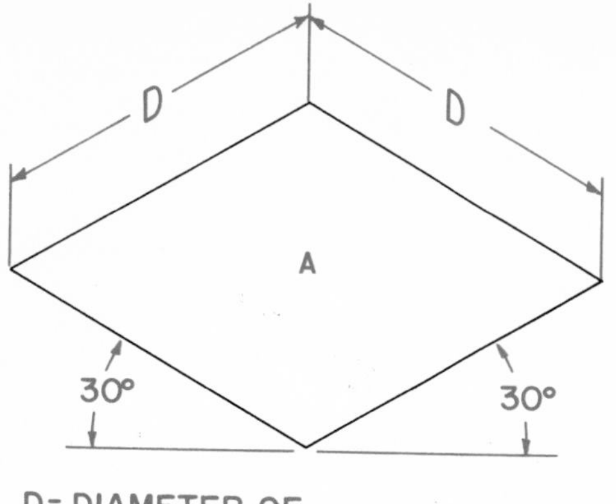

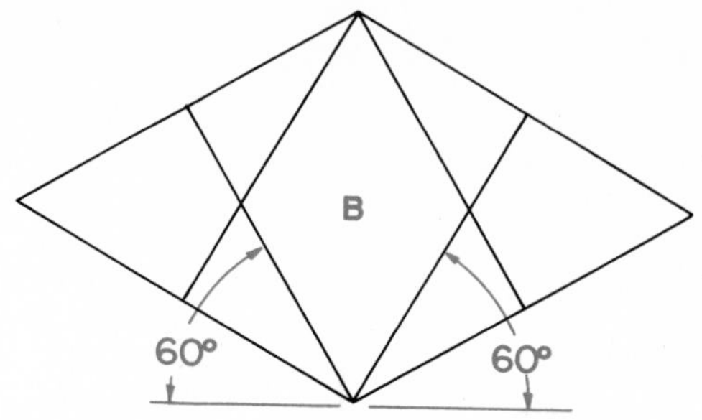

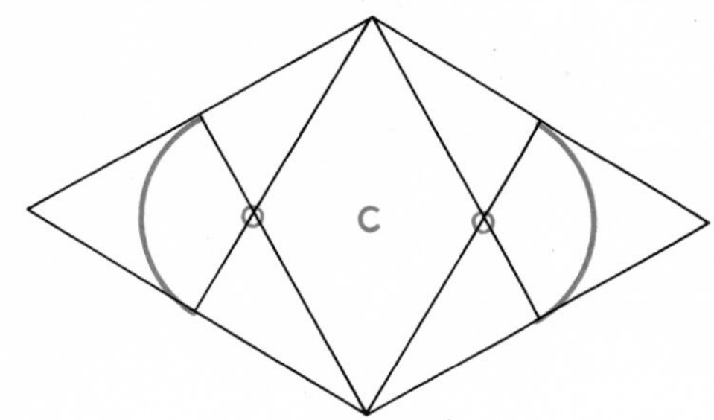

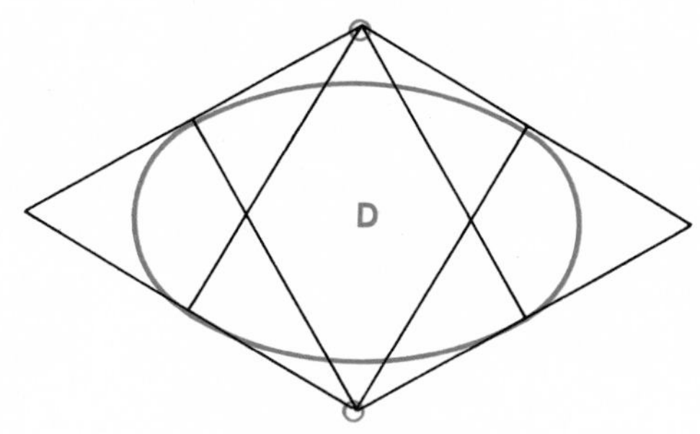

Fig. 11-8. Drawing isometric circles with a compass. Above. On a flat plane. Below. On a vertical plane.

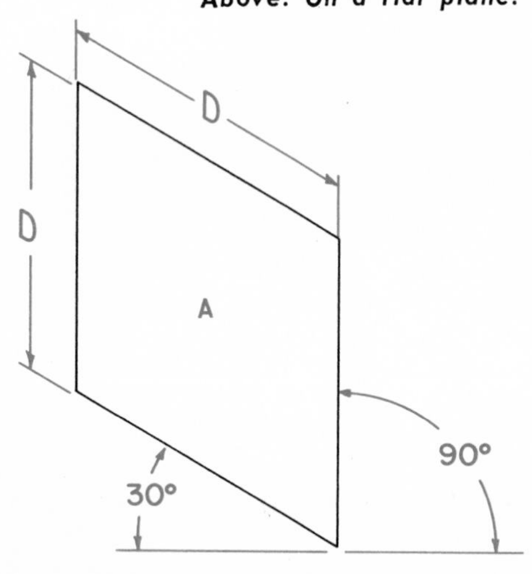

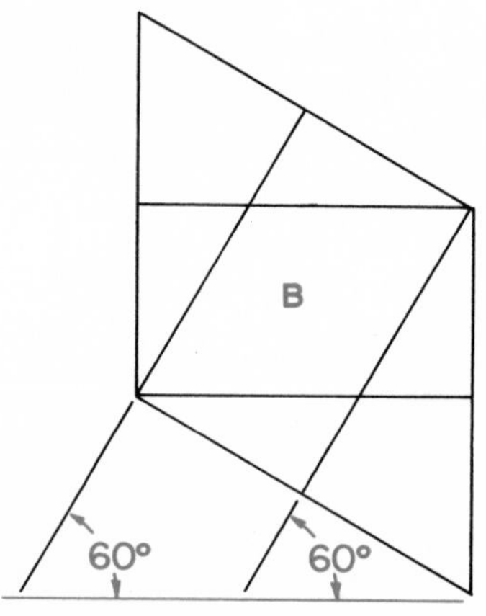

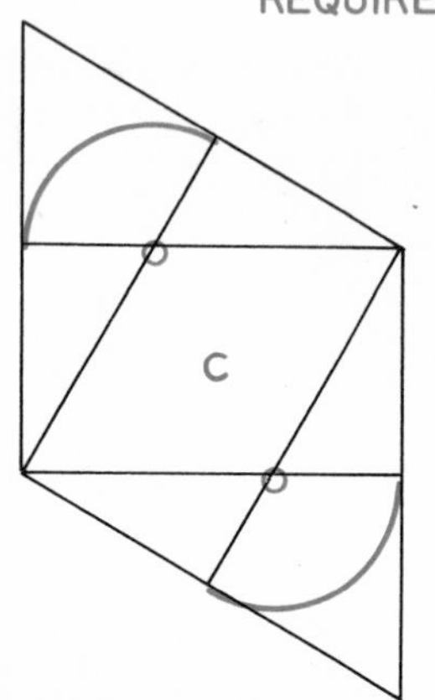

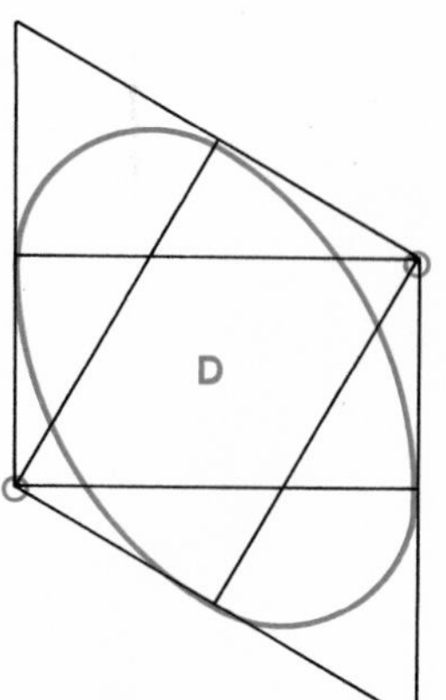

using a compass and a 60 deg. triangle, as shown in the two examples, Fig. 11-8. First, make an isometric square into which the isometric circle is to be drawn, Detail A. Using a 60 deg. triangle, draw lines as in Detail B. This provides four centers from which arcs may be drawn tangent (line which contacts other lines at only one point) to sides of isometric square, Details C and D.

It is well to know how to draw isometric circles using drafting tools, but an easier way to draw such circles is to use an ellipse template of appropriate size, Fig. 11-9.

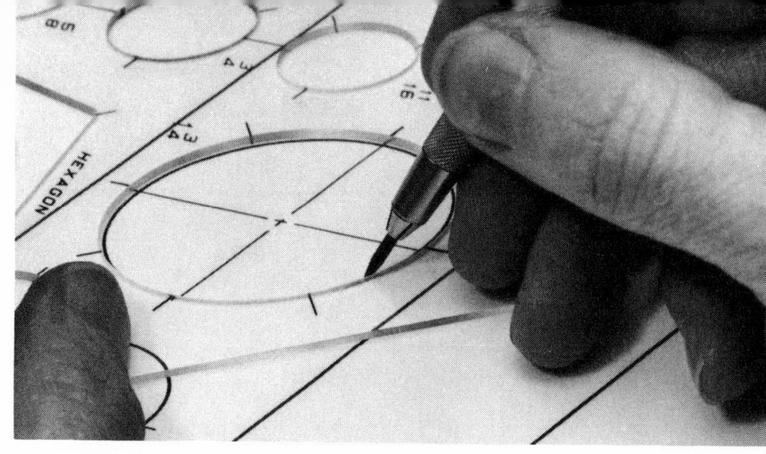

Fig. 11-9. Using an isometric circle template.

OBLIQUE PICTORIAL DRAWINGS

An OBLIQUE DRAWING is a pictorial drawing similar to an isometric drawing, except one surface, the longest dimension or front, is parallel to the picture plane (is shown in true shape and size). Three types of oblique drawings are shown in Fig. 11-10. Each type shows the front of the object as it appears in multiview form. The angle of the depth axis (construction lines drawn at an angle) may be any angle, but 15, 30, or 45 deg. are generally used.

CAVALIER OBLIQUE. An oblique drawing in which the depth axis lines are full scale (full size).

CABINET OBLIQUE. Depth axis lines are drawn one-half scale.

GENERAL OBLIQUE. Depth axis lines vary from one-half to full scale.

DIMENSIONING PICTORIAL DRAWINGS

To dimension isometric and oblique drawings, draw dimension lines parallel to the corresponding planes of the object (lines of the drawing). See Fig. 11-11. Acceptable

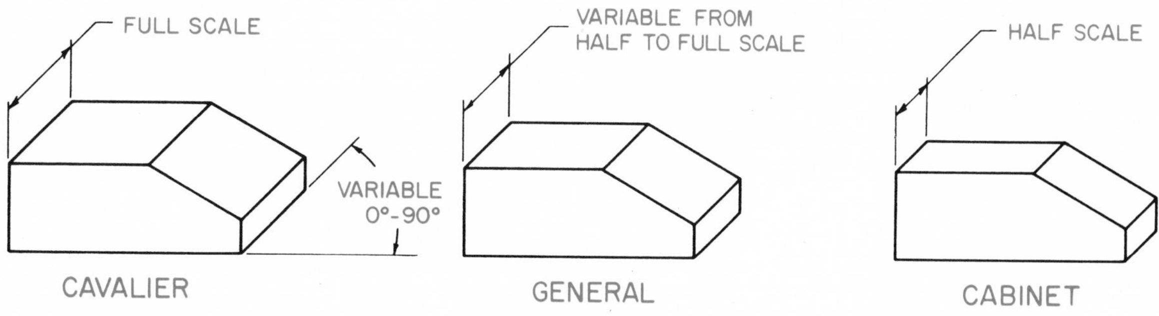

Fig. 11-10. Above. Types of oblique drawings. The CAVALIER and CABINET drawings are most frequently drawn. Fig. 11-11. Below. Dimensioning pictorials.

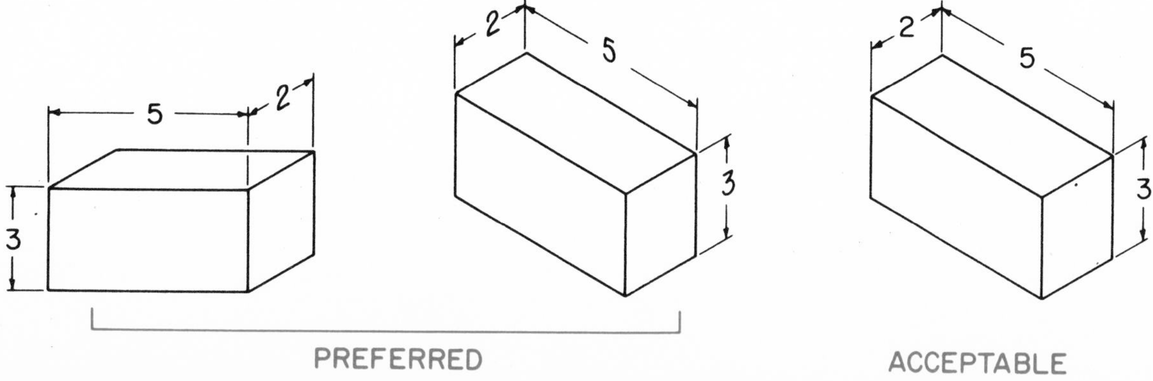

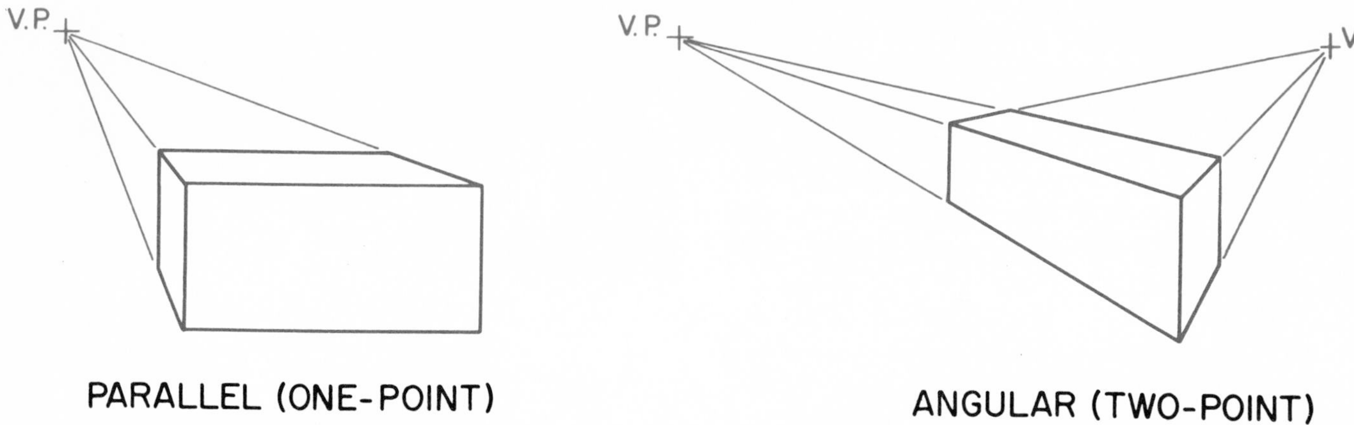

Fig. 11-12. Perspective drawings.

alternate dimensioning is also shown in Fig. 11-11.

PERSPECTIVE DRAWINGS

A PERSPECTIVE DRAWING is a drawing used by an architect, artist or draftsman, to show an object as it would appear to the eye from a certain location.

To understand how a perspective is formed, let's assume for a moment, that you are standing between two railroad tracks, in the exact center of the space, and that the tracks run straight toward the horizon as far as you can see. As you look down the tracks, the tracks appear to converge (run together) at an eye level point on the horizon. The point where the tracks meet is called the VANISHING POINT (V.P.). Lines above eye level seem to recede down to the horizon, while those below eye level appear to rise upward to the horizon. In making a perspective drawing, lines used to show depth or shape of the object if continued will converge to form a V. P. See Fig. 11-12.

To make a simplified ONE-POINT or PARALLEL PERSPECTIVE DRAWING, Fig. 11-13, the front view is drawn in its true shape in full or scale size. Draw the horizon line and select the VANISHING POINT (V.P.) at random. If at first you do not obtain a pleasing effect, try another vanishing point location. Information on locating vanishing points more accurately may be obtained from a text covering advanced drafting procedures.

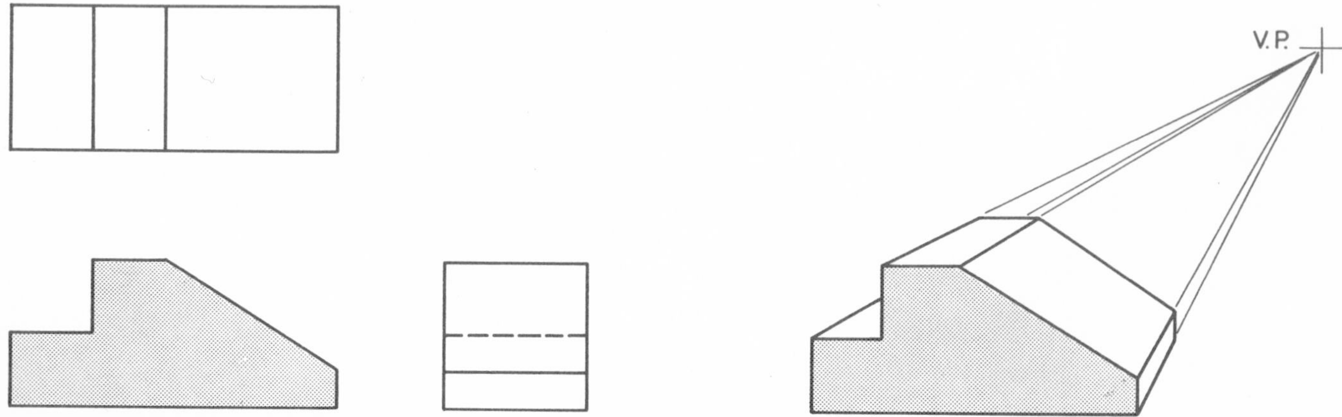

Fig. 11-13. On PARALLEL or ONE-POINT PERSPECTIVE DRAWINGS, the front view is shown in its true shape and in full or scale size.

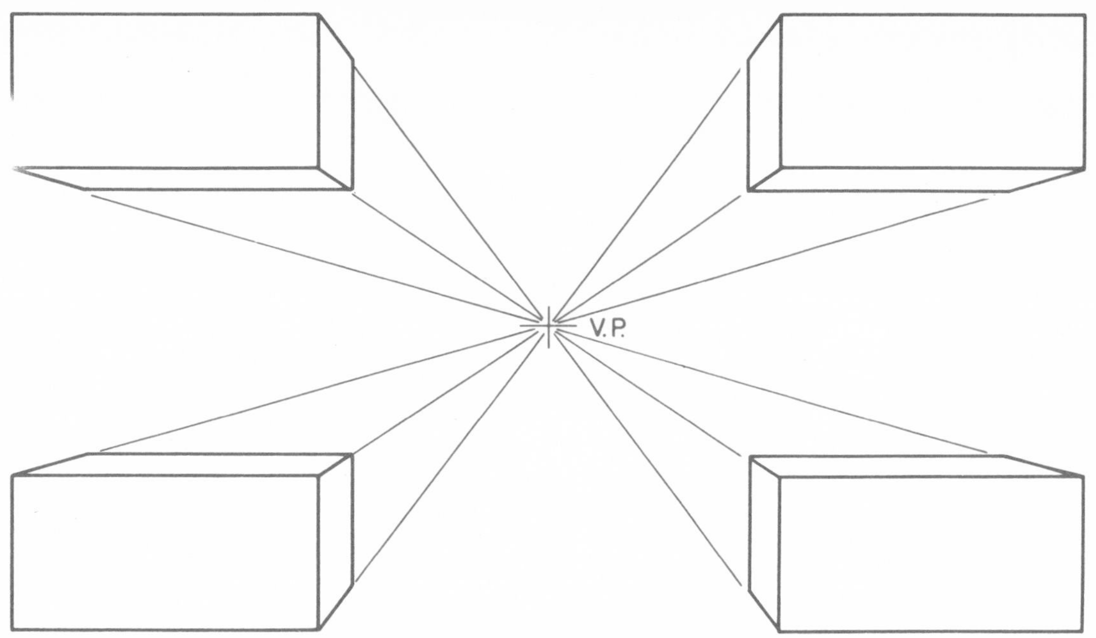

Fig. 11-14. The perspective may be located in any position relative to the vanishing point (V.P.)

The perspective may be located in any position relative to the V. P., Fig. 11-14. Project construction lines from each point on the front view to the vanishing point, Fig. 11-15.

Simple ANGULAR or TWO-POINT PERSPECTIVES may be drawn as shown Fig. 11-16.

In drawing perspectives, considerable

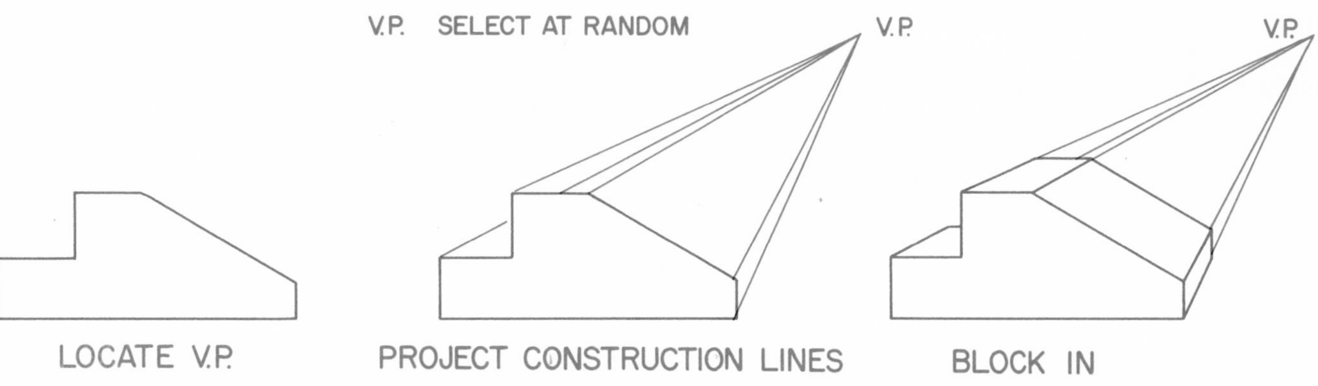

Fig. 11-15. Drawing a one-point perspective.

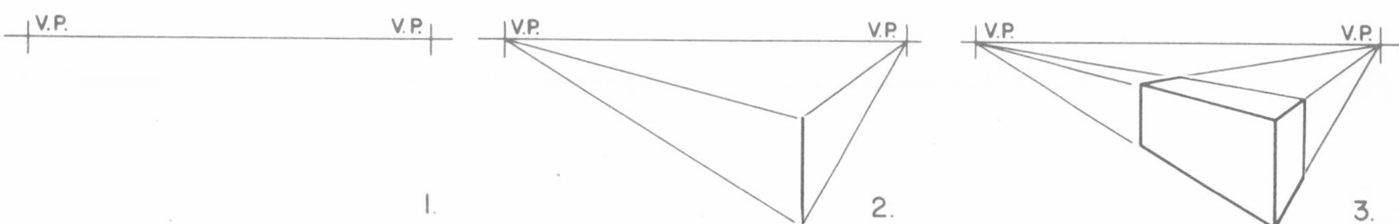

*Fig. 11-16. Drawing a simple **ANGULAR** or **TWO-POINT PERSPECTIVE**: 1–Locate the position of the horizon and two vanishing points. 2–Locate and draw a vertical line the full or scale height of the object being drawn. 3–Draw construction lines from the vertical line to the vanishing points. For simple two-point perspectives like this, it is permissible to locate the length and depth in a position that gives the most pleasing effect. Darken the necessary lines.*

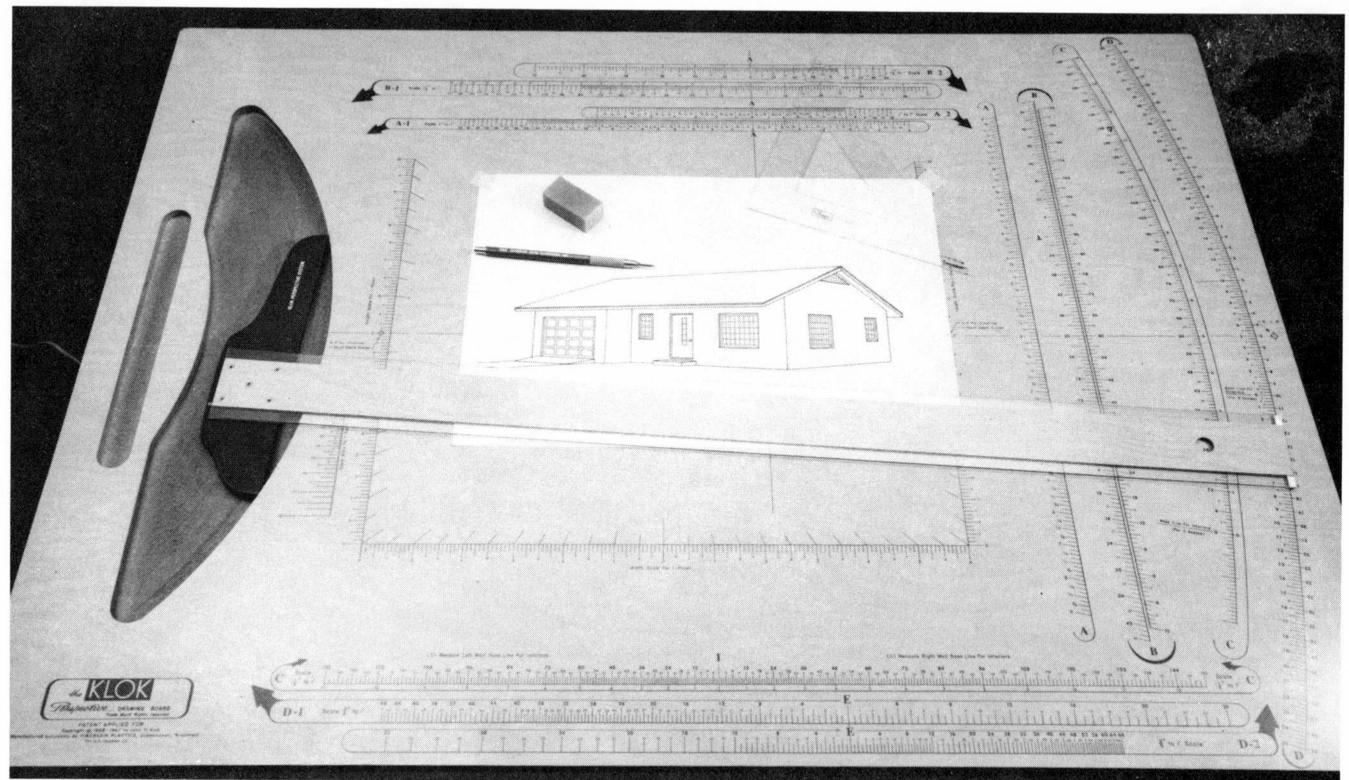

Fig. 11-17. Klok perspective drawing board.

saving of time and additional accuracy will result, if you use a perspective drawing board, as shown in Fig. 11-17. This type board is available from concerns that sell school and drafting supplies.

EXPLODED ASSEMBLY DRAWINGS

EXPLODED ASSEMBLY DRAWINGS are easy to read and may be understood without having an extensive knowledge of blueprint reading. See Fig. 11-18.

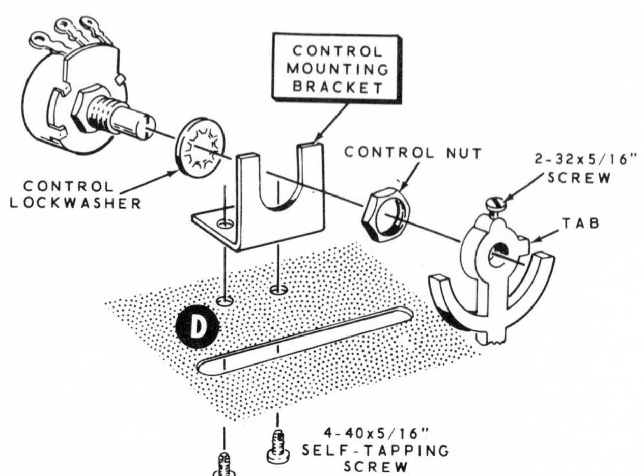

Fig. 11-18. An exploded pictorial drawing. (Heath Co.)

Such drawings are used extensively by industry and in many hobby areas. You are probably quite familiar with this type of drawing if you build model cars, planes, boats or rockets.

Exploded assembly drawings are nothing more than a series of pictorial drawings (usually isometrics) showing the parts that make up the object. They are located in proper relation to each other.

HOW TO CENTER PICTORIAL DRAWINGS

There are several ways to center pictorial drawings on the drawing sheet.

One method, which is widely used, is called the "eye balling it" method. You estimate by sight the approximate location of the starting point and develop your drawing from this point.

Another method requires the drawing of the pictorial on a second piece of paper. When it has been checked for accuracy, measure its total width and height. With this information, locate the starting point of the drawing so it will be centered on the drawing sheet which is to be evaluated.

A third method, one that will work on most problems, is shown in Fig. 11-19.

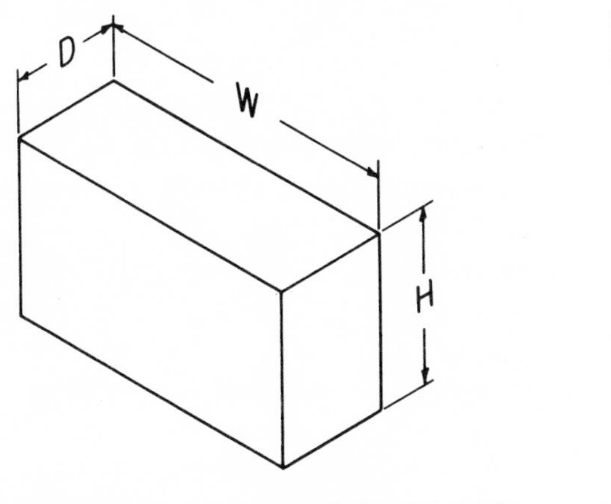

1. DESCRIBING AN ISOMETRIC VIEW.

2. LOCATE CENTER OF DRAWING AREA.

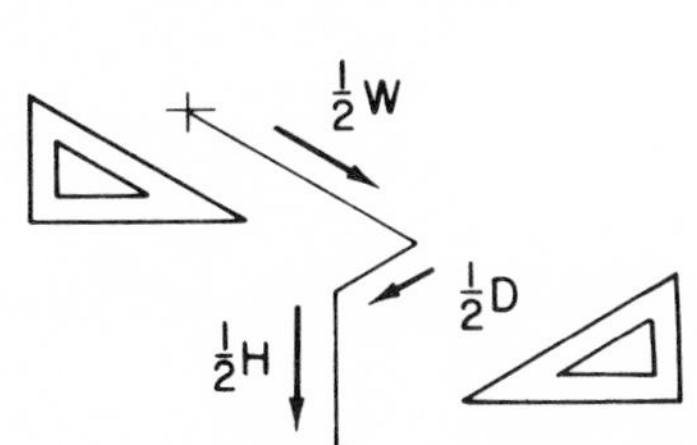

3. PLOT STARTING POINT OF DRAWING.

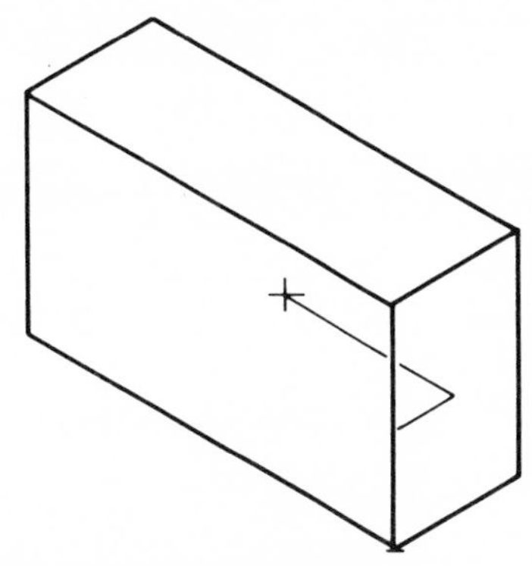

4. COMPLETE DRAWING.

Fig. 11-19. Above. An easy way to center isometric drawings. Below. An easy way to center oblique drawings.

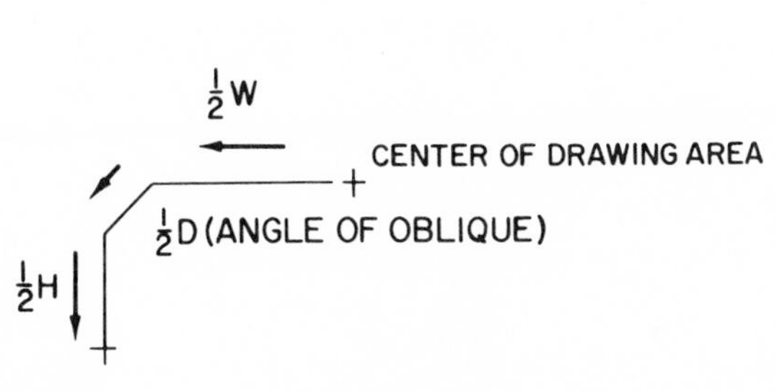

1. PLOT STARTING POINT OF DRAWING.

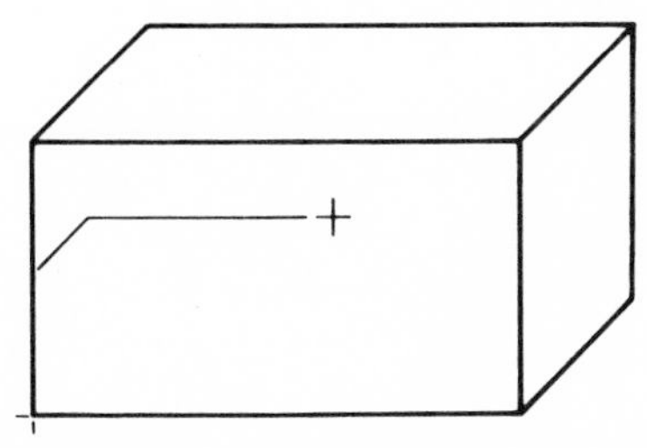

2. COMPLETE DRAWING.

TEST YOUR KNOWLEDGE - UNIT 11

1. Pictorial drawing is a method of showing an object as:
   a. It would look on the drawing.
   b. It would appear to the eye.
   c. Another form of multiview drawing.
   d. All of the above.
   e. None of the above.
2. Why are pictorials often used?
3. List 4 types of pictorial drawings:
   a. ____________________
   b. ____________________
   c. ____________________
   d. ____________________
4. Isometric pictorial drawings are drawn about three base lines. Make a sketch of these base lines.
5. The lines that represent angles in an isometric drawing are known as ________ lines.
6. Prepare a sketch that shows the difference between a cavalier and a cabinet pictorial drawing.
7. Oblique drawings may be drawn at any angle, but ________________ angles are usually used.
8. What is an "exploded assembly drawing?"

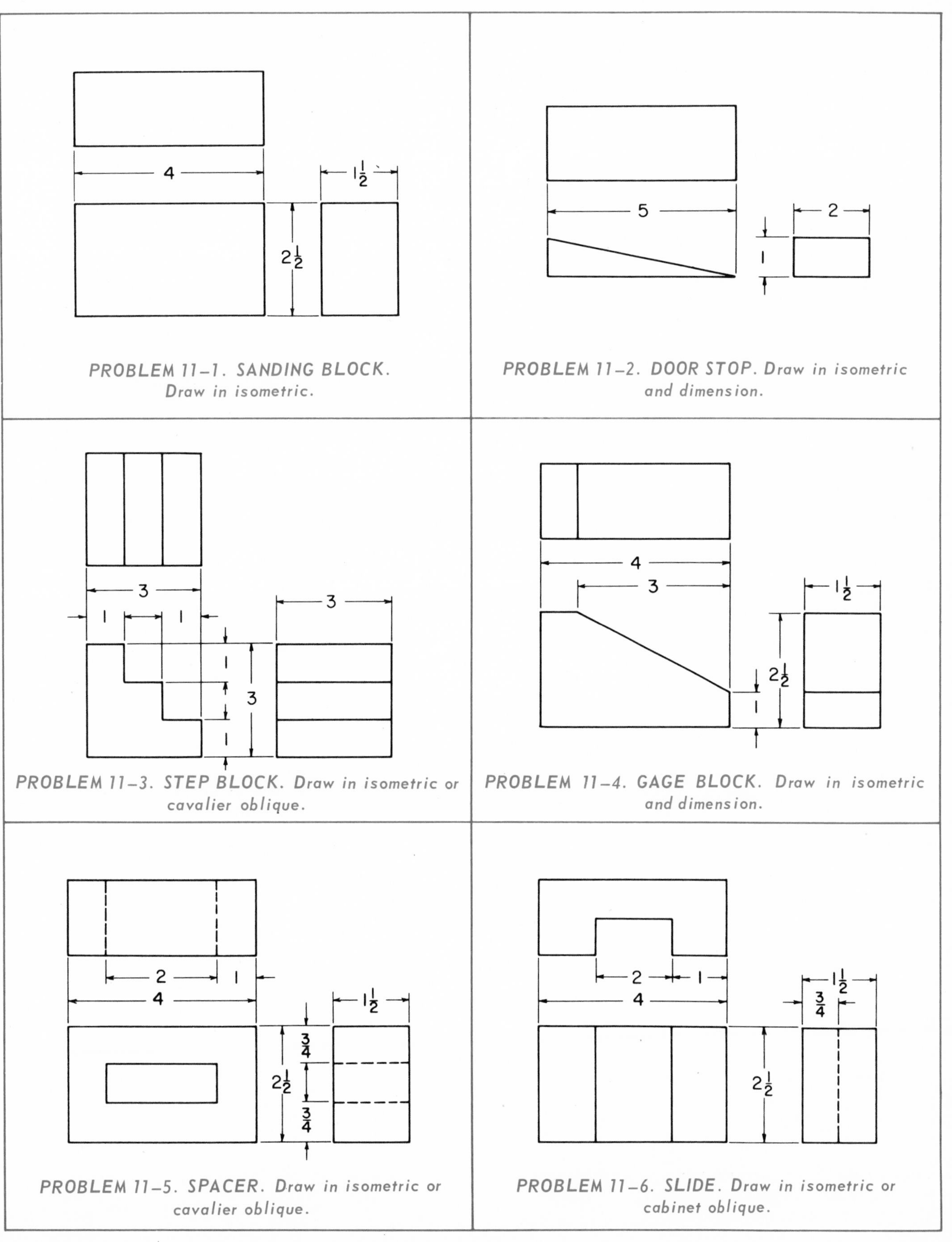

PROBLEM 11–1. SANDING BLOCK. Draw in isometric.

PROBLEM 11–2. DOOR STOP. Draw in isometric and dimension.

PROBLEM 11–3. STEP BLOCK. Draw in isometric or cavalier oblique.

PROBLEM 11–4. GAGE BLOCK. Draw in isometric and dimension.

PROBLEM 11–5. SPACER. Draw in isometric or cavalier oblique.

PROBLEM 11–6. SLIDE. Draw in isometric or cabinet oblique.

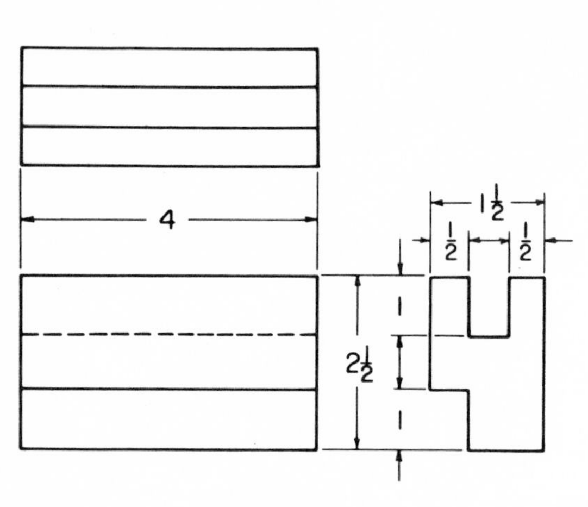

PROBLEM 11–7. ***GUIDE.*** *Draw in isometric.*

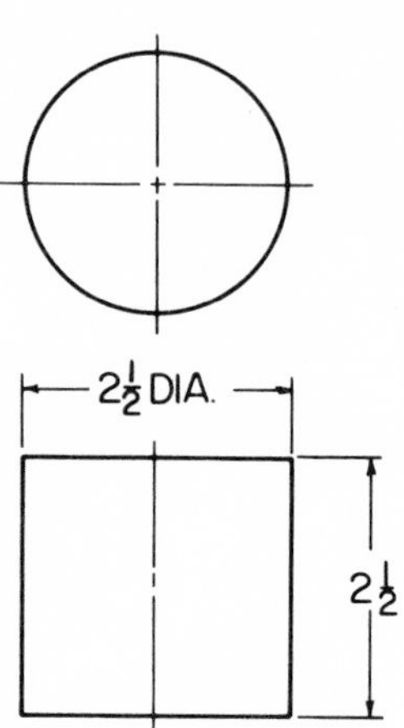

PROBLEM 11–8. ***CANNISTER.*** *Draw in isometric.*

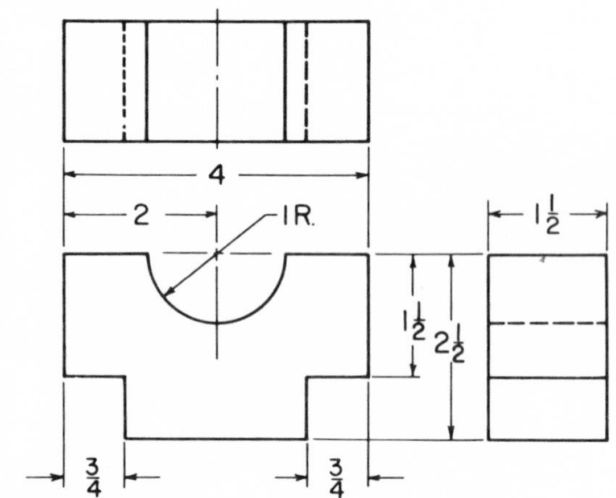

PROBLEM 11–9. ***SUPPORT.*** *Draw in cabinet or cavalier oblique and dimension.*

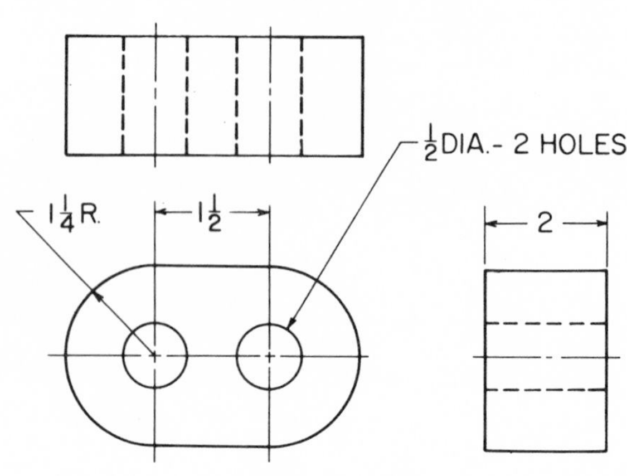

PROBLEM 11–10. ***LINK.*** *Draw in cabinet oblique.*

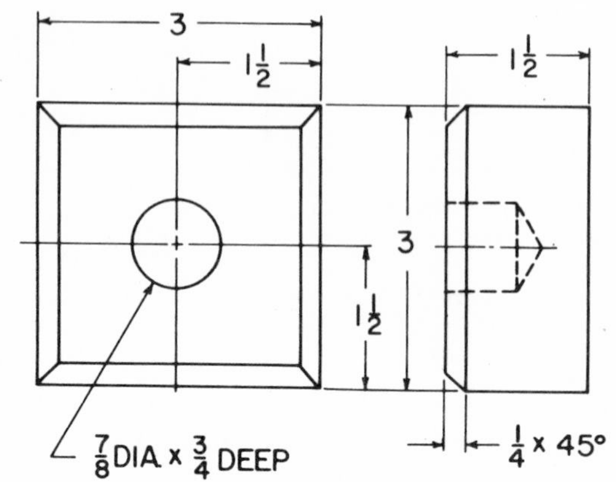

PROBLEM 11–11. ***MODERN CANDLESTICK HOLDER.*** *Draw in isometric.*

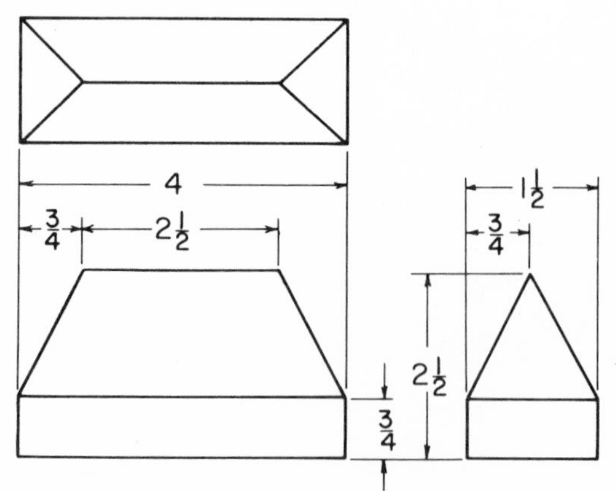

PROBLEM 11–12. ***BALANCE WEDGE.*** *Draw in isometric and dimension.*

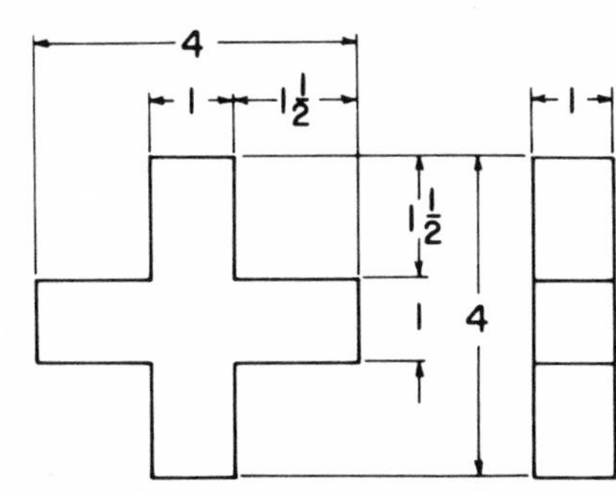

PROBLEM 11–13. **CROSS.** *Draw in one-point perspective.*

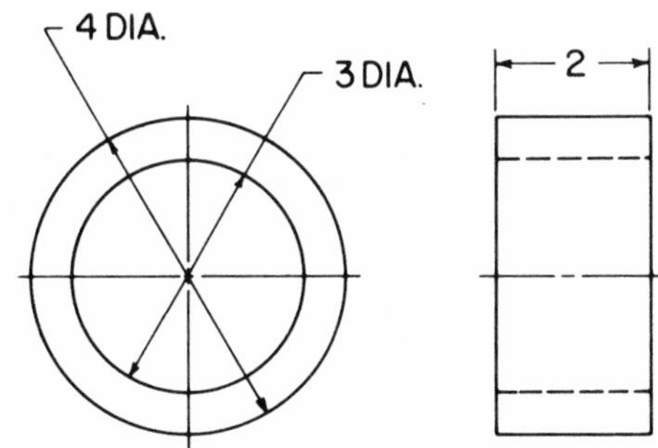

PROBLEM 11–14. **BEARING.** *Draw in isometric or one-point perspective.*

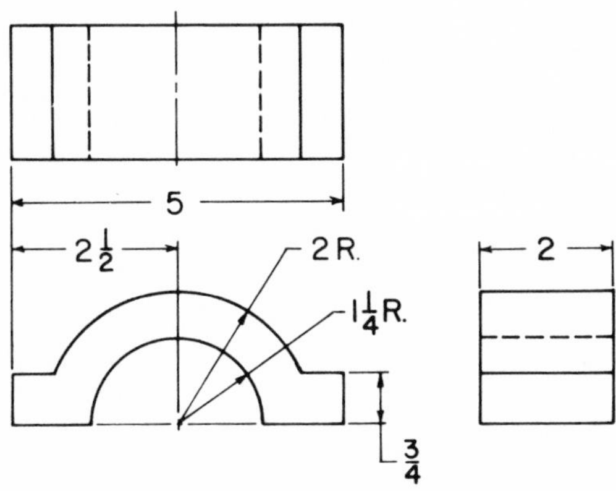

PROBLEM 11–15. **BEARING CAP.** *Draw in isometric and dimension.*

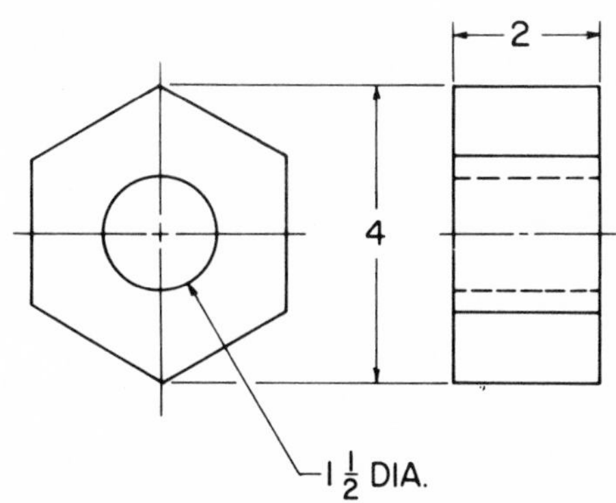

PROBLEM 11–16. **HEXAGONAL BASE.** *Draw in isometric or cabinet oblique.*

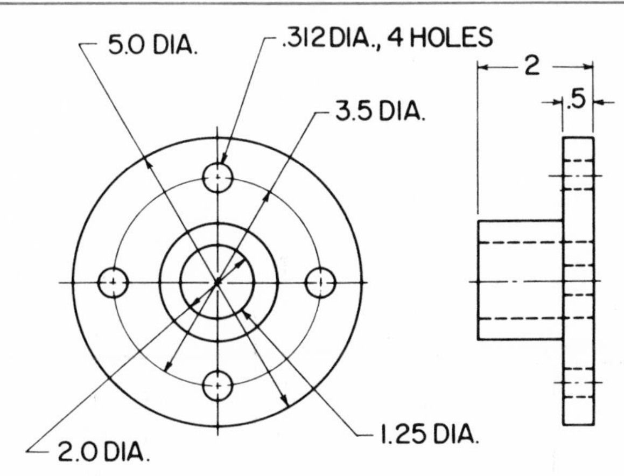

PROBLEM 11–17. **FACE PLATE.** *Draw in isometric.*

PROBLEMS:

DESIGN AND DRAW PICTORIALLY

1. BOOK RACK
2. SHOE SHINE BOX
3. MODERN BIRD HOUSE
4. COFFEE TABLE
5. END TABLE

DESIGN AND DRAW AS EXPLODED DRAWING

1. PICTURE FRAME
2. BOOK CASE
3. JEWELRY BOX
4. WALL STORAGE CABINET
5. DESK

PROBLEM 11–18. **SPECIAL DESIGN PROBLEMS.**

Unit 12
PATTERN DEVELOPMENT

A pattern is a full-size drawing of the various surfaces of an object stretched out on a flat plane, Fig. 12-1. A pattern is frequently called a STRETCHOUT.

Patterns or stretchouts are produced utilizing a form of drafting called PATTERN DEVELOPMENT. The method is also known as SURFACE DEVELOPMENT and SHEET METAL DRAFTING.

Pattern development is important to many occupations and hobbies.

Patterns were required to make the clothing and shoes you wear. Pattern development plays an important part in the fabrication of sheet metal ducts and pipes needed in the installation of heating and air conditioning units. Stoves and refrigerators are fabricated from many sheet metal parts. Accurate patterns had to be developed for the parts before the appliance could be put into production.

Draftsmen in the aerospace industry must be familiar with pattern development techniques. Manufacturing sheet metal parts for airplanes and rockets, Fig. 12-2, involve the use of many patterns.

If you have ever built a flying model airplane, Fig. 12-3, you know that many patterns are needed to cut the parts to shape for assembly.

Even wallets and handbags were cut to shape using patterns as guides.

As you can see, patterns play an important part in the manufacture of many items. How many can you name?

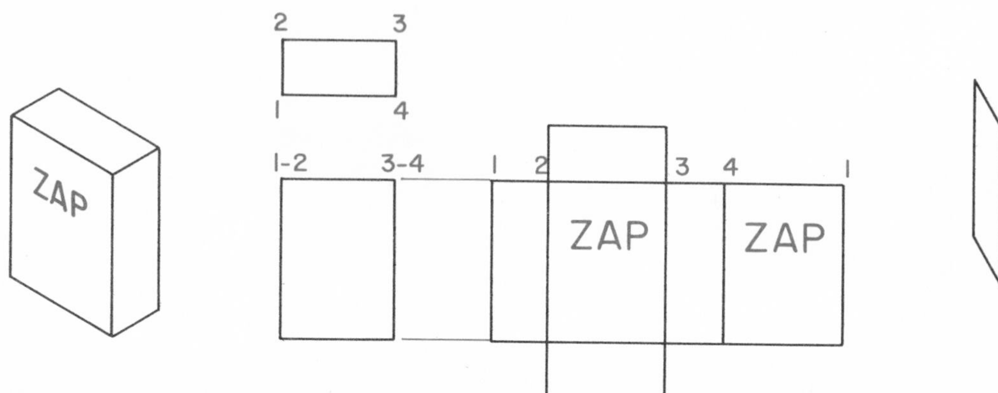

Fig. 12-1. The pattern or stretchout of a cereal type box. Check this layout against a cereal box you have cut open.

Fig. 12-2. Draftsman in the aviation industry must be familiar with pattern development. (Gates–Learjet)

HOW TO DRAW PATTERNS

Regular drafting techniques as described elsewhere in this text, are used to draw patterns. Basic pattern development falls into two categories:

PARALLEL LINE DEVELOPMENT. The technique used to make patterns for prisms and cylinders.

RADIAL LINE DEVELOPMENT. Patterns for regular tapering forms (cones, pyramids, etc.) are developed utilizing this method.

Combinations and variations of the basic developments are used to draw patterns for more complex geometric shapes.

While regular drafting techniques are used in pattern development, some lines have additional meanings. Sharp folds or bends are indicated on the stretchout by a visible object line (see PATTERN DE-

Fig. 12-3. Patterns for wing ribs, fuselage formers and tail sections must be developed before a model airplane, like this radio controlled model, can be constructed.

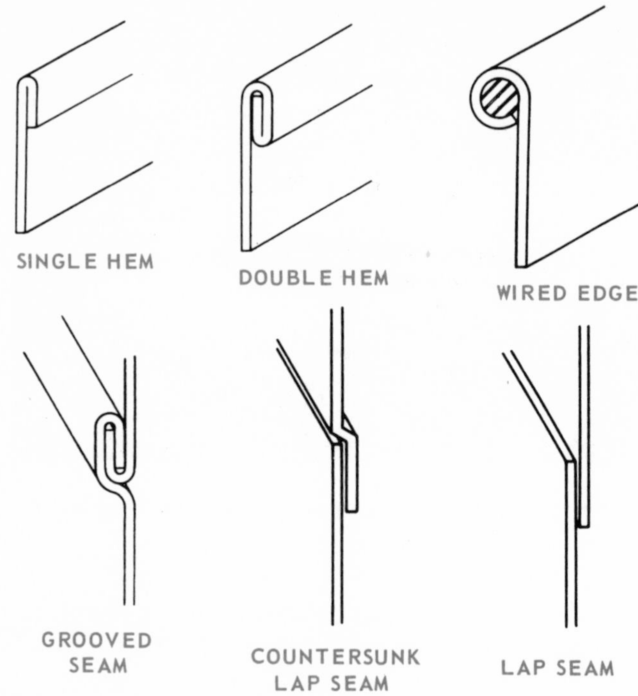

Fig. 12-4. Typical hems, edges and seams used to join and give rigidity to sheet metal. Extra material is required.

VELOPMENT OF A REGULAR PRISM, page 155, for an example of this type).

Curved surfaces are shown on the pattern by construction lines or center lines (an example of this type is shown in PATTERN DEVELOPMENT OF A CYLINDER, page 154).

Stretchouts are seldom dimensioned.

When developing a pattern or stretchout, allow additional metal for HEMS, EDGES and SEAMS, Fig. 12-4.

HEMS are used to strengthen the lips of sheet metal objects. They are made in standard fractional sizes, 3/16 in., 1/4 in., 3/8 in., etc.

The WIRED EDGE gives extra strength and rigidity to sheet metal edges.

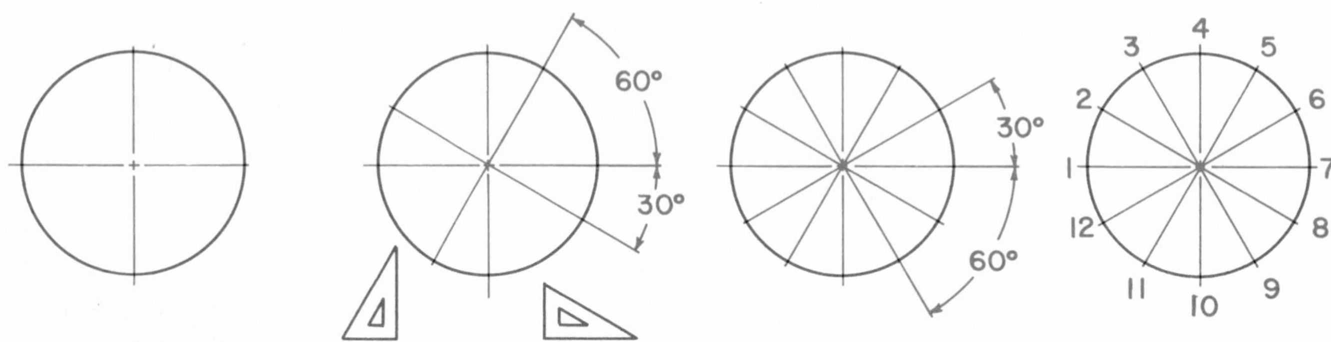

Fig. 12-5. How to divide a circle into twelve (12) equal parts.

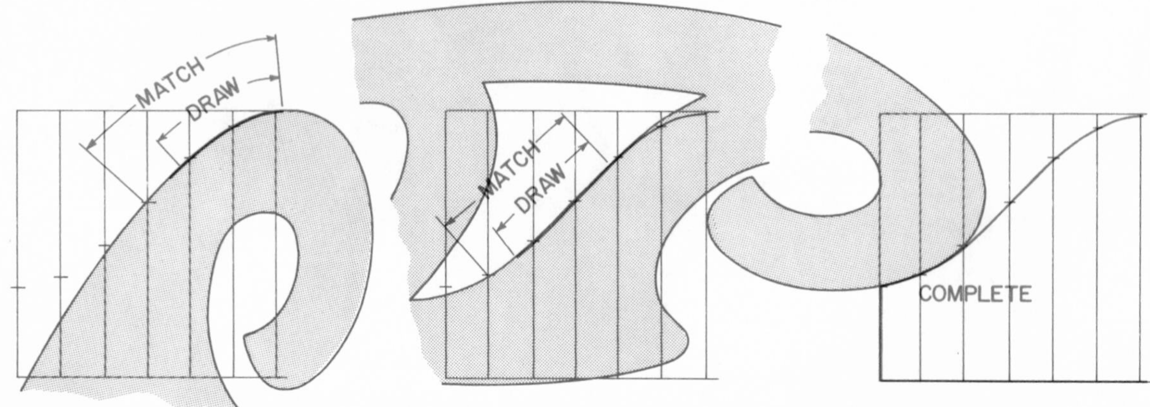

Fig. 12-6. Using a French Curve to draw an irregular curve.

SEAMS make it possible to join sheet metal sections. They are usually finished by soldering and/or riveting.

Reference lines and points are needed when developing the stretchout of a circular object. To provide these, the circle is divided into twelve (12) equal parts, Fig. 12-5.

Irregular curves are drawn using a FRENCH CURVE. The points of the irregular curve are plotted. The points may be connected by a lightly sketched line. Match the French Curve to the sketched line or points taking care to make the curve of the line flow smoothly. Fig. 12-6 shows how this is done.

TEST YOUR KNOWLEDGE-UNIT 12

1. Pattern development is a form of drafting that ______________________________
______________________________.
2. Pattern development is also known as:
   a. ____________________.
   b. ____________________.
3. Patterns are also known as ____________.
4. List five uses of patterns.
   a. ____________________.
   b. ____________________.
   c. ____________________.
   d. ____________________.
   e. ____________________.
5. In pattern development, a heavy solid line (visible object line) indicates that ______
______________________________.
6. Very light lines (construction lines) and center lines indicate that ____________.
7. Make a sketch of a wired edge, a lap seam, and a grooved seam.
8. Irregular curves are drawn with a ______
____________________.

PATTERN DEVELOPMENT OF A CYLINDER

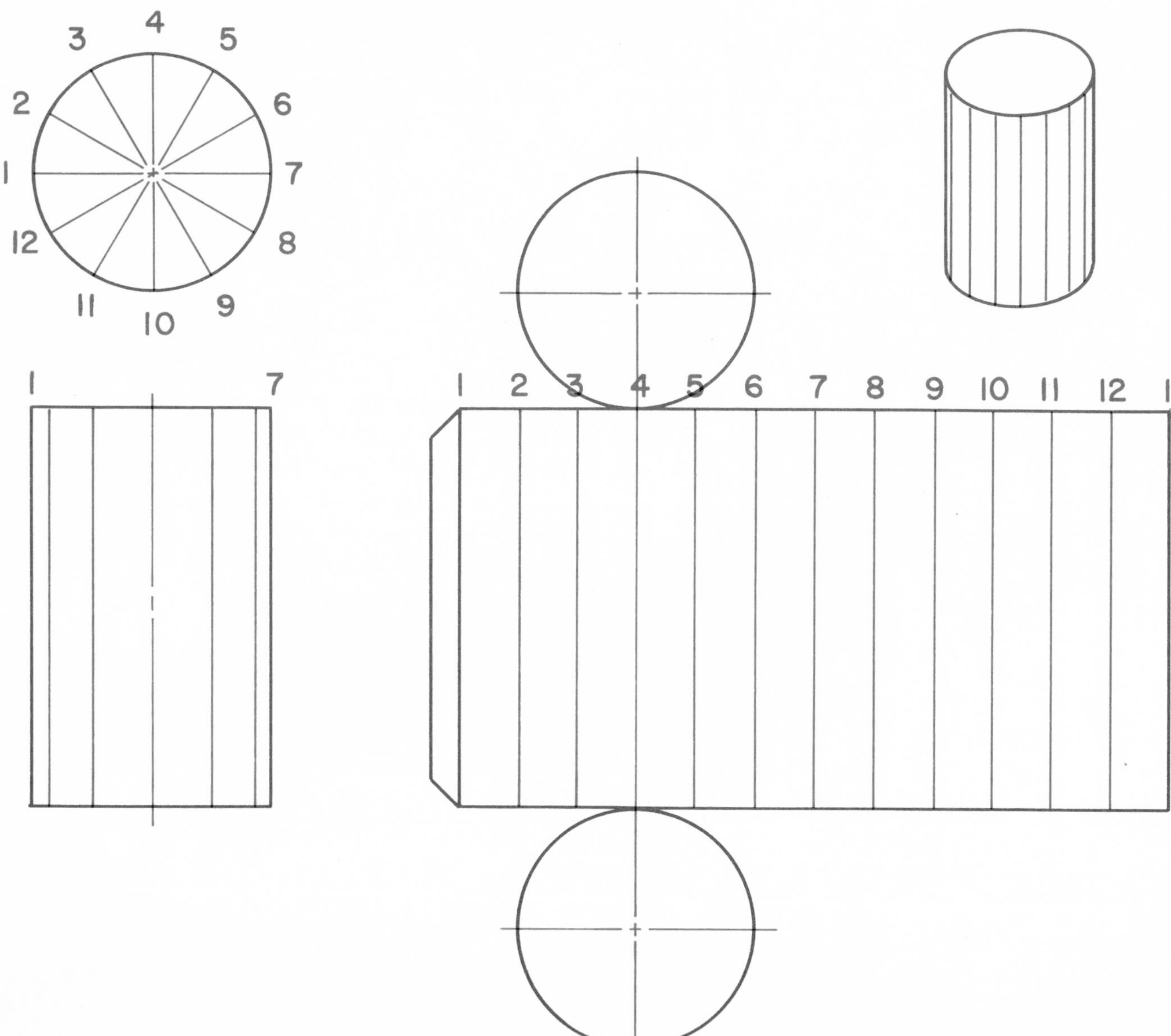

1. Draw front and top views of required cylinder. Divide top view into twelve (12) equal parts and number as shown.
2. The height of the pattern or stretchout is the same as the height of the front view. Project construction lines from the top and bottom of the front view.
3. Allow sufficient space (one-inch is adequate) between the front view and the pattern, and draw a vertical line. This will locate line 1 of the pattern.
4. Set your compass or divider from 1 to 2 (the points where the division lines intersect the circle) on the top view. Transfer this distance to the extended lines of the pattern to locate reference lines 1, 2, 3, - 12, 1.
5. Draw the top and bottom tangent to the extended lines.
6. Allow 1/4 in. for seams, and go over all outlines with visible object lines. The lines that represent the curves or circular lines are drawn in color, or, are left as construction lines.

PATTERN DEVELOPMENT OF A RECTANGULAR PRISM

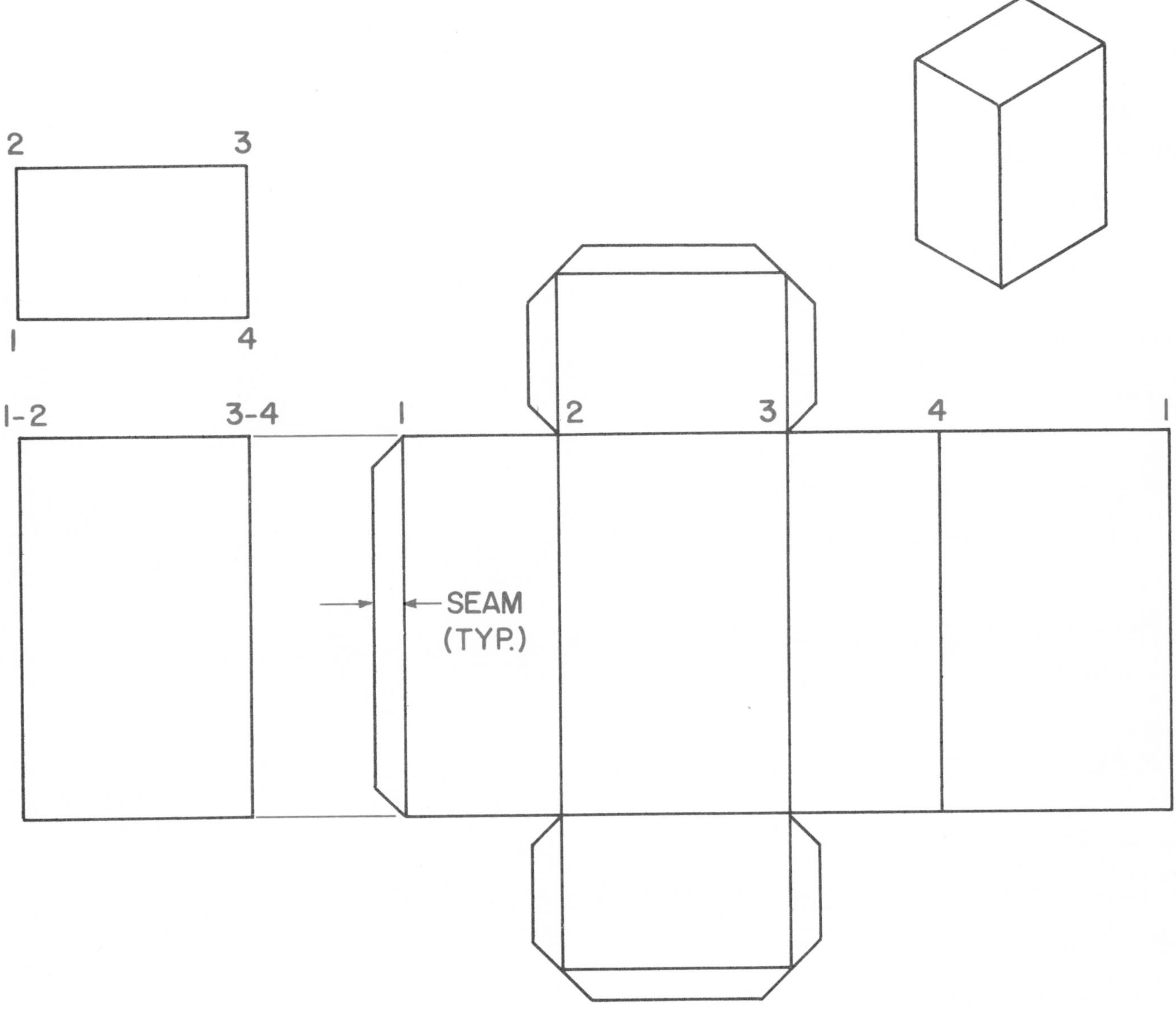

1. Draw the front and top views.
2. The height of the pattern is the same as the height of the front view. Project construction lines from the top and bottom of the front view.
3. Measure over one inch from the front view and draw a vertical line between the extended lines to locate line 1.
4. Set your compass or divider from 1 to 2 on the top view, and transfer this distance to the extended lines. Locate the other distances in the same manner.
5. Construct the top and bottom as shown.
6. Allow 1/4 in. for seams, and go over all outlines and folds with visible object lines. The pattern may be cut out, folded to shape and cemented together using rubber cement.

PATTERN DEVELOPMENT OF A TRUNCATED PRISM

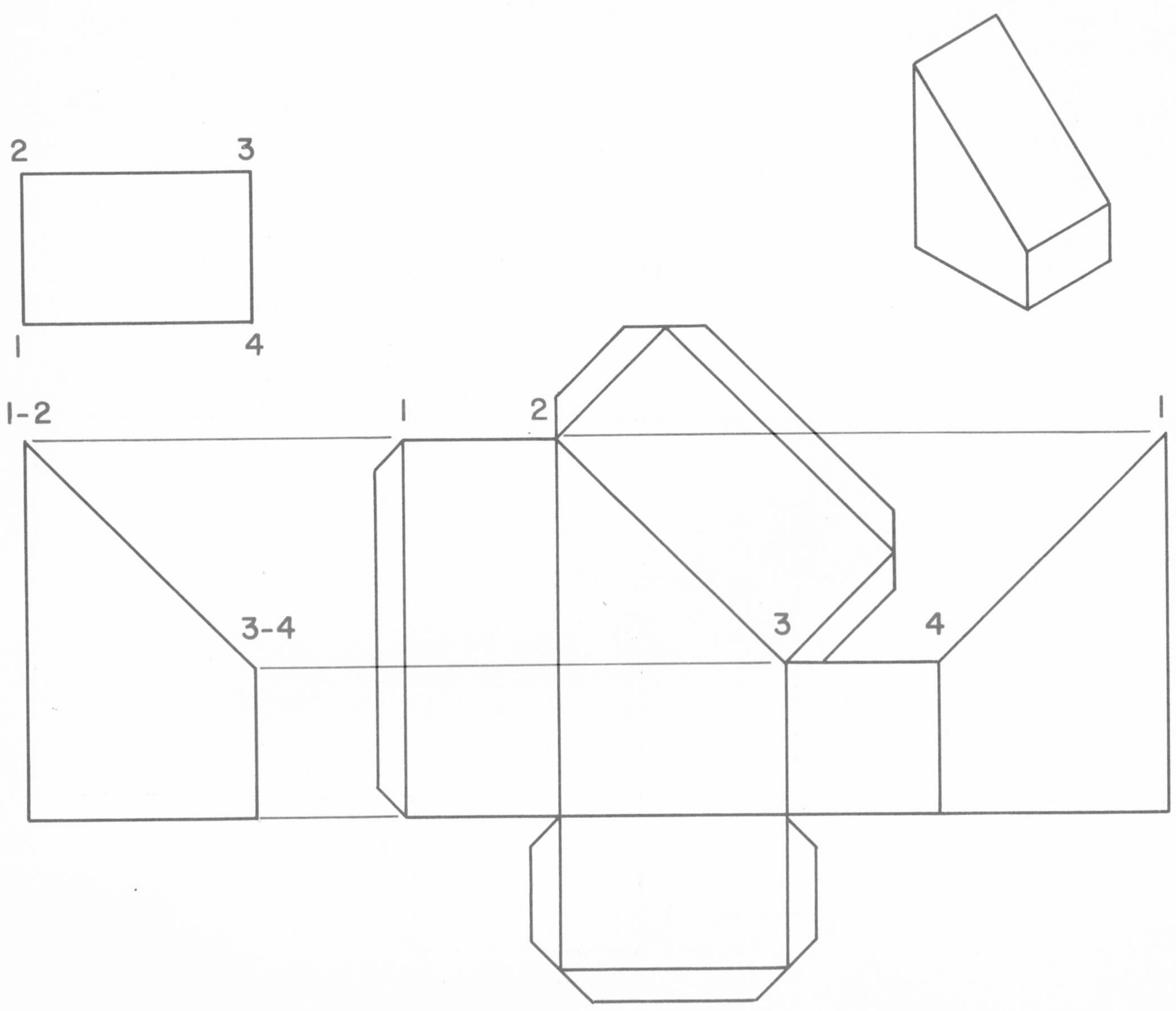

1. Draw front and top views. Number the points as shown.
2. Proceed as in previous examples of pattern development.
3. Mark off and number the folding points. Project point 1 on the front view, to line 1 of the stretchout. Repeat with points 2, 3 and 4 to lines 2, 3 and 4.
4. Connect the points 1 to 2, 2 to 3, 3 to 4 and 4 to 1.
5. Draw the top and bottom in position.
6. Allow material for seams, and go over the outline and fold lines with visible object lines.

PATTERN DEVELOPMENT OF A TRUNCATED CYLINDER

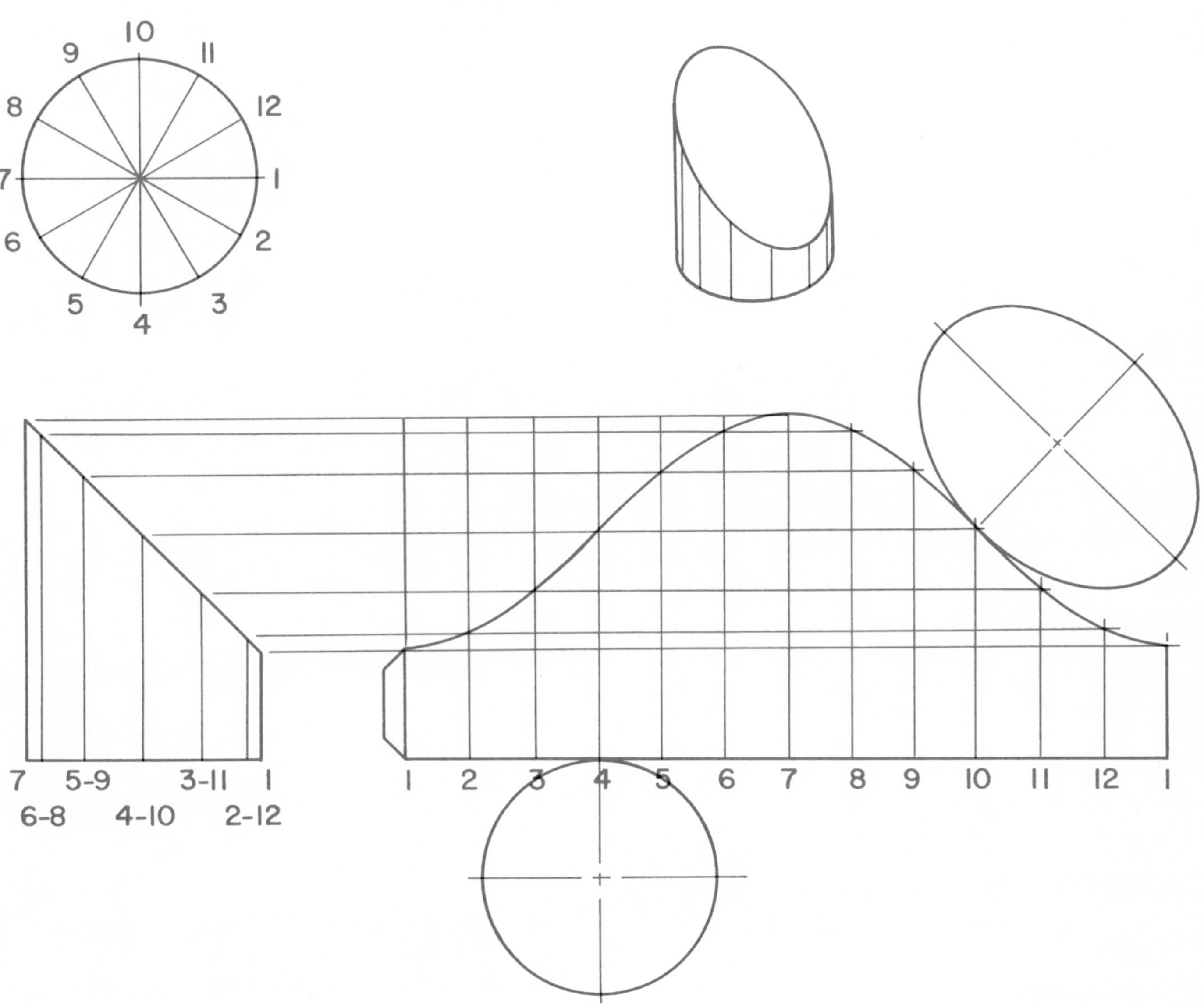

1. Draw the front and top views. Divide and number as shown.
2. Extend lines from the top and bottom of the front view.
3. Allow one inch between the front view and the pattern and draw line 1.
4. Set your compass or dividers from 1 to 2 on the top view and step off twelve (12) equal divisions on the extended lines of the pattern. Number them.
5. Draw vertical construction lines at each of the above divisions.
6. The curve of the pattern is developed by projecting lines from the points on the front view. Point 1 is projected over until it intersects lines 1 on the pattern; points 2-12 intersect lines 2 and 12; etc. When all of the points are located they are connected with a curved line drawn with a French Curve.
7. Complete by adding the top and bottom. The top is developed as an auxiliary view.

PATTERN DEVELOPMENT OF A PYRAMID

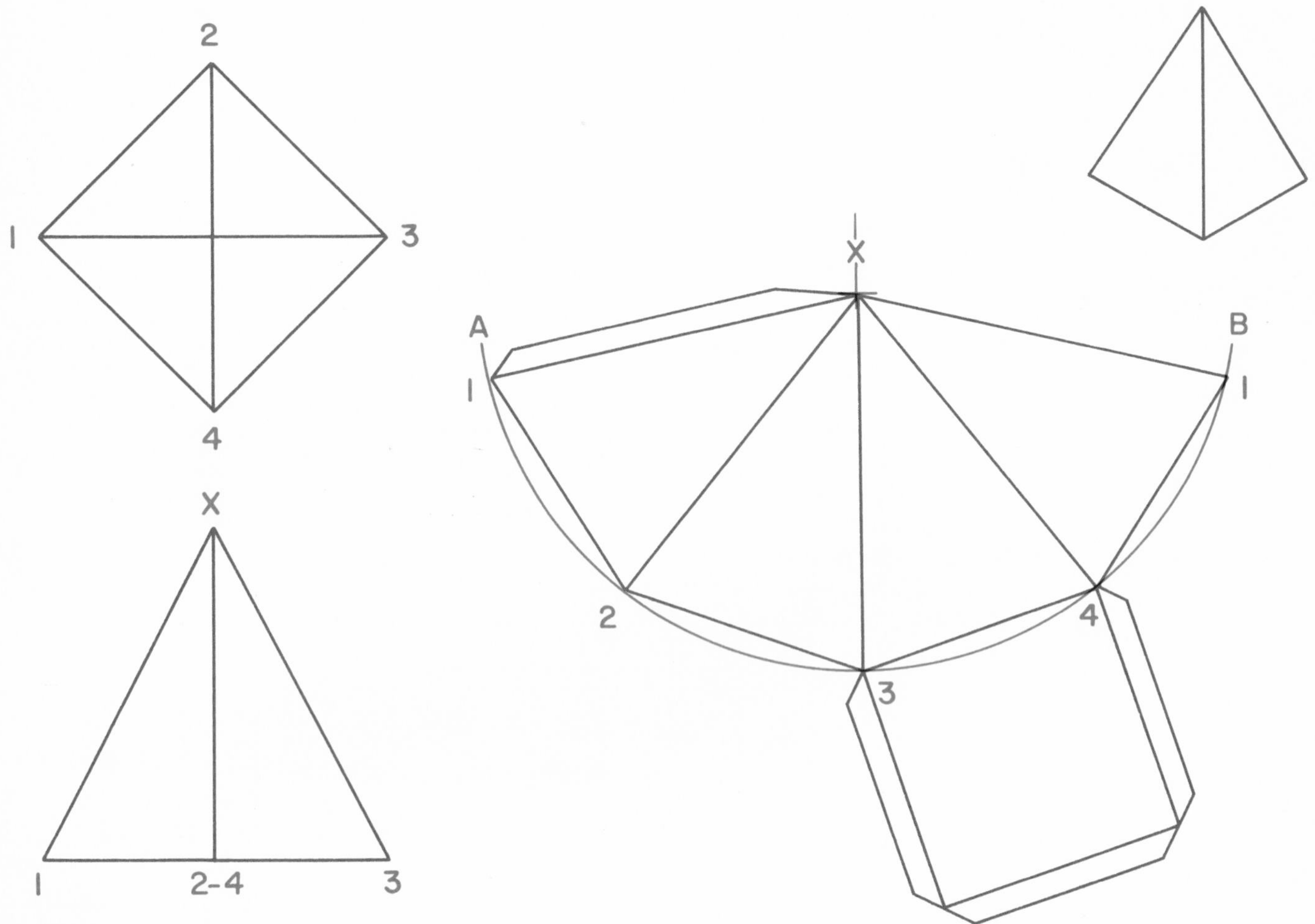

1. Draw the front and top views. Number as shown.
2. Locate center line X of the stretchout.
3. Set your compass to a radius equal to X-1 on the front view and using the above center, draw arc A-B.
4. Draw a vertical line through center X and arc A-B.
5. Set compass from 1 to 2 on top view and at point where vertical line intersects the arc as the starting point, step off two (2) divisions on each side of the line. (Points 1-2-3-4-1 on the stretchout.)
6. Connect the points, draw the bottom in place, and go over the outline and folds with object lines.

PATTERN DEVELOPMENT OF A CONE

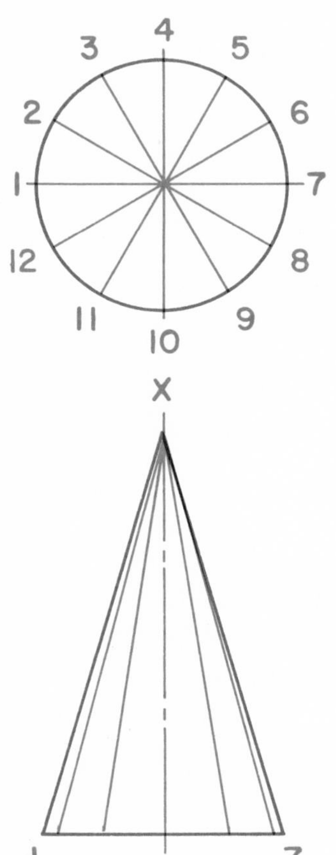

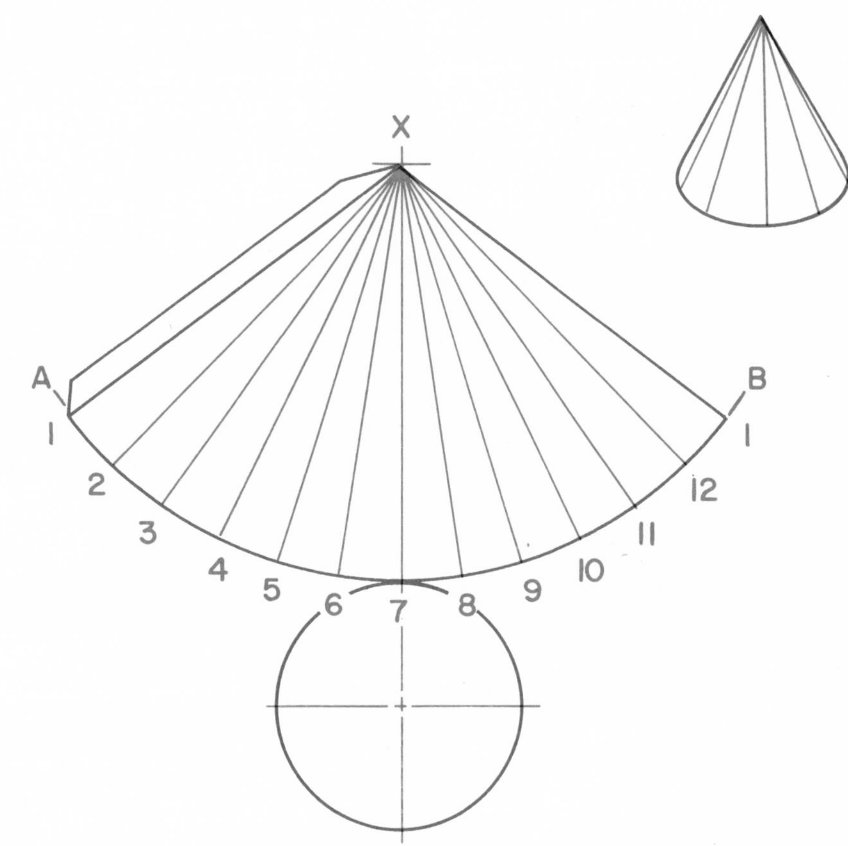

1. Draw the front and top views. Divide the top view into twelve (12) equal parts. Number as shown.
2. Locate center line X of the stretchout.
3. Set your compass from X to 1 on the front view and with X of the stretchout as the center, draw arc A-B.
4. Draw a vertical construction line through center line X and the arc.
5. Set your compass from 1 to 2 on the top view, and with the point where the vertical line intersects the arc as the starting point, step off six (6) divisions on each side of the line. (Points 1-2-3-4 etc. on the pattern.)
6. Go over the outline carefully with object lines.
7. The lines that represent the curved portion may be drawn in color, or, they may be left as construction lines.

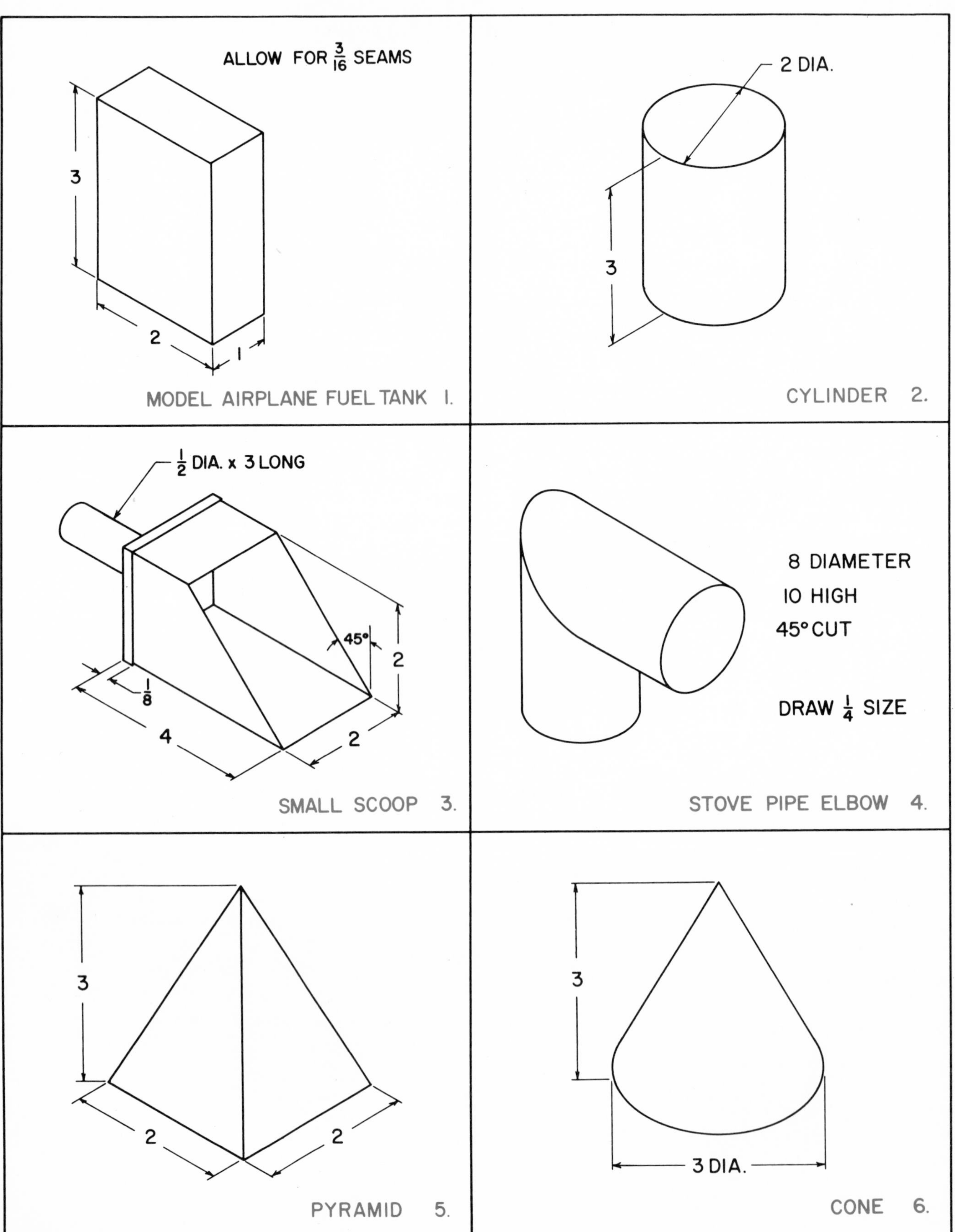

PROBLEMS 12–1 to 12–6. Develop patterns for the objects above. Use the dimensions provided in the problems.

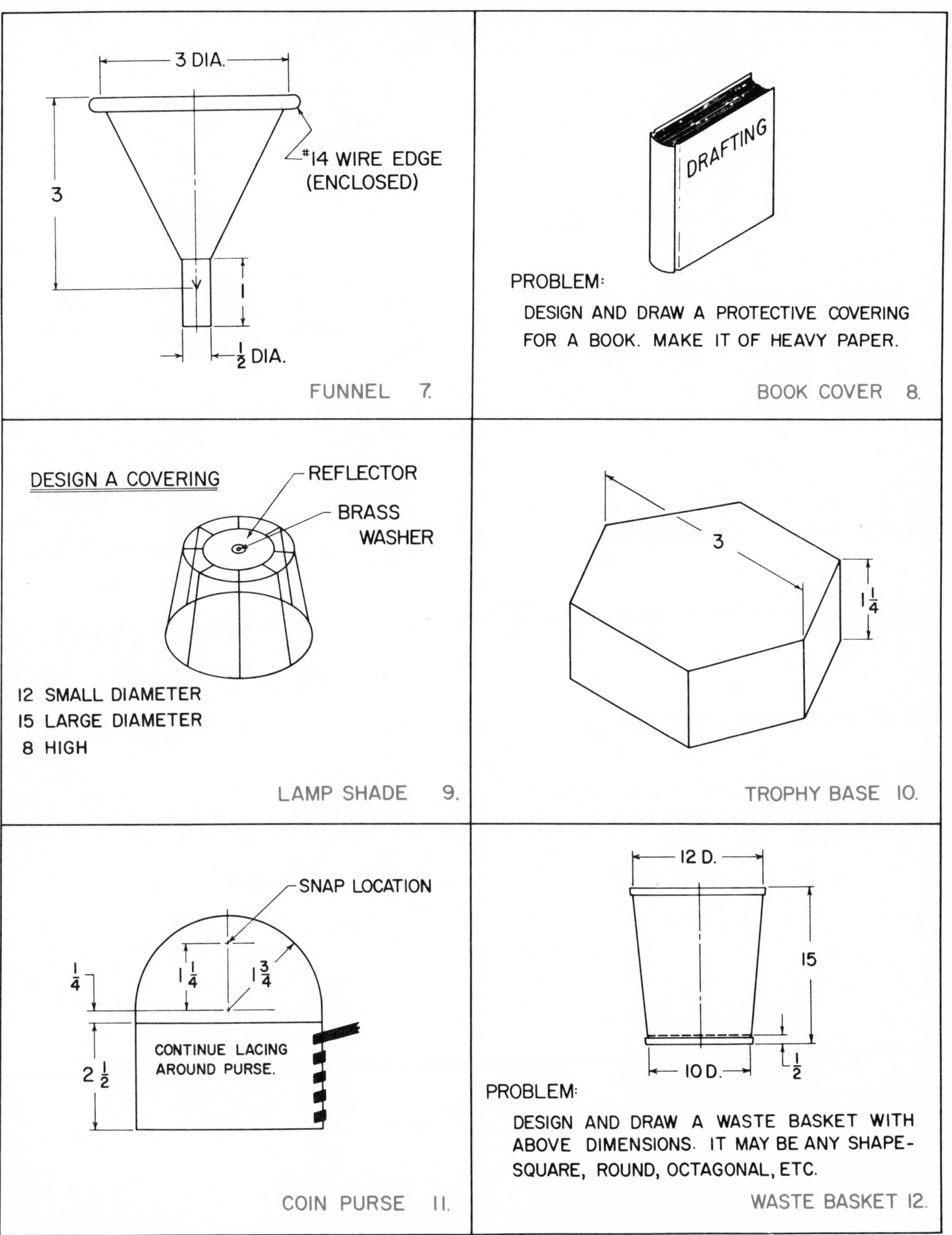

PROBLEMS 12–7 to 12–12. Develop patterns for the objects above. Use the dimensions provided in the problems.

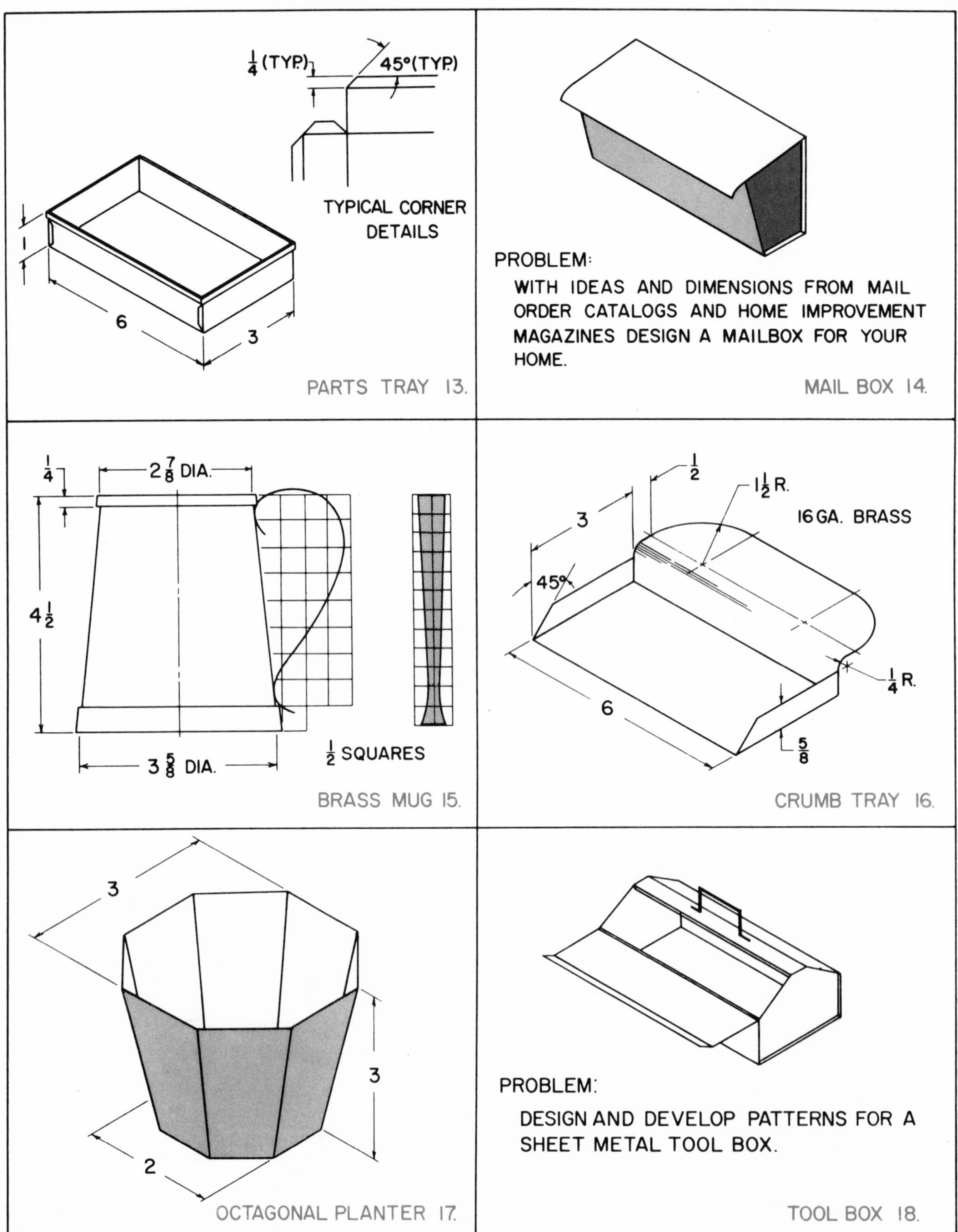

PROBLEMS 12–13 to 12–18. *Develop patterns for the objects above. Use the dimensions provided in the problems.*

Pattern Development Problems

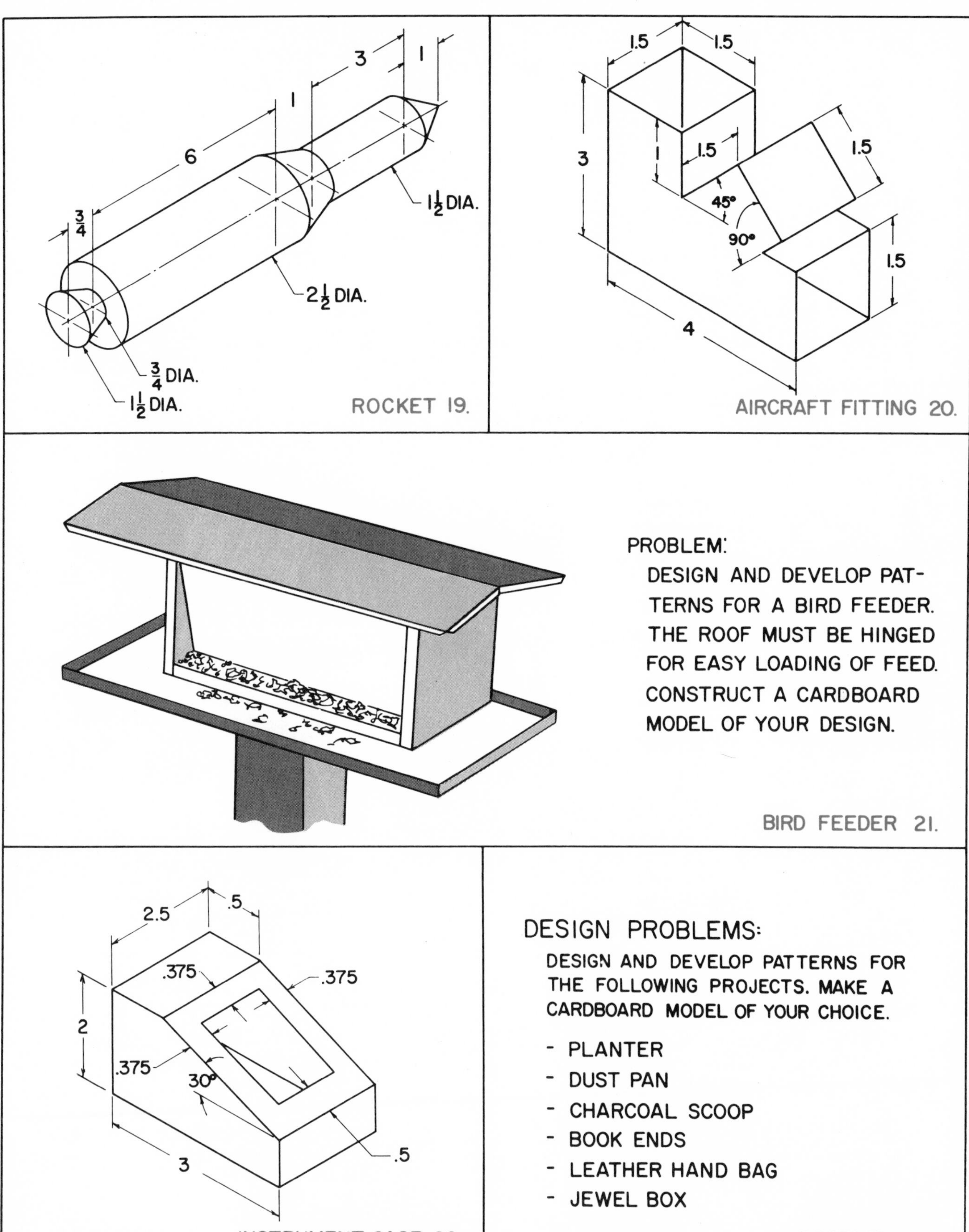

PROBLEMS 12–19 to 12–23. Develop patterns for the objects above. Use the dimensions provided in the problems.

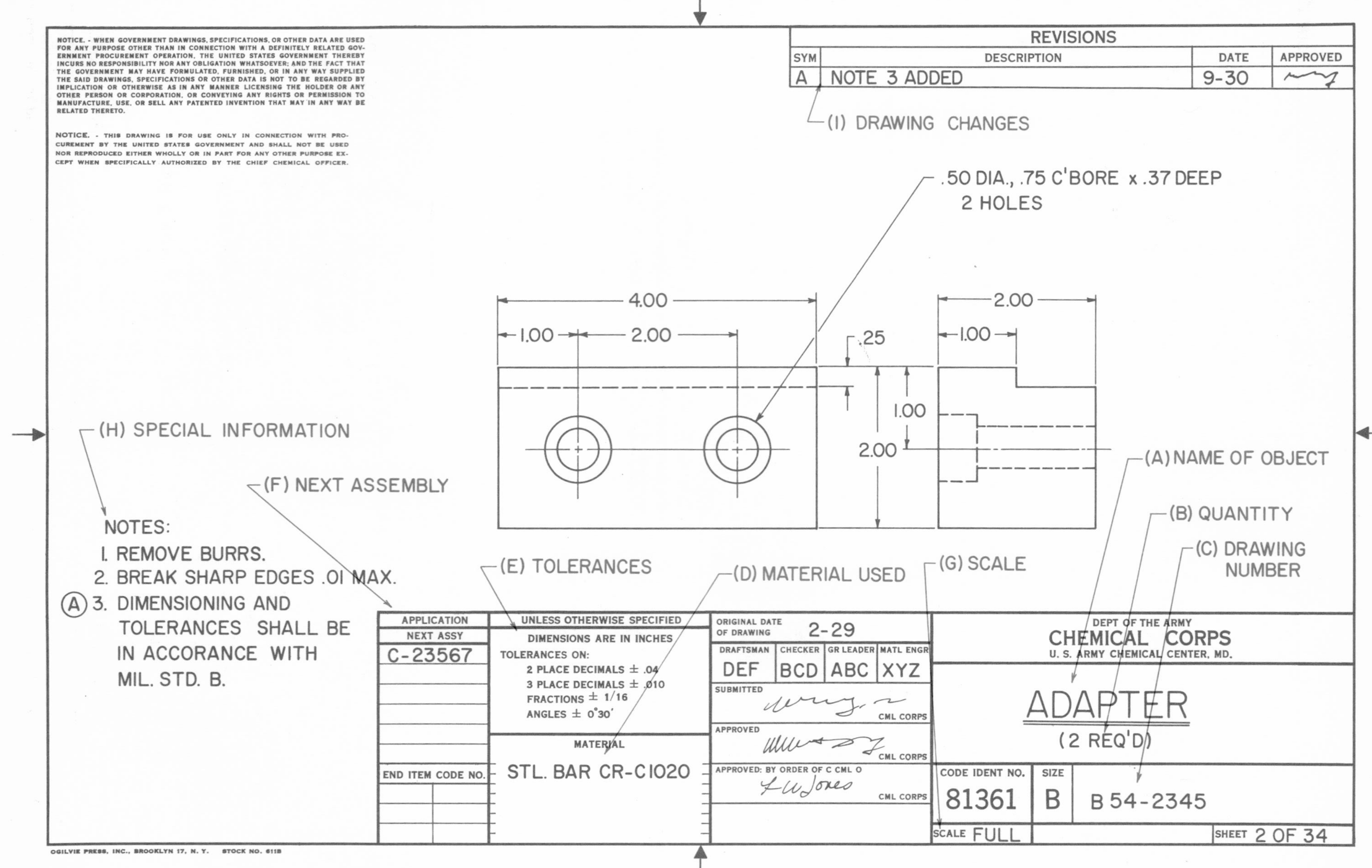

Fig. 13-1. A working drawing.

Unit 13
WORKING DRAWINGS

It would not be practical for industry to manufacture a product without using drawings which provide complete manufacturing details.

Drawings required range from a single freehand sketch for a simple job, to many thousands of drawings needed to manufacture a complex product like the helicopter shown in Fig. 1-1, page 7. The drawings which tell the craftsman what to make and establish the standards to which he must work are called WORKING DRAWINGS. See Fig. 13-1.

Working drawings fall into two categories, DETAIL DRAWINGS and ASSEMBLY DRAWINGS. An example of a DETAIL DRAWING is given in Fig. 13-2. It includes a view

Fig. 13-2. A detail drawing.

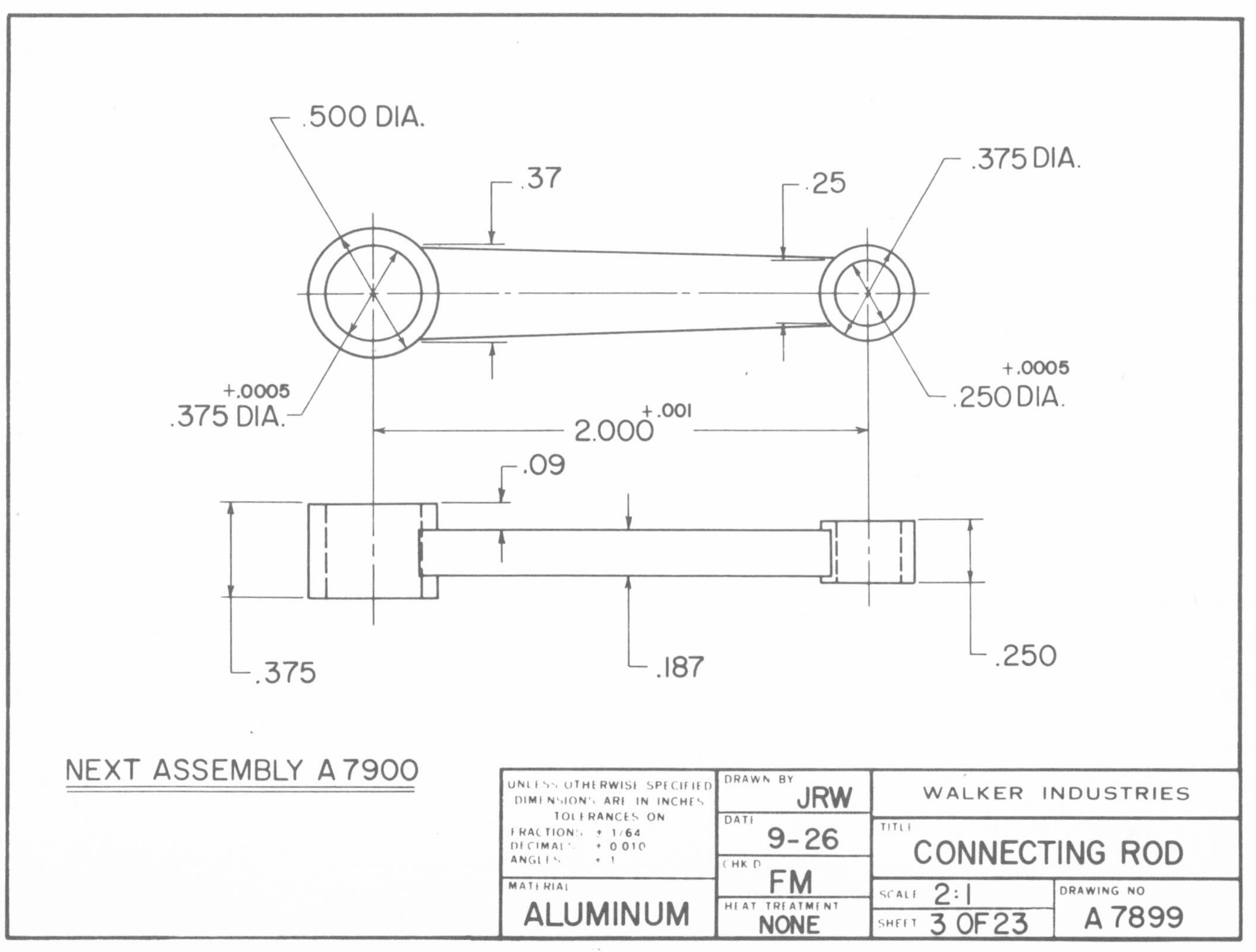

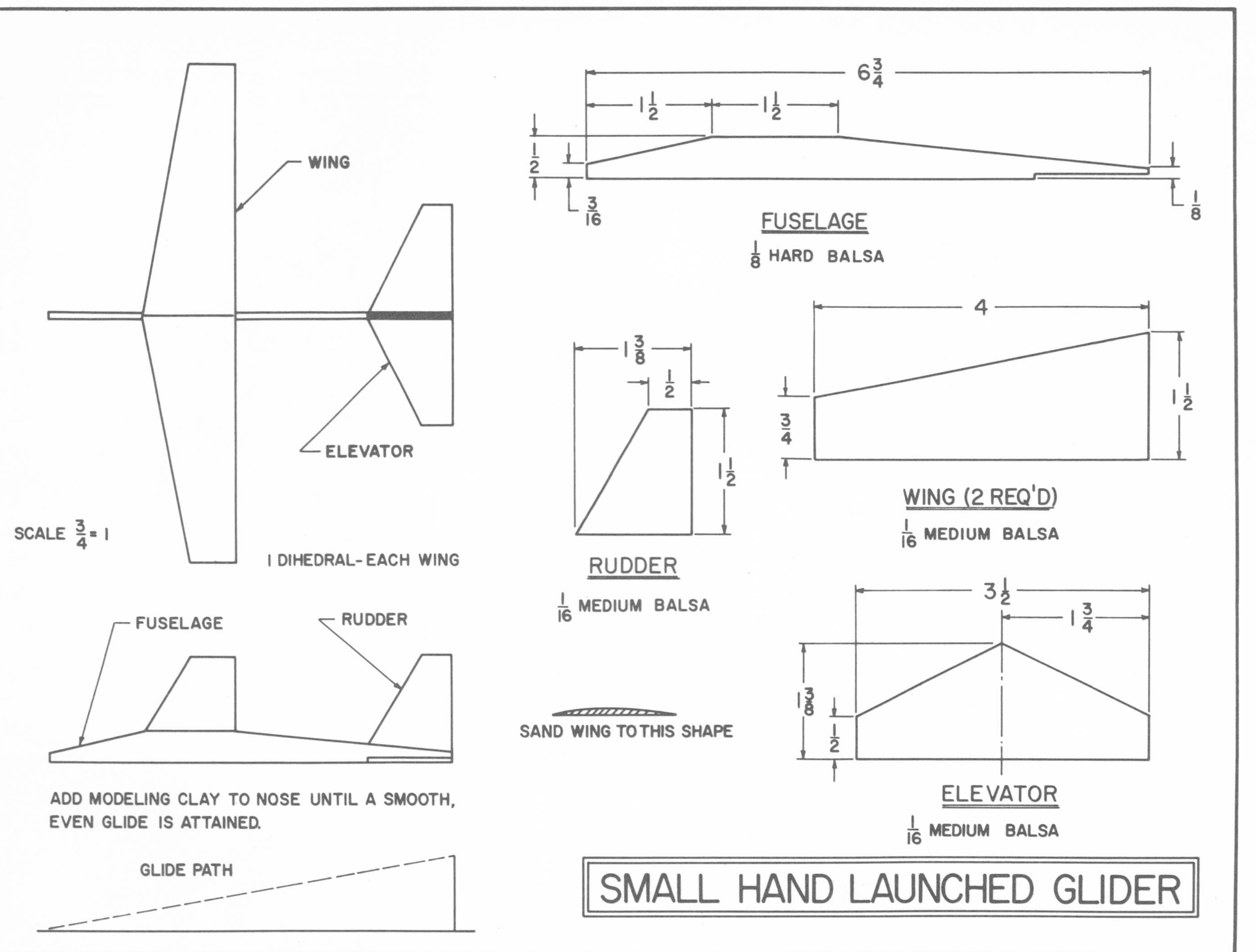

Fig. 13-3. If the item is small enough, it is permissible to draw the details and assembly on the same drawing sheet.

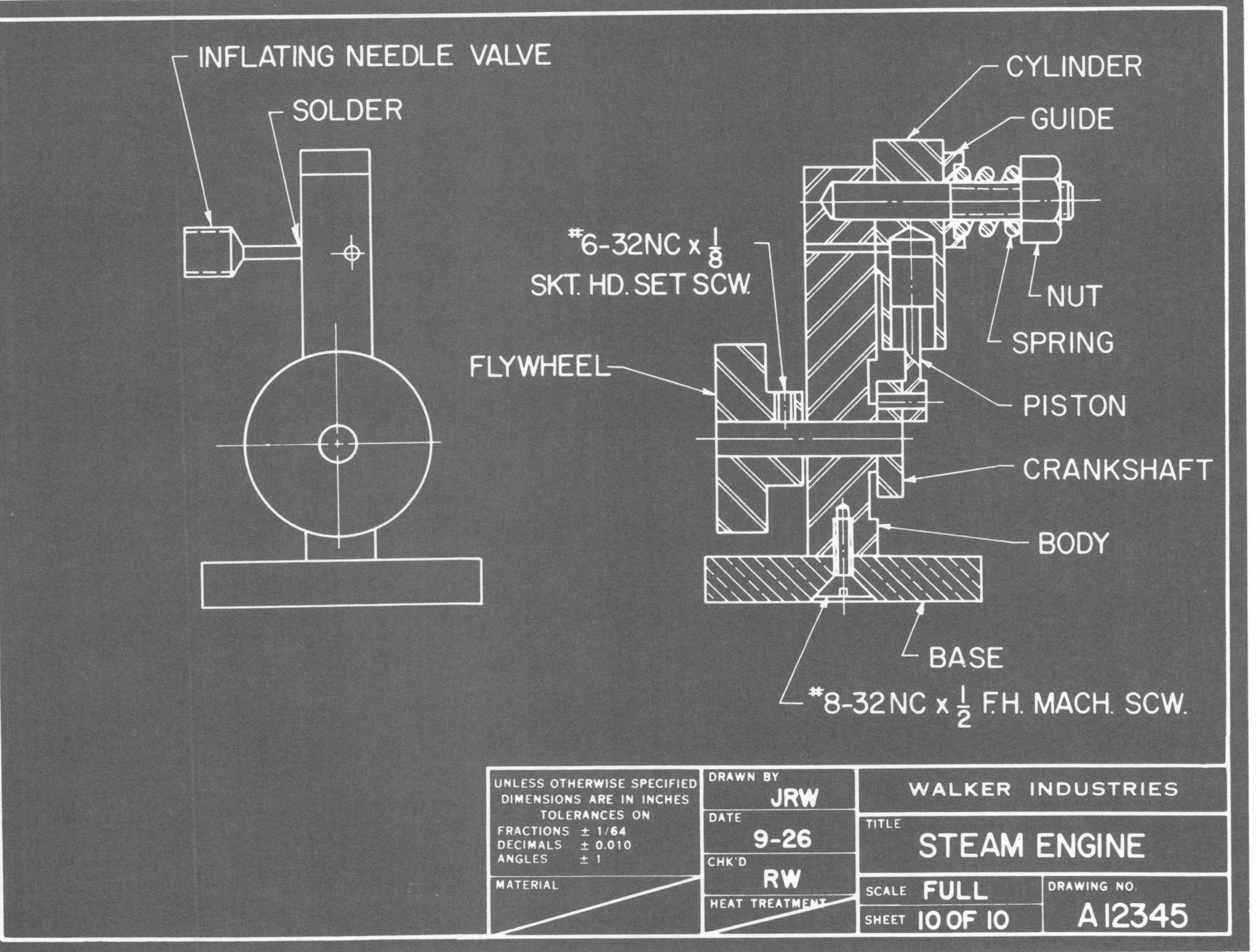

Fig. 13-4. An assembly drawing.

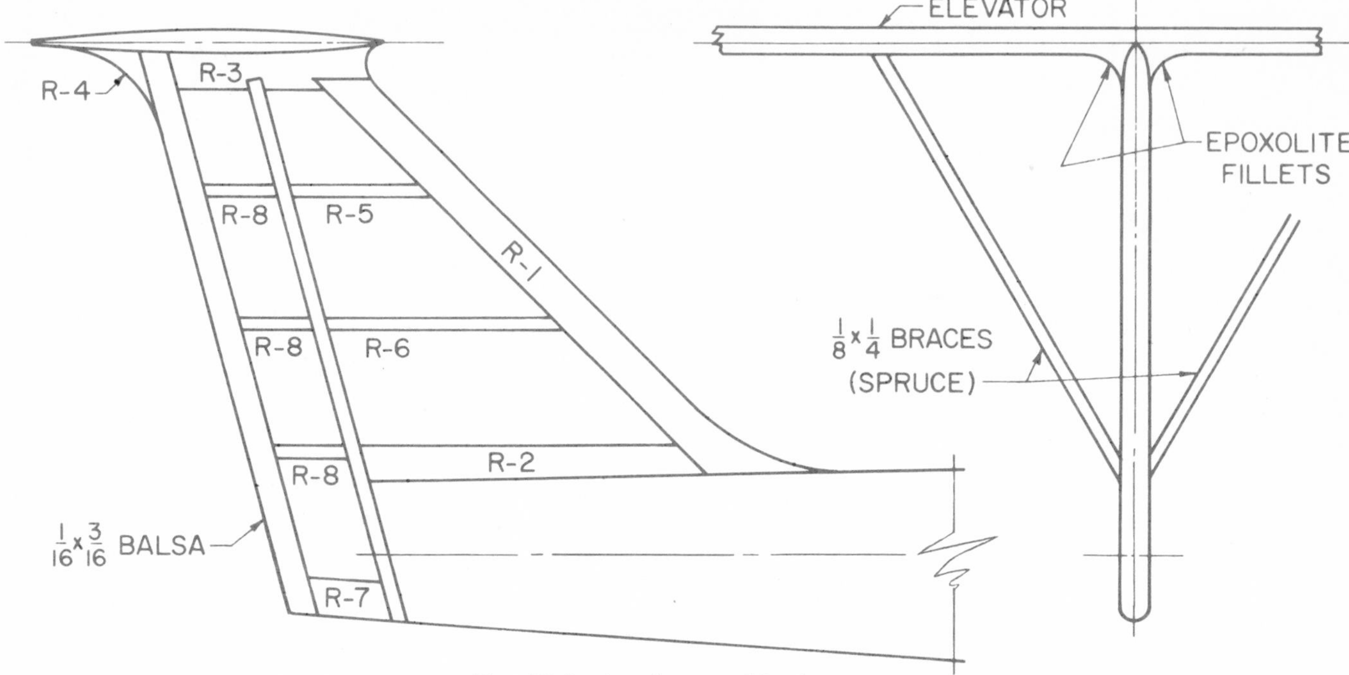

Fig. 13-5. A sub-assembly drawing.

(or views) of the product, with dimensions and other pertinent information required to make the part. This information includes: (Refer to Fig. 13-1).

A. NAME OF THE OBJECT. The name assigned to the part. This should appear on all drawings making reference to the piece.

B. QUANTITY. The number of units needed in each assembly.

C. DRAWING NUMBER. Number assigned to the drawing for filing and reference.

D. MATERIAL USED. The exact material specified to make the part.

E. TOLERANCES. The permissible deviation, either oversize or undersize, from a basic dimension.

F. NEXT ASSEMBLY. The name given to the major assembly on which the part is to be used.

G. SCALE. Drawings made other than full size are called scale drawings.

H. SPECIAL INFORMATION. Information pertinent to the correct manufacture of the part that is not included on the various views of the object.

I. REVISIONS. Changes that have been made on the original drawing.

In most instances, the detail drawing provides information on a single part. However, it is permissible to draw all of the parts and the assembly of a small or simple mechanism on the same sheet, Fig. 13-3.

An example of the second type of working drawing, an ASSEMBLY DRAWING, is given in Fig. 13-4. As its name implies, this contains views that show where and how the various parts fit into the assembled product. They also show how the complete object will look. When the product is large or complex it may not be possible to present all of the information on one sheet. A SUB-ASSEMBLY DRAWING, Fig. 13-5, is utilized when this problem arises. This type of drawing illustrates the assembly of only a small portion of the complete object.

Fig. 13-6. Parts list.

PARTS LIST		
No.	*Name*	*Quan.*
1	CRANKCASE	1
2	CRANKSHAFT	1
3	CRANKCASE COVER	1
4	CYLINDER	2
5	PISTON	2

$\frac{1}{4}$

$\frac{1}{2}$ $3\frac{1}{2}$ $\frac{3}{4}$

Under certain situations, a drawing may include a PARTS LIST, Fig. 13-6, and/or a BILL OF MATERIALS, Fig. 13-7. The information on the parts list is read from the top down when the list is located at the upper right corner of the drawing sheet, or, read from the bottom up when located above the title block. Either position is acceptable. The parts are usually listed according to size or in their order of importance.

An IDENTIFYING NUMBER, Fig. 13-8, is placed on each sheet for convenience in filing or rapid location of the drawing.

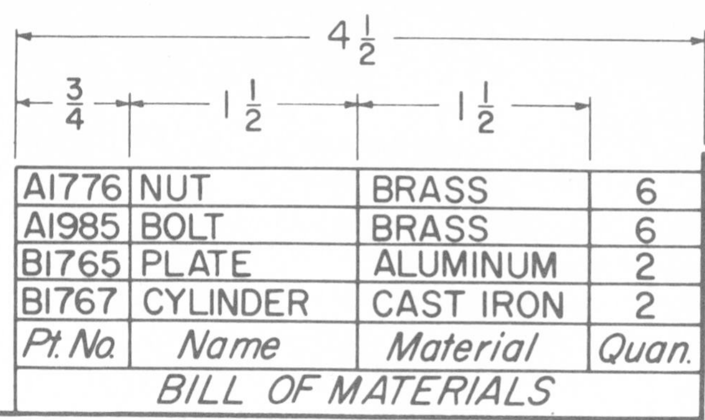

Pt. No.	Name	Material	Quan.
A1776	NUT	BRASS	6
A1985	BOLT	BRASS	6
B1765	PLATE	ALUMINUM	2
B1767	CYLINDER	CAST IRON	2

BILL OF MATERIALS

Fig. 13-7. Bill of materials.

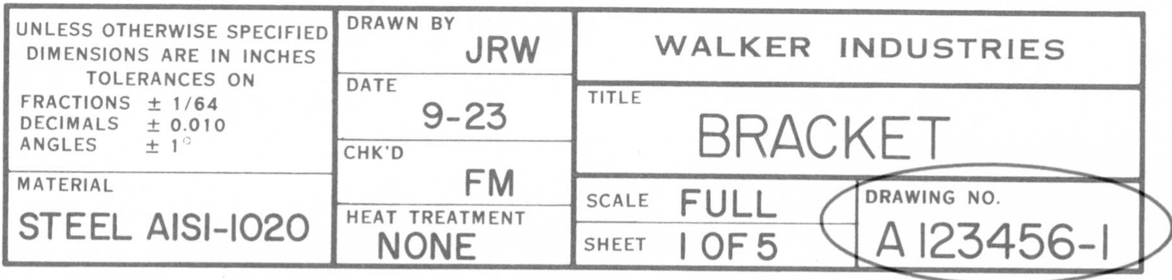

UNLESS OTHERWISE SPECIFIED DIMENSIONS ARE IN INCHES
TOLERANCES ON
FRACTIONS ± 1/64
DECIMALS ± 0.010
ANGLES ± 1°
MATERIAL STEEL AISI-1020
DRAWN BY JRW
DATE 9-23
CHK'D FM
HEAT TREATMENT NONE
WALKER INDUSTRIES
TITLE BRACKET
SCALE FULL
SHEET 1 OF 5
DRAWING NO. A 123456-1

Fig. 13-8. For convenience in filing and locating drawings, industry gives each plate an identifying number.

TEST YOUR KNOWLEDGE-UNIT 13

1. Why are working drawings used?
2. The ____________ drawing has a picture of the part with dimensions and other information needed to make the object.
3. The ____________ drawing shows where and how the object described in the above drawing fits into the complete assembly of the object.
4. On large or complex objects ________ ____________ drawings are frequently used. These drawings show the assembly of only a small portion of the complete object.

The drawing described in Question No. 2 includes much information necessary to manufacture a part. Some of this information is described below. Place the letter of the correct description in the blank before the words below.

5. ______ QUANTITY
6. ______ TOLERANCES
7. ______ MATERIAL USED
8. ______ SCALE
9. ______ REVISIONS

A. The exact material specified to make the part.
B. Drawings made other than full size are called this.
C. The number of units needed in each assembly.
D. The permissible deviation, either oversize or undersize, from a basic dimension.
E. Changes that have been made on the original drawing.

OUTSIDE ACTIVITIES

1. From a local industry, try to obtain samples of detail drawings.
2. Try to obtain samples of assembly drawings.
3. Try to obtain samples of sub-assembly drawings.
4. Prepare a bulletin board display using the theme WORKING DRAWINGS.
5. Design a suitable title block for the drawing sheets of a company that you are planning to start.
6. Secure the piston and rod assembly from a small single cylinder gasoline engine and prepare detail drawings for them. Indicate all dimensions as decimals. It may be necessary for you to learn how to read a micrometer to make accurate measurements.
7. Design a foot stool and prepare the drawings necessary to mass-produce it in the school shop. Make your drawings on tracing vellum so that a number of prints can be made.
8. Design a coffee or end table. Prepare the drawings needed to make it in the school shop.

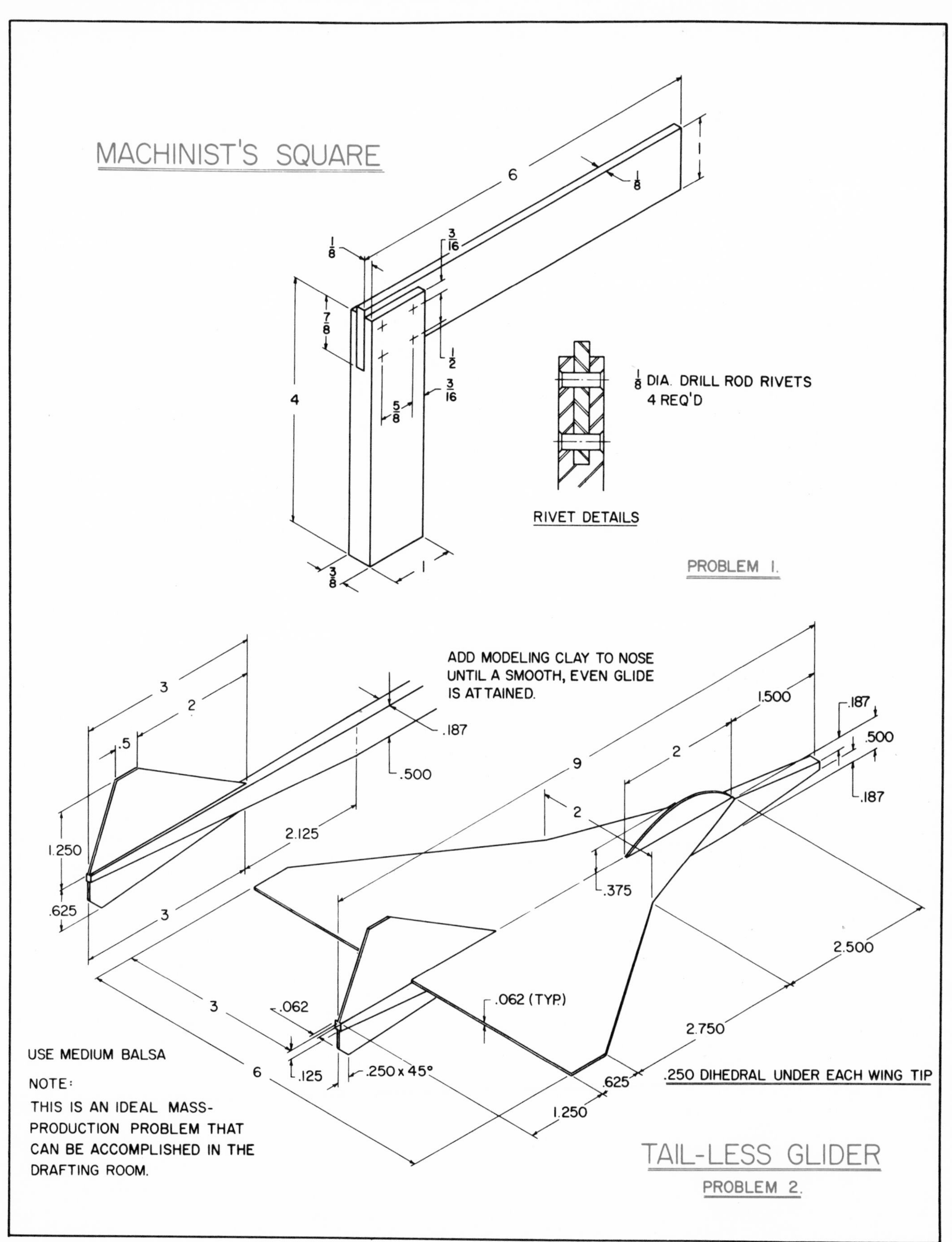

***PROBLEM 13–1. Above. MACHINIST'S SQUARE.** Make detail and assembly drawings. **PROBLEM 13–2. Below. TAIL-LESS GLIDER.** Prepare detail and assembly drawings. Design the necessary patterns, fixtures, etc., needed to mass-produce the glider.*

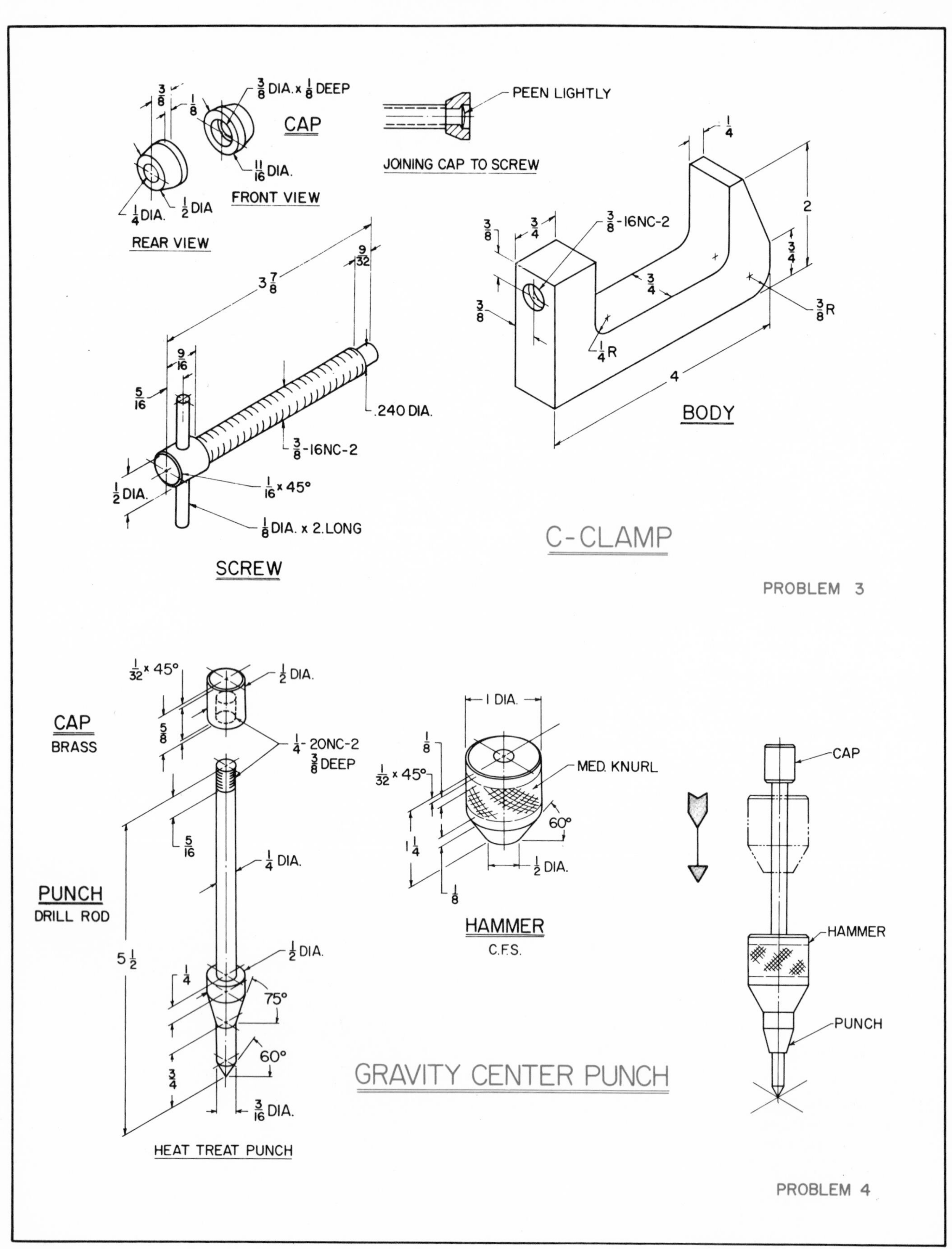

PROBLEM 13–3. Above. C–CLAMP. Prepare detail and assembly drawings. PROBLEM 13–4. Below. GRAVITY CENTER PUNCH. Make detail drawings.

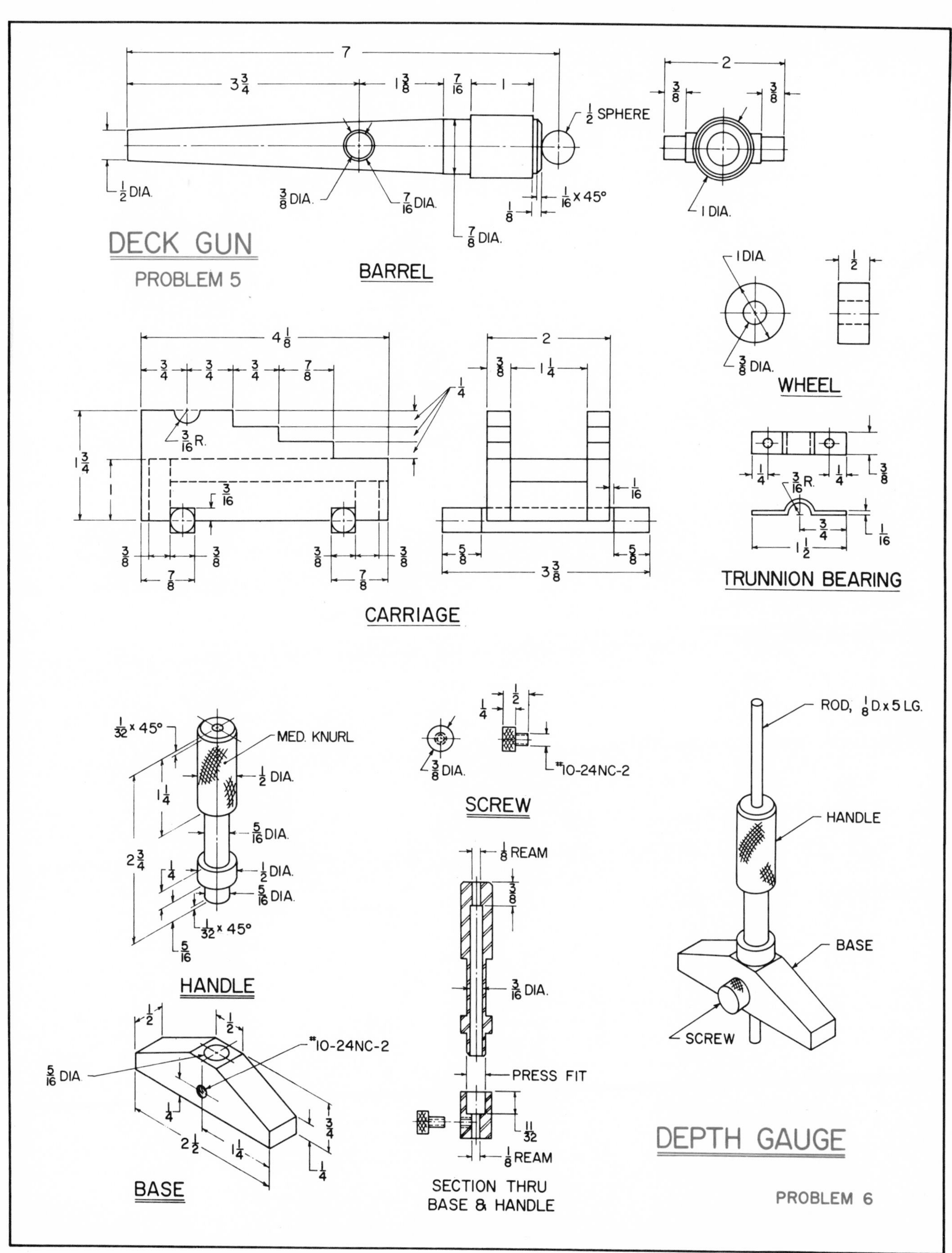

*PROBLEM 13–5. Above. **DECK GUN.** Prepare an assembly drawing. PROBLEM 13–6. Below. **DEPTH GAUGE.** Make detail and assembly drawings.*

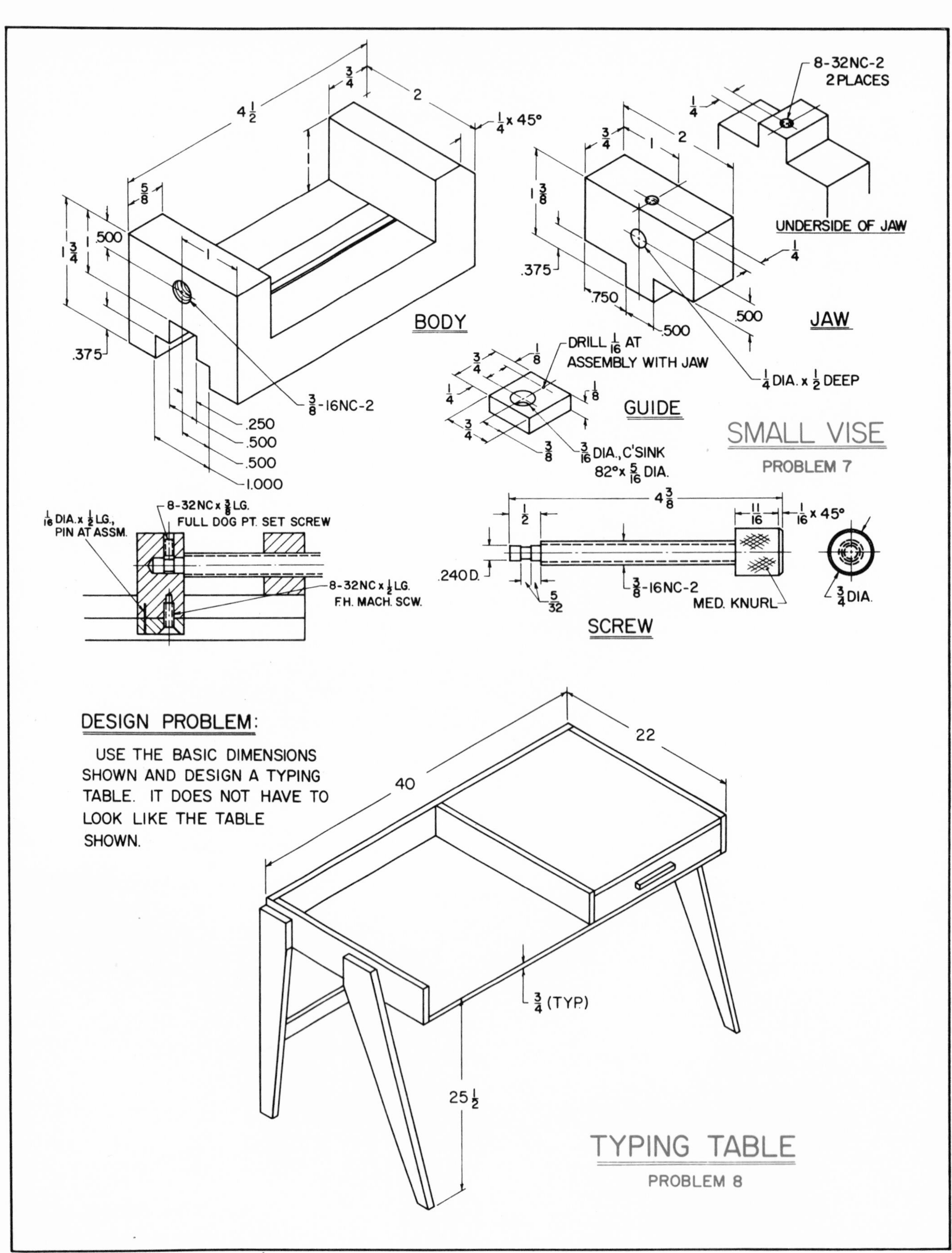

PROBLEM 13–7. *Above.* **SMALL VISE.** *Prepare detail and assembly drawings.* **PROBLEM 13–8.** *Below.* **TYPING TABLE.** *Design a table using the dimensions given. Prepare the drawings that will be needed to make your table.*

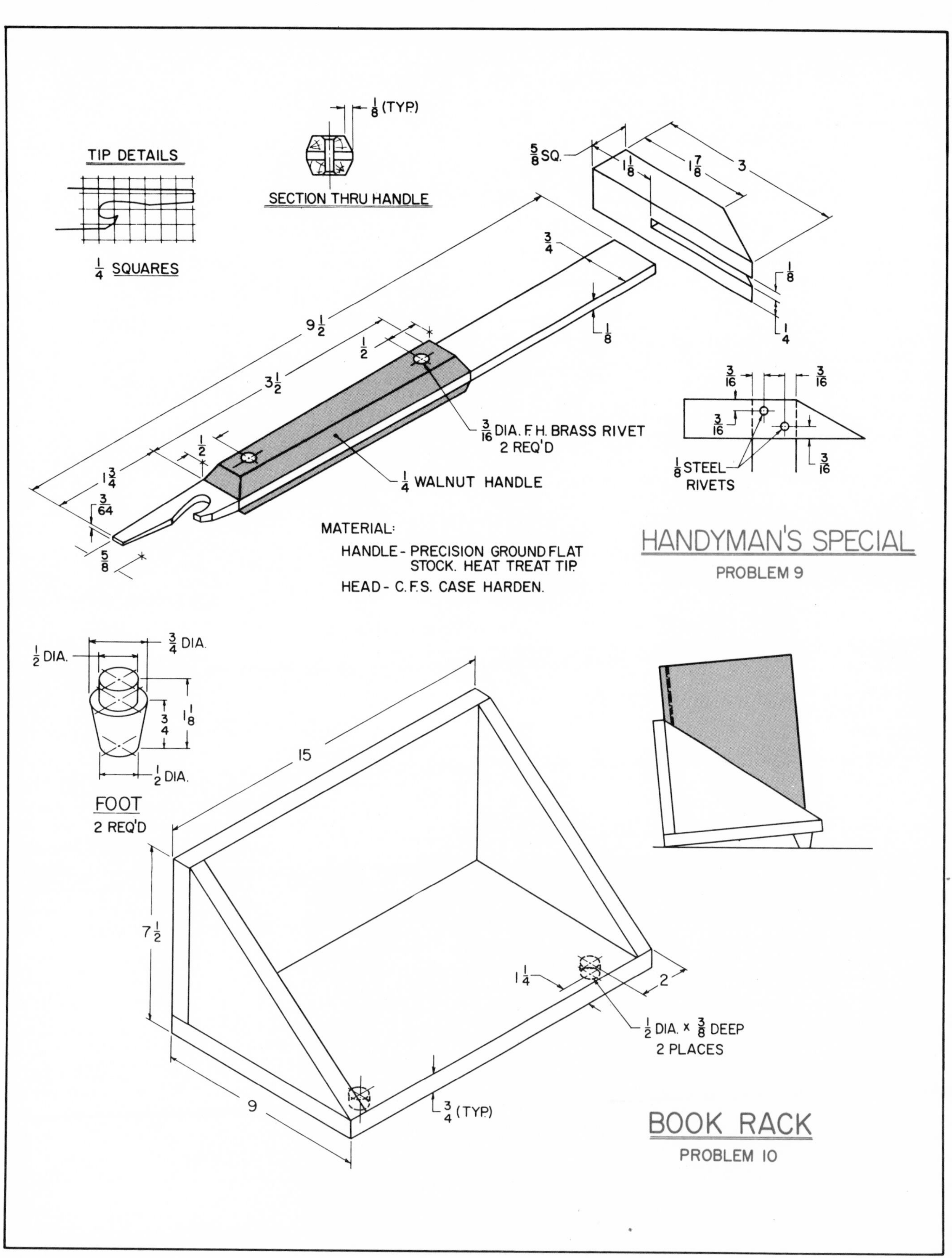

***PROBLEM 13–9.** Above. **HANDYMAN'S HELPER.** Prepare detail and assembly drawings.*
***PROBLEM 13–10.** Below. **BOOK RACK.** Make a fully dimensioned assembly drawing.*

Unit 14
PRINT MAKING

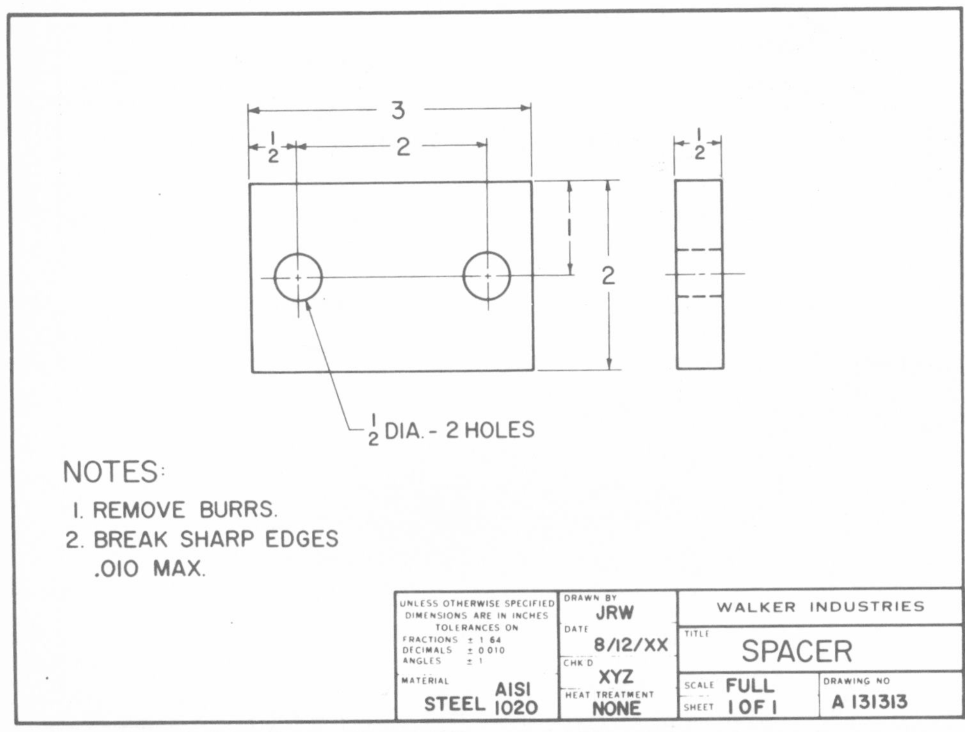

POSITIVE PRINT

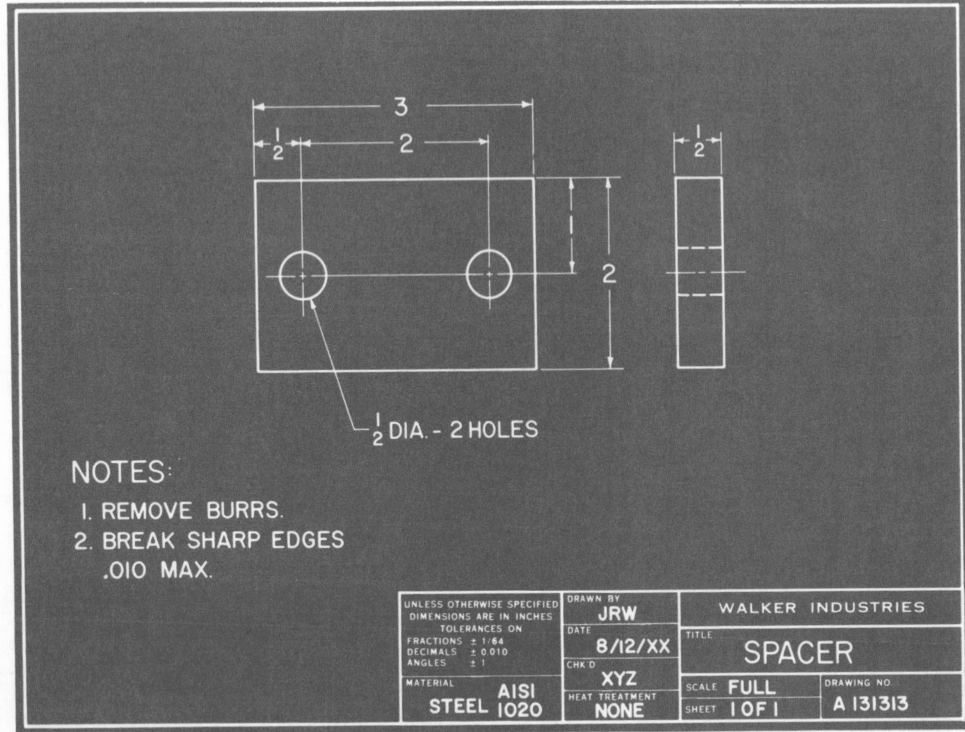

BLUEPRINT

Fig. 14-1. The true blueprint is a print with white lines on a blue background. The term "blueprint" is commonly used when referring to all types of prints regardless of the color of the lines or the background.

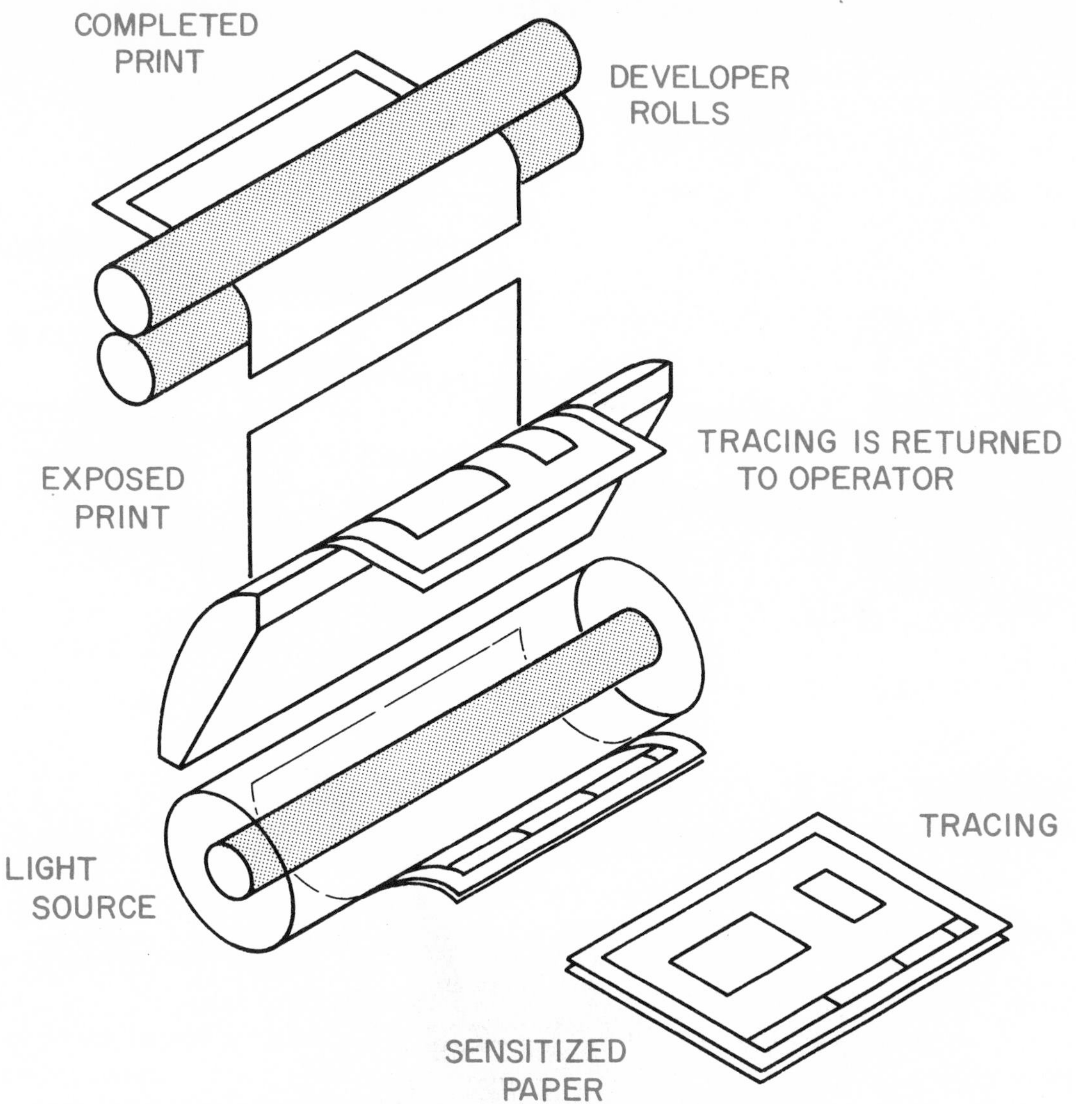

Fig. 14-2. The diazo process. On most machines the print is made in one continuous process.

Original drawings are seldom used on a job because they would soon become worn and soiled and difficult to read. Also, several sets of identical drawings are frequently needed at the same time because craftsmen at different locations are working on the product or structure.

High cost makes it impractical to draw up a set of plans for each person who needs them and to replace plans damaged or ruined.

When several sets of plans are needed, the original drawings are reproduced or duplicated. The technique used must produce accurate copies of the originals, must not destroy the original drawings and must be speedy and reasonable in cost.

Several of the more important duplicating processes will be described in this Unit. The original drawing to be reproduced is usually made on semitransparent (translucent) material such as tracing paper, cloth or film.

BLUEPRINTS

The BLUEPRINT PROCESS, Fig. 14-1, is the oldest of the techniques employed to duplicate drawings. The print consists of white lines on a blue background.

To make a print by the blueprint method, the tracing is placed in contact with a sensitized paper (unexposed blueprint paper).

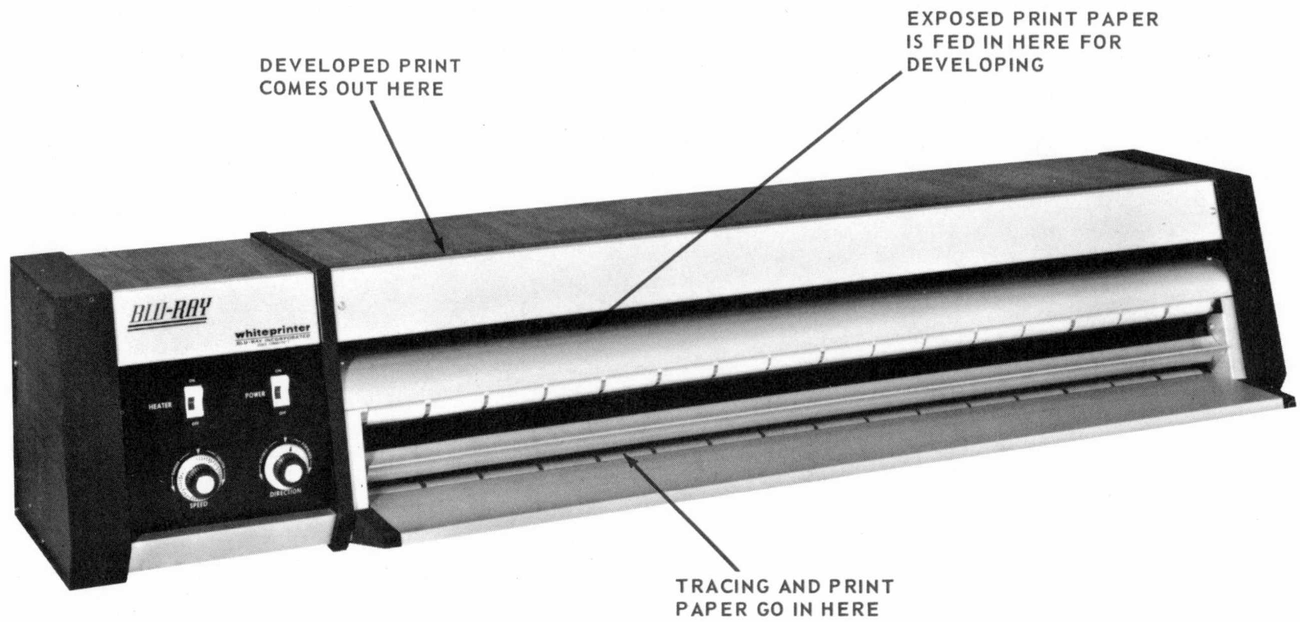

Fig. 14-3. A combination printer and developer type diazo copying machine. (Blue-Ray, Inc.)

The two sheets are exposed to a bright light. The light cannot penetrate the opaque lines of the tracing and exposure takes place only where the light strikes the sensitive coating on the print paper.

The exposed print is developed by washing in water. The lines are "fixed" by washing in a solution of potassium dichromate crystals dissolved in water to prevent the lines from fading. After a final washing the print is dried and pressed.

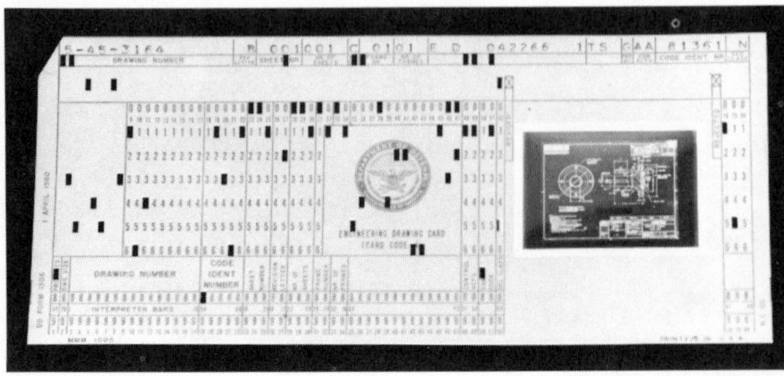

Fig. 14-4. The small negative on the microfilm aperture card is enlarged by a photographic process to the desired print size. The card is easier to store than full size prints and when needed can be retrieved by computerized techniques. Some microfilm methods utilize the film in roll form.

DIAZO PROCESS

The DIAZO PROCESS, Fig. 14-2, is a versatile copying technique for making direct positive copies (dark lines on light background) of anything that is drawn, written or typed on semitransparent paper, cloth or film. The process produces the copies quickly and is inexpensive.

The print is made by placing the tracing in contact with sensitized copying material (paper, cloth or film) and exposing it to light. After the tracing has been removed, the exposed copy is developed by passing it through ammonia vapors.

With most modern diazo copying machines, Fig. 14-3, the print is made in one continuous process. In addition to turning out the finished copy in seconds, it is clean, odorless and quiet.

Using this process black-on-white, black-on-color, color-on-white or color-on-color prints can be produced.

THE MICROFILM PROCESS

The MICROFILM PROCESS, Fig. 14-4, was originally designed to reduce storage facilities and to protect prints from loss. With this process, the drawing is reduced by photographic means. Finished negatives can be stored in roll form, or on cards adaptable to computerized storage and retrieval. To produce a working print, the microfilm image is retrieved from the files and enlarged (called BLOW-BACKS) on photographic paper. The print is discarded or destroyed when it is no longer needed. Microfilms can also be viewed on a READER to check details, Fig. 14-5.

HOW TO MAKE A TRACING

The first step in making a tracing for reproduction is to draw it on a semi-transparent or translucent material. There are many types of this material available in sheet or roll form.

TRACING PAPER is frequently used because it is inexpensive. Tracing paper is used extensively for making preliminary plans, sketching and the like.

TRACING CLOTH is excellent for making original tracings that must withstand severe handling. Changes and alterations can be made without damaging the drawing quality of the cloth.

Tracing cloths are more expensive than papers but retain their drawing and reproduction qualities almost indefinitely. They are made from specially treated, fine grade linen cloth. One surface has a matte finish to take pencil or ink.

TRACING FILMS made from acetate, Mylar or polyester are ideally suited for work requiring a high degree of dimensional stability (it will not expand or contract in heat, cold or high humidity). They have high transparency and make prints of maximum contrast. One surface has a matte finish that will take pencil or ink. Corrections are made with an eraser or damp cloth.

Tracing films are tough and tear resistant. They do not discolor or become brittle with age.

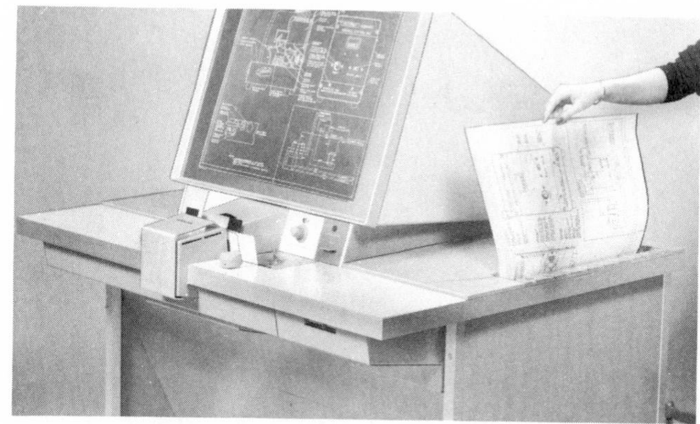

Fig. 14-5. A microfilm reader. The enlarged print can be studied on a view screen in this unit and then made into a print of the required size. (RECORDAK)

PENCIL TRACINGS

Sharp, clear reproductions can be made from pencil tracings with modern reproduction equipment and materials. At present most drawings used by industry are done in pencil.

Many draftsmen make original drawings on the tracing material that will be used to produce the copies. For complex jobs, the layout is frequently traced from the original drawing onto other sheets which are used in making copies (this is where we get the term TRACING).

Tracings for reproduction are made the same way as the conventional drawings you have been making. Block in the needed views with a 4H or 6H pencil. If they are drawn lightly enough they will not have to be erased. Complete the drawing and add notes and dimensions using a 2H pencil. Keep the lines uniformly sharp and dense as this type line reproduces best.

Every effort must be made to keep your drawings clean. Working with clean hands

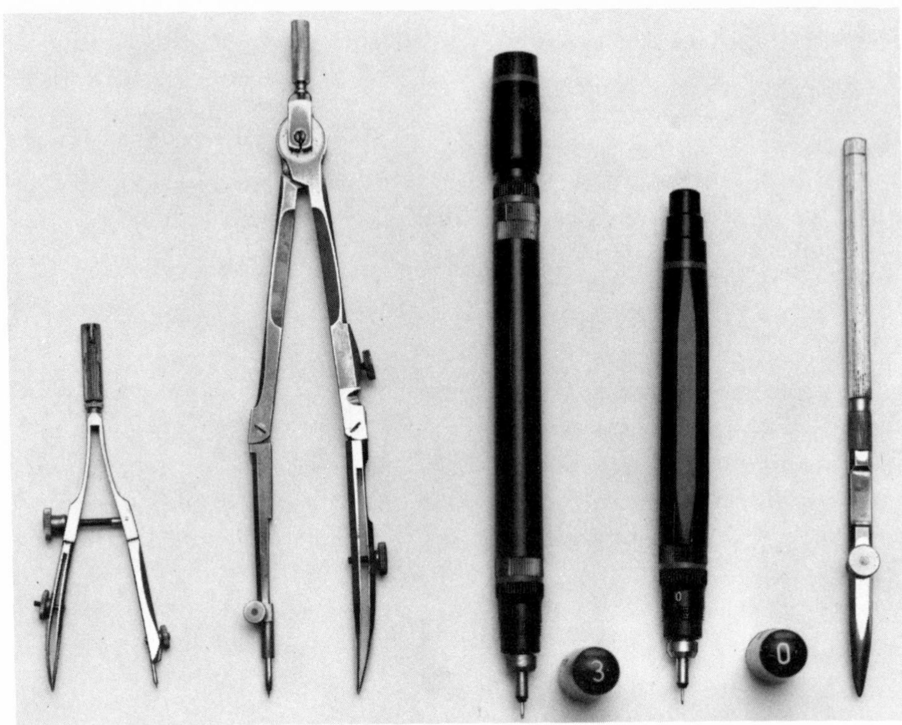

Fig. 14-6. Drafting equipment used in inking: Compasses, Technical Fountain Pens, and Ruling Pen.

and a piece of paper under your hand when you letter dimensions and notes will help.

INKED TRACINGS

Inked tracings are more difficult to prepare than tracings in pencil. A knowledge of inking techniques, and much patience and practice are required to produce acceptable inked tracings.

A dense black, waterproof ink (called INDIA INK) is used. Lines are drawn with a RULING PEN or a TECHNICAL FOUNTAIN PEN, Fig. 14-6. The ruling pen can be adjusted to make different width lines, but different size points are needed on a technical fountain pen to vary line width.

Hold the pen in much the same manner you would hold a pencil. Practice drawing

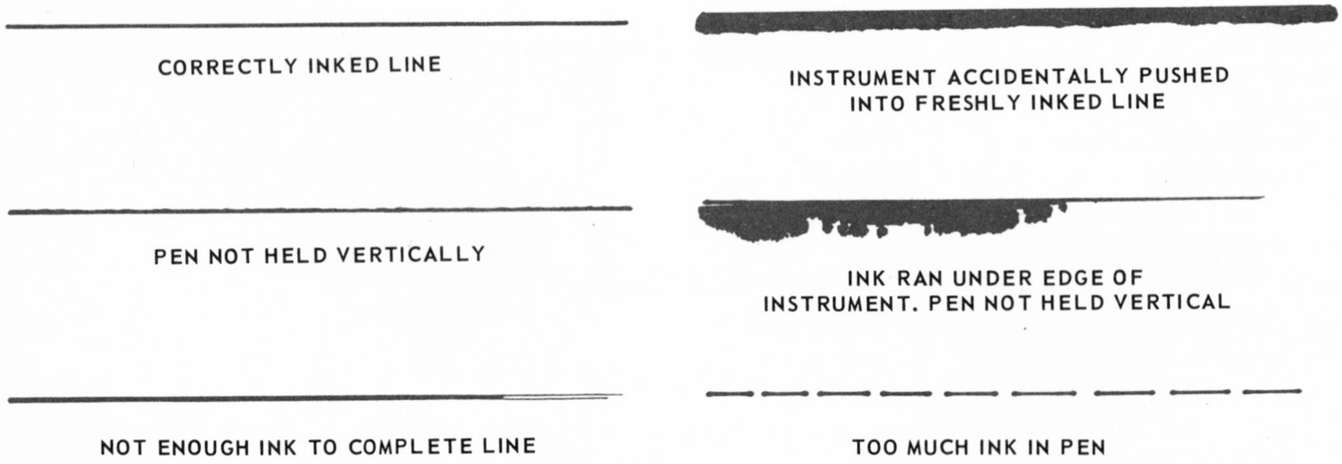

CORRECTLY INKED LINE

INSTRUMENT ACCIDENTALLY PUSHED INTO FRESHLY INKED LINE

PEN NOT HELD VERTICALLY

INK RAN UNDER EDGE OF INSTRUMENT. PEN NOT HELD VERTICAL

NOT ENOUGH INK TO COMPLETE LINE

TOO MUCH INK IN PEN

Fig. 14-7. Common mistakes made in inking.

lines until they are uniform in width. Make sure the ink does not run under the edge of the T-square or triangle.

Most of the difficulties you will encounter when starting to ink are shown in Fig. 14-7. Many of them can be avoided, or their occurance greatly reduced, if you keep the pen clean and fill the pen with the quill on the ink bottle cap, Fig. 14-8, (NEVER dip the pen in the ink). Use great care so that you do not accidentally slide your instruments or hand into freshly inked lines.

When inking a tracing, the experienced draftsman uses a definite order of working. The order preferred by the author is to ink:

1. All circles, arcs and tangents, starting from the smallest to the largest.
2. Hidden circles and arcs.
3. Irregular curved lines.

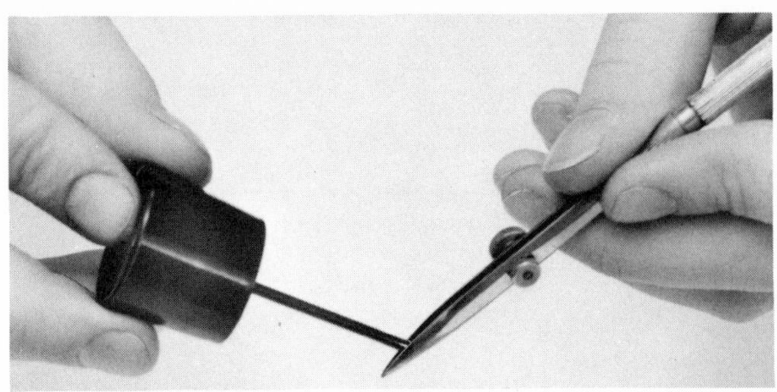

Fig. 14-8. The correct way to fill a ruling pen. NEVER dip the pen in the ink bottle.

4. Straight lines in this order:
   a. Horizontal.
   b. Vertical.
   c. Inclined.
5. Hidden straight lines in the same order listed in 4.
6. All center, extension and dimension lines.
7. Border and section lines.
8. Lettering, dimensions and arrowheads.

TEST YOUR KNOWLEDGE - UNIT 14

1. Why must drawings be duplicated?
2. A blueprint is composed of __________.
3. List the sequence for making blueprints.
4. The diazotype process is often called a __________ print process.
5. How is the diazotype print developed?
6. In microfilming the original drawing is reduced in size __________.
7. Enlarged prints made from microfilm are called __________.
8. List three basic types of tracing media.
9. When making a tracing, it must be remembered that only a _______, _______ line will print well.
10. When making an ink tracing, the ink used is called __________ ink.
11. __________ is the most usual difficulty encountered when inking.

OUTSIDE ACTIVITIES

1. Make a pencil tracing of one of your drawings and duplicate it by one of the duplicating processes - blueprint, moist print, diazotype print, etc.
2. Make an ink tracing of the same drawing and duplicate it by the same reproduction process. Compare the results.
3. Secure an example of each of the reproduction processes described in this unit. Prepare a bulletin board display around them.
4. Obtain an example of a microfilm aperture card and print made from a microfilm master.

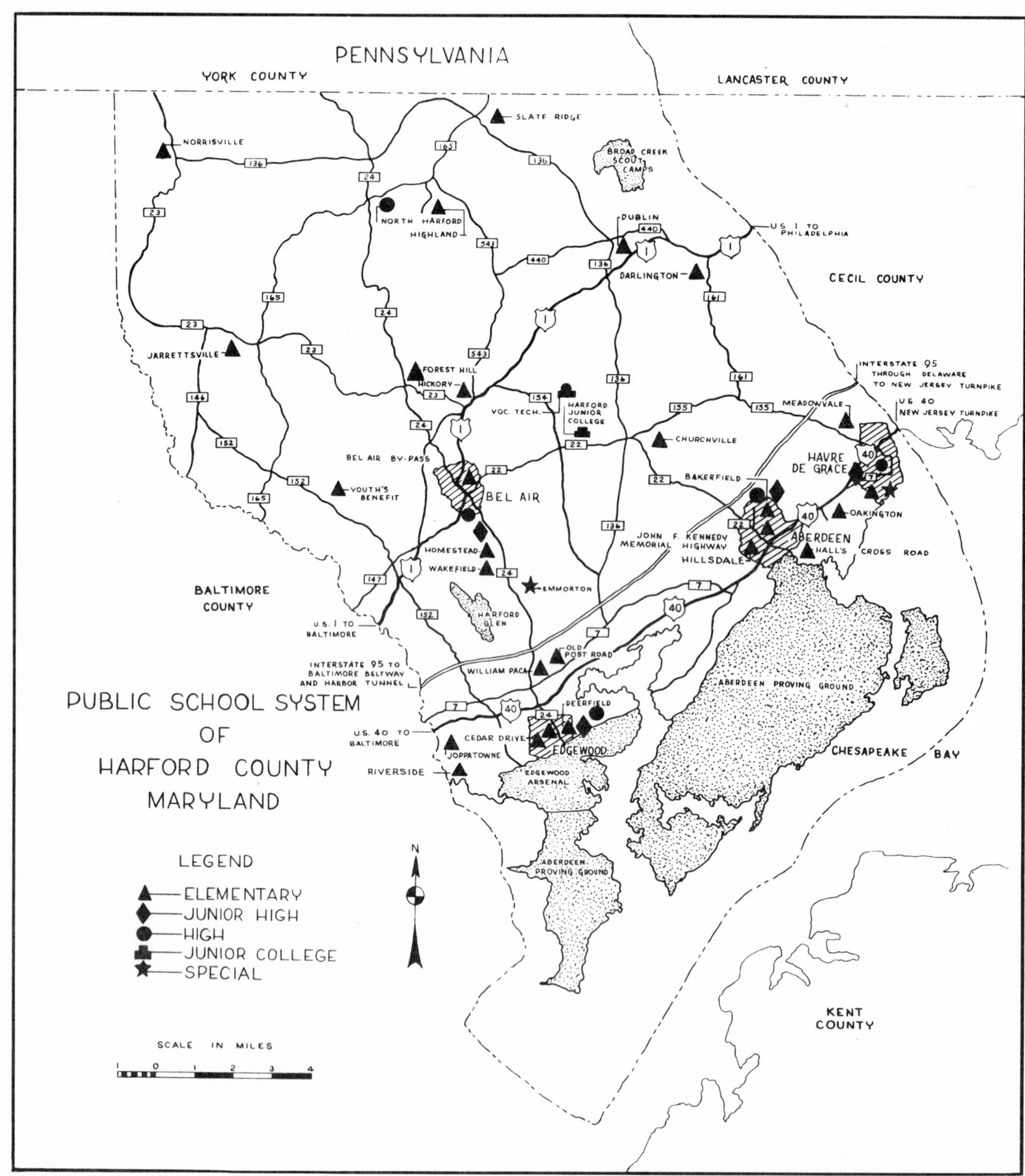

Fig. 15-1. Special purpose map showing the location of schools in Harford County, Maryland (home of the author).

Unit 15
MAPS

What is a map? We often think of a map as being a graphic picture showing a portion of the earth's surface. However, there are many types of maps. Aeronautical charts or maps provide information for fliers. Airports are shown, as well as compass directions, restricted areas and radio station frequencies needed for navigation. Weather maps let us know what to expect weatherwise. An example of a special purpose map is given in Fig. 15-1.

All of us have used a map of some sort. It may have been a road map to plan a vacation trip, or to find the shortest route from one town to another. Or, it may have been in school when studying history or geography.

A professional map maker is called a CARTOGRAPHER. He has the specialized skills required to design and draw maps.

There are times when the draftsman is called upon to draw maps. This unit will describe the maps he would be expected to prepare.

Although aerial photography has greatly reduced the amount of time needed to prepare maps, a great deal of "leg work" is necessary if a map is to be accurate. A SURVEYOR, using an instrument called a TRANSIT, Fig. 15-2, establishes the boundaries of the area being mapped. Distances are measured with a tape and the BEARINGS (angles the lines make with due north and south) are recorded in the form of FIELD NOTES. This information is used to draw a map as in Fig. 15-3. North is toward the top of the sheet and is so indicated.

Maps are always drawn to scale because it is not possible to draw them full size. The size of the land area being mapped will determine the scale. A drawing of a town lot (a PLAT) may be drawn 1 in. = 30 ft. A scale of 1:62500 (approximately 1 in. = a mile) is employed by the United States Geological Survey for mapping large land areas.

Fig. 15-2. Students setting up a transit.

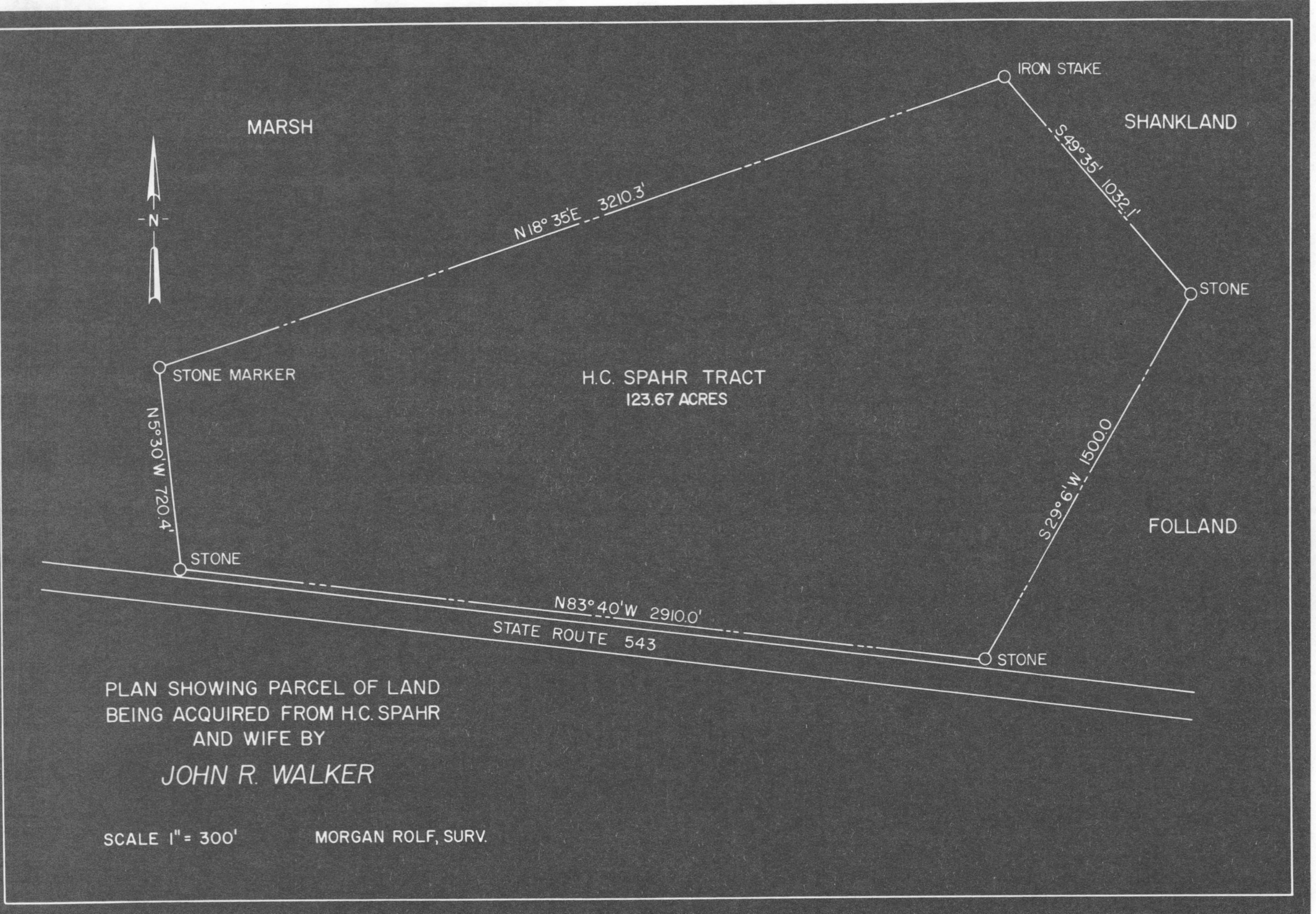

Fig. 15-3. Copy of a typical land survey plan.

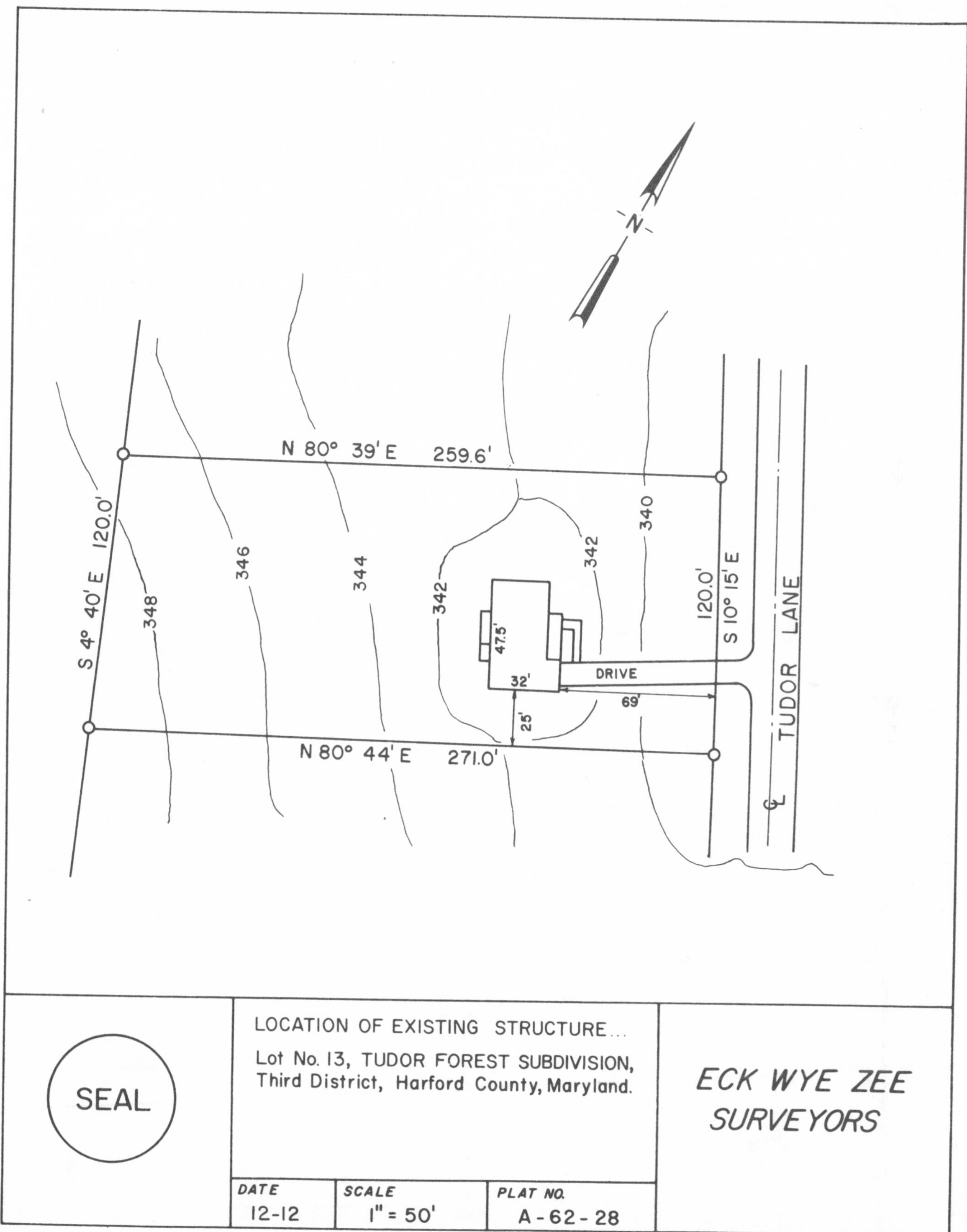

Fig. 15-4. Plat plan.

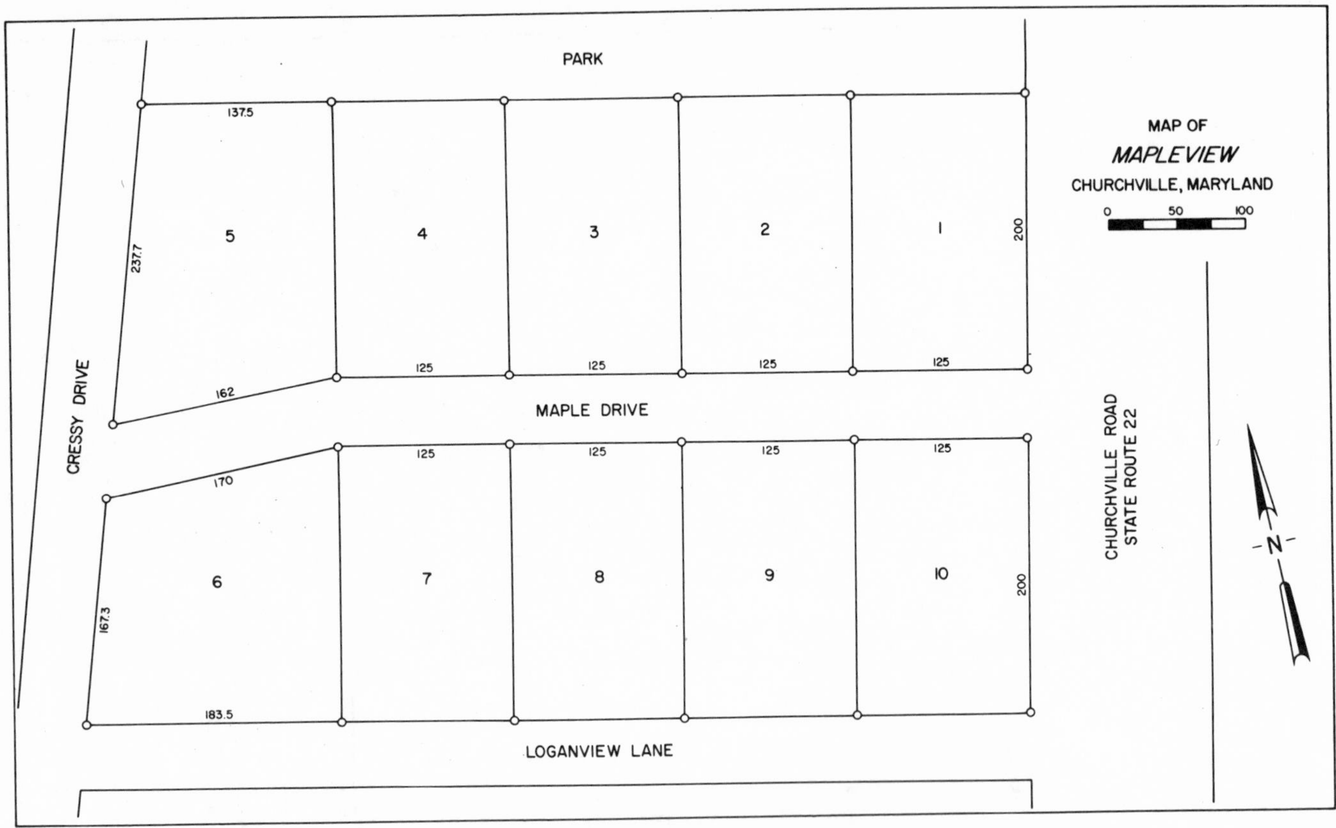

Fig. 15-5. Map of a section of a city showing the layout of streets and lots.

PLAT PLAN

The map of a town lot is called a PLAT or PLOT PLAN, Fig. 15-4. The exact location of the lot and its size is given, as well as how the house or structure is situated on it. The elevation or height of the land above sea level may also be shown as a series of irregular lines (called CONTOUR LINES) drawn on the map at predetermined differences in elevation.

CITY MAP

Street and lot layouts are shown on a CITY MAP, Fig. 15-5. Such maps employ a GRAPHIC SCALE such as those shown in Fig. 15-6, to indicate the scale of the map.

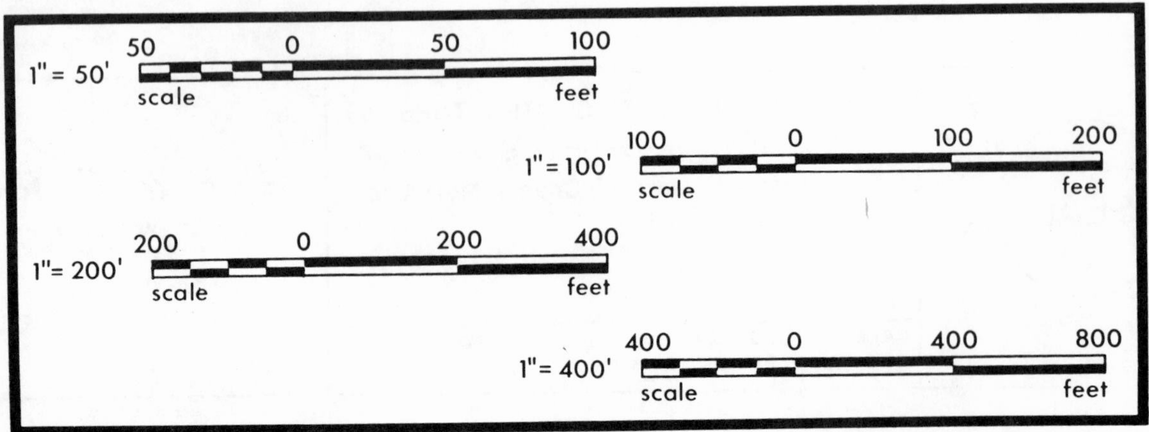

Fig. 15-6. Graphic scales.

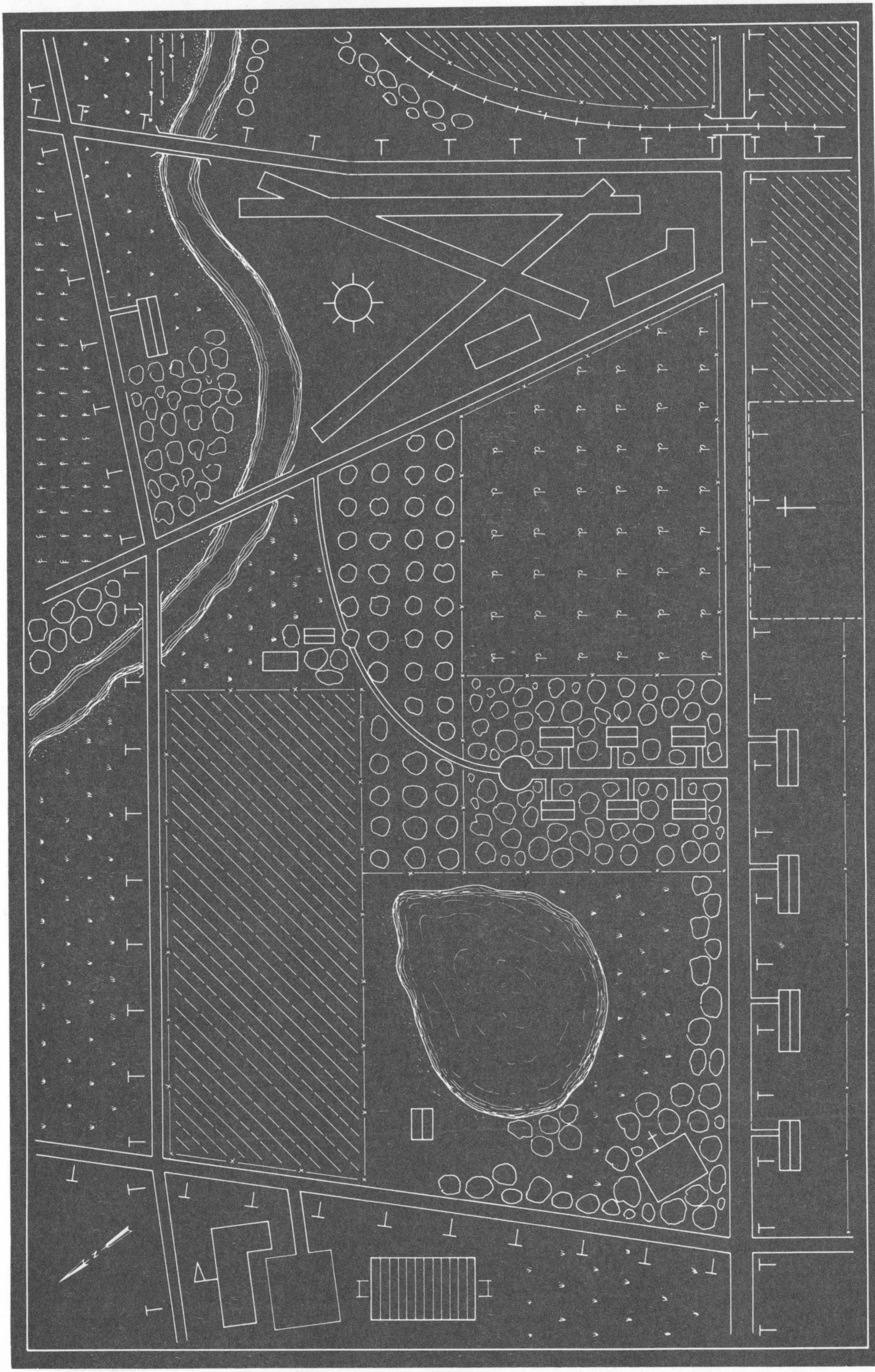

Fig. 15-7. Topographical map.

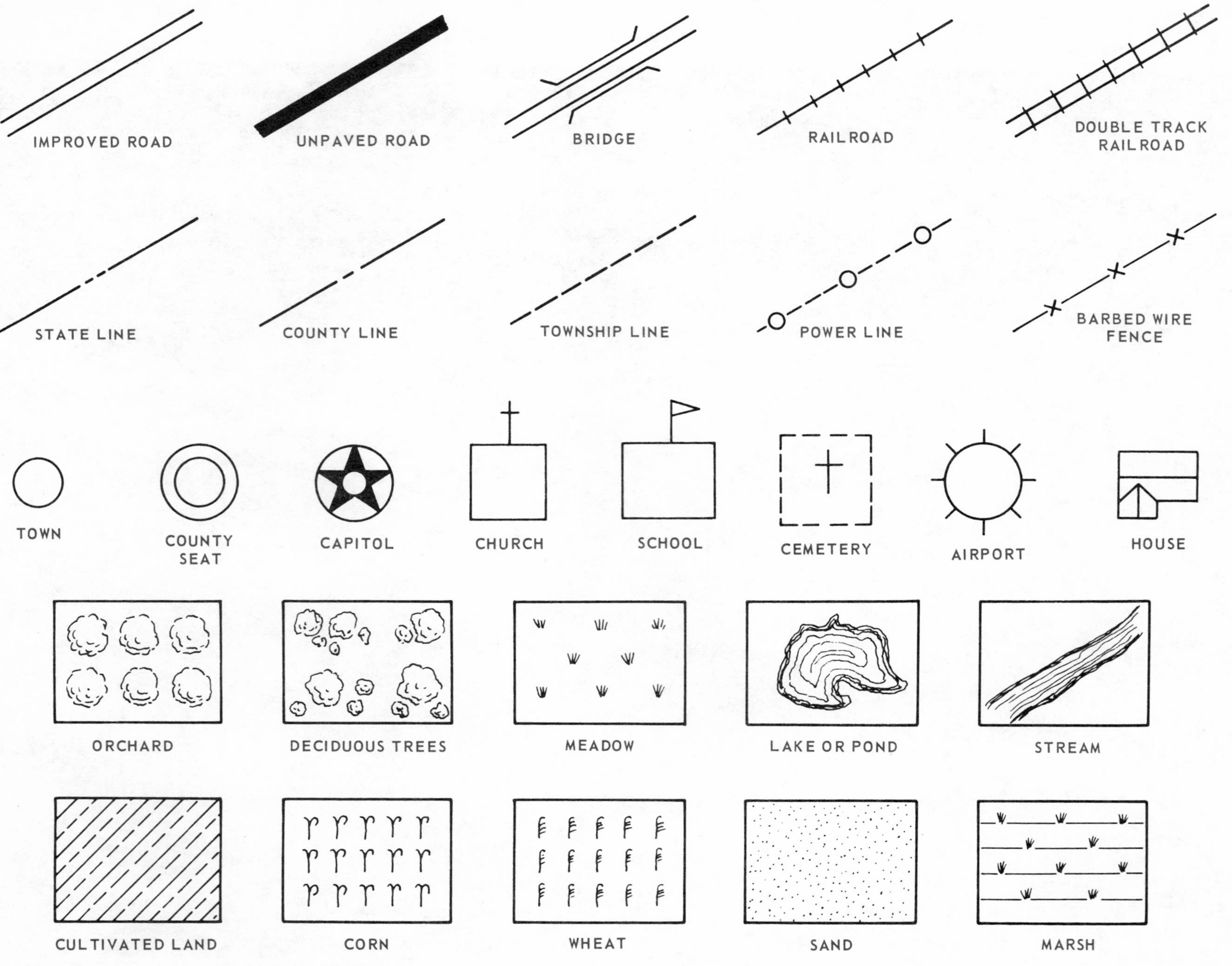

Fig. 15-8. Symbols used on maps.

TOPOGRAPHIC MAP

Information about the physical features of land is included on a TOPOGRAPHICAL MAP, Fig. 15-7. Orchards, streams, forests, airports, harbors, roads, bridges, buildings and other land characteristics can be shown. Quite frequently contour lines are included on a topographic map.

SYMBOLS

By using SYMBOLS, Fig. 15-8, it is possible to include a large amount of information on a map. The symbol is often a stylized drawing (has a resemblance to) of the feature it represents.

TEST YOUR KNOWLEDGE - UNIT 15

1. List some uses you have made of maps.
2. A map is usually thought of as a ________ ________________.
3. A ____________uses an instrument called a transit in preparing maps.
4. Why is it impractical to draw a map full size or half size?
5. A plat plan is a map of ______________.
6. A city map shows ________________.
7. A topographic map shows ____________ ________________.
8. Topographic features are represented on a map by special symbols. Draw the symbols that represent the following:
   a. Church.
   b. House.
   c. School.
   d. Power line.
   e. Barbed wire fence.
   f. Orchards.
   g. Corn.
   h. Wheat.
   i. Sand.

OUTSIDE ACTIVITIES

1. Secure samples of special purpose maps.
2. Prepare a map showing the school grounds.
3. Make a plat plan of the property on which your home is located.
4. Prepare a map of your neighborhood.
5. Draw a map showing the route you take in coming to school.
6. CLASS PROJECT: Draw a map of the surrounding community and have each student draw in the route he travels in coming to school.
7. Secure a road map of your state. Plot the shortest route between your home town and the capitol of your state. If you live in the capitol city, plot the shortest route between it and the next largest city in the state.
8. Prepare a special map that will show the locations of the various schools in your school district.
9. MULTI-GROUP PROJECT: If survey equipment is available, have each group survey the school grounds and compare their results. Compare the results with an actual survey of the grounds.

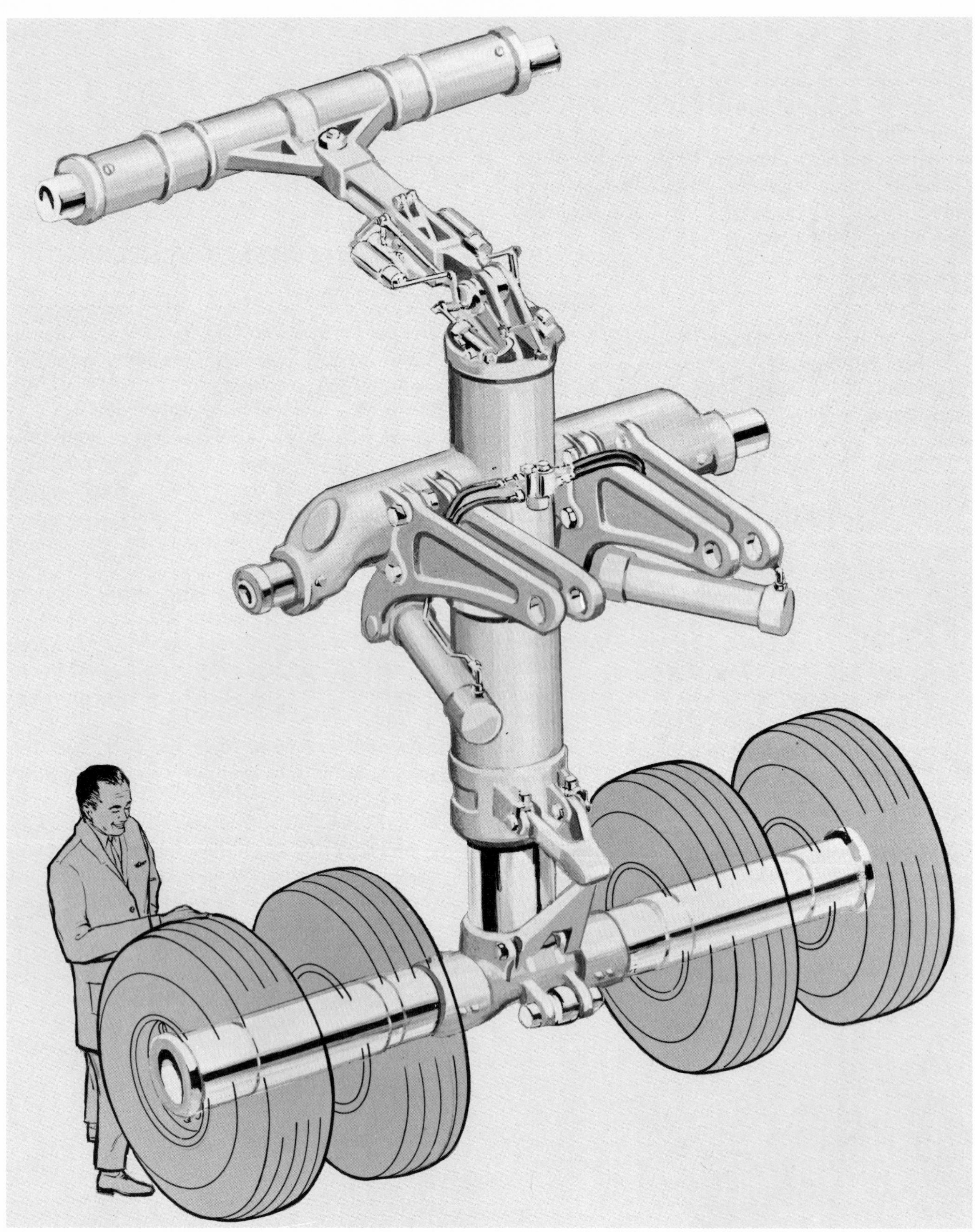

Industry drawing, showing the size of a nose landing gear used on a modern airplane. (Bendix)

Unit 16
CHARTS AND GRAPHS

Industry and education have many uses for charts and graphs. A few of the more widely used kinds are described and shown in this unit.

By using charts and graphs it is possible to show trends, make comparisons, and measure progress quickly without studying a mass of statistics.

A draftsman is often asked to prepare charts and graphs. He takes the statistical information (figures) and decides upon the most effective way to present this material in an interesting and easily understood manner.

GRAPHS

Graphs frequently used are:

1. Line Graph.
2. Bar Graph.
3. Circle or Pie Graph (sometimes called an Area Graph).

LINE GRAPH

The LINE GRAPH may be used to make comparisons, Fig. 16-1. Lines that present the information are called CURVES. When only one curve appears on the graph, it should be drawn as a solid line.

Fig. 16-1. A ***LINE GRAPH*** *that compares the times needed by three different cars to reach 60 mph.*

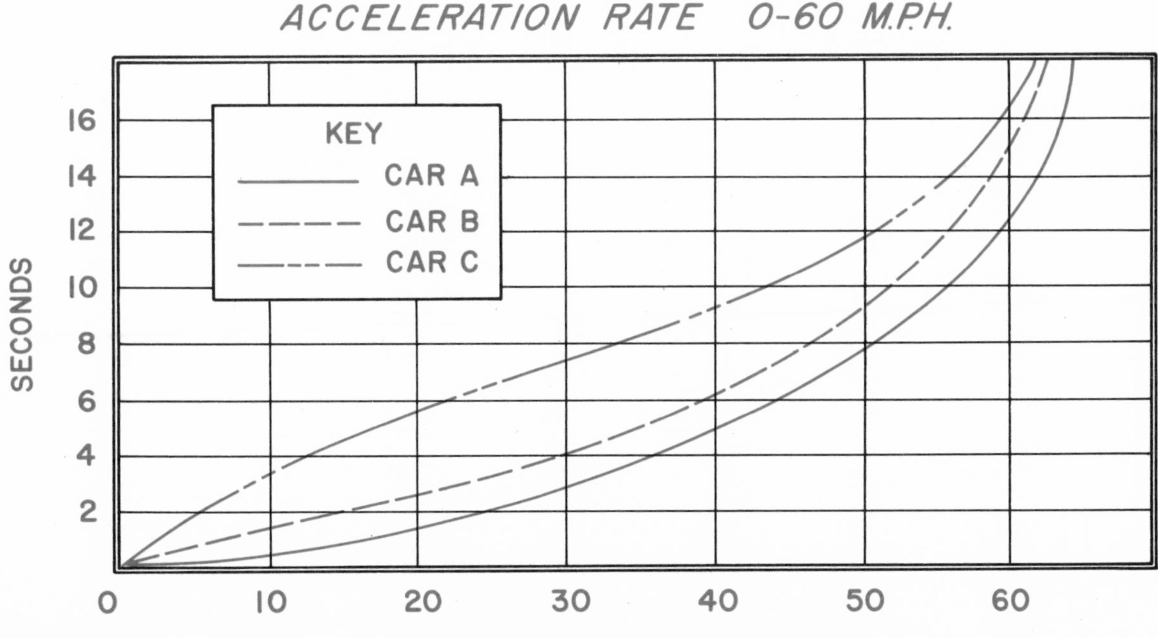

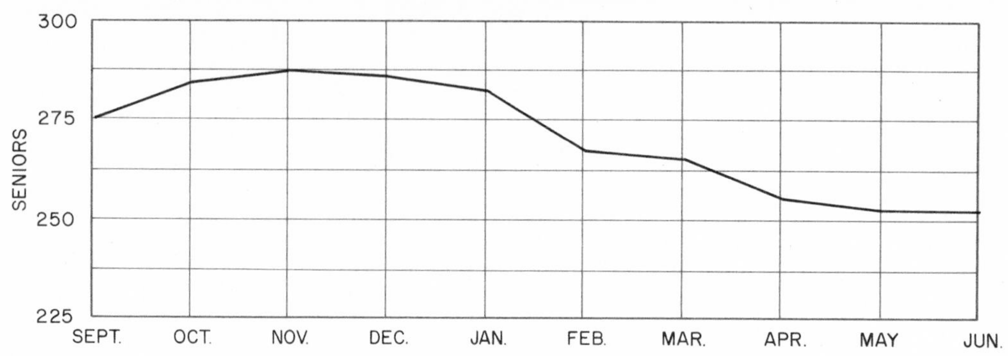

Fig. 16-2. ***LINE GRAPH*** *showing trends.*

When more than one curve is employed, each line should be clearly labeled and a KEY should be included with the graph to show what each curve represents.

A line graph may be used to show trends, that is, what has happened or what may happen. A line graph is illustrated in Fig. 16-2.

BAR GRAPHS

Comparisons between quantities or conditions can also be made using BAR GRAPHS.

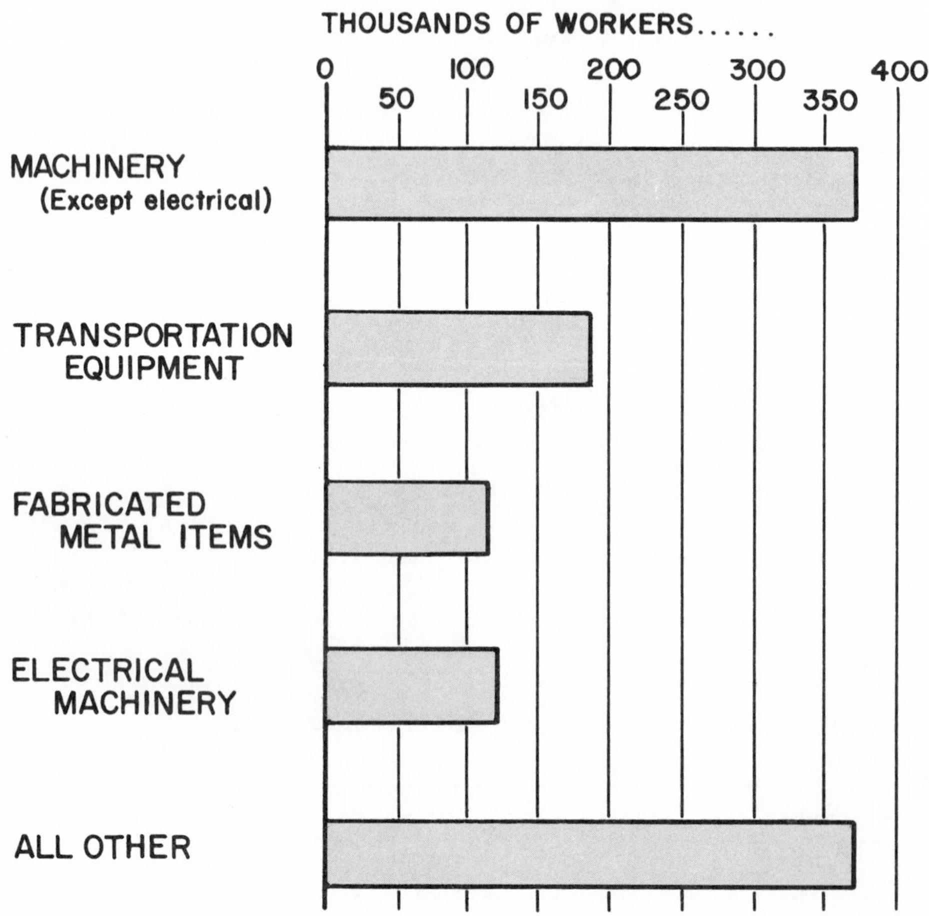

Fig. 16-3. ***A HORIZONTAL BAR GRAPH*** *that shows where machinists are employed.*

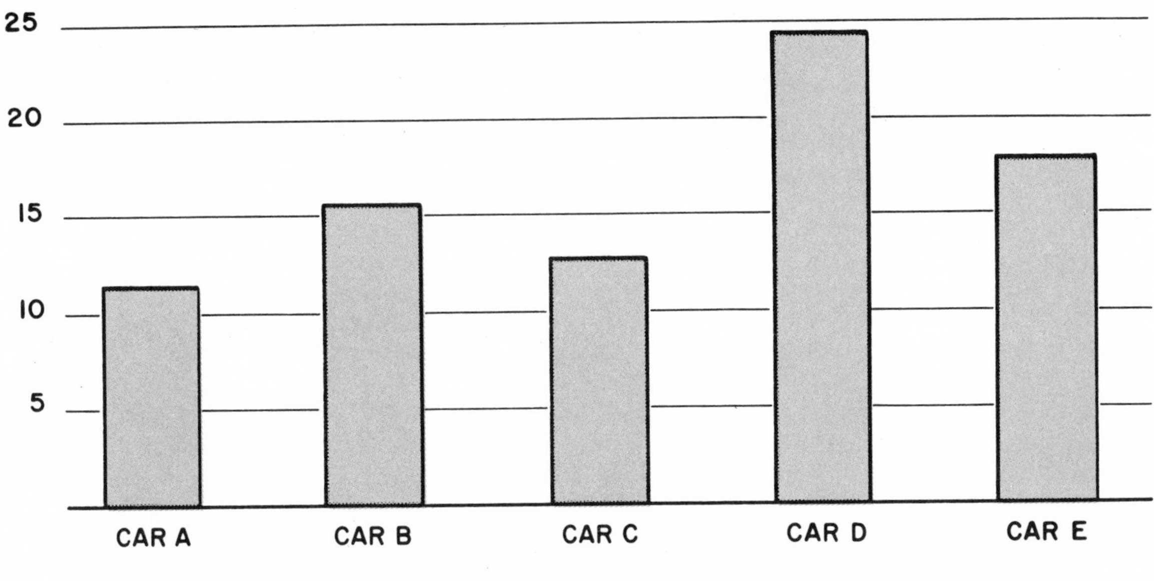

Fig. 16-4. Car economy is shown in this VERTICAL BAR GRAPH.

Several different forms of the bar graph are available to the graph maker.

The HORIZONTAL BAR GRAPH, see Fig. 16-3, presents information on a horizontal plane.

With the VERTICAL BAR GRAPH, Fig. 16-4, information is given in a vertical or upright position.

The COMPOSITE BAR GRAPH, Fig. 16-5, can be drawn in either a vertical or hori-

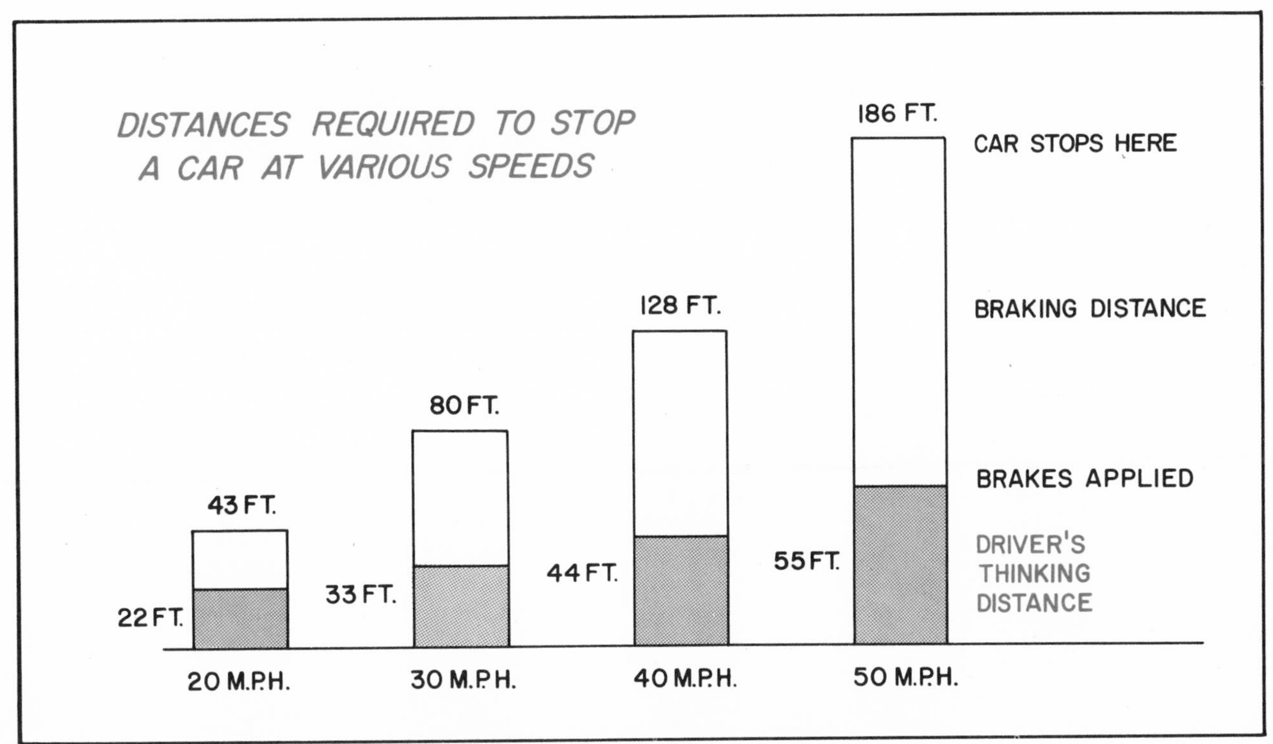

Fig. 16-5. COMPOSITE BAR GRAPH.

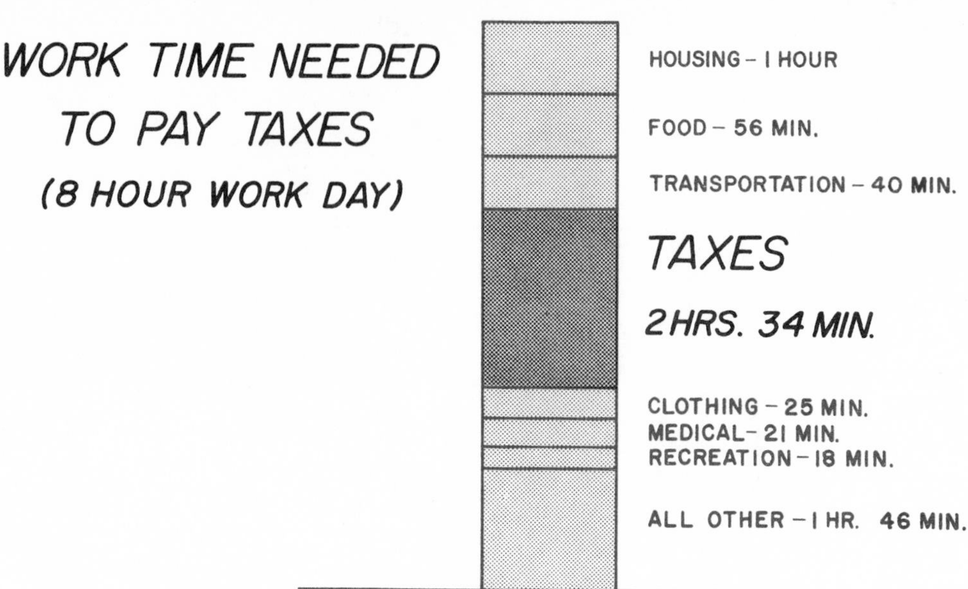

Fig. 16-6. A ***100 PERCENT BAR GRAPH*** *which indicates amount of time one person worked to pay his taxes.*

zontal position. This compares several items of information in the same graph.

A 100 PERCENT BAR GRAPH, Fig. 16-6, consists of a singular rectangular bar. In this type graph information is presented on a percentage basis.

The PICTORIAL BAR GRAPH, Fig. 16-7, is a variation of the bar graph. It uses pictures to represent the information instead of bars. Pictures tend to make the graph more interesting.

CIRCLE OR PIE GRAPH

The CIRCLE or PIE GRAPH is shown in Fig. 16-8. This is composed of a segmented circle and shows the entire unit divided into comparable parts.

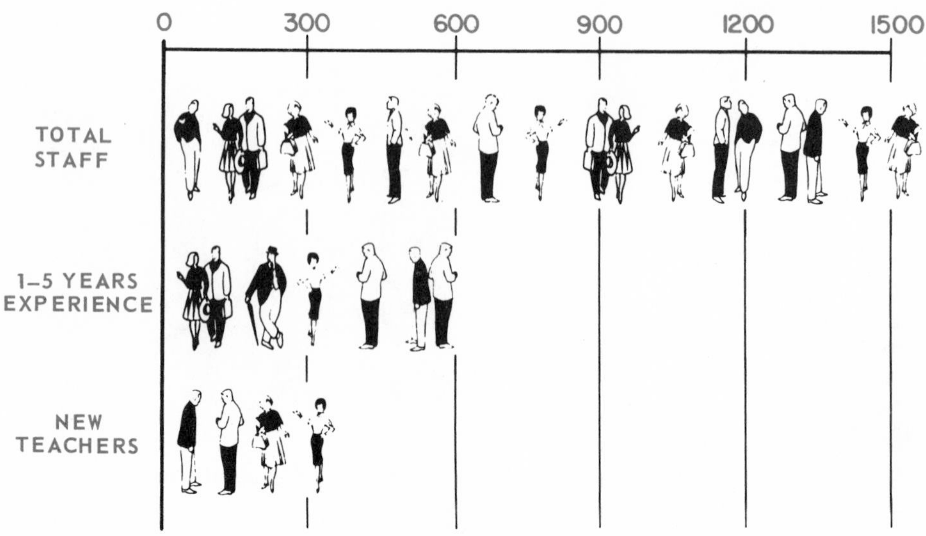

Fig. 16-7. A ***PICTORIAL BAR GRAPH*** *showing experience of a teaching staff in Harford County, Maryland.*

The circle graph is also known as an AREA GRAPH.

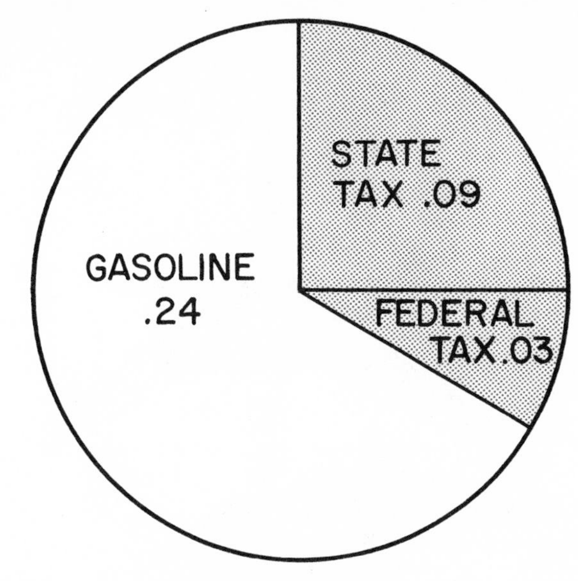

*Fig. 16-8. The **CIRCLE** or **PIE GRAPH** is sometimes called an **AREA GRAPH**. This one shows that one-third the cost of a gallon of gasoline is state and federal taxes.*

CHARTS

CHARTS are another means employed to convey information rapidly. The most familiar of these is the ORGANIZATION CHART, Fig. 16-9. This shows the order of responsibility and the relation of persons and/or positions in an organization.

FLOW CHARTS, Fig. 16-10, may be used to show the sequence, or order of operations of how a product is manufactured and/or distributed.

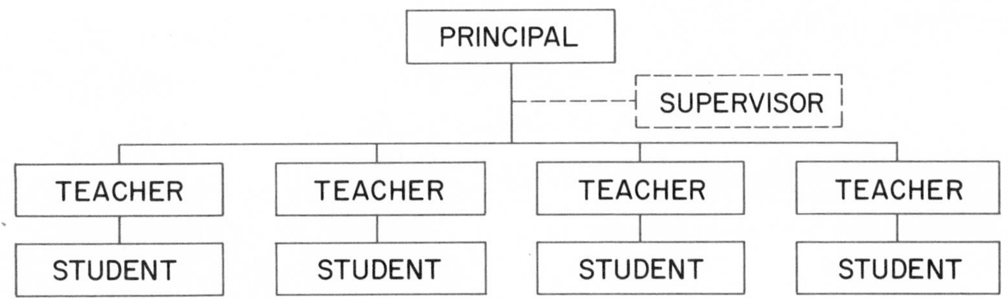

*Fig. 16-9. Above. An **ORGANIZATION CHART**. Fig. 16-10. Below. This **FLOW CHART** illustrates the sequence a spacecraft used to dock with the lunar module.*

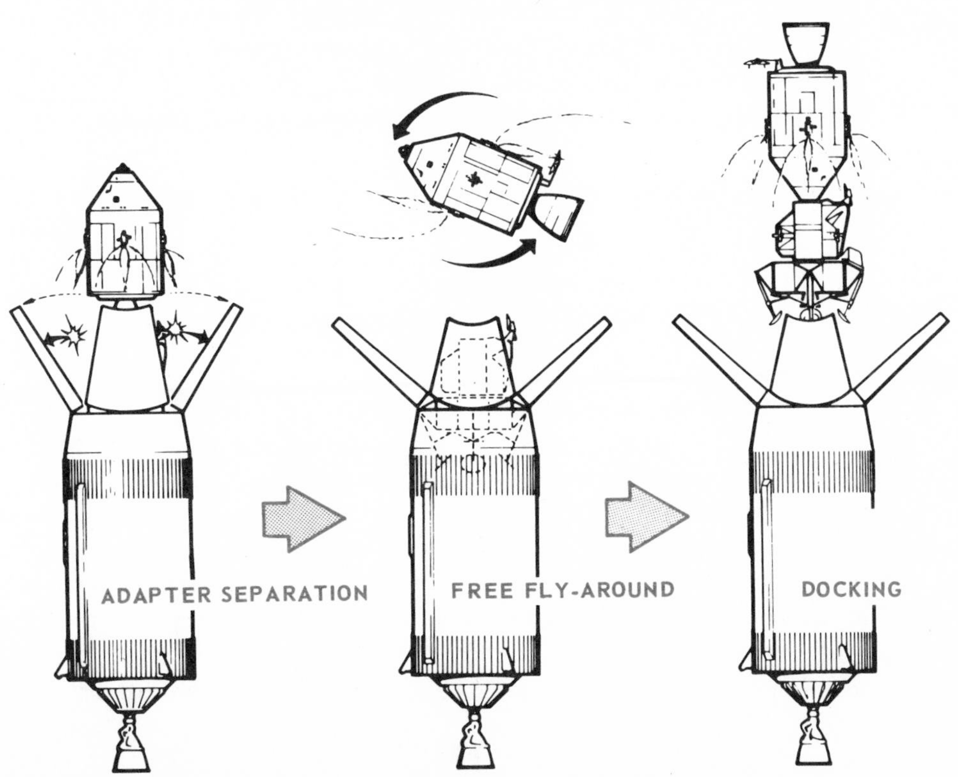

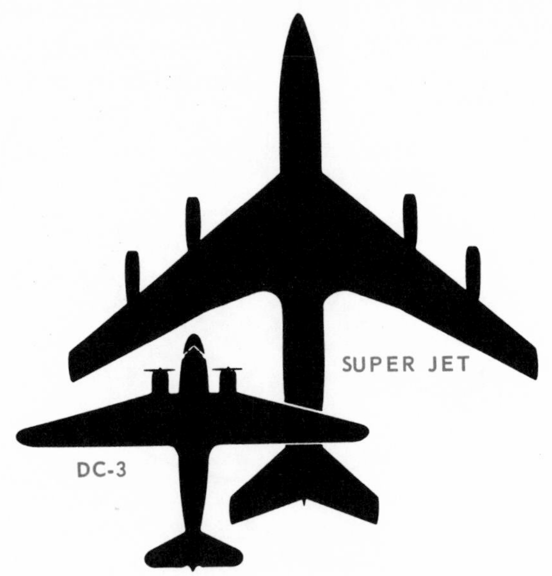

Fig. 16-11. PICTORIAL CHART comparing the commercial aircraft used in the early 1940's and the commercial jet of today. (Trans World Airlines)

The PICTORIAL CHART, Fig. 16-11, is an ideal way of presenting comparisons in an interesting and easily understood manner.

PREPARING CHARTS AND GRAPHS

There are many variations of the graphs presented in this unit. It is up to the ingenuity of the person preparing the material to decide which type will do the best job.

Do not start a graph or chart until ALL of the information has been gathered. Then a rough drawing should be prepared to determine what size presentation will be best.

Use color whenever possible to brighten up your presentation and to make the information stand out.

TEST YOUR KNOWLEDGE - UNIT 16

1. Why are charts and graphs used?
2. List the three most commonly used kinds of graphs.
   a. ______________.
   b. ______________.
   c. ______________.
3. The line that presents the information on a ________________ graph is called ________________.
4. Of what use is the KEY that should be included when more than one line is used on a graph?
5. Briefly describe each of the following graphs:
   Horizontal Bar Graph.
   Vertical Bar Graph.
   Composite Bar Graph.
   100 Percent Bar Graph.
   Pictorial Bar Graph.
6. The __________ or __________ graph is also known as an Area Graph.
7. Briefly describe each of the following charts:
   Flow Chart.
   Organization Chart.
   Pictorial Chart.
8. Why should color be used on a chart or graph?

OUTSIDE ACTIVITIES

1. Develop a pie graph of how you spend your allowance.
2. Prepare a bar graph to show the approximate increase in horsepower in a particular make of automobile from 1940 to the present time. Use five year steps.
3. Prepare a line graph showing how the price of the automobile used in 2 has risen in the same periods.
4. Make a 100 percent bar graph showing a breakdown of the cost of a gallon of gasoline - including cost of the gasoline, state, federal and local taxes.
5. Draw a pictorial bar graph showing how far an automobile will travel after the brakes are applied at 25 mph, 45 mph, 60 mph, and 75 mph.
6. Make an organization chart of the pupil

personnel system used in your drafting room or Industrial Arts shops.

7. Develop a picture chart showing how the size of a particular make of automobile has increased in the past 20 years.
8. Make a picture graph showing the enrollment in each grade of your school. Let each symbol represent 25 students.
9. Design a flow chart showing how a simple stool could be manufactured in the Industrial Arts shop.
10. Make a line graph showing the enrollment of the Industrial Arts classes in your school.

Unit 17
MANUFACTURING PROCESSES

Metals, and other materials are available in a variety of shapes and sizes. See Fig. 17-1. Since the purpose of most drawings is to describe a part or product that is to be manufactured, it is important that the draftsman have an understanding of how the materials can be cut, shaped, formed and fabricated. He should also know which material is best suited for a particular application or product.

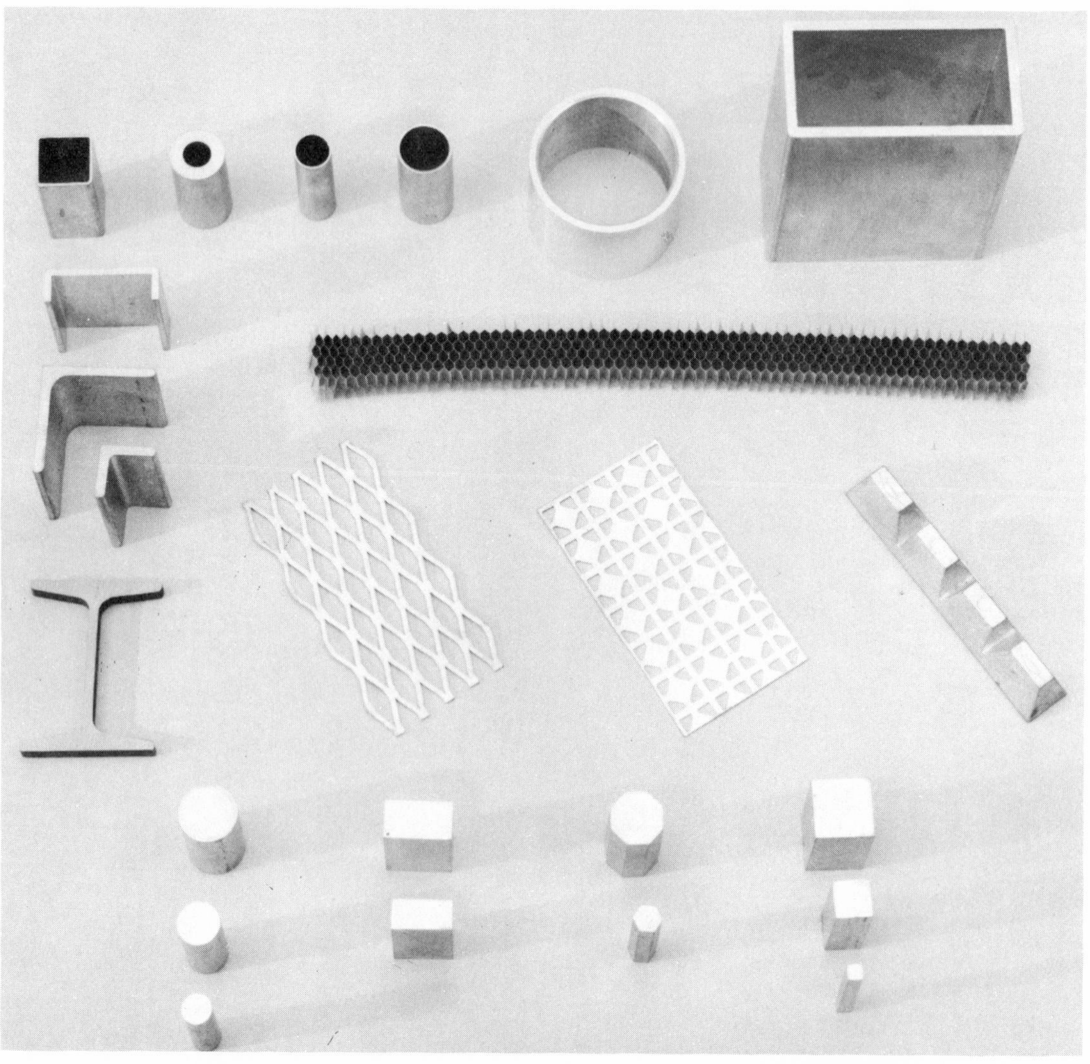

Fig. 17-1. Many kinds and shapes of metal are available.

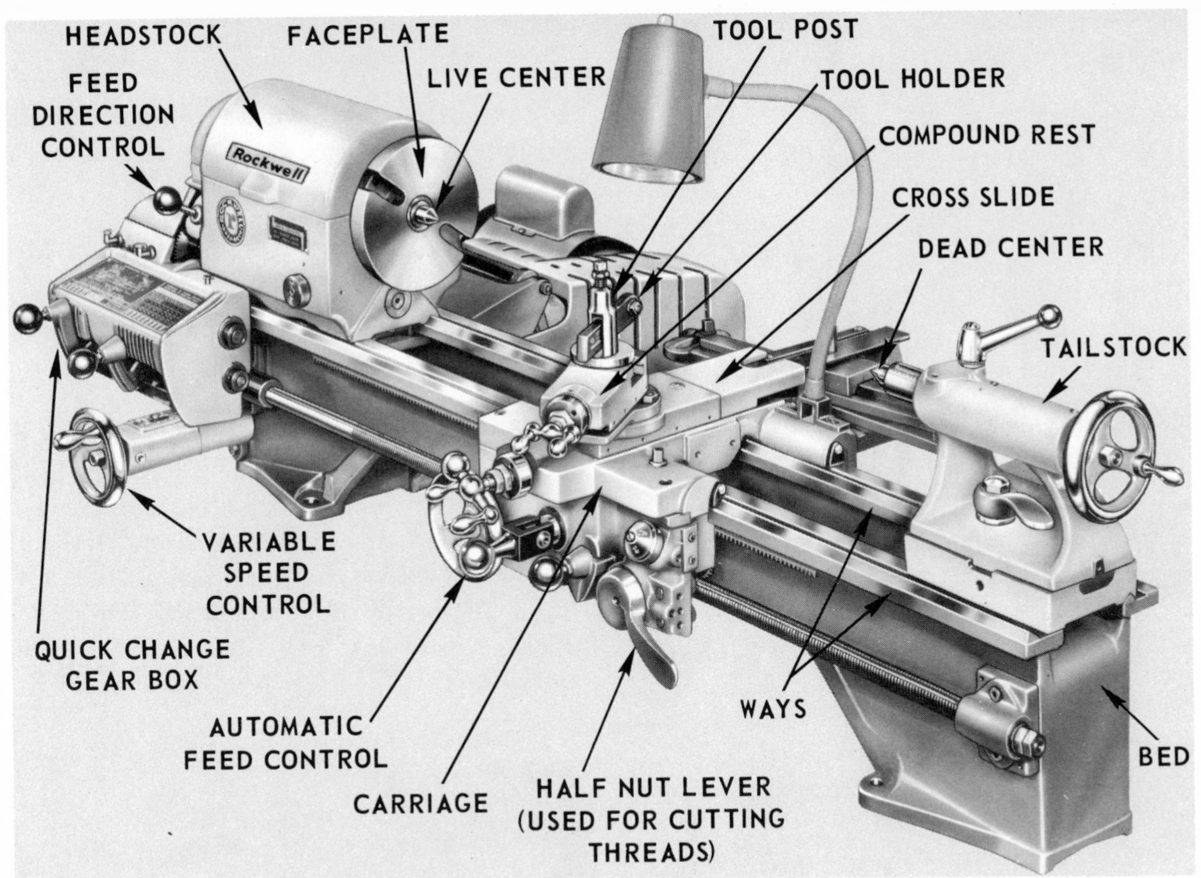

Fig. 17-2. Metal cutting lathe. (Rockwell Mfg. Co.)

MACHINE TOOLS

A machine tool is a power driven tool, not hand portable, used to shape or form metal by cutting, impact, pressure, electrical techniques, or a combination of these processes.[1]

The world of today could not exist without MACHINE TOOLS. They produce the accurate and uniform parts needed for the assembly line and make mass-production techniques possible.

Machine tools are manufactured in a large range of styles and sizes. Only basic tools and manufacturing techniques will be covered in this unit.

LATHE

The LATHE, Fig. 17-2, is one of the oldest and most important of the machine tools. It operates on the principle of the work being rotated against the edge of the cutting tool, Fig. 17-3. The cutting tool can

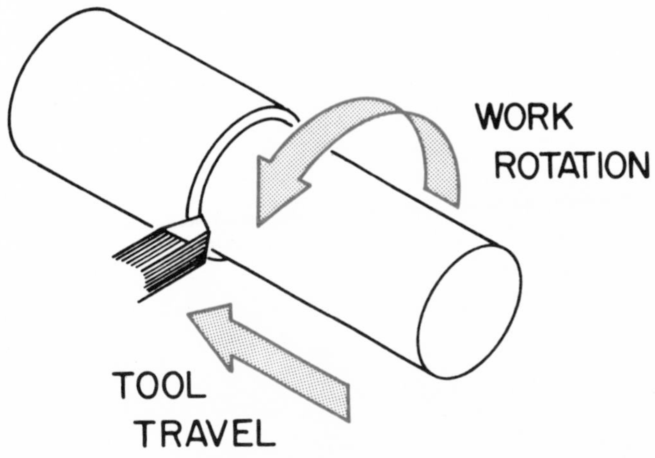

Fig. 17-3. The operating principle of the lathe.

[1] National Machine Tool Builder's Association
2139 Wisconsin Avenue, N.W.
Washington, D. C. 20007

Fig. 17-4. Boring (machining an internal surface). (Clausing)

be controlled and can be moved lengthwise and across the face of the material being machined (turned).

Operations other than turning can be performed on the lathe. It is possible to drill, bore, Fig. 17-4, ream, and cut threads and tapers.

There are many variations of the basic lathe. The TURRET LATHE, Fig. 17-5, is used when a number of identical parts must be machined. It is a conventional lathe fitted with a six-sided tool holder called a TURRET. Different cutting tools fitted in the turret rotate into position for the machining operations.

Fig. 17-5. Turret lathe. (Clausing)

Fig. 17-6. Partial assembly of a plate and gear-train of a wrist watch being held by tweezers. The gear shafts and pins were made on a lathe. (Bulova Watch)

Other lathes range in size from the small lathe used by the instrument and watchmaker, Fig. 17-6, to the large lathes that machine the forming rolls for steel mills, Fig. 17-7.

DRILL PRESS

The DRILL PRESS, Fig. 17-8, is probably the best known of the machine tools. On it, a cutting tool called a TWIST DRILL is rotated against the work with sufficient pressure to cut its way through the material, Fig. 17-9.

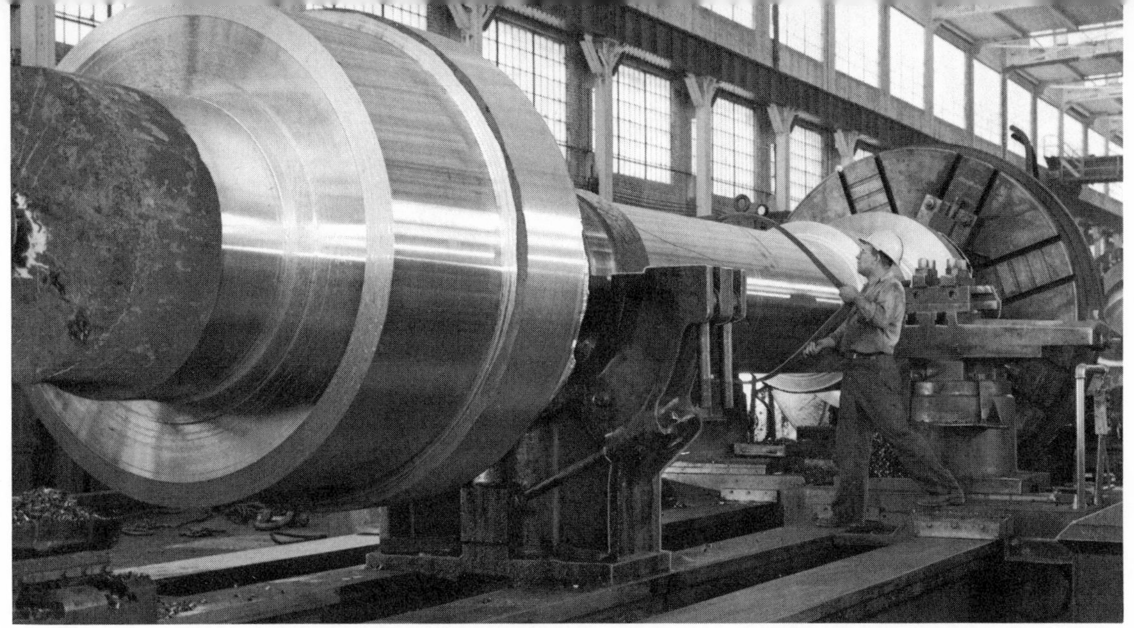

Fig. 17-7. Lathe used to machine forming rolls for a steel mill.

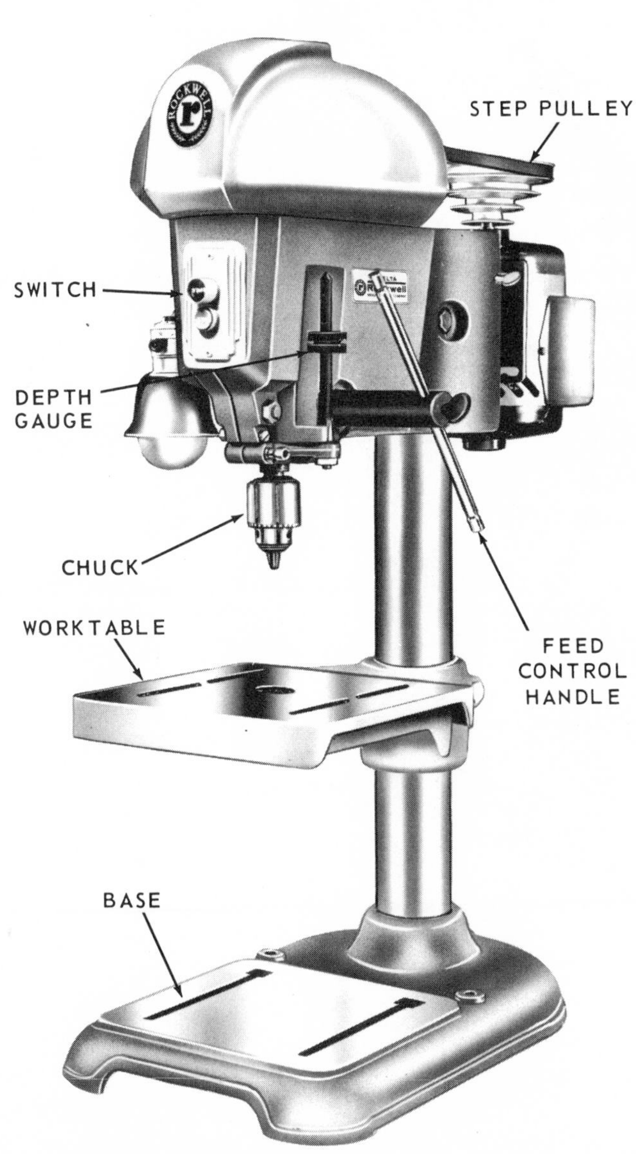

Fig. 17-8. A bench model drill press.

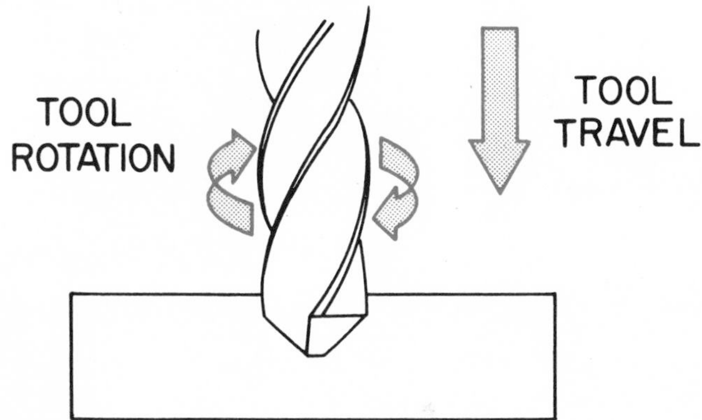

Fig. 17-9. The operating principle of the drill press.

Other operations which can be performed on the drill press include: REAMING, Fig. 17-10, finishing a drilled hole to close

Fig. 17-10. Reaming produces a more accurate hole than drilling. The hole must be drilled slightly smaller than the required finished hole size.

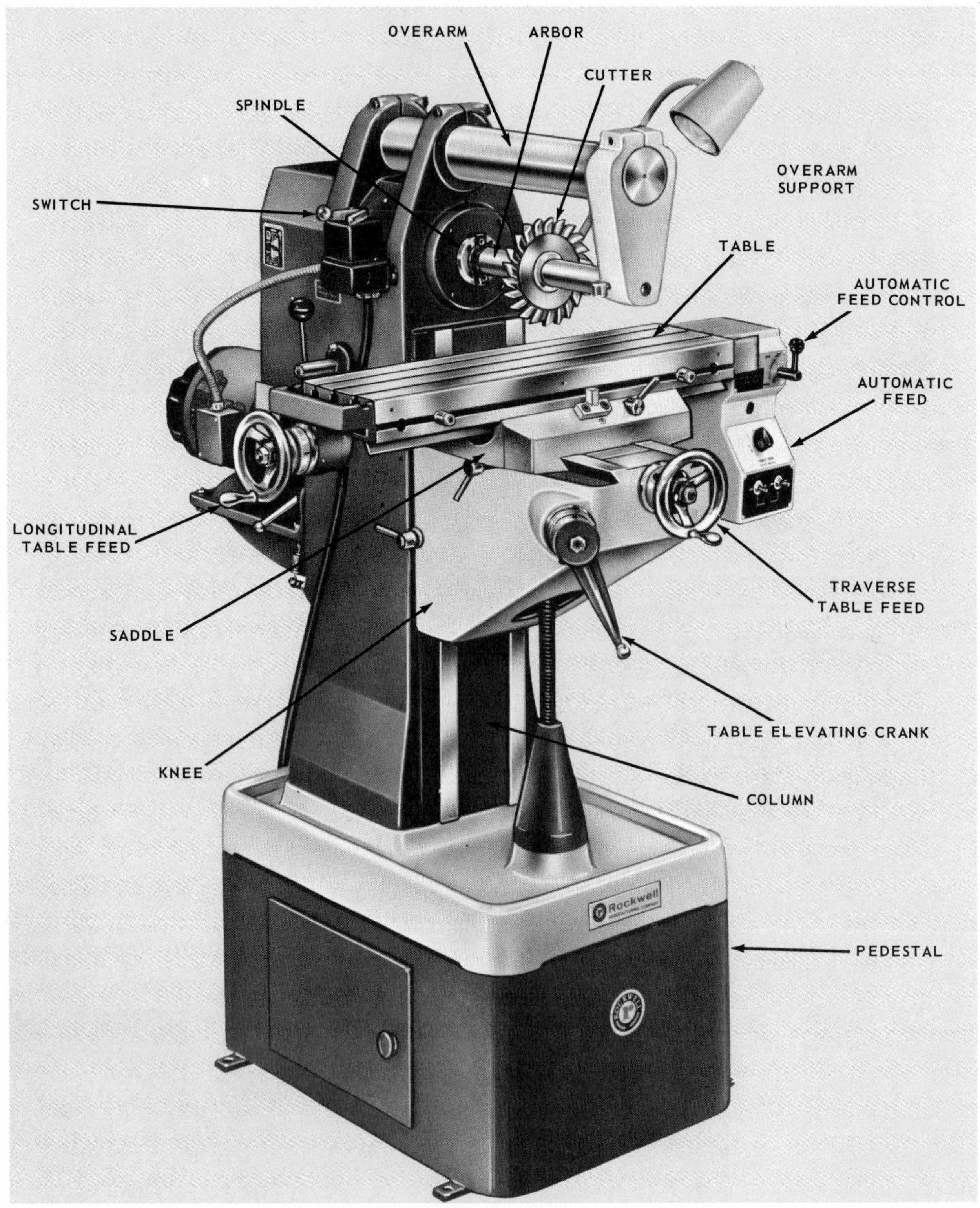

Fig. 17-11. Horizontal milling machine. (Rockwell)

tolerances; COUNTERSINKING, cutting a chamfer on a hole so a flat head fastener can be inserted; TAPPING, the process of cutting internal threads.

MILLING MACHINE

The MILLING MACHINE, Fig. 17-11, is quite a versatile machine. It can be used to

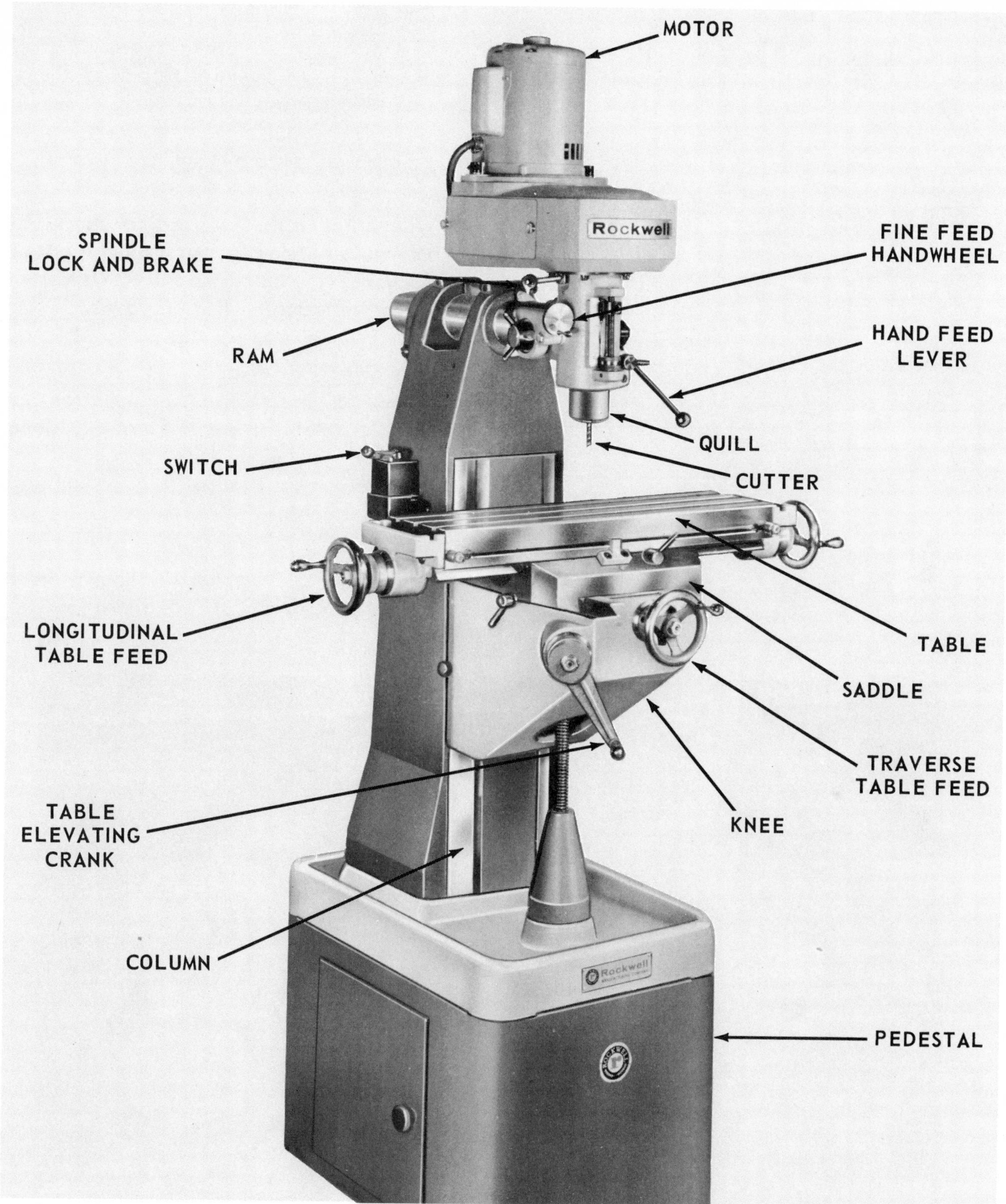

Fig. 17-12. The vertical milling machine. (Rockwell Mfg. Co.)

machine flat and irregularly shaped surfaces, drill, bore and cut gears.

The VERTICAL MILLING MACHINE, Fig. 17-12, differs from the HORIZONTAL MILLING MACHINE shown in Fig. 17-11, in that the cutter is mounted vertical to the worktable.

Metal is removed by means of a rotating cutter that is fed into the moving work, Fig. 17-13.

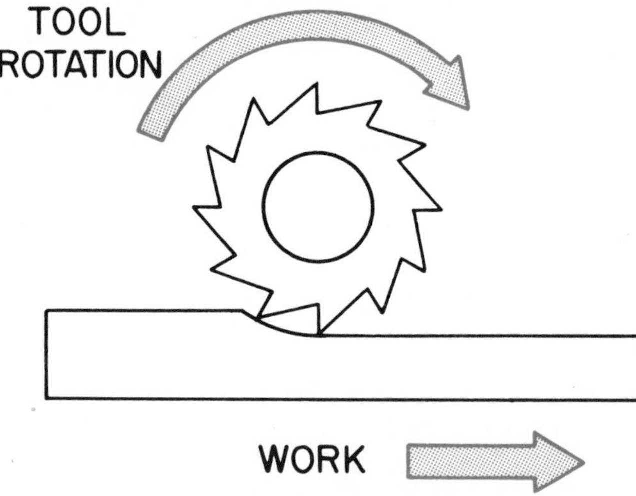

Fig. 17-13. The operating principle of the milling machine.

There are also other types of milling machines. They vary in size and by the types of automatic controls.

SHAPER, PLANER AND BROACH

Where the lathe is used to shape metal to a circular shape, planing machine tools - the SHAPER, PLANER and BROACH, are employed to machine flat surfaces.

The SHAPER, Fig. 17-14, is considered too slow for most mass-production operations. Much of the work formerly done on the shaper is being done faster and more economically on the milling machine and broaching machine.

The SHAPER works as shown in Fig. 17-15. The PLANER differs from the shaper

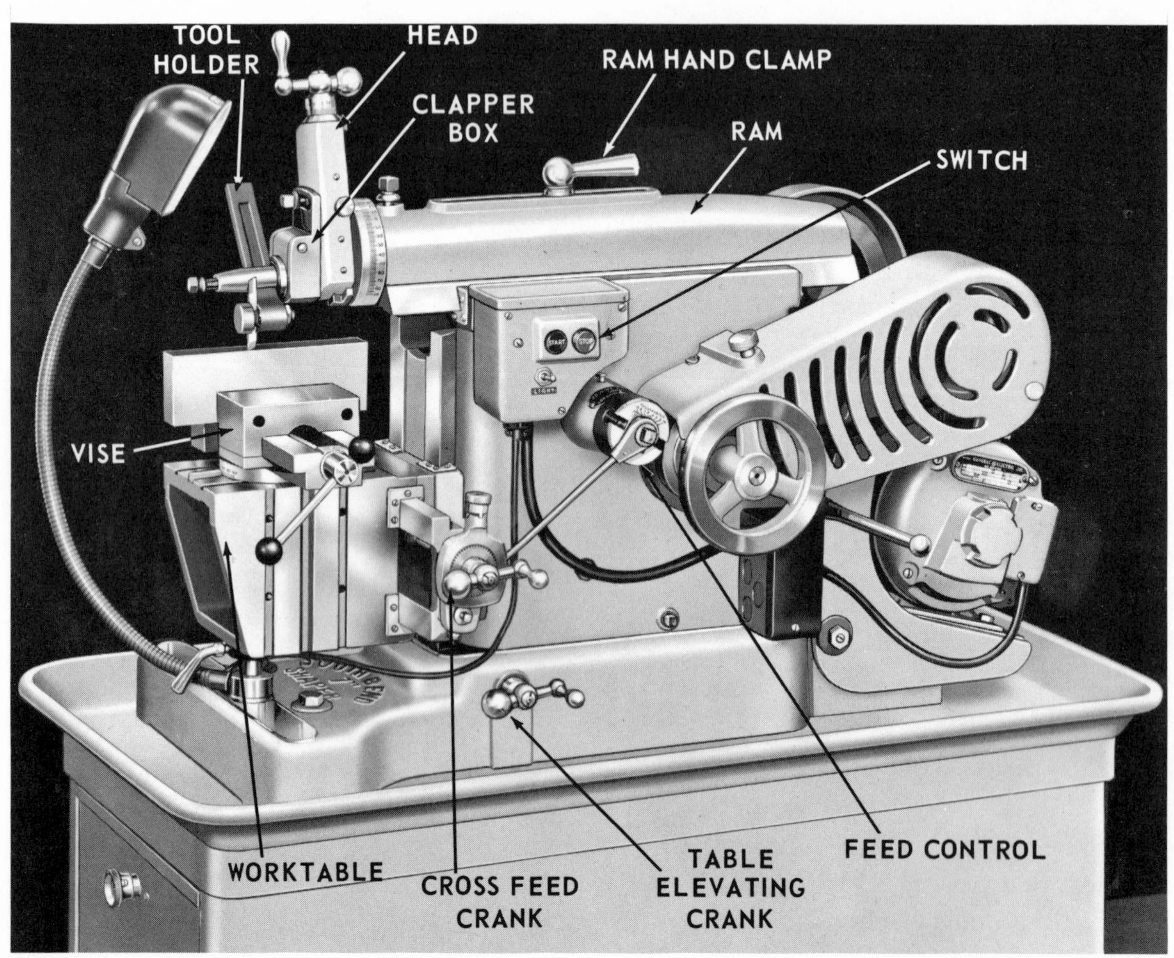

Fig. 17-14. The shaper. (South Bend Lathe)

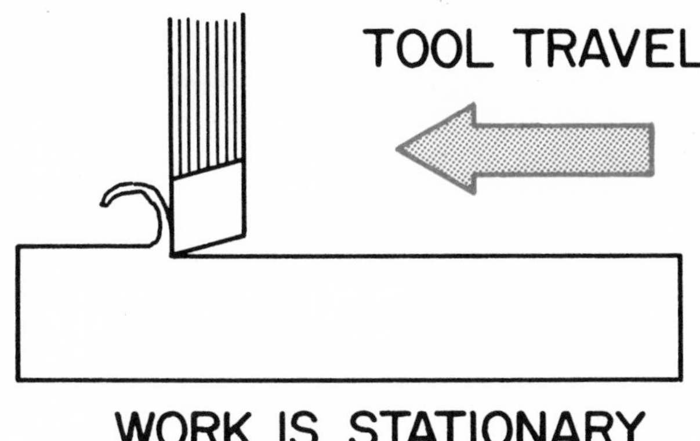

Fig. 17-15. The operating principle of the shaper.

in that the work travels back and forth while the cutter remains stationary, Fig. 17-16.

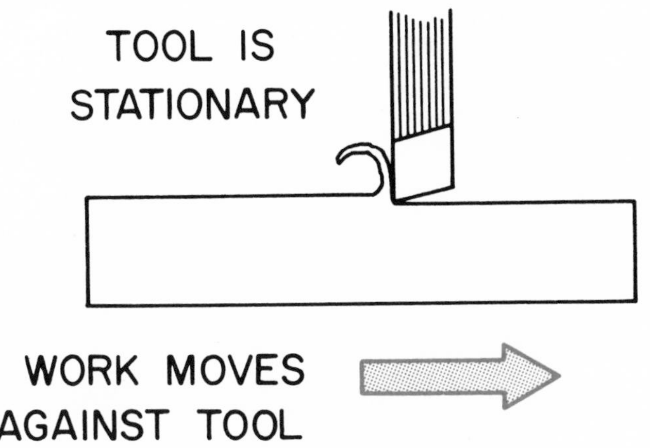

Fig. 17-16. The operating principle of the planer.

BROACHING, Fig. 17-17, is similar to shaping, but instead of a single cutting tool advancing after each stroke across the work, the broach is a long tool with many cutting teeth. Each tooth has a cutting edge a few thousandths of an inch higher than the one before and increases in size to the exact finished size required. The broach is pushed or pulled across the work.

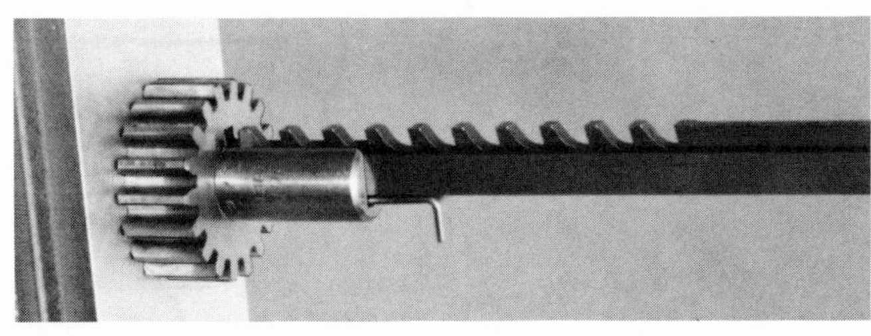

Fig. 17-17. A broach being used to cut a keyway in a gear. Note the number of teeth on the tool.

The broach is ideal for producing keyways and irregular shaped openings.

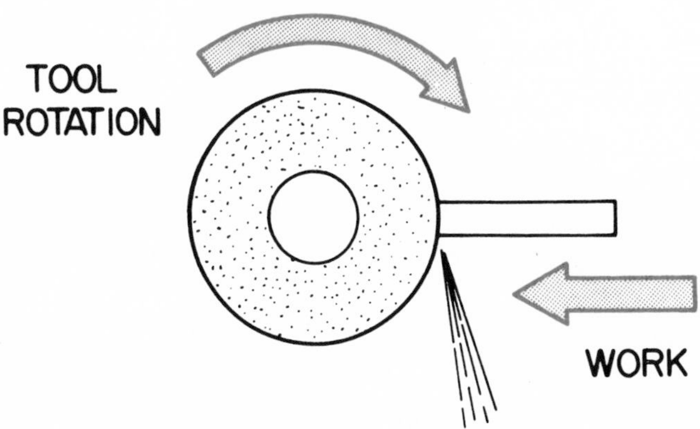

Fig. 17-18. The operating principle of the grinder.

GRINDING

GRINDING, Fig. 17-18, is an operation that removes material by rotating an abrasive wheel against the work. The PEDESTAL GRINDER, Fig. 17-19, is the simplest and most widely used grinding machine.

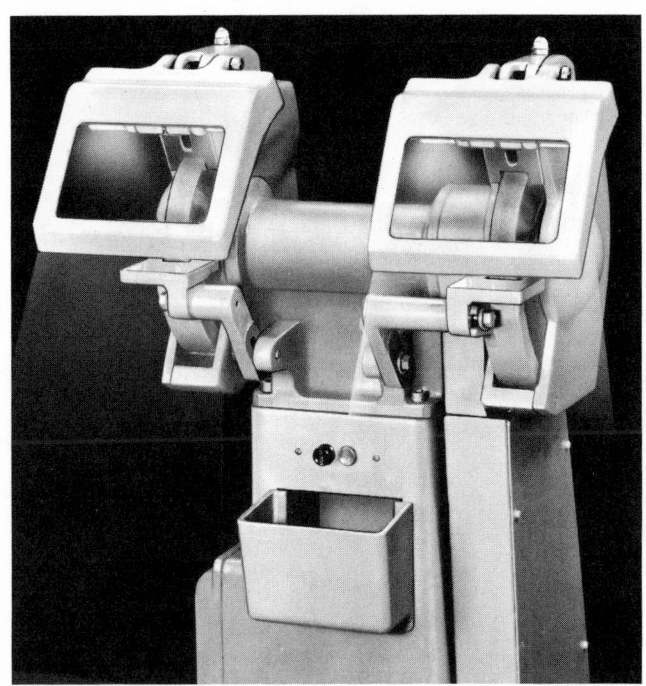

Fig. 17-19. Pedestal grinder. (South Bend Lathe)

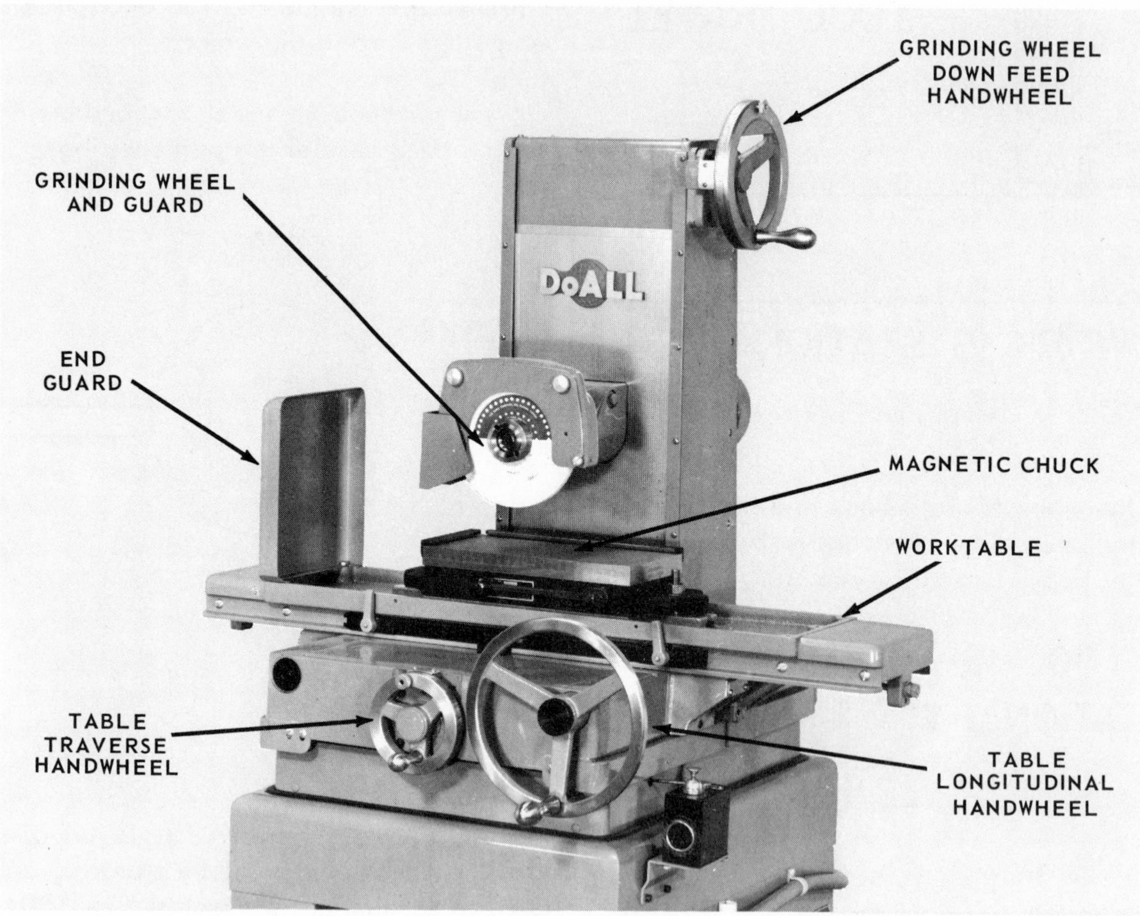

Fig. 17-20. Manually operated surface grinder. (The DoALL Co.)

Fig. 17-21. A cylindrical grinder. Closeup showing part being ground and abrasive wheel. (Landis Tool Co.)

A grinding machine like the SURFACE GRINDER, Fig. 17-20, is used to put fine finishes on flat surfaces and to finish them to close tolerances.

With the CYLINDRICAL GRINDER, Fig. 17-21, it is economically feasible to machine hardened steel parts to tolerances of 1/100000 (0.00001) in. with extremely fine surface finishes.

SHEARING AND FORMING TECHNIQUES

SHEARING is a process where the material (usually in sheet form) is cut to shape using action similar to cutting paper with scissors.

STAMPING is divided into two separate classifications: CUTTING and FORMING.

The cutting operation is also known as BLANKING, Fig. 17-22. It involves cutting flat sheets to the shape of the finished part.

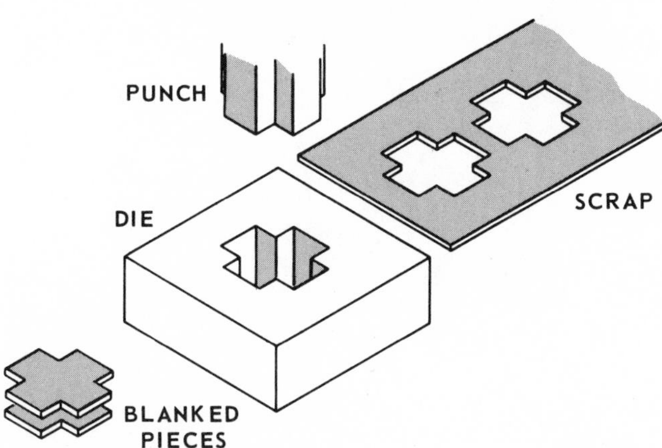

Fig. 17-22. Blanking operation. Cutting flat sheets.

FORMING is a process where flat metal blanks are given three-dimensional form, Fig. 17-23.

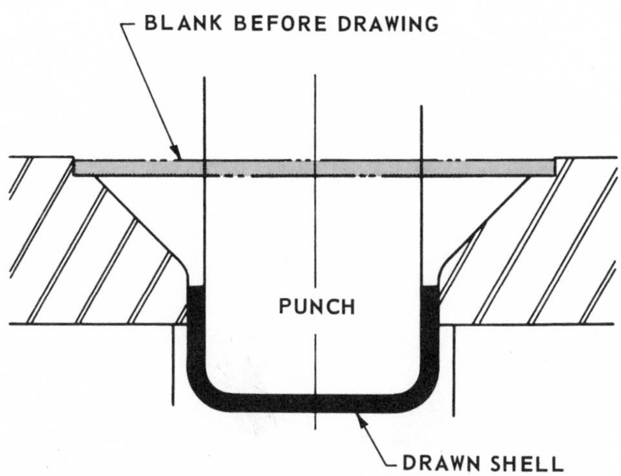

Fig. 17-23. Forming is a process where flat metal blanks are given three-dimensional form.

With PRESSING, metal is shaped by using pressure and sometimes heat to form it to the required shape.

CASTING TECHNIQUES

Materials can also be given shape and form by reducing it to a molten state and pouring it into a mold of the desired shape. This is called casting.

SAND CASTING

In the SAND CASTING PROCESS, Fig. 17-24, the mold that gives the molten metal shape is made of sand. It is one of the oldest metal forming techniques known to man.

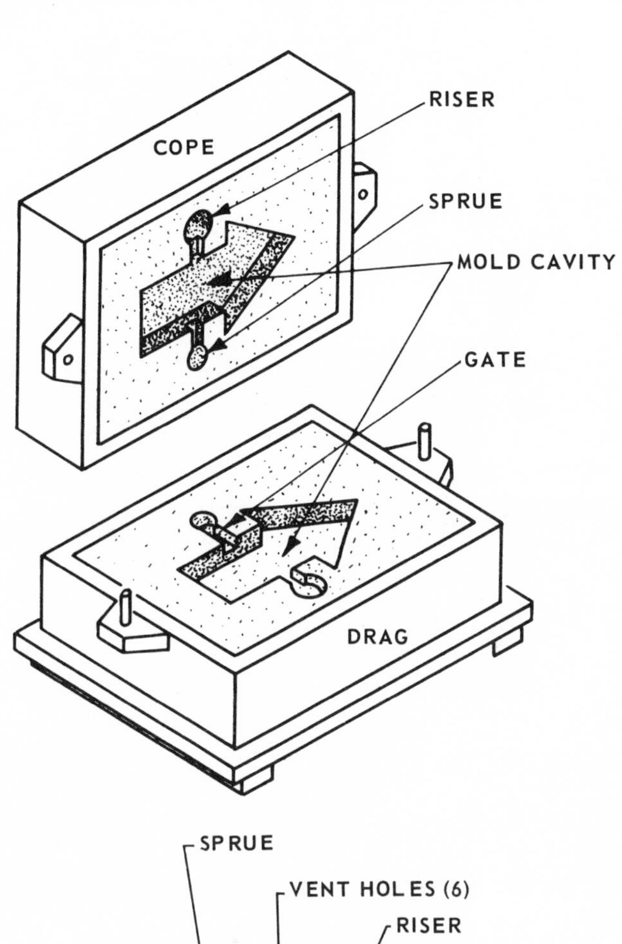

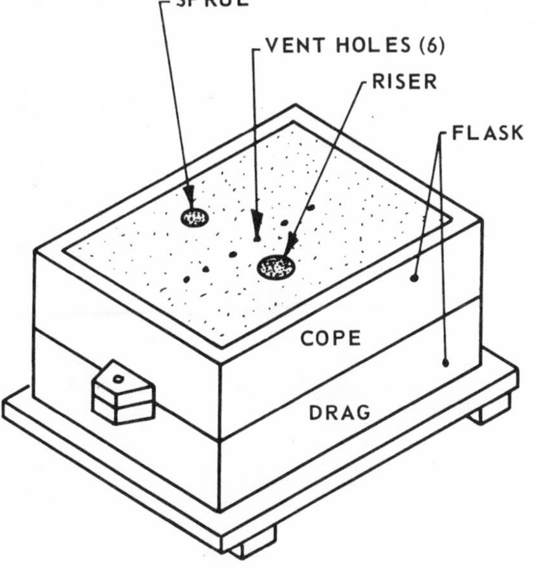

Fig. 17-24. Parts of a sand mold.

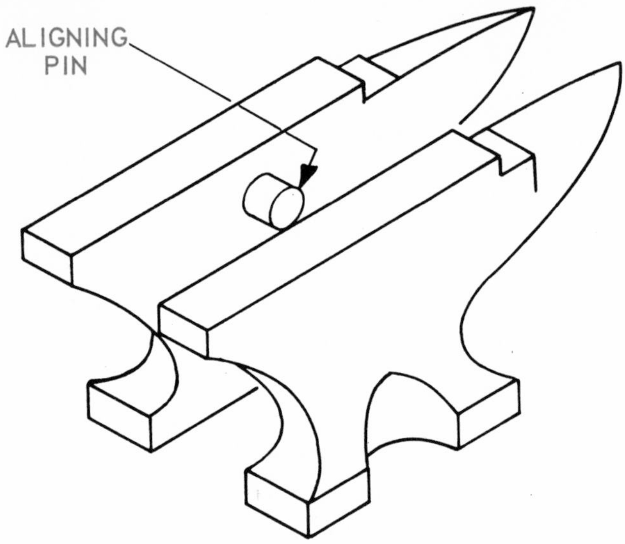

Fig. 17-25. Typical two-piece pattern used to make cavity in mold.

Fig. 17-26. This piston was cast in a permanent mold in the same way you would cast toy soldiers and fishing sinkers.

The sand mold is made by packing sand in a box called a FLASK, around a PATTERN of the shape to be cast. When the pattern has been DRAWN (removed) from the mold and the mold halves, called the COPE and the DRAG, are reassembled, a cavity remains.

Before assembling the mold, openings called SPRUES, RISERS and GATES are

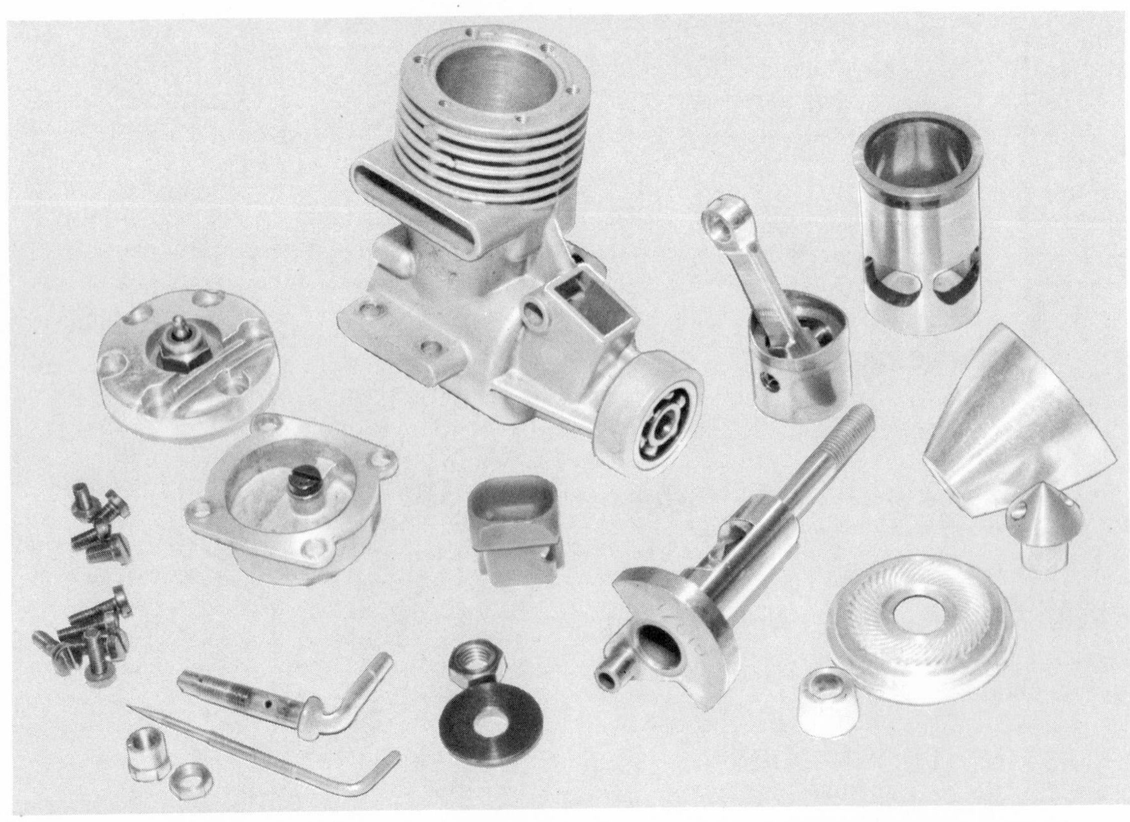

Fig. 17-27. Small gas engine. The crankcase, cylinder head and crankcase cover were die cast.

made in the mold. The sprue is the opening into which the molten metal is poured. The risers allow the hot gasses to escape. The gates are trenches that run from the sprues and risers to the mold cavity and permit the molten metal to reach the cavity.

Because metal contracts as it cools, PATTERNS, Fig. 17-25, must be made oversize to allow for shrinkage.

PERMANENT MOLD CASTING

Some molds used for metal casting are made from metal. These molds are called PERMANENT MOLDS because they do not have to be destroyed to remove the casting. The process produces castings with a fine surface finish and a high degree of accuracy, Fig. 17-26.

Fishing sinker and toy soldier molds are familiar examples of products made using the permanent mold process.

DIE CASTINGS

DIE CASTING, Fig. 17-27, is a process where molten metal is forced into a DIE or MOLD under pressure. The pressure is maintained until the metal solidifies at which time the mold is opened and the casting is removed, Fig. 17-28. The mold or die is made of metal.

Die cast parts have good dimensional accuracy and require little finish machining.

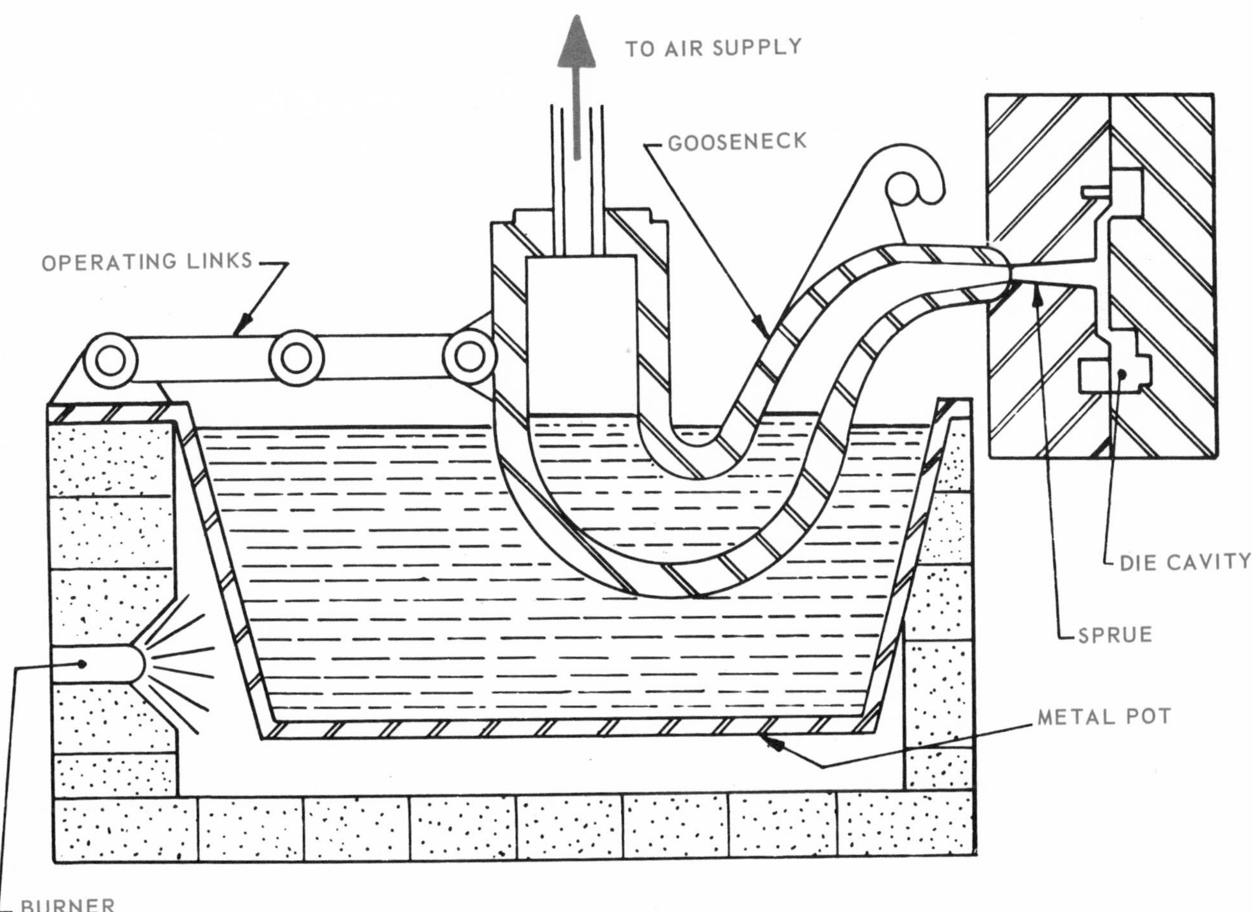

Fig. 17-28. Diagram of a die casting machine. The metal is forced into the die cavity (mold) under pressure to make a casting that is denser than a sand casting.

TEST YOUR KNOWLEDGE - UNIT 17

1. Make a sketch showing how a lathe operates.
2. Make a sketch showing how a drill press operates.
3. Make a sketch showing how a milling machine operates.
4. What is a machine tool?
5. What are some other operations that can be performed on a drill press?
   a. ____________________.
   b. ____________________.
6. List four machines used to machine flat surfaces.
   a. ____________________.
   b. ____________________.
   c. ____________________.
   d. ____________________.
7. Make a sketch showing how the surface grinder operates.
8. A ____________ is used to make the mold cavity in a sand mold.
9. It is made slightly larger than the finished casting because metal ________.
10. How do sand castings differ from castings made in permanent molds and die castings?
11. Make a sketch showing a typical sand mold.

OUTSIDE ACTIVITIES

1. Prepare a bulletin board using clippings which illustrate basic machine tools.
2. Secure samples of work made on a lathe.
3. Secure samples of work machined on a drill press.
4. Secure samples of work machined on a milling machine.
5. Secure samples of work machined on a surface grinder.
6. Secure samples of work cast in a sand mold.
7. Secure samples of work made by die casting.
8. Bring into class an example of a permanent mold such as a sinker mold or toy soldier mold.

Unit 18
WELDING DRAWINGS

WELDING is a widely used industrial technique for fabricating metal pieces. It is a method of joining by heating metals to a high temperature which causes them to melt and fuse together. The high temperatures needed are generated by electricity (arc welding) or by burning gases (usually oxyacetylene). See Fig. 18-1.

Fig. 18-1. In the manufacturing, construction and service industries, welding is a widely used metal fabricating technique. (Union Carbide Corp., Linde Div.)

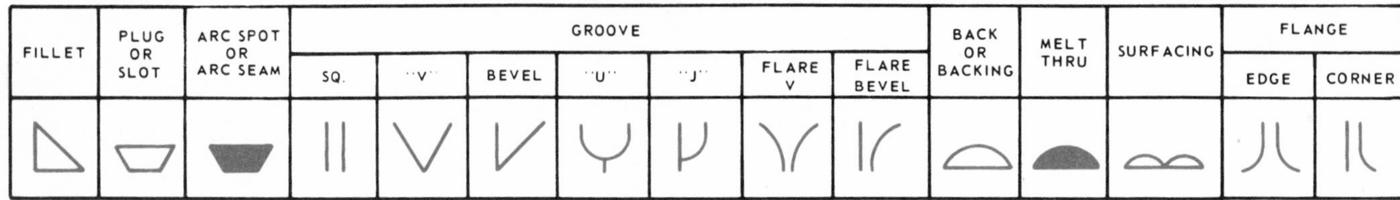

BASIC ARC AND GAS WELD SYMBOLS

TYPE OF WELD			
RESISTANCE SPOT	PROJECTION	RESISTANCE SEAM	FLASH OR UP SET

BASIC RESISTANCE WELD SYMBOLS

WELD ALL AROUND	FIELD WELD	CONTOUR	
		FLUSH	CONVEX

SUPPLEMENTARY WELD SYMBOLS

Fig. 18-2. Welding symbols.

The American Welding Society (AWS) has developed a series of symbols which tell the welder exactly what to do.

The symbols are included on drawings where the assemblies require welding. Basic arc and gas welding symbols are shown in Fig. 18-2.

USING WELDING SYMBOLS

Before welding symbols can be used effectively the draftsman must be familiar with the various types of joints used in welding, Fig. 18-3.

The location of the arrow with respect to the joint is very important when using the welding symbol to specify the required weld. The side of the joint indicated by the arrow is considered the ARROW SIDE. The side opposite the arrow side of the joint is considered the OTHER SIDE of the joint.

When the weld is to be made on the ARROW SIDE of the joint, the weld symbol is placed on the reference line so it is TOWARD THE READER, Fig. 18-4. A weld on the OTHER SIDE of the joint is specified by placing the weld symbol on the reference line AWAY FROM THE READER, Fig. 18-5. Welds on both sides of the joint are specified by placing the symbol on BOTH SIDES of the reference line, Fig. 18-6.

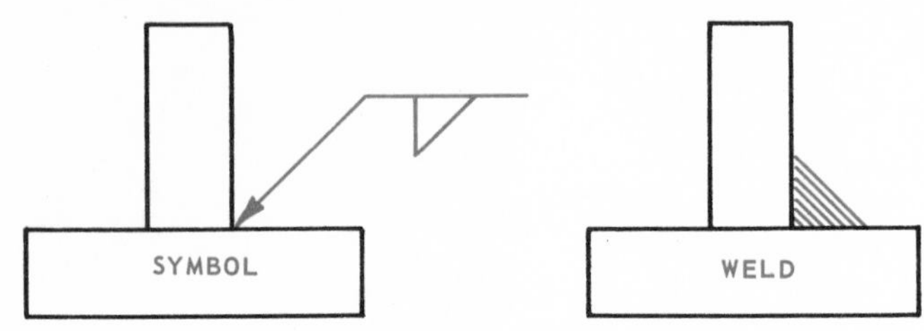

*Fig. 18-4. Weld symbol that indicates that the weld is to be made on the **ARROW SIDE** of the joint. Note the weld symbol is **TOWARD** the reader.*

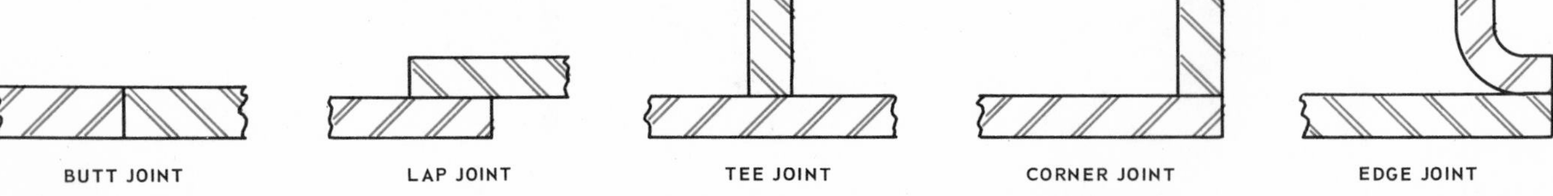

Fig. 18-3. Basic joints used in welding.

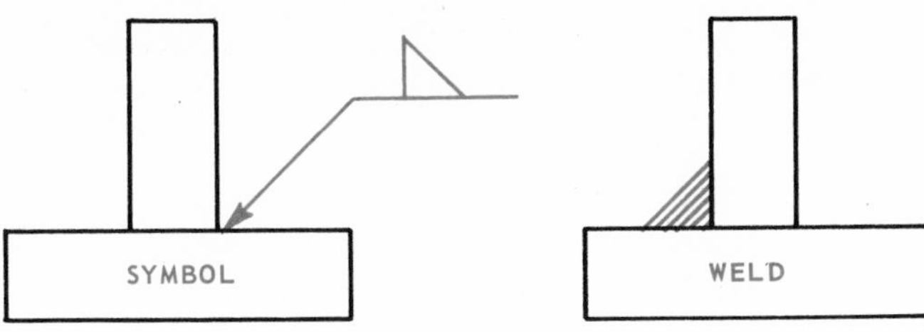

Fig. 18-5. Weld symbol indicates that the weld is to be made on the ***OTHER SIDE*** *of the joint.*

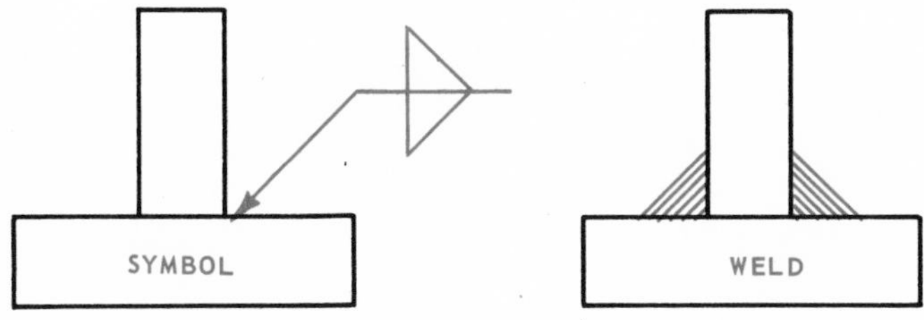

Fig. 18-6. Weld symbol indicates that the weld is to be made on ***BOTH SIDES*** *of the joint.*

A weld that is made all of the way around the joint is specified by drawing a circle at the point where the arrow is bent, Fig. 18-7. Field welds (welds not made in the shop) are specified as shown in Fig. 18-8.

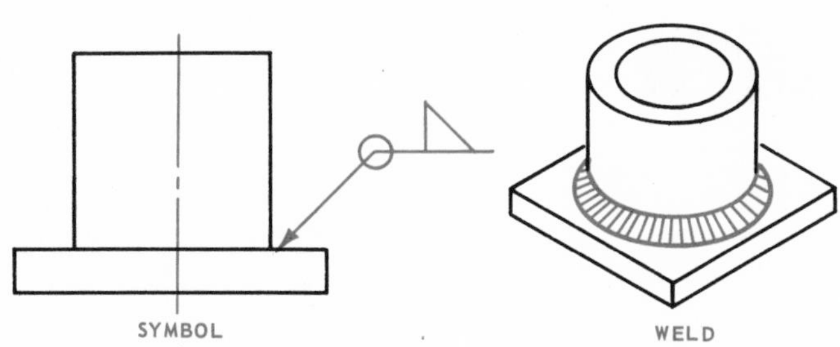

Fig. 18-7. Weld symbol that indicates a weld to be made ***ALL AROUND*** *the joint.*

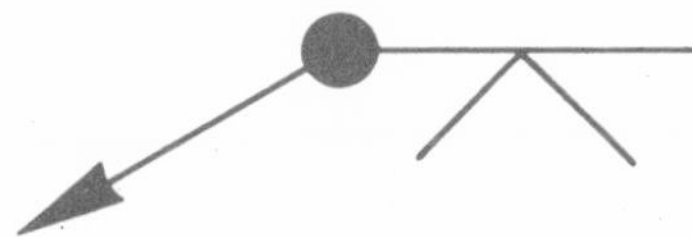

Fig. 18-8. Field welding symbol.

Weld size is placed to the left of the symbol, Fig. 18-9. The figure indicating the length of the weld is placed to the right of the symbol.

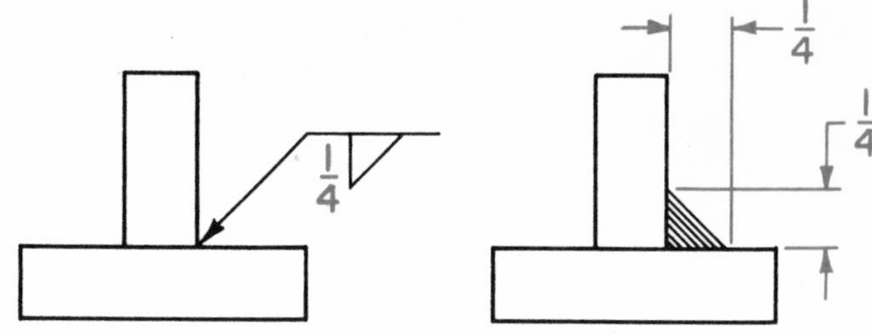

Fig. 18-9. How weld size is indicated.

DRAWINGS FOR ASSEMBLIES TO BE FABRICATED BY WELDING

Items that are to be assembled by welding are usually composed of several different pieces, Fig. 18-10. Because the individual pieces are seldom cut to size, welded, and machined in the same general area,

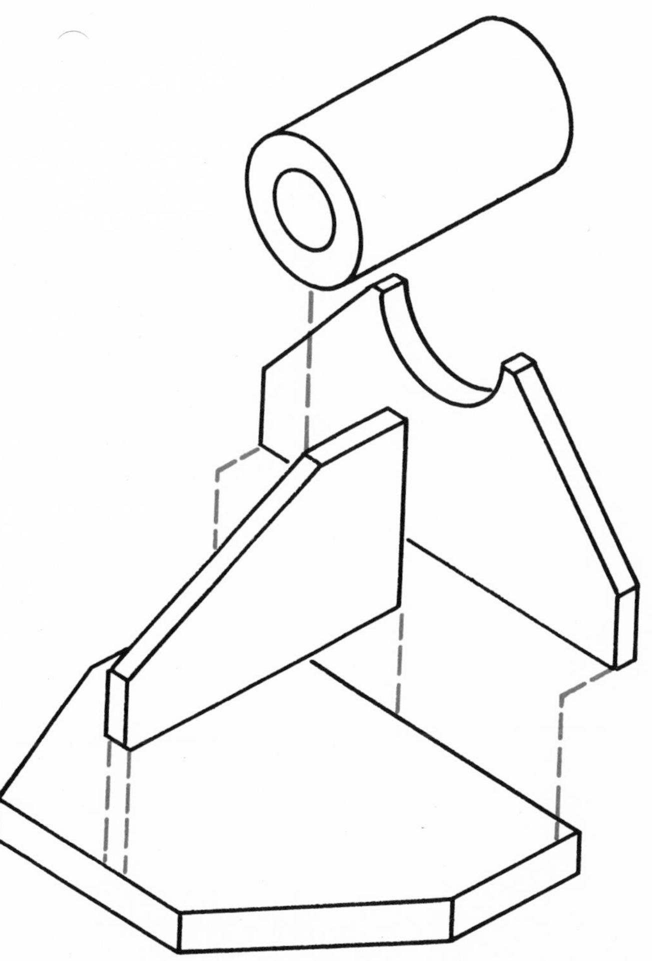

Fig. 18-10. A welded assembly is composed of several different pieces.

several drawings are usually required - each of which provides information on a specific operation.

For simple jobs, all of the needed information can often be included on a single drawing, Fig. 18-11.

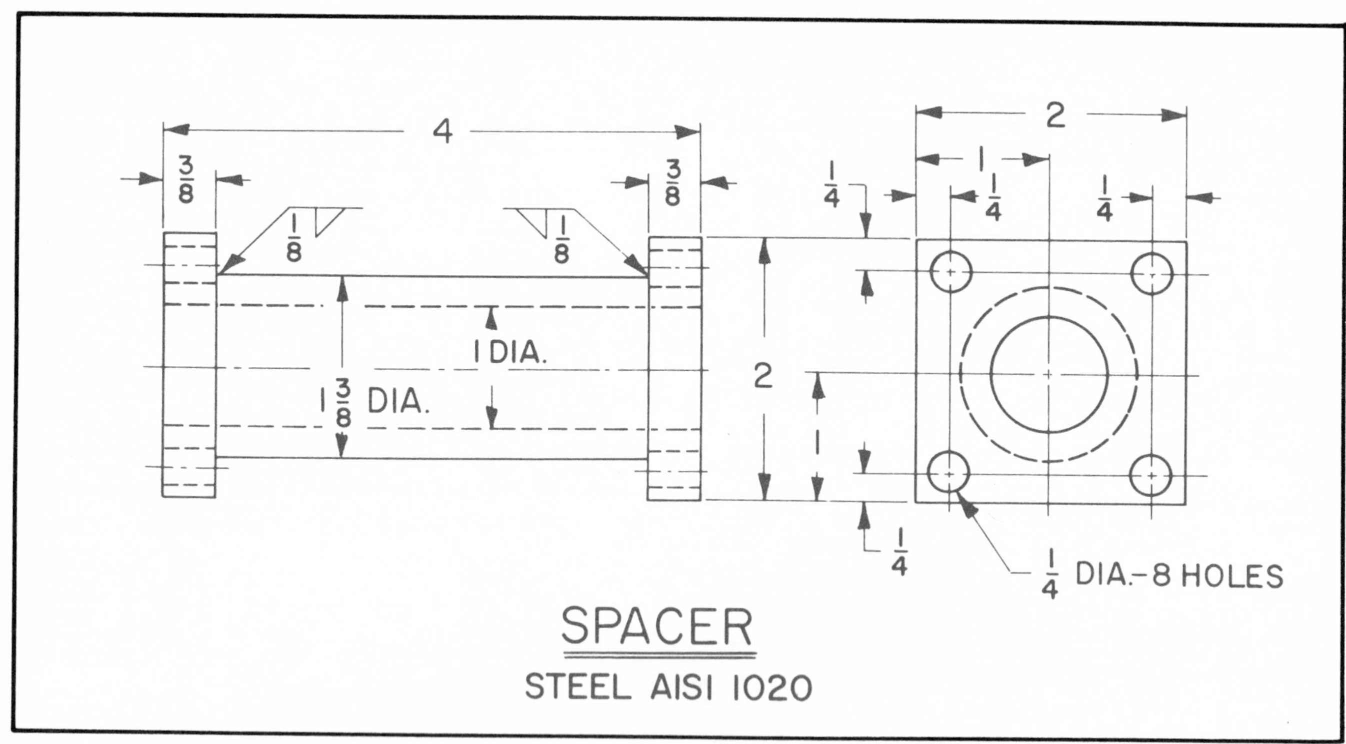

Fig. 18-11. Cutting, welding and machining information is included on this drawing.

TEST YOUR KNOWLEDGE - UNIT 18

1. How are the high temperatures needed for welding generated?
2. Welding symbols were developed to:
   a. Lessen the possibilities of the wrong type of weld being made.
   b. Tell the welder exactly what to do.
   c. Eliminate "hit or miss" welds.
   d. All of the above.
3. Sketch the symbol for a fillet weld on the arrow side of the joint.
4. Sketch the symbol for a fillet weld on the other side of the joint.
5. Sketch the symbol for a fillet weld on both sides of the joint.
6. Sketch the symbol for a fillet weld all of the way around the joint.
7. What is a field weld?
8. Why are several different drawings needed for a job that is made up of several pieces and assembled by welding?

OUTSIDE ACTIVITIES

1. Secure samples of welded joints and mount them on a display board. Label the samples and add the proper drafting symbol to the display.
2. Visit a local industry that makes extensive use of welding and get samples of the work they produce.
3. Invite a professional welder to the school shop to demonstrate the safe and proper way to gas weld and to electric arc weld.

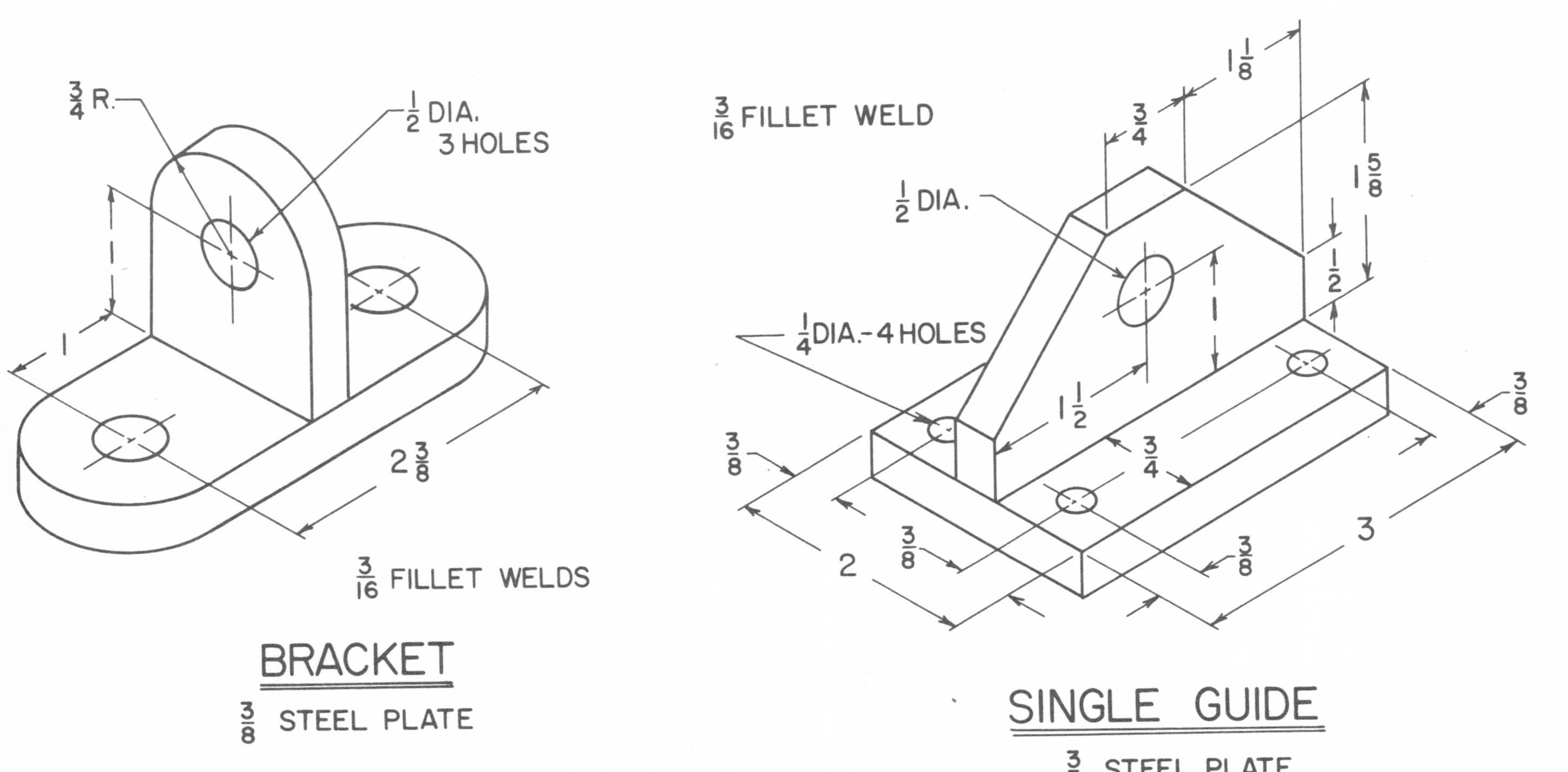

PROBLEM 18–1. Left. BRACKET. *Prepare a drawing showing the necessary views to fabricate it. Use the appropriate welding symbol.* **PROBLEM 18–2. Right. SINGLE GUIDE.** *Prepare a drawing with the information necessary to fabricate it by welding.*

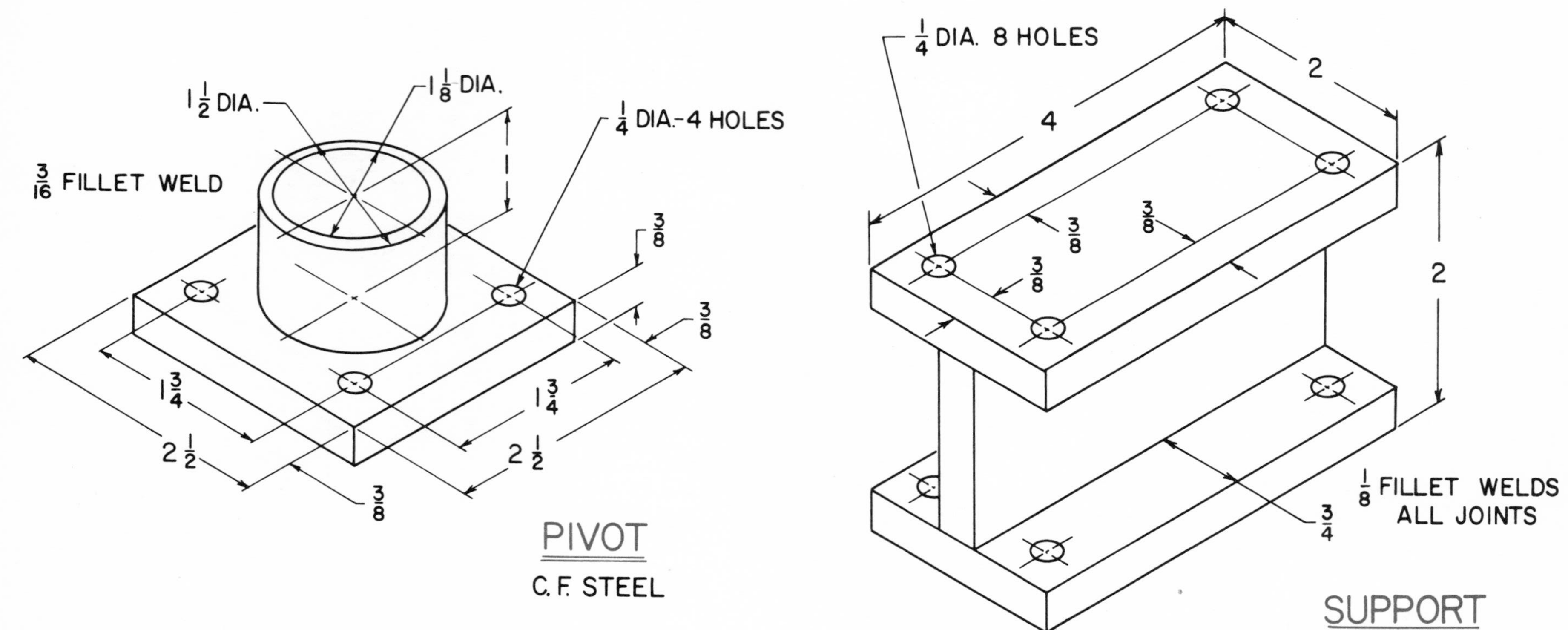

PROBLEM 18–3. Left. PIVOT. The fillet weld is to be made all around the joint. PROBLEM 18–4. Right. SUPPORT. Prepare a drawing showing the views necessary to fabricate the BRACKET.

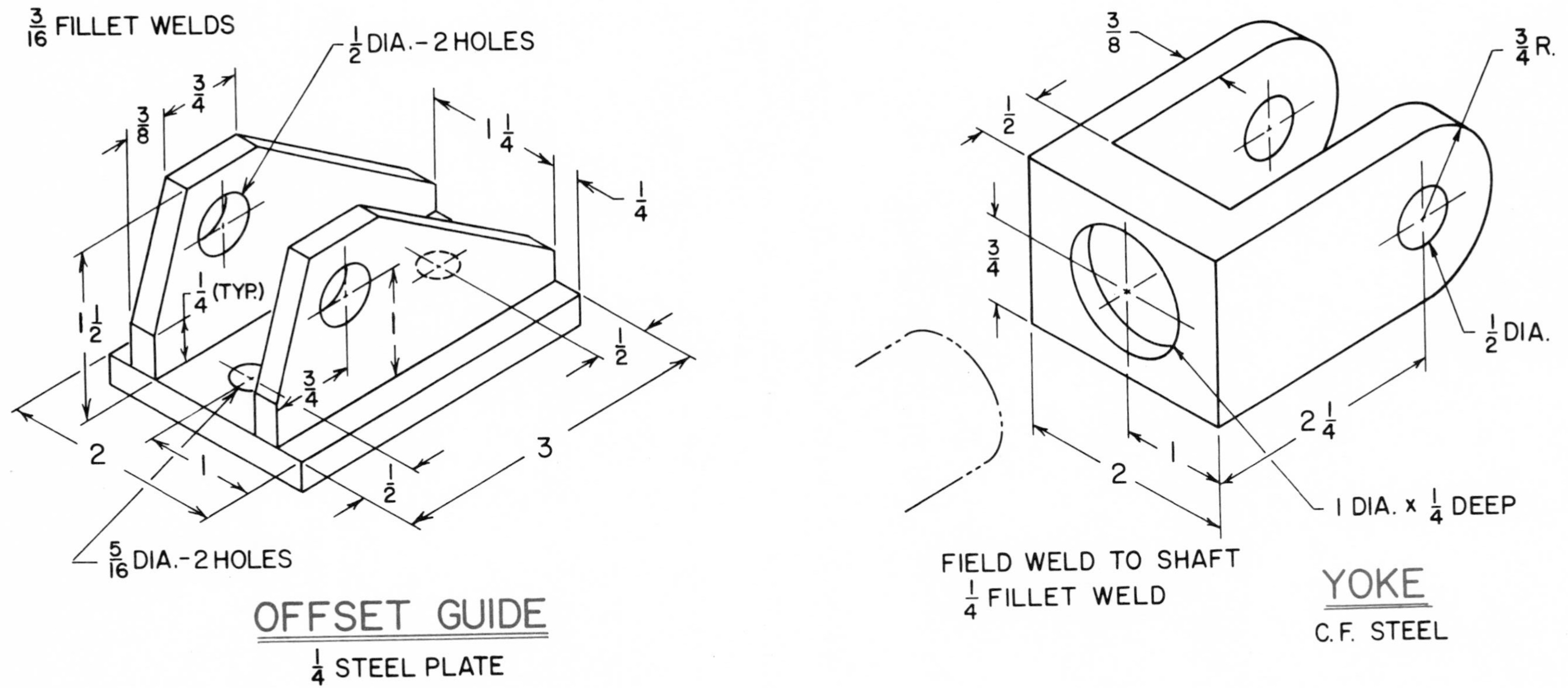

***PROBLEM** 18–5. **Left. OFFSET GUIDE.** The weld is to be made on both sides of each vertical piece.*
***PROBLEM** 18–6. **Right. YOKE.** This part is to be welded to a 1 in. diameter shaft at the job site.*

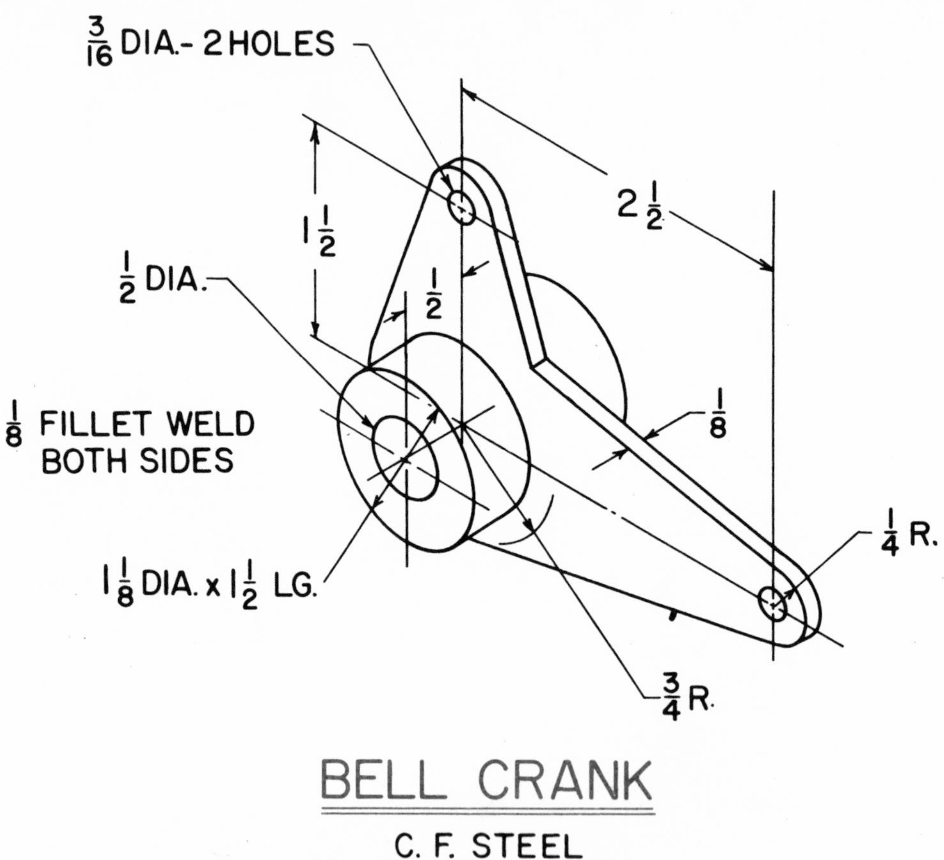

***PROBLEM 18–7. BELL CRANK.** A 1/8 fillet weld is specified on both sides of the joint.*

Fig. 19-1. A few of the many thousands of different types and sizes of fasteners used by industry.

Unit 19
FASTENERS

A fastener is a device used to hold two objects or parts together. Fasteners include screws, rivets, nails, nuts and bolts, etc., Fig. 19-1. They are used extensively to assemble manufactured items. Because of the importance of fasteners, draftsmen and mechanics should be familiar with them, know what type to use for particular applications and how to draw them.

THREADED FASTENERS

Threads have many applications. They are used to:

1. Make adjustments.
2. Transmit motion.
3. Assemble parts.
4. Apply pressure.
5. Make measurements.

How many examples of these applications can you name?

Fasteners that fall into the threaded category use the wedging action of the threads to hold things together.

While many different thread profiles (shapes) have been developed to do these many jobs, the thread profile shown in Fig. 19-2 has been standardized in the United States. It is known as the AMERICAN NATIONAL THREAD SYSTEM. The National Coarse (NC) and the National Fine (NF) are

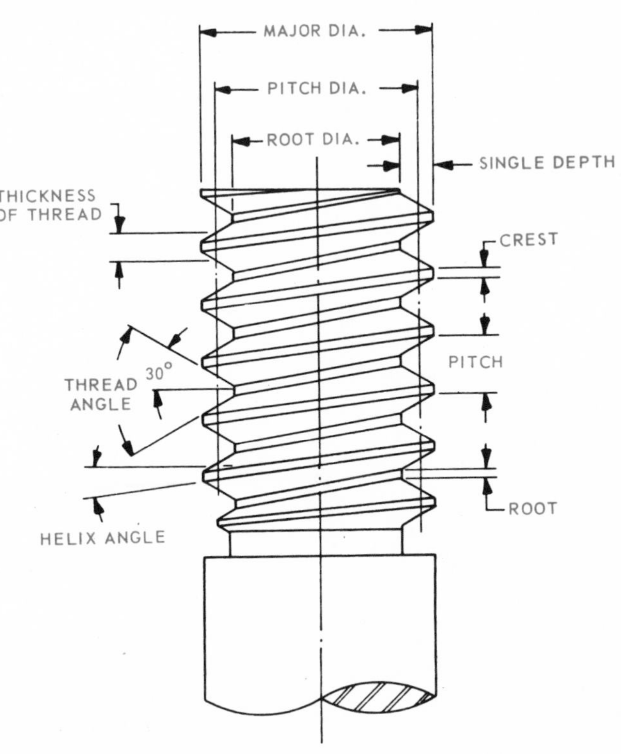

Fig. 19-2. The parts of a screw thread.

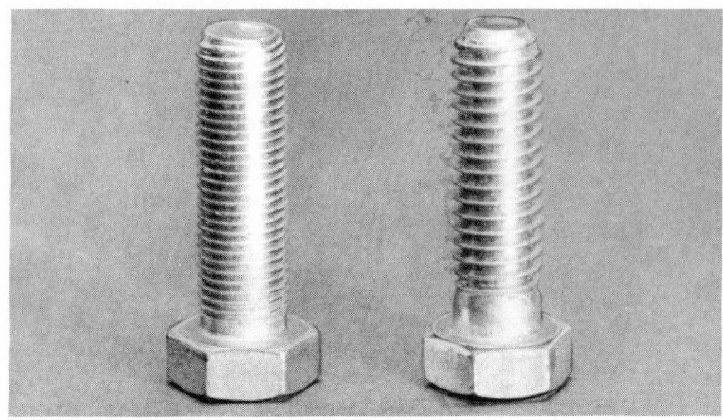

Fig. 19-3. A comparison between National Fine (NF) threads (left) and National Coarse (NC) threads. Both bolts are the same diameter.

the most widely used thread series in the American National Thread System. The NC series has fewer threads per inch of length for a given diameter than the NF series, Fig. 19-3.

It is time consuming to show threads on the drawing as they would actually appear. For this reason the SCHEMATIC and SIMPLIFIED representation of threads, Fig. 19-4, are generally used.

The DETAILED representation looks like the actual screw thread. It is sometimes employed where confusion might result if the simplified representation were employed.

The SCHEMATIC representation is easier to draw. It should not be used for hidden threads or sections of external threads.

The SIMPLIFIED representation of the screw thread is a fast and easy way to draw threads. For this reason it is widely employed in drafting. It should be avoided where there is a possibility of this representation being confused with other details on the drawing.

Regardless of which thread representation the draftsman decides to draw, the thread size must be noted on the drawing. An accepted way of presenting this information is shown in Fig. 19-5, and means:

1/2 - The major or large diameter of the thread.
13 - Represents number of threads per inch.

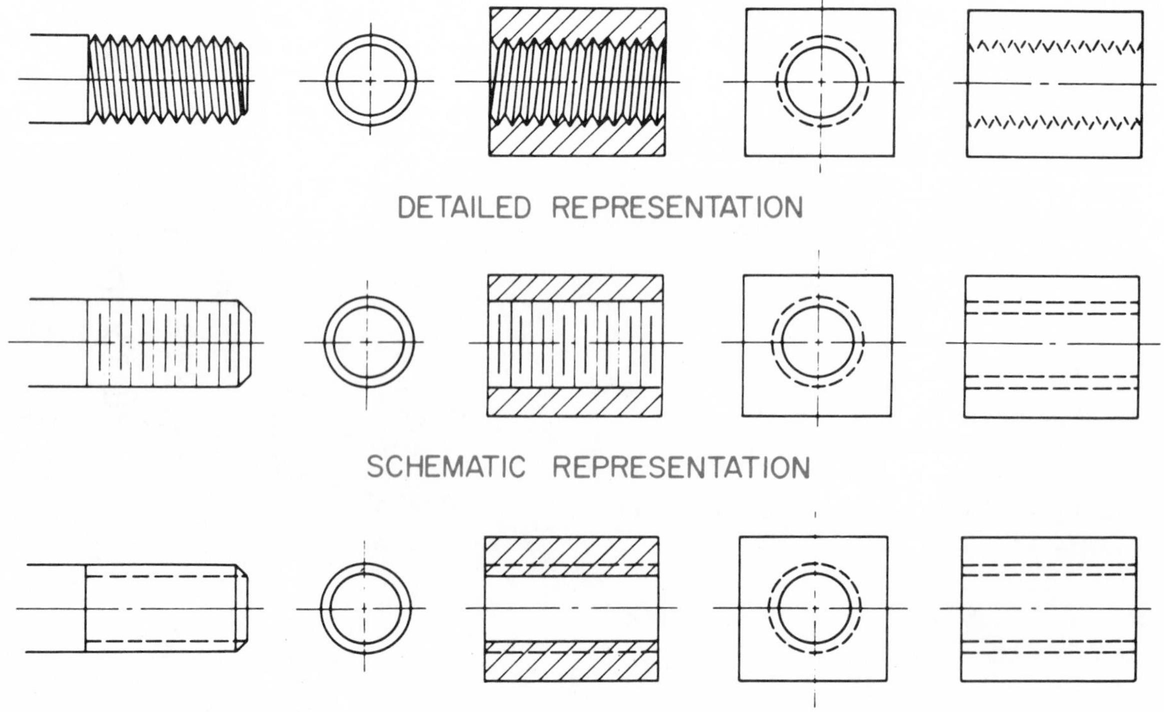

Fig. 19-4. Approved ways of representing threads on a drawing.

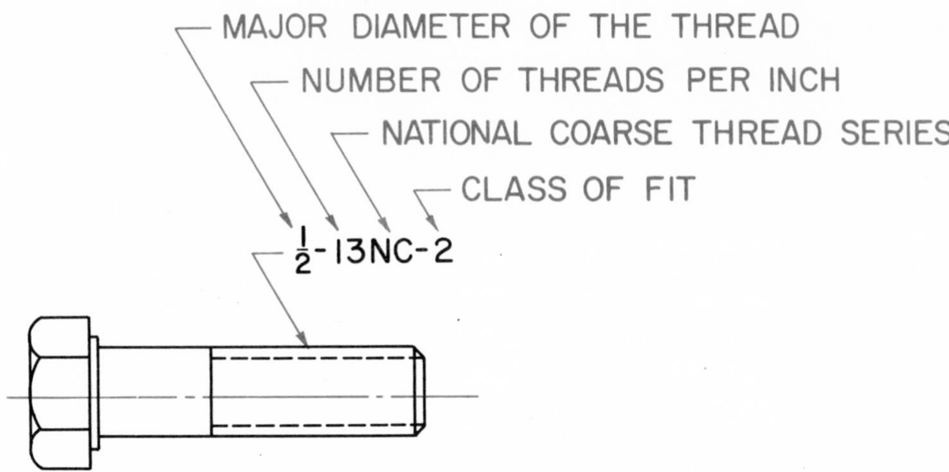

Fig. 19-5. How thread size is noted and what each term means.

NC - Means the thread is an American Standard thread of the National Coarse Series.

2 - The class of fit of the thread (how snug or loose the thread is when it is screwed into a mating part).

- Class 1 - A loose fit and is seldom used today.
- Class 2 - A free fit and is employed for most regular work.
- Class 3 - A medium fit and is used for more precise work such as automobile engines.
- Class 4 - A very snug fit that is utilized in aerospace applications, jet engines, etc.

Threads are understood to be right-hand threads. Left-hand threads are represented by the letters LH.

TYPES OF THREADED FASTENERS

The fasteners described in this unit can usually be found in the school shop or a typical hardware store. The majority of them are made of steel, brass or aluminum. Special applications may require them to be made of other materials.

MACHINE SCREWS

MACHINE SCREWS, Fig. 19-6, are available with single slotted or cross slotted (Phillips) heads in round, flat, fillister, pan and oval head styles. Nuts (square or hexagonal) are not furnished with machine screws and must be purchased separately. They have many applications in general assembly work where screws less than 1/4 in. in diameter are needed. Machine screws are

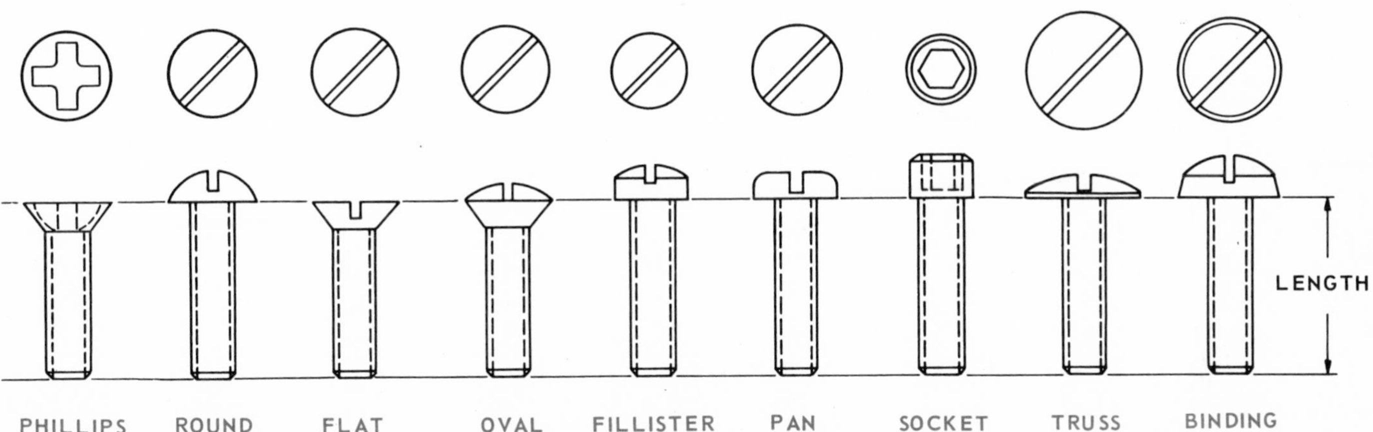

Fig. 19-6. Types of machine screws.

made with both NC and NF threads. Size of machine screws and other fasteners described in this unit can be found in the Tables, starting on page 300.

MACHINE BOLTS

MACHINE BOLTS are manufactured with square and hexagonal heads, Fig. 19-7. They are used to assemble machinery and other items that do not require close tolerance fasteners.

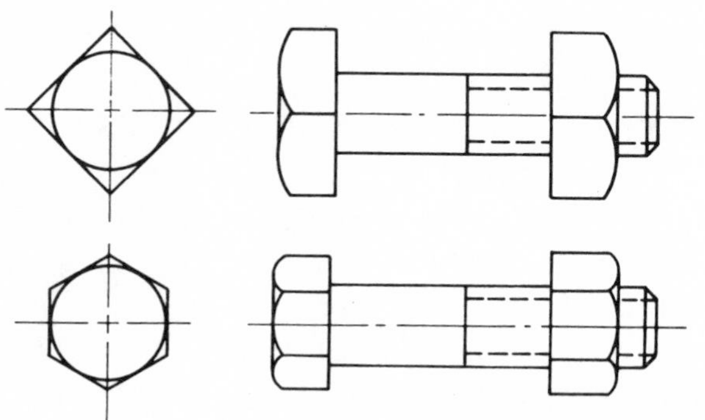

Fig. 19-7. Machine bolts.

CAP SCREWS

CAP SCREWS, Fig. 19-8, are used when the assembly requires a stronger, more precise and better appearing fastener. They are primarily employed to bolt two pieces together. The screw passes through a CLEARANCE HOLE in one part and screws into a THREADED HOLE in the other, Fig. 19-9. They are made with coarse (NC) and fine (NF) threads and are extensively utilized in the assembly of machine tools.

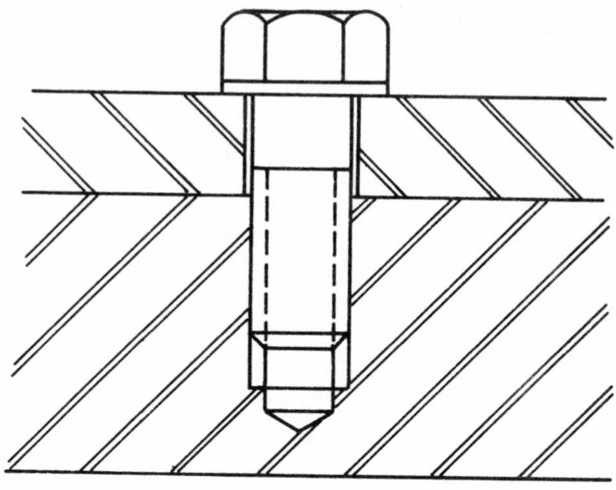

Fig. 19-9. How a cap screw works.

SETSCREWS

SETSCREWS are usually made of steel and are heat-treated to make them stronger, Fig. 19-10. Major use is to prevent slippage of pulleys on shafts. Setscrews are available in many different head and point styles.

STUD BOLT

STUD BOLTS, Fig. 19-11, are threaded on both ends. One end is threaded into a tapped

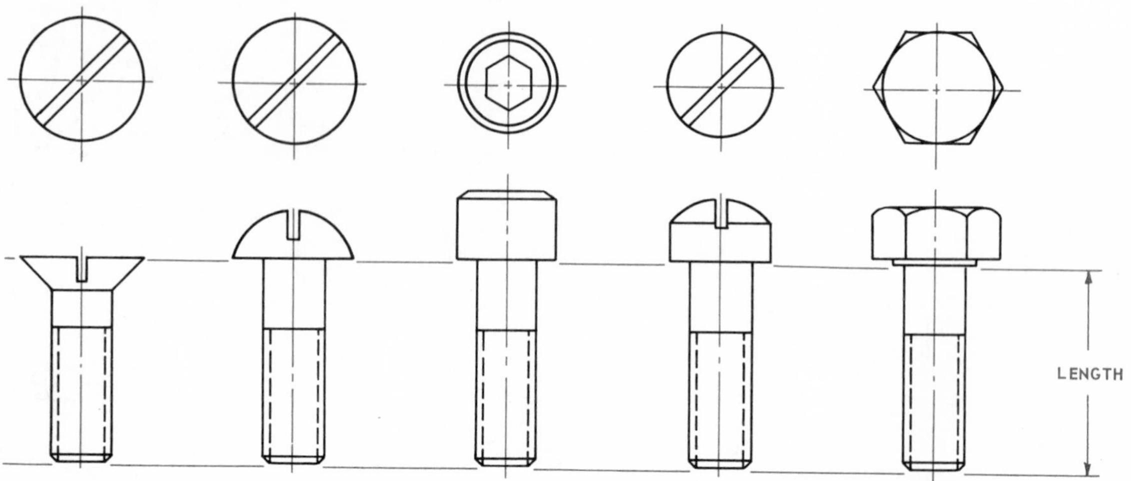

Fig. 19-8. Types of cap screws.

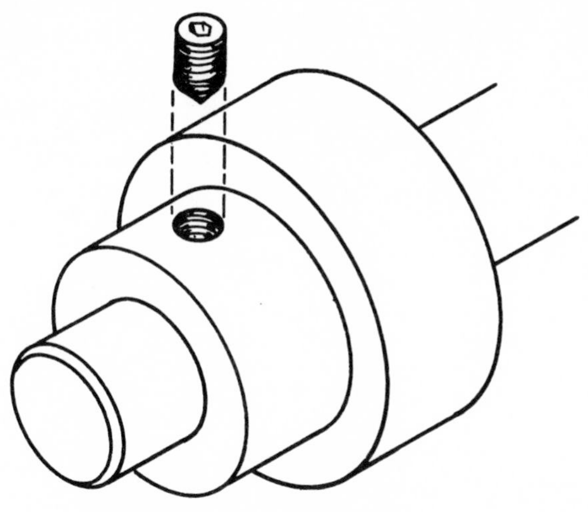

Fig. 19-10. A typical setscrew application.

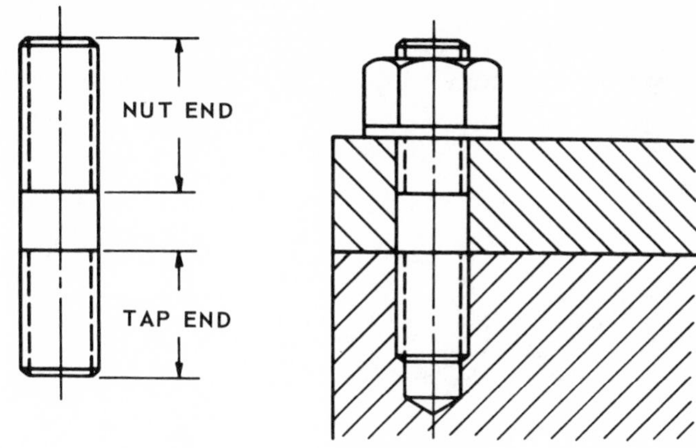

Fig. 19-11. Stud bolt.

hole, the piece to be clamped is fitted over the stud, and the nut is screwed on to clamp the two pieces together.

NUTS

NUTS are screwed down on bolts or screws to tighten or hold together the objects through which they pass. A few of the many types available are shown in Fig. 19-12.

WASHERS

WASHERS are used with nuts and bolts to distribute the clamping pressure over a larger area, and to prevent the fastener

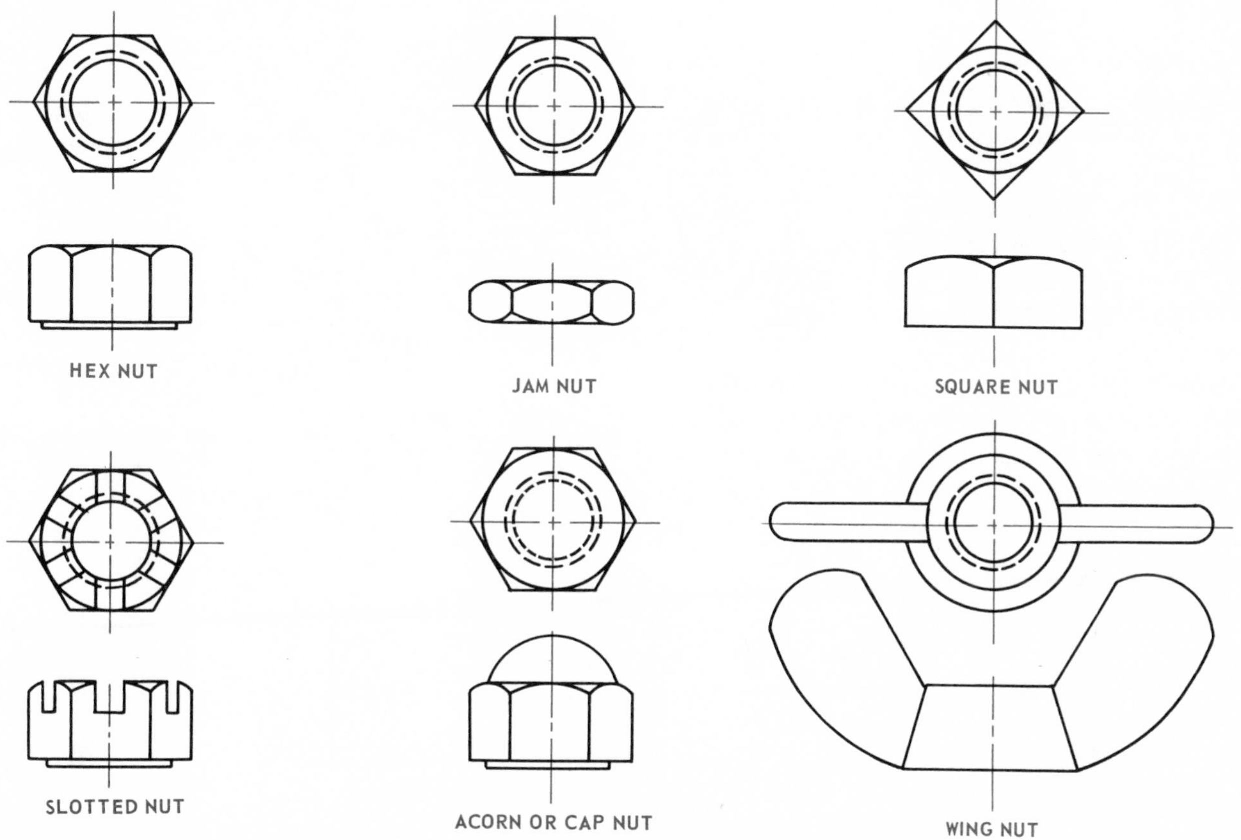

Fig. 19-12. Common nut styles.

from marring the work surface when it is tightened. Many types are manufactured, Fig. 19-13.

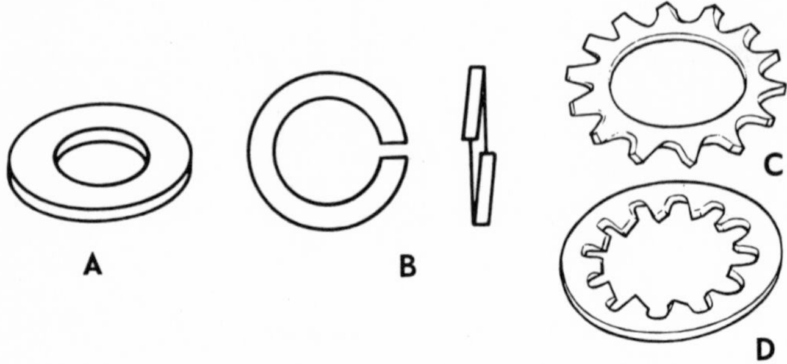

Fig. 19-13. Washer types. A–Plain washer. B–Split type lock washer. C–External type lock washer. D–Internal type lock washer.

NONTHREADED FASTENERS

Nonthreaded fasteners comprise a large group of holding devices.

RIVETS

Permanent assemblies are made with RIVETS, Fig. 19-14. Two or more pieces of material are held together by these headed pins. Holes are drilled in the material to be riveted. The shank of the rivet is passed through the hole. After aligning the pieces, the plain end of the rivet is upset or "headed" by hammering to form a second head.

Some rivets are expanded by a small explosive charge or by other forms of pressure. The parts are drawn together by the heading operation.

COTTER PIN

The COTTER PIN, Fig. 19-15, is fitted into a hole drilled crosswise in a shaft and prevents the parts from slipping or turning off.

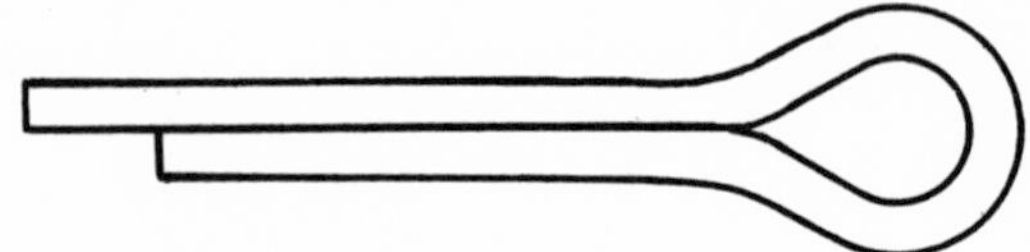

Fig. 19-15. Cotter pin.

KEYS, KEYWAYS AND KEYSEATS

A KEY, Fig. 19-16, is a small piece of metal partially fitted into the shaft and partially in the hub to prevent the rotation of the gear or pulley on the shaft.

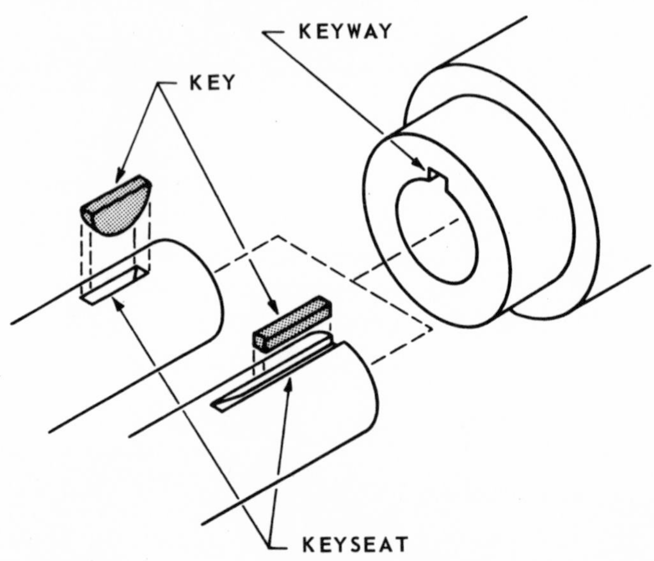

Fig. 19-16. Two types of keys are shown. The half-round key is called a Woodruff key. The other is a rectangular key.

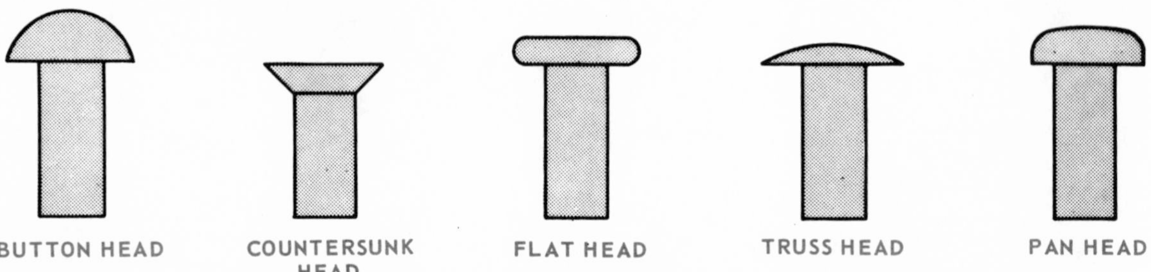

Fig. 19-14. Rivet head styles.

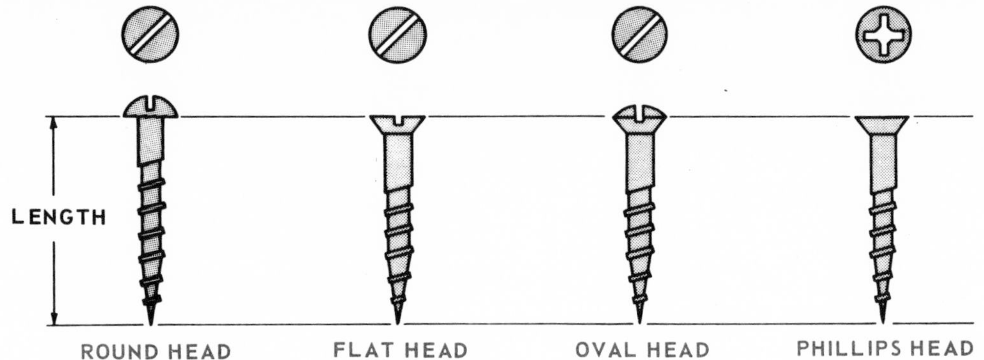

Fig. 19-17. How the various types of wood screws are measured.

The KEYSEAT is machined in the shaft. The KEYWAY is cut in the hub of the mating part.

Different styles of keys have been devised for special applications.

FASTENERS FOR WOOD

The most common fasteners used in wood are nails and screws.

NAILS

Using NAILS is an easy way to fasten wood pieces together. Nails are usually made of mild steel, but some are made of aluminum. For exterior work, nails made from mild steel are given a galvanized (zinc) coating.

Nail size is given as "penny" and is abbreviated with a lower case letter "d."

WOOD SCREWS

WOOD SCREWS are manufactured from several different kinds of metal. Screw size is indicated by the diameter of the shank and the length. Head styles and how each is measured for length is shown in Fig. 19-17. Wood screws are available in lengths from 1/4 in. to 6 in.

HOW TO DRAW BOLTS AND NUTS

The method shown in Fig. 19-18 is an approximation but is acceptable for many drafting applications. Information needed is: (1) bolt diameter, (2) bolt length and (3) the type of head or nut.

Fig. 19-18. Information needed to draw a bolt and nut.

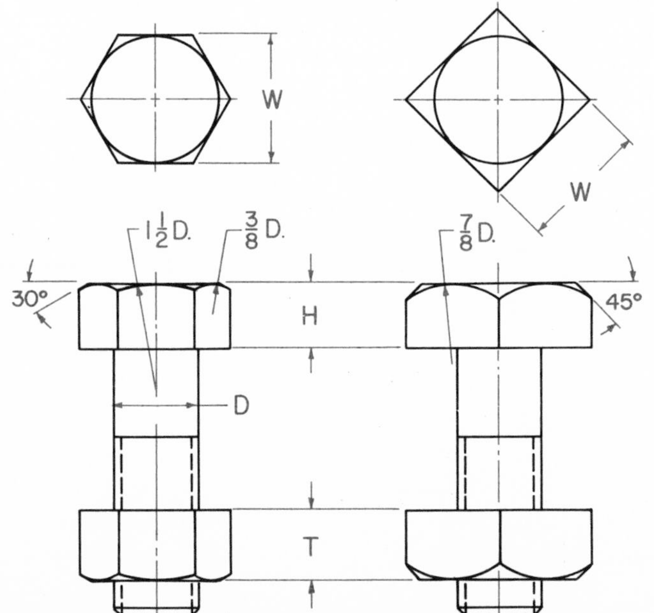

BOLT & NUT FORMULAS
$W = 1\frac{1}{2}$ D.
$H = \frac{3}{4}$ D.
$T = \frac{7}{8}$ D.

To draw square and hexagonal bolts and nuts, as in Fig. 19-19, this procedure is recommended:

1. Draw center lines and lines representing the diameter (D).

2. On center line, draw a circle (diameter = 1 1/2D).

3. Using triangles, circumscribe the hexagon (or square) about the circle.

4. Develop side view.

5. Draw arcs in bolt head and nut using radii given in Fig. 19-18.

6. Complete by drawing the chamfers on the nut and bolt head. Draw threads (either simplified or schematic) on bolt.

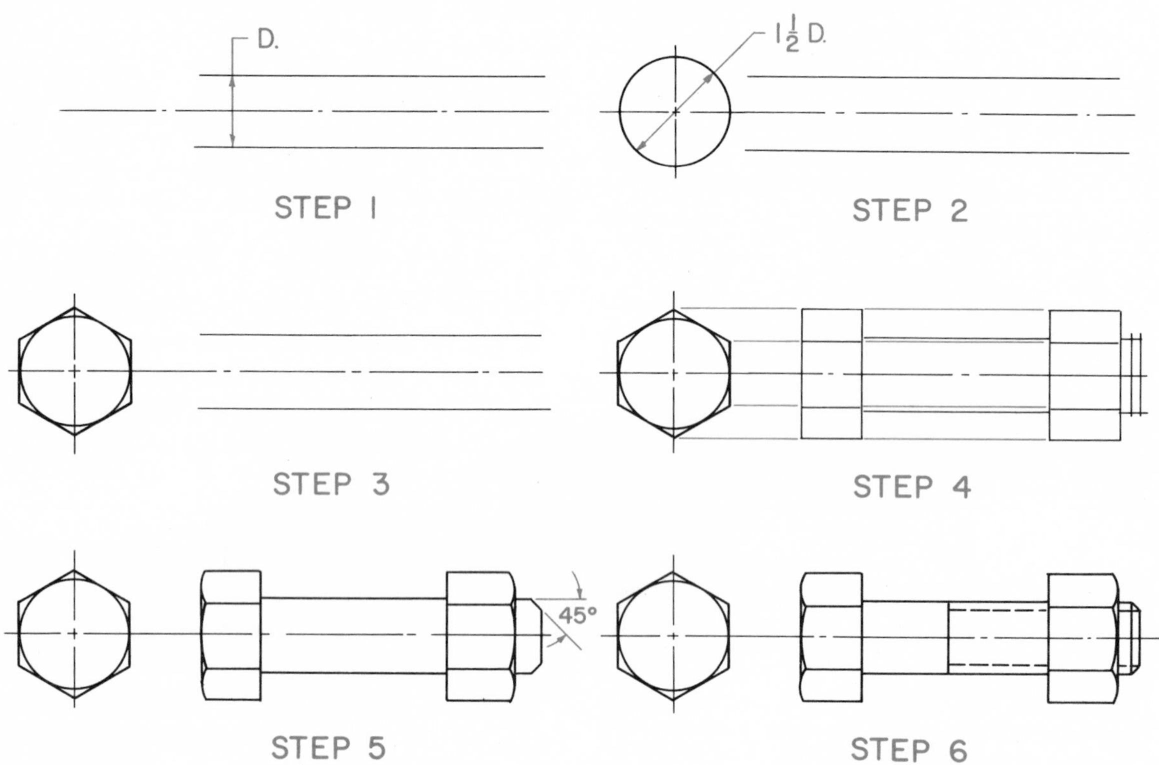

Fig. 19-19. Steps in drawing a bolt and nut. The same procedure is followed whether the nut and bolt is to have a square or hexagonal shaped head.

TEST YOUR KNOWLEDGE - UNIT 19

1. Identify 10 different fasteners. Underline the threaded fasteners.
2. What are fasteners?
3. Give five (5) applications that make use of fasteners.
4. List some uses made of screw threads.
   a. ____________________.
   b. ____________________.
   c. ____________________.
5. What is the difference between a National Coarse (NC) thread and a National Fine (NF) thread?
6. Make sketches of a detailed representation of a screw thread, a schematic representation and a simplified representation.
7. Explain the meaning of the following screw thread size: 1/2 - 20NF - 3LH.
8. What fasteners are most commonly used to join wood?
9. Place the letter of the correct definition in the blank to the left of the fastener.

___Machine screw	A. Threaded on both ends.
___Cap screw	B. Fitted on a bolt.
___Machine bolt	C. Distributes the clamping pressure of the nut or bolt.
___Setscrew	
___Stud bolt	
___Nut	D. Prevents slippage of a pulley on a shaft.
___Washer	

E. Used for any general assembly work where screws less than 1/4 in. in dia. are needed.
F. Used when close tolerances are not required.
G. Used when assembly requires a stronger, more precise and better appearing fastener.

OUTSIDE ACTIVITIES

1. Draw 3/4-10NC-2 x 4 in. long hexagonal and square head bolts and nuts on the same sheet. Allow 3 in. between the drawings. Use a simplified thread representation.
2. Draw 1-8NC-2 x 3 in. long hexagonal and square head bolts and nuts on the same sheet. Allow 3 in. between. Use a schematic thread representation.
3. Secure samples of machine screws, cap screws, machine bolts, setscrews, stud bolts, nuts, washers, rivets, cotter pins, nails and wood screws. Make a display for the drafting room.
4. Get sample showing a key, keyseat and keyway in use.
5. Secure an example of a setscrew application.

Unit 20
ELECTRICAL AND ELECTRONICS DRAFTING

Our world, as we know it today, could not exist without electricity and the electronic devices it powers. You will realize how true this is by just listing the electrical devices you depend upon in your home - TV set, transistor radio, washer, refrigerator, stereo, range and the lighting to name but a few.

There are many other electronic devices that may not be as familiar such as the computers that aid the astronauts in their flights to the moon; the devices that control the machines that make, inspect, assemble and test so many of the products we use, Fig. 20-1. Many hobbies are electronically oriented, Fig. 20-2.

Fig. 20-1. Modern machine tools, like this machining center, are controlled electronically. Note the various cutting tools. (Kearney and Trecker)

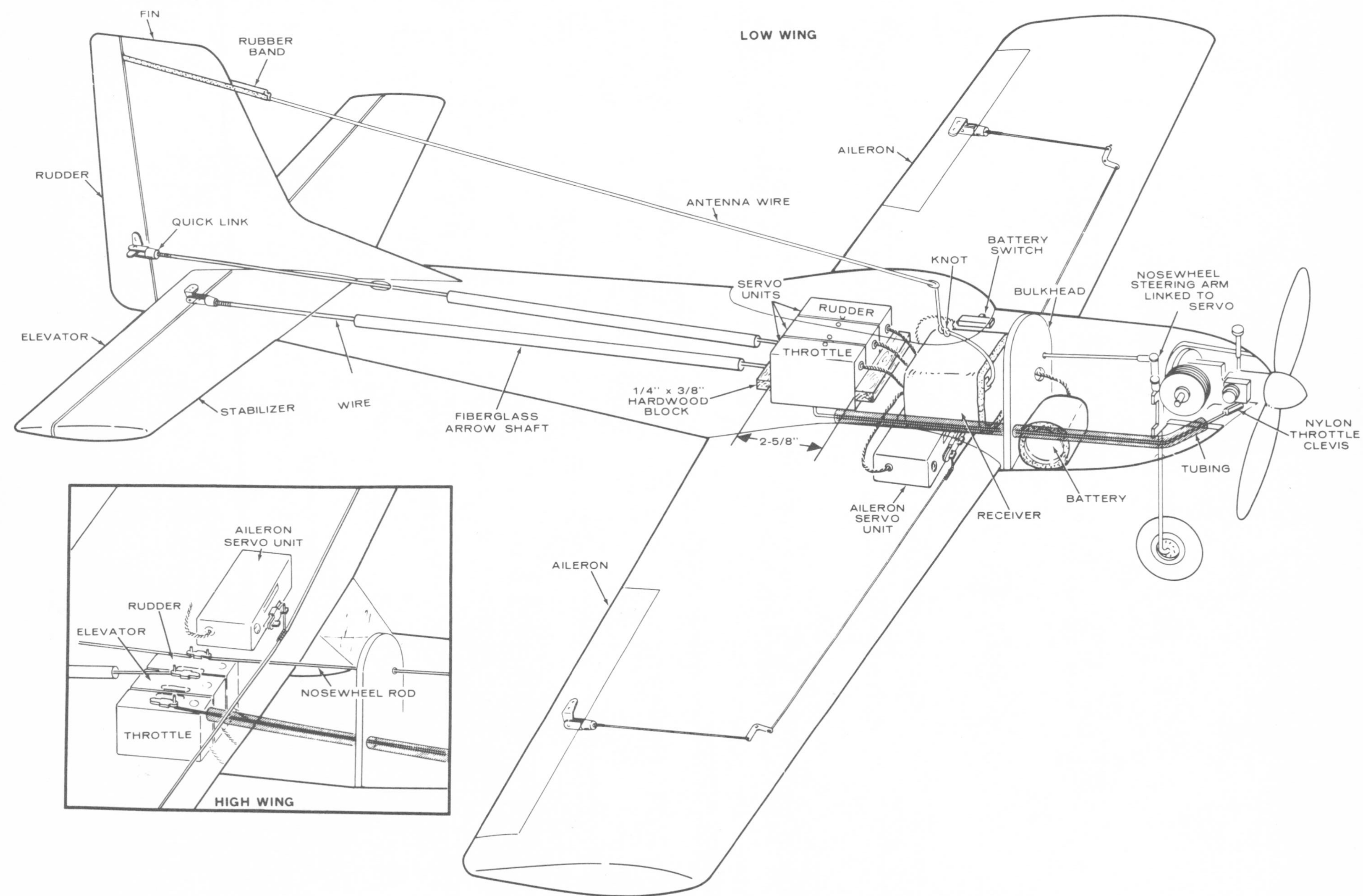

Fig. 20-2. Radio control in a model airplane. Many hobbies are electronically oriented. (Heath Company)

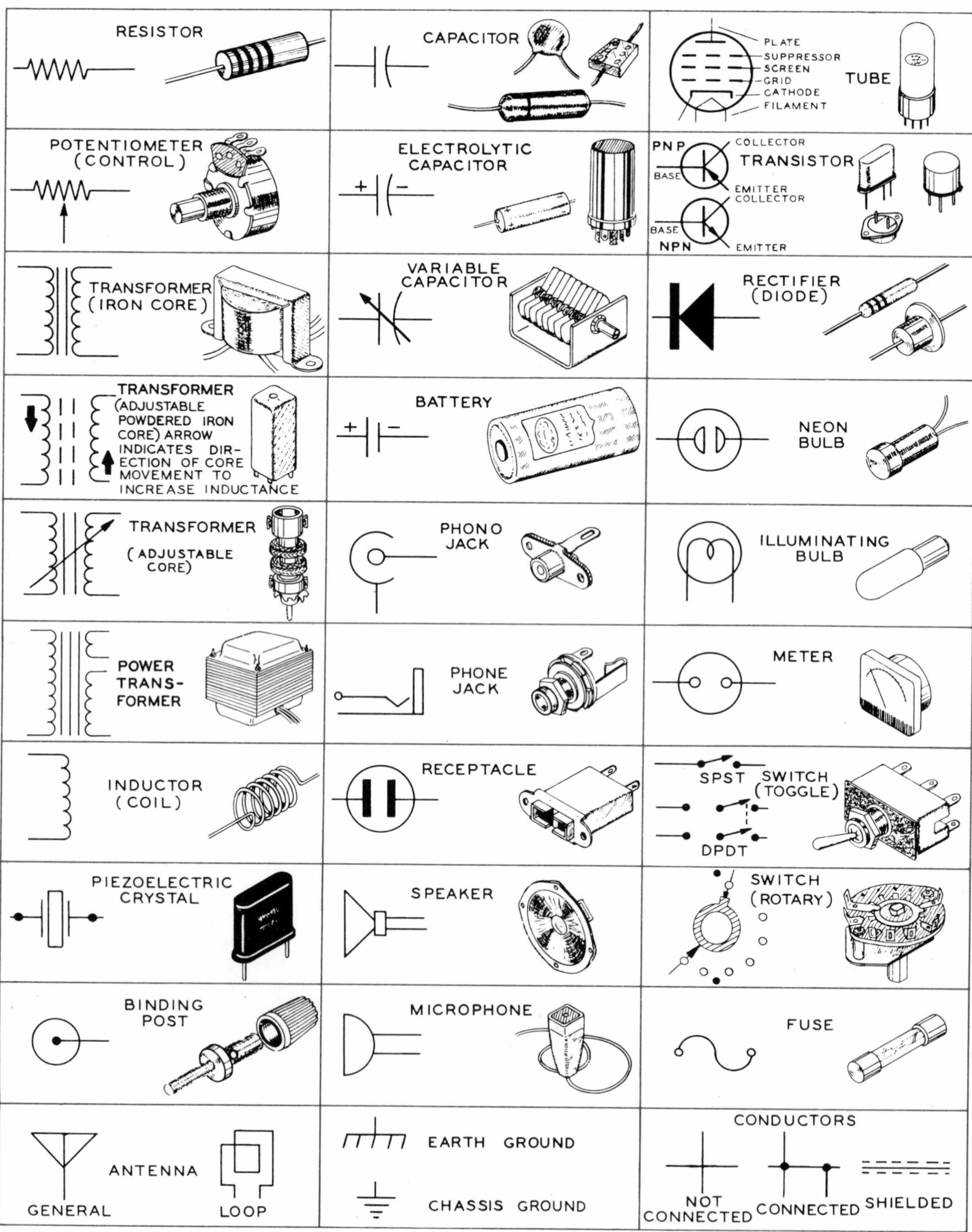

Fig. 20-3. Typical electronic symbols and related illustrations.

ELECTRICAL AND ELECTRONIC DRAFTING

Electrical and electronic drafting is done in much the same manner as conventional drafting. The same type equipment is needed.

However, instead of using regular multi-view drawings, a large portion of electrical and electronic drafting is diagrammatic in character. That is considerable use is made of symbols. The symbols, Fig. 20-3, represent the various components and wires that make up the electronic circuit. They are easier and quicker to draw than the actual part. Symbols are combined on a diagram that shows the function and relation of each component in the circuit.

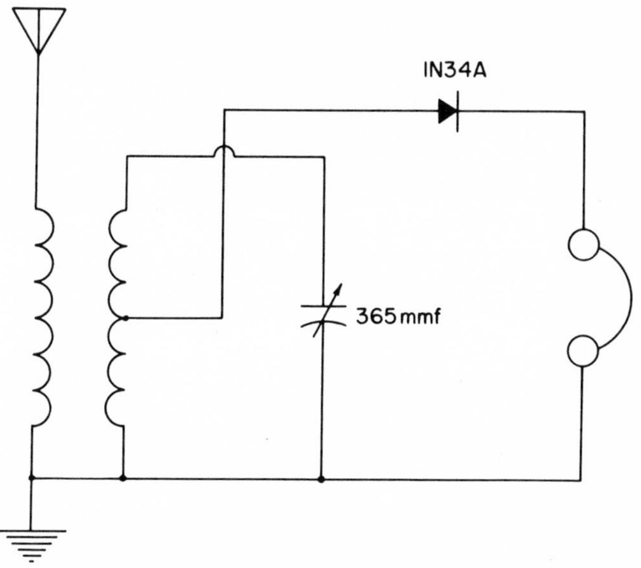

Fig. 20-4. A schematic diagram of a crystal radio.

TYPES OF DIAGRAMS USED IN ELECTRICAL AND ELECTRONIC DRAFTING

The draftsman working in this area of drafting must be familiar with the different types of diagrams.

SCHEMATIC DIAGRAM, Fig. 20-4. A drawing using symbols and single lines to show the electrical connections and functions of a specific electronic circuit. The various components that make up the circuit are drawn without regard to their actual physical size, shape or location.

CONNECTION OR WIRING DIAGRAM, Fig. 20-5. While commonly used to show the distribution of electricity on architectural drawings, this type of diagram may be drawn to show

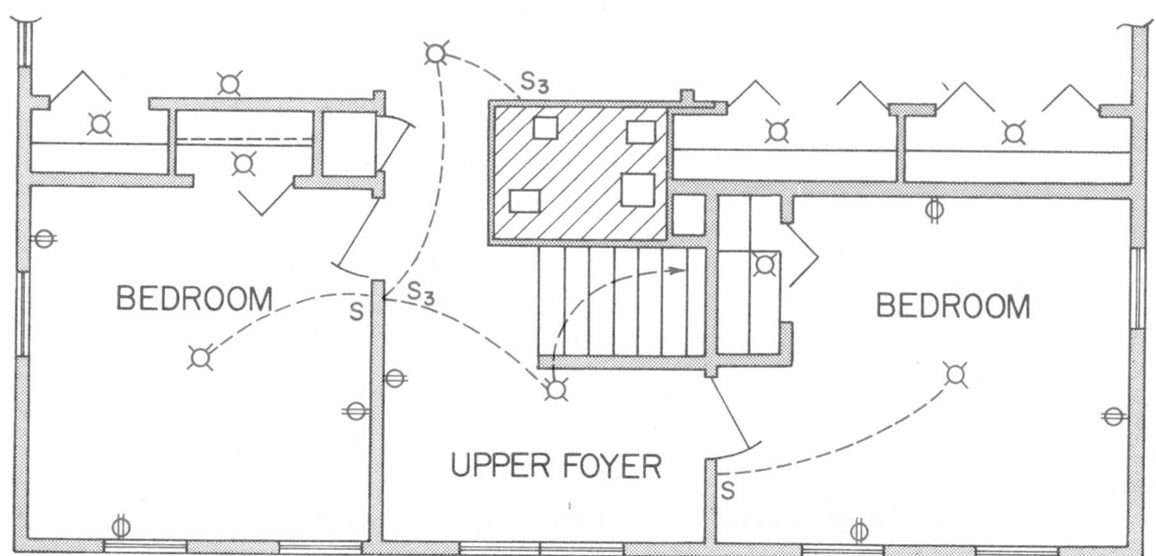

Fig. 20-5. A wiring diagram showing the location of switches, plugs and lighting in a modern home.

the general physical arrangement of the transistors, diodes, resistors, switches, etc., that make up an electronic circuit, Fig. 20-6.

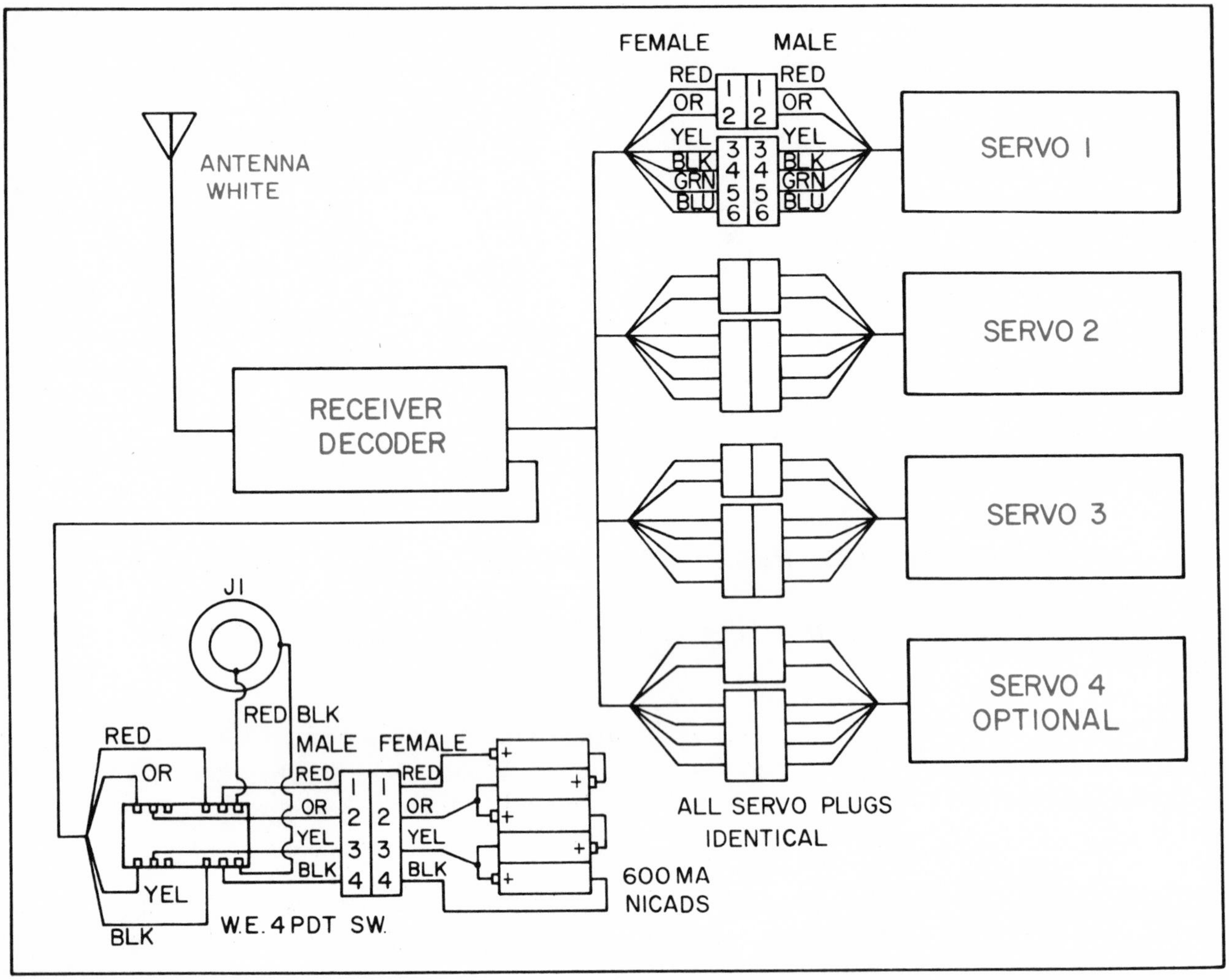

Fig. 20-6. A wiring diagram of a radio control installation similar to the unit shown in Fig. 20-2.

BLOCK DIAGRAM, Fig. 20-7. A simplified way to show the operation of an electronic device. This utilizes "blocks" (squares and/or rectangles) that are joined by a single line. It reads from left to right.

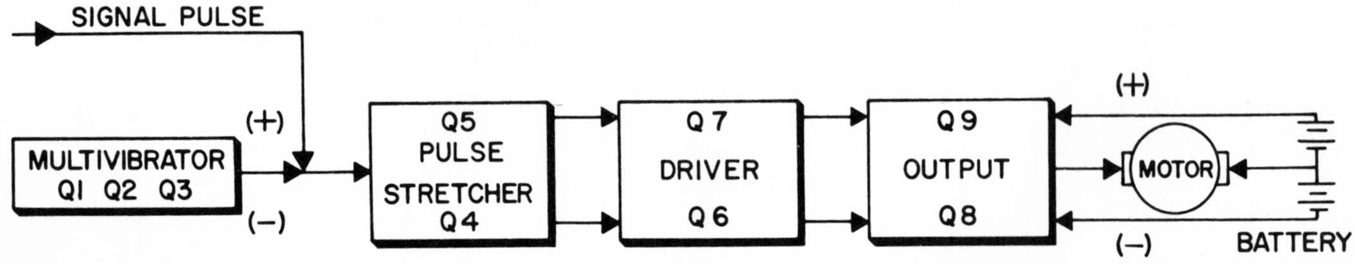

Fig. 20-7. A block diagram of servo that may be used to activate the controls of a radio controlled model airplane, boat or car.

PICTORIAL DIAGRAM, Fig. 20-8. In this diagram, the components are drawn in pictorial form and in their proper location. The pictorial diagram is used extensively by electronic kit manufacturers because it is so easy to understand.

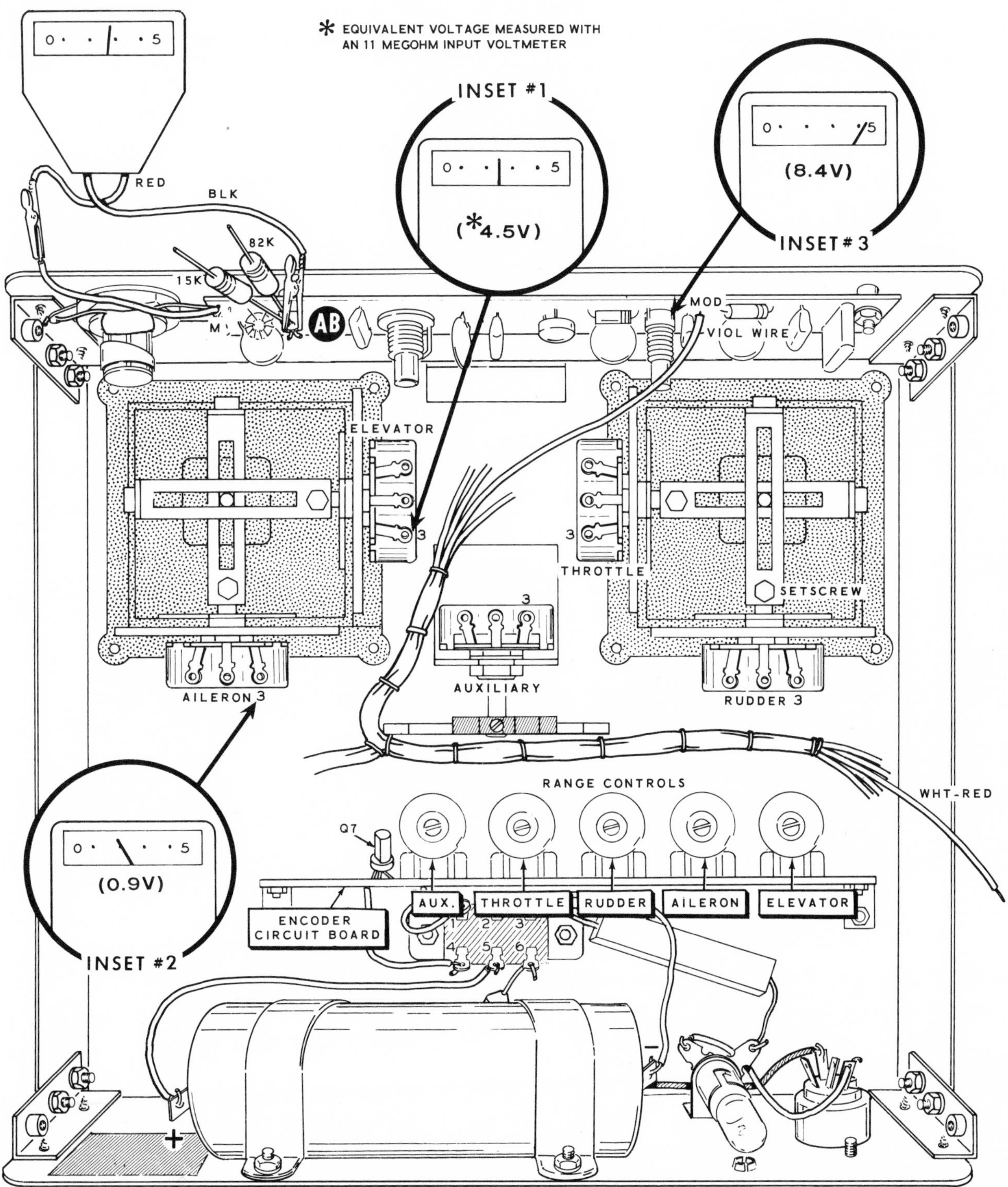

Fig. 20-8. A pictorial diagram in which electrical components are drawn in pictorial form. (Heath Company)

DRAWING ELECTRICAL AND ELECTRONIC SYMBOLS

Symbols in circuit diagrams need not be drawn to any particular scale. However, they should be shaped correctly, large enough to be seen clearly and in proportion, Fig. 20-9.

Symbols and lines are drawn the same weight as a visible object line. A darker line may be used when a portion of the diagram must be emphasized.

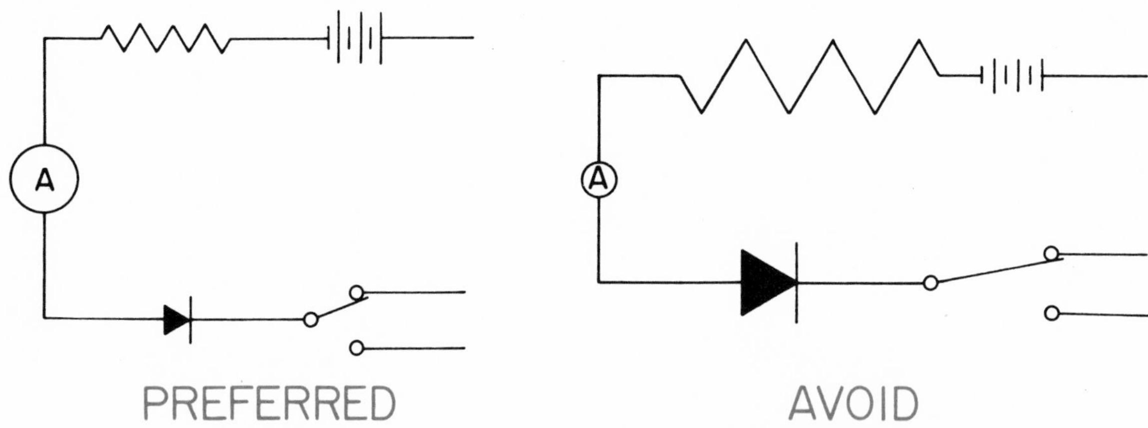

Fig. 20-9. Draw symbols in proportion.

TEST YOUR KNOWLEDGE - UNIT 20

1. How does electrical and electronic drafting differ from conventional drafting?
2. Why are symbols used to represent electrical and electronic components on drawings?
3. A SCHEMATIC DIAGRAM is a drawing ______________________________.
4. How does a BLOCK DIAGRAM differ from a WIRING DIAGRAM?
5. What is a PICTORIAL DIAGRAM?
6. Symbols and lines used in electrical and electronic drafting are drawn the same weight as a __________ __________ line.
7. The electrical diagrams used on house plans are called ________________ diagrams.

OUTSIDE ACTIVITIES

1. Make a schematic diagram of a desk lamp.
2. Draw the following electronic symbols: (space them uniformly on the sheet).

   a. battery (single cell)
   b. battery (three cell)
   c. antenna
   d. crystal
   e. crystal diode
   f. fuse
   g. ground
   h. headphones
   i. jack
   j. lamp
   k. capacitor (variable)
   l. line plug
   m. loud speaker
   n. microphone
   o. variable resistor
   p. switch
   q. transistor
   r. resistor
   s. wires connected
   t. wires not connected
3. Prepare a schematic diagram of a two cell flashlight.

4. Make a wiring diagram of your bedroom.
5. Secure a small battery powered toy and examine how it operates. Make a suitable diagram showing your findings.
6. Prepare a pictorial diagram of a crystal radio set.
7. Make a schematic diagram of a single tube radio receiver.
8. Make a suitable diagram showing four batteries, switch and lamp wired in series.
9. Make a suitable diagram showing four batteries, switch and lamp wired in parallel.

Fig. 21-1. Typical small home. Plans for this home will be given in this Unit. (National Plan Service, Inc.)

Unit 21
ARCHITECTURAL DRAFTING

Plans which provide craftsmen with the information needed to construct buildings in which people live, work and play are called ARCHITECTURAL DRAWINGS. See Fig. 21-1.

Buying a home is probably one of the largest investments you will make in your lifetime. An understanding of the basic principles of architectural drafting will be a great help if you plan to design or construct a new home, modernize an existing home, or judge the soundness and value of a home offered for sale.

The ability to read and interpret architectural drawings is essential to those in the construction industry such as carpenters, masons, plumbers, electricians, roofers, etc. It is also useful to workers in lumber yards, hardware and building supply stores.

BUILDING A HOME

To make sure your home is built to your specific requirements, it is important that you have a good plan and a well defined contract with your builder.

Plans provide the vast amount of information needed to construct a modern home. They should incorporate applicable aspects of the local and state building codes.

BUILDING CODES

Building codes are laws which provide for the health, safety and general welfare of the people in the community.

Building codes are based on standards developed by government and private agencies.

BUILDING PERMIT

In most communities, the building contractor or owner must file a formal application for a building permit. Plans and specifications are submitted for the proposed structure. They are reviewed by building officials to determine whether they meet local building code requirements.

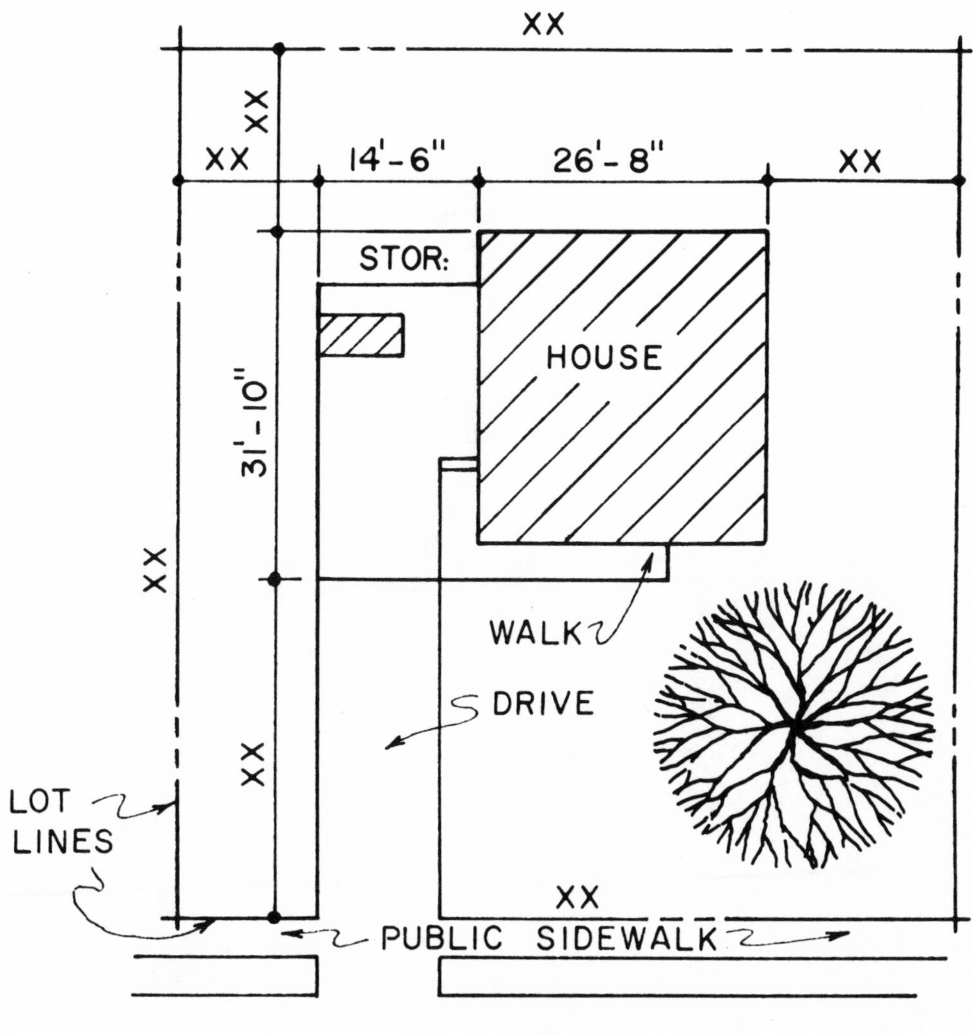

PLOT PLAN

Fig. 21-2. A plot plan for the house shown in Fig. 21-1.

An inspection card is usually posted on the building site and the work is inspected by local building officials as construction progresses. The card is signed as each phase of construction is approved.

PLANS

Because it is not possible to include all construction details on a single sheet, a set of typical house plans will include a plot plan, foundation and/or basement plan, floor plans, elevations (front, rear and side views of the home), wall sections, and built-in cabinet and fireplace details.

SCALE

The plans are generally drawn to a one-fourth inch scale, (1/4" = 1'-0"). This means that 1/4 in. on the drawing equals 1 ft. in the building being constructed.

A larger scale, 1" = 1'-0", is generally used when greater detail is needed on a structural part. Framing plans are often drawn to 1/8" = 1'-0" scale.

PLOT PLAN

The PLOT PLAN, Fig. 21-2, shows the location of the buildings on the building site, as well as walks, driveways and patios. Overall building and lot dimensions are included. Contour lines (lines which indicate the slope of the surface) are sometimes shown.

ELEVATIONS

ELEVATIONS are the front, rear and side views of the house, Fig. 21-3. They are made of lines which are visible when the building is viewed from various positions. Included on the elevations are the floor levels, grade lines, window and door heights, roof slope and the types of materials to be used on the walls and roof. Foundation and footing lines below grade are indicated with hidden lines.

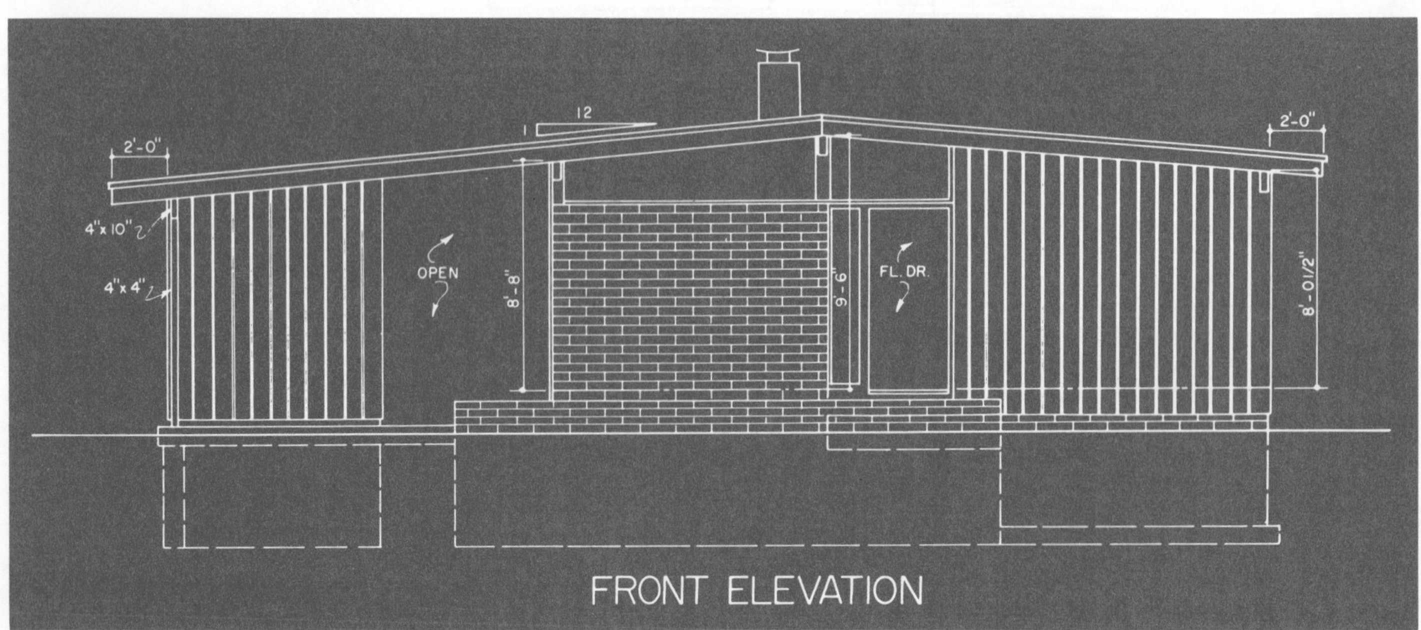

Fig. 21-3. Elevation drawing.

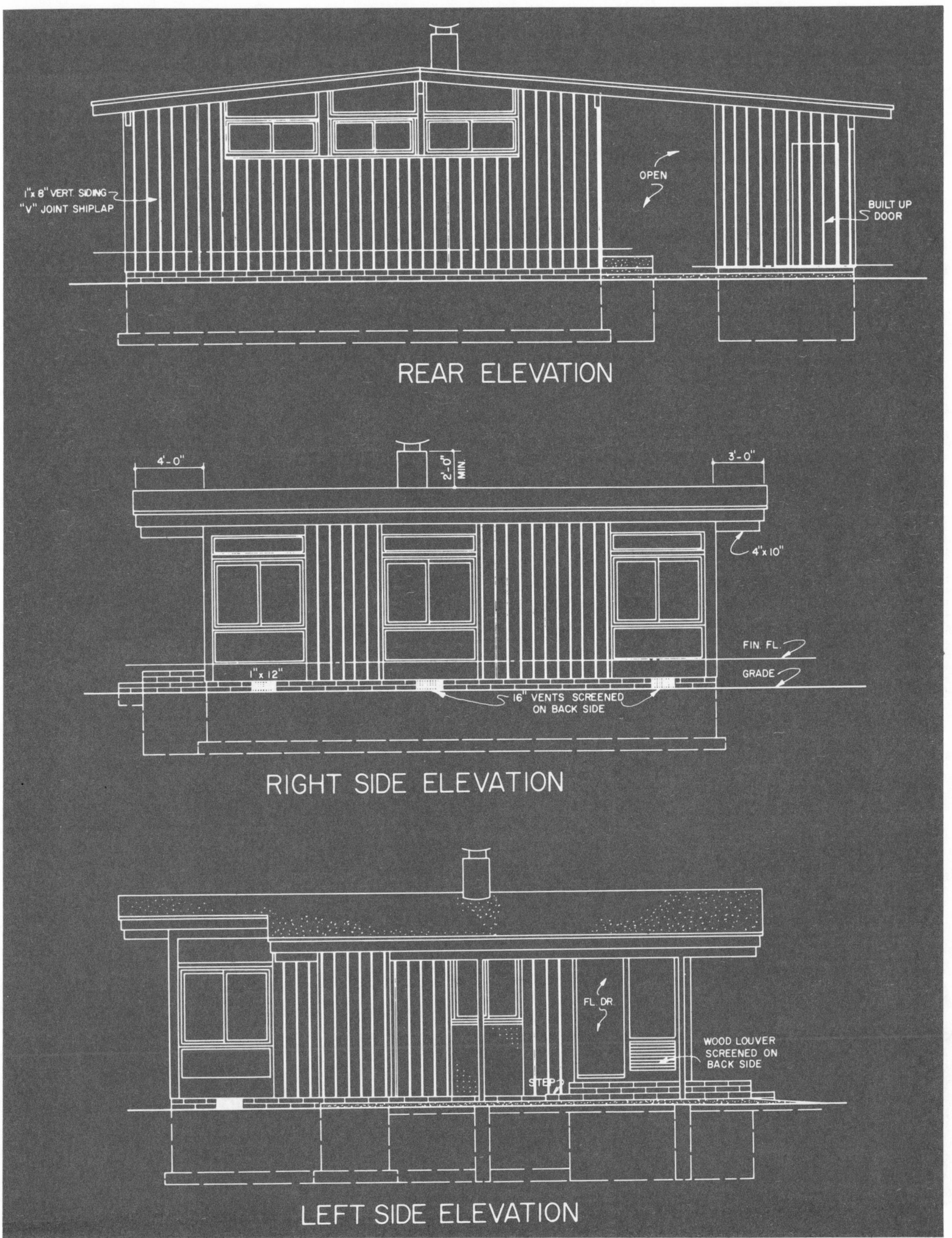

Fig. 21-3. (Continued)

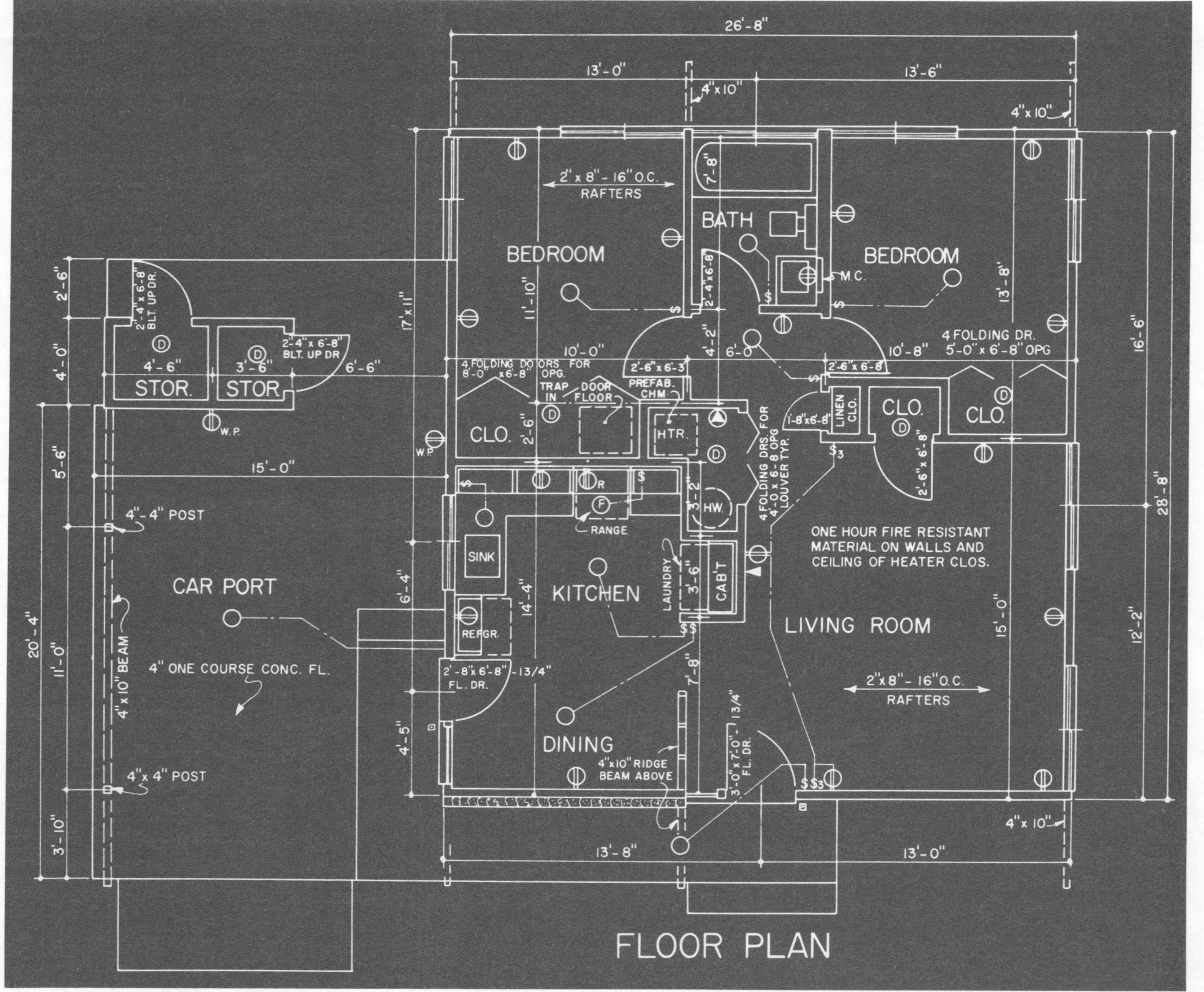

Fig. 21-4. ***Floor plan.*** *(National Plan Service, Inc.)*

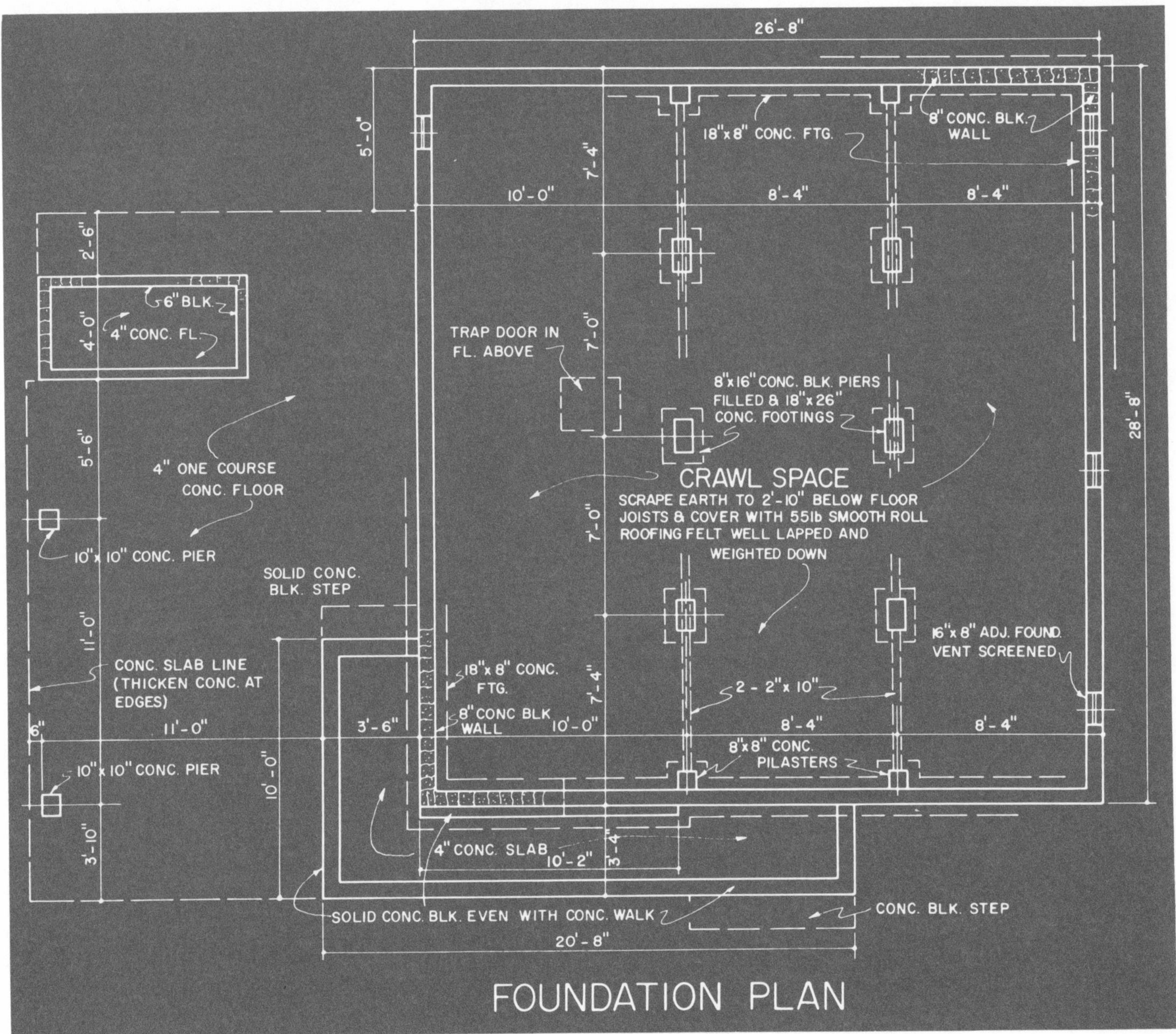

Fig. 21-5. Foundation plan.

FLOOR PLANS

The size and shape of the building, as well as the interior arrangement of the rooms are shown on the FLOOR PLAN, Fig. 21-4. Additional information such as the location and sizes of the interior partitions, doors, windows, stairs and utility installations (plumbing, electrical, etc.) is also included.

Foundation and basement plans are frequently combined on a single sheet, Fig. 21-5.

SECTIONS

SECTIONAL VIEWS, Fig. 21-6, are used to give construction details of the structure from basement or foundation footing, to the ridge of the roof. Break lines are frequently incorporated into the section to reduce the size of the drawing and save time in drawing the plan.

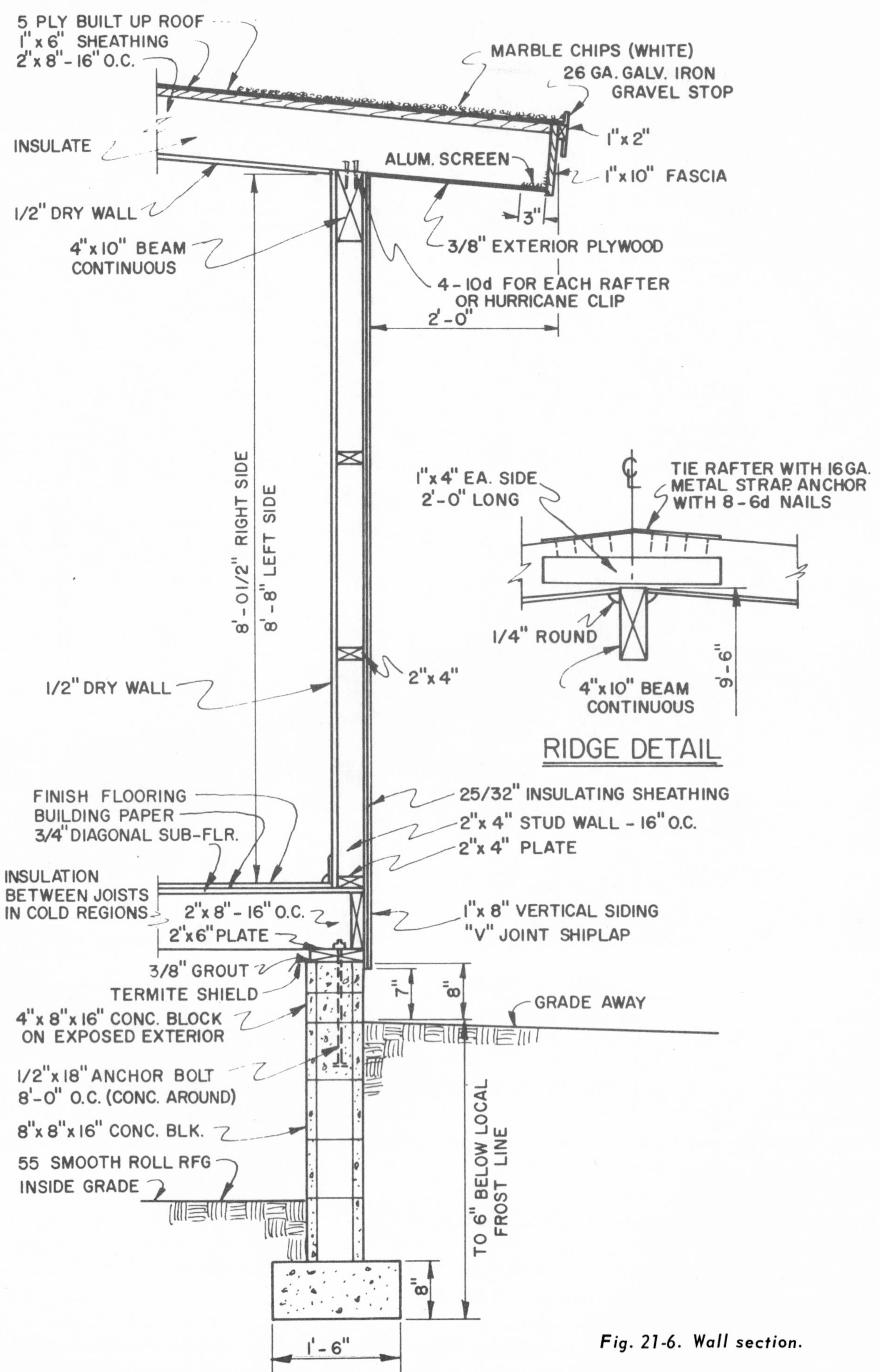

Fig. 21-6. Wall section.

Along with the sizes of framing materials, the types and kinds of materials to be used for sheathing, insulation, interior and exterior wall surfaces and the like are also indicated.

DETAILS

Information to assist the craftsmen in constructing such things as built-in cabinets, fireplaces and other pertinent information are given on DETAIL SHEETS, Fig. 21-7.

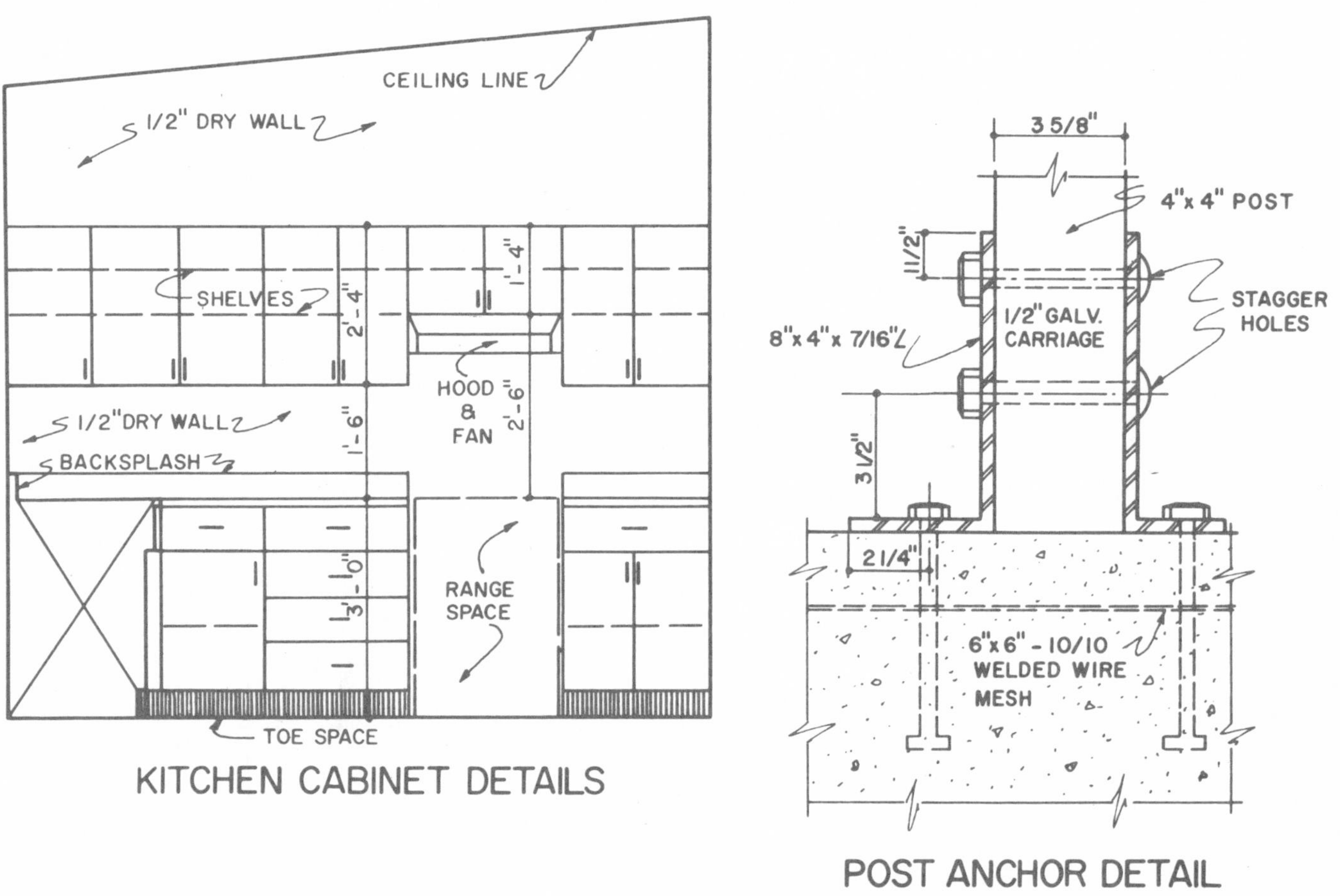

Fig. 21-7. Details found on a typical architectural drawing.

ARCHITECTURAL DRAFTING TECHNIQUES

In general, most conventional drafting techniques will apply to architectural drafting.

SYMBOLS, "SPEC" SHEETS

Extensive use of SYMBOLS, Fig. 21-8, will be noted in architectural drafting. Symbols are employed because it is not practical to show on the plans items such as doors, windows, plumbing fixtures, etc., as they would actually appear in the structure.

Sheets of specifications usually accompany house plans. These "spec sheets" describe in writing things that cannot be easily indicated on the drawings such as quality of materials, how installation of specific items are to be made, etc.

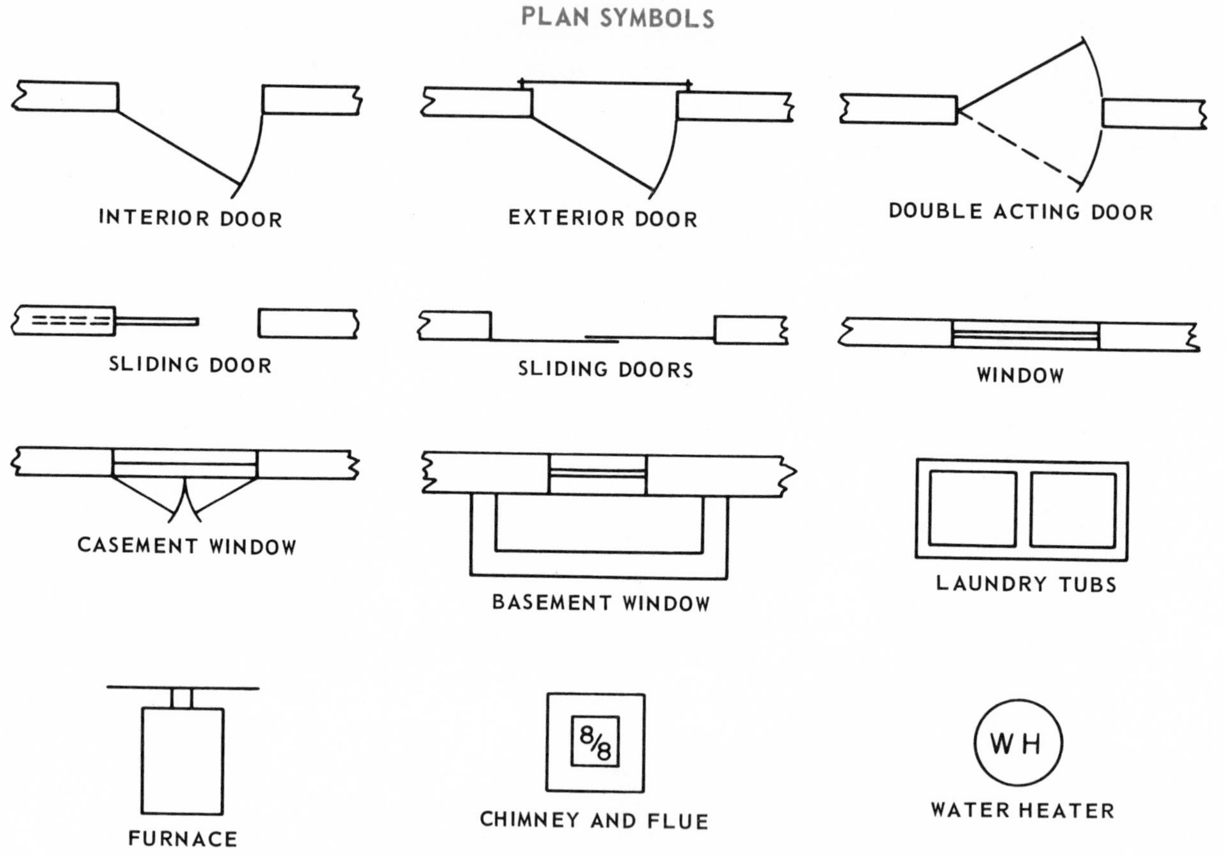

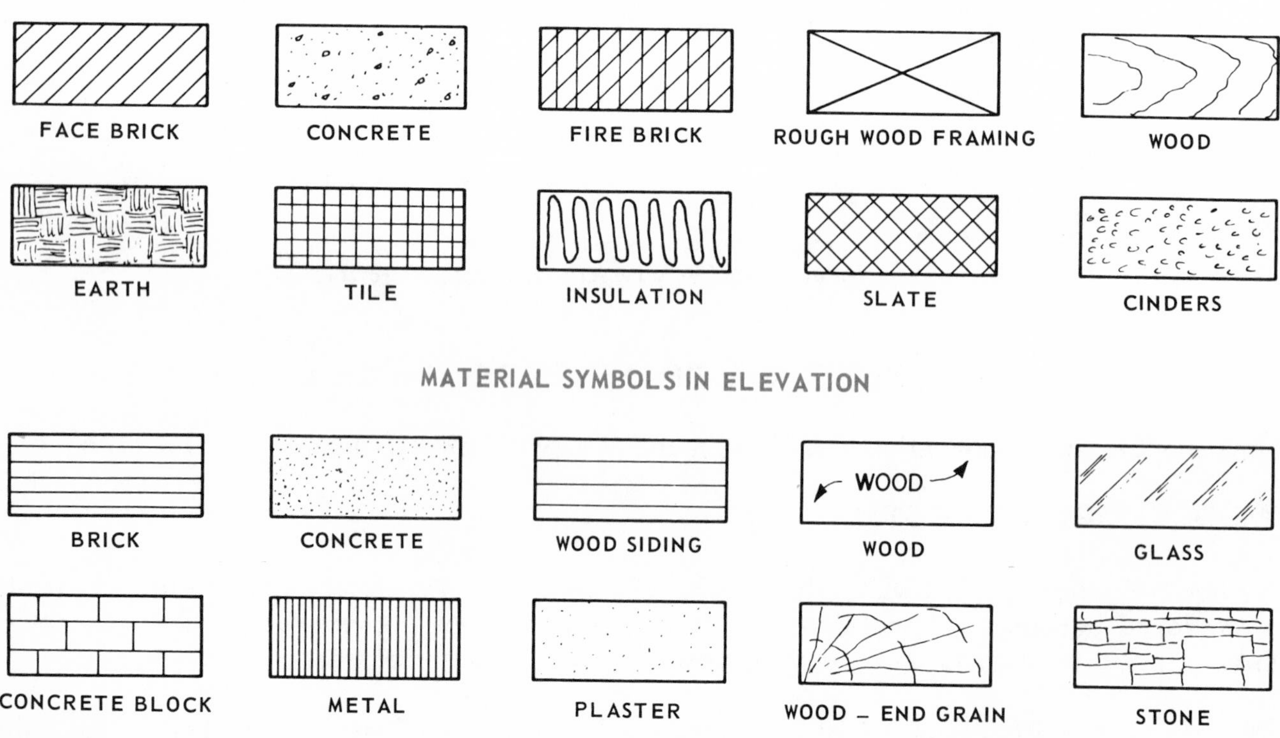

Fig. 21-8. Symbols used to indicate materials and fixtures on drawings.

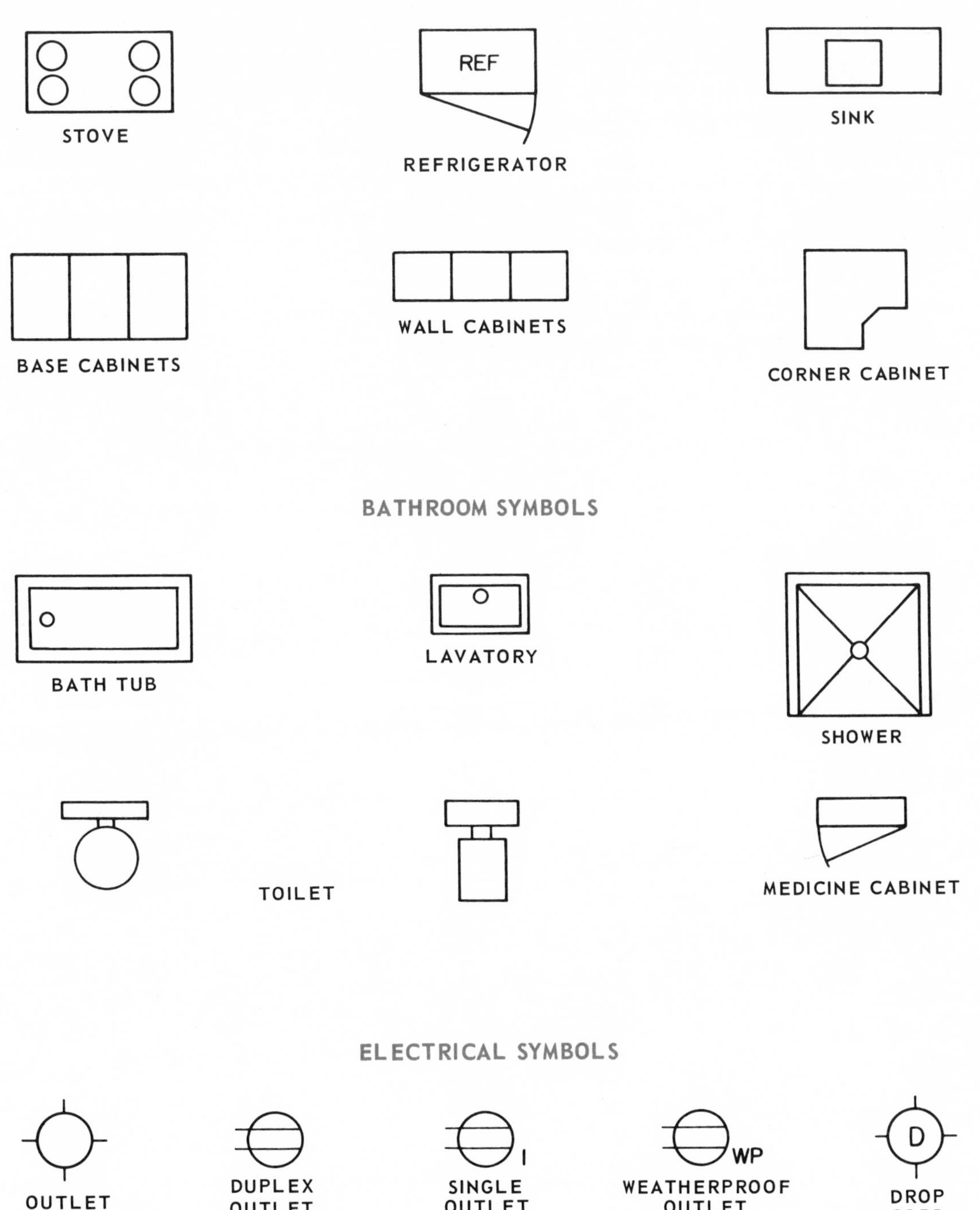

Fig. 21-8. (Continued)

SCALE

With few exceptions, architectural plans are drawn to scale. When drawn to scale, the views presented on the drawing are shown in an accurate proportion of the full-sized structure. A 1/4" = 1'-0" scale would mean that 1/4 in. on the drawing equals 1 ft. in the structure to be built.

Scale drawings are relatively easy to make if an ARCHITECT'S SCALE is used. This scale is divided into various major units, each unit representing a distance of one foot. Shown in Fig. 21-9 is a 1/4 " = 1'-0" unit of the scale.

This unit and the other units on the scale are further subdivided into equal parts or multiples of twelve to represent inches, Fig. 21-10.

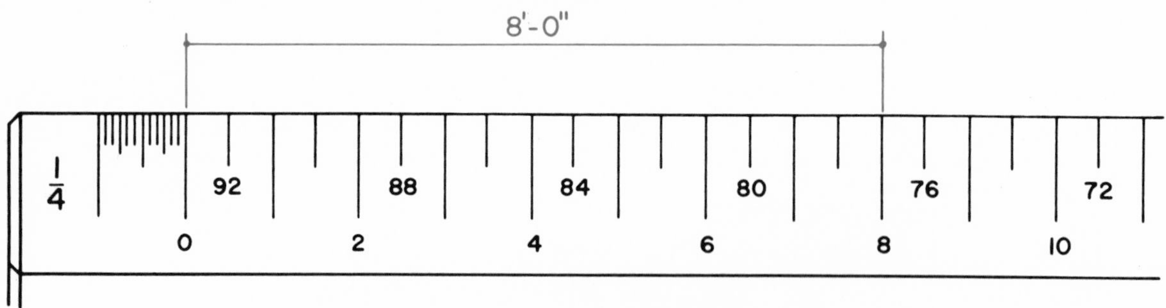

Fig. 21-9. Measuring feet on architect's scale.

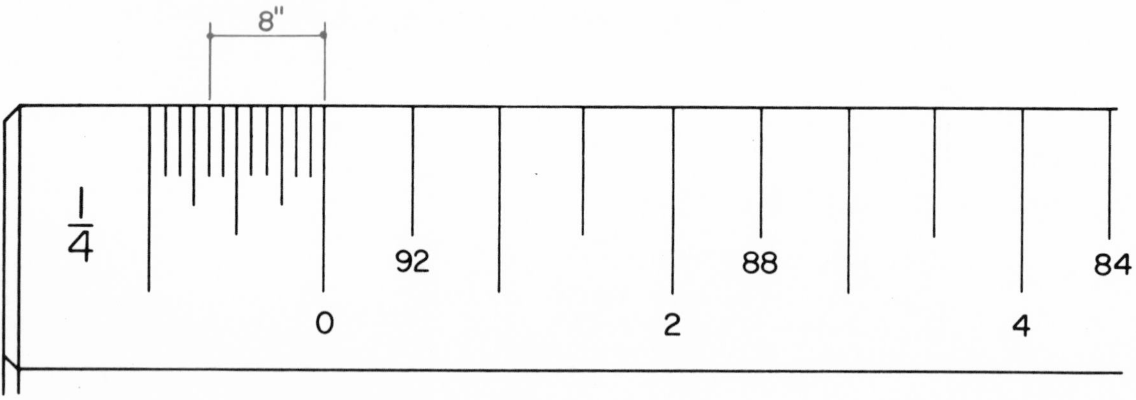

Fig. 21-10. Measuring inches on architect's scale.

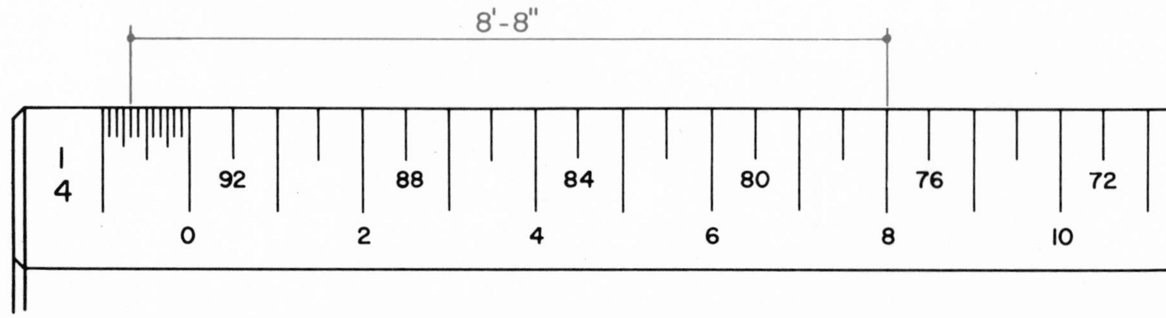

Fig. 21-11. Measuring both feet and inches on architect's scale.

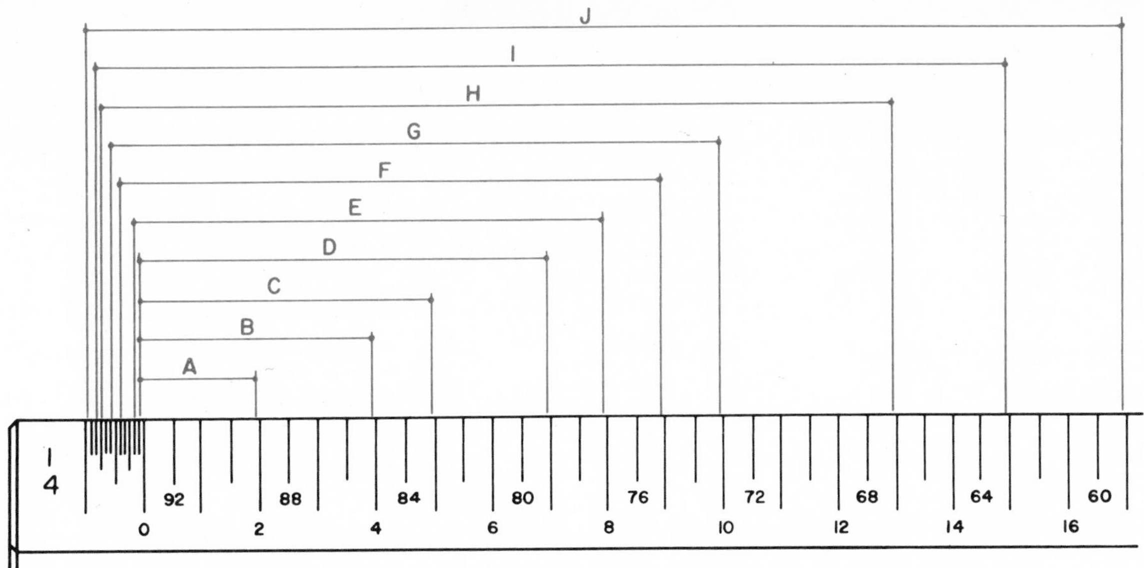

Fig. 21-12. How many of these dimensions can you read correctly?

When using the architect's scale to make a measurement, for example, 8'-8" (eight feet eight inches), start at the 0 and go to the right to locate 8' (eight feet). Then, moving in the opposite direction from 0 locate the 8" (eight inches). The combined measurement would be 8'-8", Fig. 21-11.

Fig. 21-12 will give you an opportunity to practice reading the architect's scale. List your answers on a separate piece of paper.

TITLE BLOCK

A TITLE BLOCK, Fig. 21-13, for architectural drawings should include the following:

1. Type of structure (house, garage, shed, etc.).
2. Where it is to be located.
3. Architect's name (and/or draftsman).
4. Date.
5. Drawing scale or scales to be used.
6. Sheet number and number of sheets making up the full set of drawings (Sheet 1 of 10, etc.).

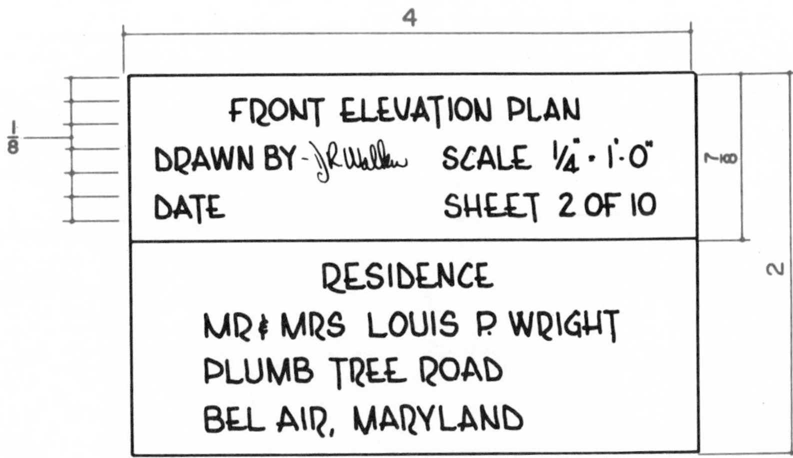

Fig. 21-13. Plan title block.

The title block may be made any convenient size; however, 2 in. by 4 in. is a suitable size for most home plans. It is usually located in the lower right corner of the drawing sheet.

LETTERING

While the conventional single stroke Gothic lettering is sometimes used on architectural drawings, most architectural lettering follows the Roman alphabet of letters and numbers, Fig. 21-14. Many architects and architectural draftsmen develop their own distinctive style

A B C D E F G H I J K L M N O P Q R S T

U V W X Y Z & 1 2 3 4 5 6 7 8 9 0

a b c d e f g h i j k l m n o p q r s t u v w x y z

Fig. 21-14. One style of architectural lettering.

of lettering. Whether you decide to use the single stroke Gothic lettering or develop a style of your own, remember that your lettering must be legible.

DIMENSIONING

There is only a slight difference between the dimensioning techniques used on architectural drawings and on other forms of drafting. The dimensions are read from the bottom and right sides of the drawing sheet. Measurements over twelve inches are expressed in feet and inches. Foot and/or inch marks (' and ") are used on all dimensions. Dimensions may be placed on all sides of the drawing and through the drawing itself. Dimension figures, usually 1/8 in. high, are written above the dimension line which may be capped with arrowheads or small darkened circles, Fig. 21-15.

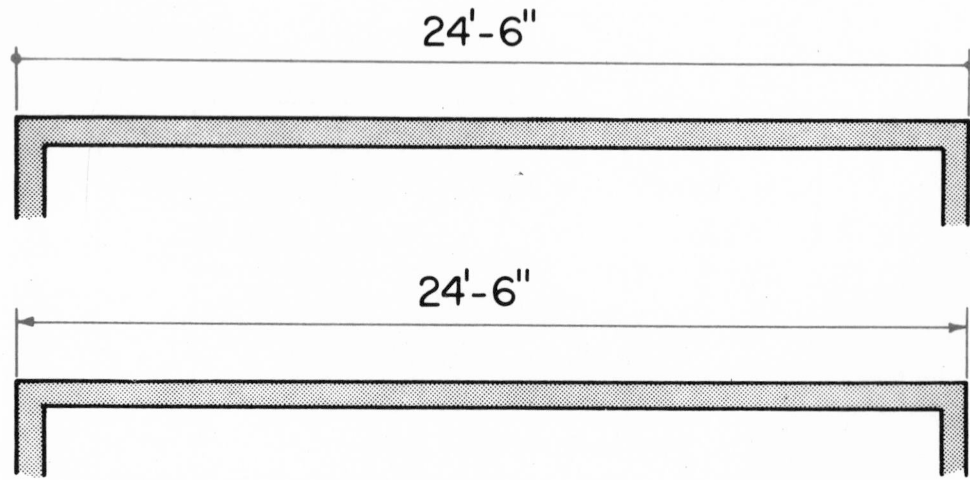

Fig. 21-15. Ends of dimension lines may be capped by arrowheads, or darkened circles may be used.

DIMENSIONING RULES

1. Make sure that the dimensioning is complete. A builder should never have to assume or measure a distance on the drawing for a dimension.
2. ALWAYS letter dimensions full-size regardless of the scale of the drawing.
3. Dimensions and notes should be at least 1/4 in. away from the view.
4. Align dimensions across the view wherever possible.
5. Place overall dimensions outside the view.
6. Overall dimensions are determined by adding detail dimensions. Do not scale the drawing.
7. Dimension partitions center to center or from center to outside wall.
8. Room size may be indicated by stating the width and length of the room.
9. Windows, doors, beams, etc., are dimensioned to their centers.
10. Window and door sizes and types are given in a SCHEDULE, Fig. 21-16.

WINDOW SCHEDULE					
MARK	REQ'D	OVERALL SASH SIZE			DESCRIPTION
		WIDTH		HEIGHT	
W-1	4	5'-4"	x	4'-0"	ALUMINUM SLIDING WINDOW WITH INSULATING SASH FIBERGLASS SCREENS
W-2	3	5'-4"	x	2'-0"	SAME AS ABOVE
W-3	1	4'-0"	x	3'-0"	"
W-4	1	2'-7"	x	5'-8"	1/4" PLATE GLASS
W-5	1	9'-6"	x	2'-4"	FIXED, SAME AS ABOVE

Fig. 21-16. Window schedule of the type found on drawings.

ARCHITECTURAL ABBREVIATIONS

Time and space can be saved when lettering architectural drawings by using abbreviations. Following are some commonly used:

Asphalt Tile - AT
Beam - BM
Bedroom - BR
Brick - BRK
Ceiling - CLG
Center Line - CL or ℄
Concrete - CONC
Drawing - DWG
Door - DR
Elevation - EL
Exterior - EXT
Floor - FLR
Footing - FTG
Grade Line - GL
Lavatory - LAV
Living Room - LR
Plaster - PL
Room - RM

OBTAINING INFORMATION

A great deal of research into catalogs and reference books must be made on the part of the student to find the many standard sizes that are essential in completing an accurate set of plans.

Additional information may be found in mail order catalogs, lumber and building supply company literature, from the Federal Housing Administration, and copies of your local building ordinances. Specialized textbooks in architecture, carpentry, plumbing and other building areas are excellent sources of information when planning a home or other structure.

PLANNING A HOME

When planning a home you must determine how much money will be available for building purposes. Some sources suggest that a home should not cost more than two and one-half times the person's yearly salary.

One way to determine the size of the home is by the square foot method. To set up a typical planning problem, assume that $18,000.00 is available to build the home.

Building costs vary from one area of the country to another area. A local contractor can give you the approximate cost per square foot of construction in your community. Let's assume the approximate cost of frame construction is $15.00 per square foot.

Divide the cost per square foot ($15.00) into the money available to build the house ($18,000.00) and you will find that a frame home of 1,200 square feet of floor area can be built. Outlines of homes having 1,200 square feet of floor area are shown in Fig. 21-17.

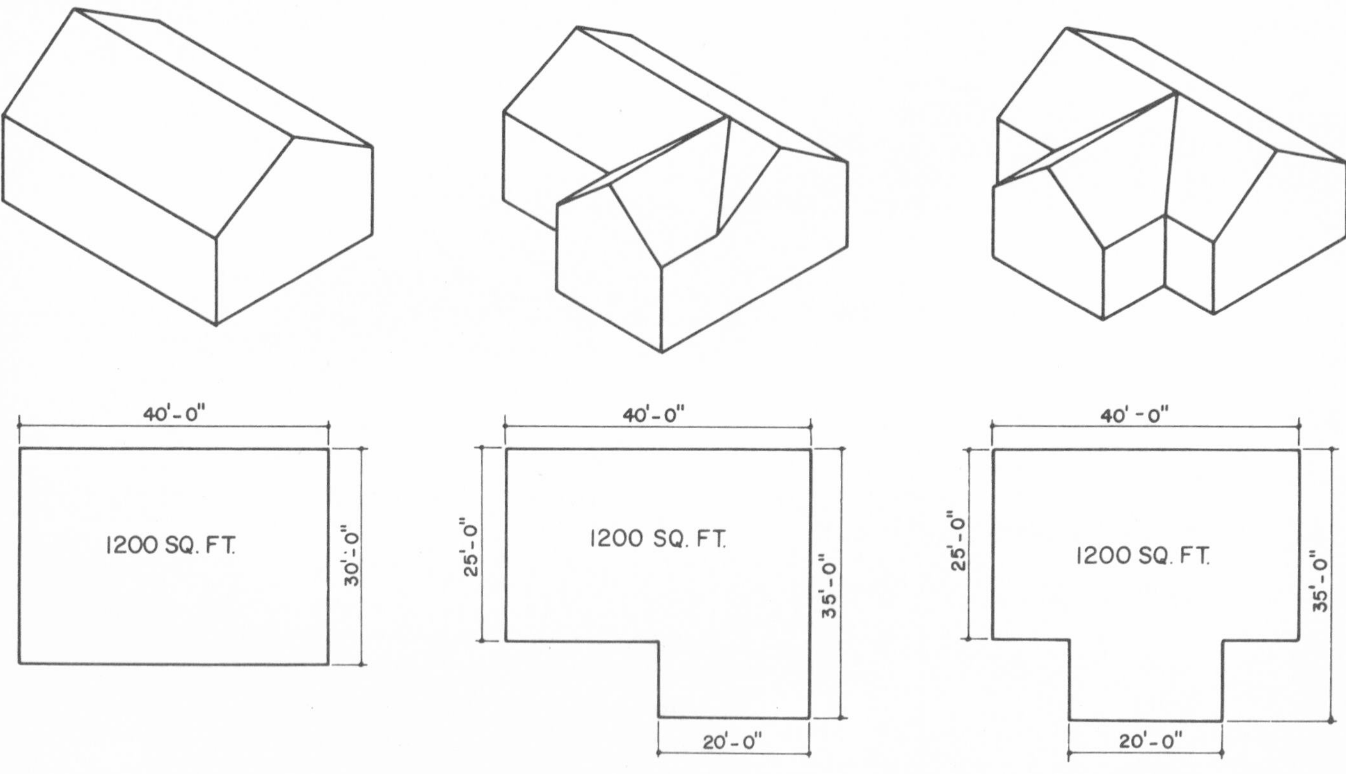

Fig. 21-17. Examples of floor areas of 1200 square feet.

After determining the approximate size and shape of the proposed house, you should try to develop a suitable floor plan. Room use and space available must be carefully considered and balanced.

Be sure to include closet and storage facilities. Using scale cutouts of the furniture that you plan to use in the various rooms will aid you in your planning. The cutouts MUST be made to the same scale as the floor plan. Prepare a scale drawing of the room showing all openings and the areas the doors take up when opened.

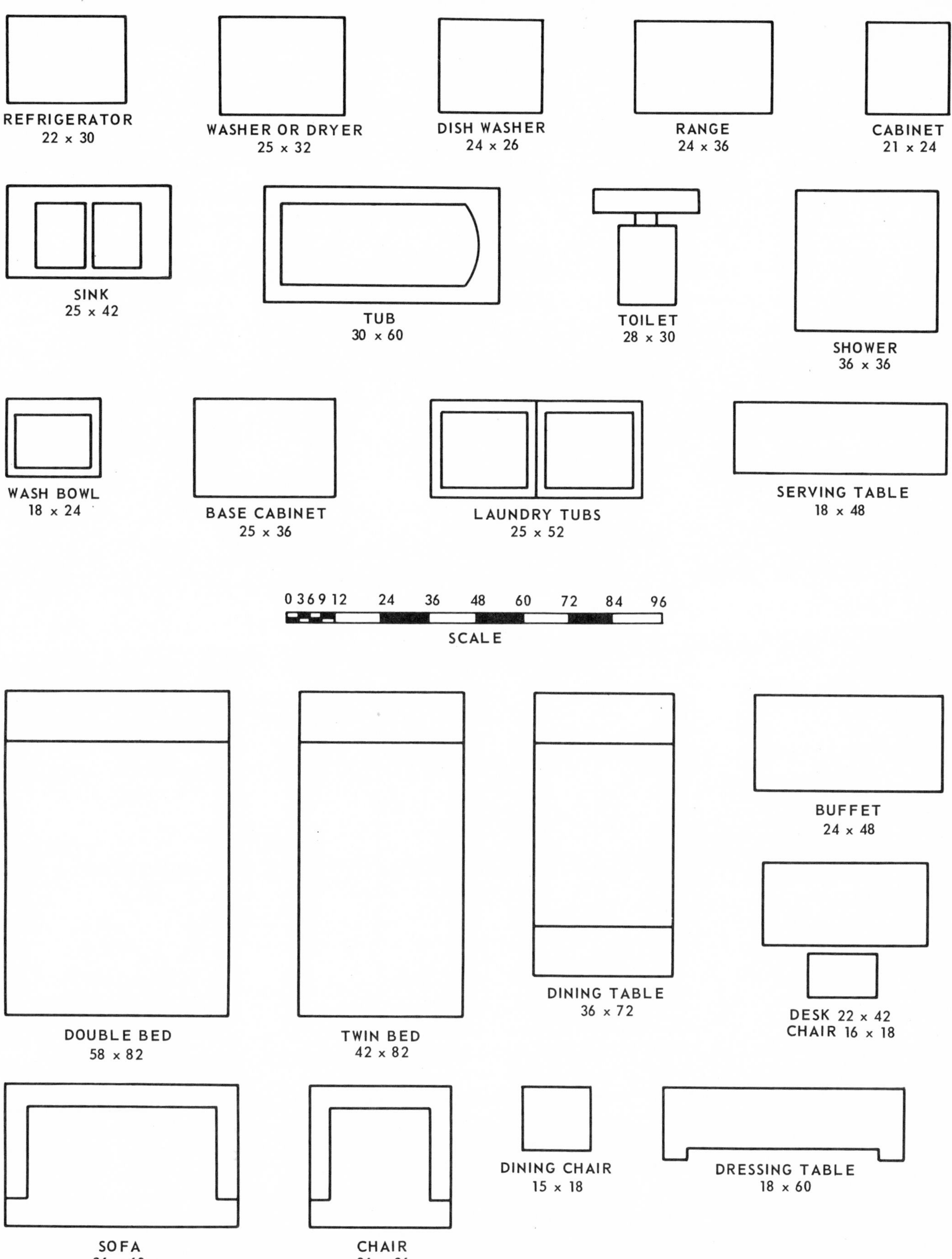

Fig. 21-18. Furniture dimensions that will aid in planning a home.

By moving the cutouts, it is easy to find out whether the room is large enough for the intended use.

Sizes of some furniture and fixtures are given in Fig. 21-18. Dimensions of items not shown may be found in manufacturer's catalogs, home magazines and mail order catalogs.

Cutout planning for a typical room is shown in Fig. 21-19.

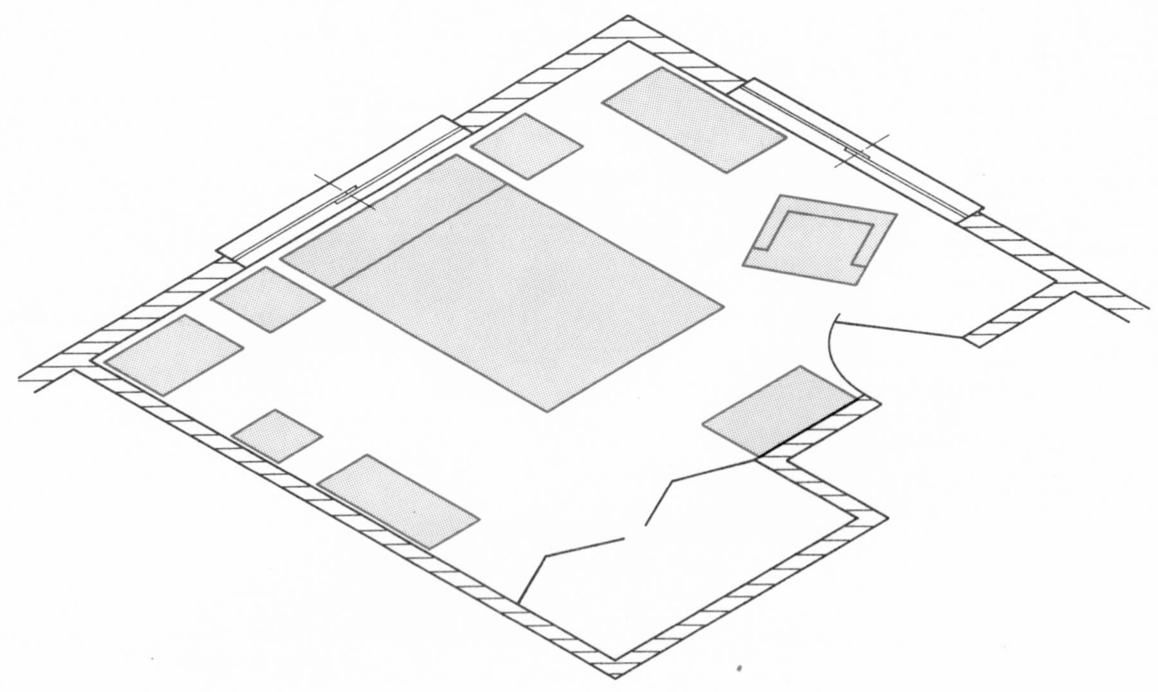

Fig. 21-19. How cutouts may be used to plan a bedroom.

TEST YOUR KNOWLEDGE - UNIT 21

1. Architectural drawings are ______________________________.
2. The ability to read and interpret architectural drawings is essential to workers such as:
   a. ____________________.
   b. ____________________.
   c. ____________________.
   d. ____________________.
   e. ____________________.
3. What are building codes?
4. Architectural drawings are generally drawn to a scale of ____________________.
5. The location of the buildings on the building sites is shown on the ______________. This drawing also shows the walks, driveways and patios.
6. Drawings that show the various views (front, rear and side views) of the structure are called ______________.
7. The floor plan shows the
   a. Size and shape of the rooms.
   b. Location and sizes of the windows and doors.

c. Location of the utilities (plumbing, heating and electrical fixtures).
d. All of the above.
e. None of the above.

8. Construction details of the structure from basement or foundation to the ridge of the roof are shown on ______________________ .
9. Make a sketch showing two recommended methods of dimensioning architectural drawings.

OUTSIDE ACTIVITIES

1. Make a scale drawing (1/4" = 1'-0") of the school drafting room. Use cutouts to determine an efficient layout of the room.
2. Prepare a drawing of your bedroom showing the location of the furniture in the room. Use a scale of 1/2" = 1'-0".
3. Draw the floor plan of a two-car garage with a workshop at one end.
4. Design and draw a full set of plans for a two room summer cottage or hunting cabin.
5. Draw the plans necessary to convert your basement into a workshop or recreation room.
6. Design and draw the plans necessary to construct a storage and tool house.
7. Design a small two bedroom house that can be constructed at minimum cost. Construct a model of it.
8. Plan a structure to house a sports car sales agency.

Unit 22
AUTOMATED DRAFTING

COMPUTER GRAPHICS is a term often applied to computer directed drafting systems. Drawing devices are used in conjunction with a computer to make written or drawn presentations. See Fig. 22-1. Computer directed systems are frequently used when the volume of graphic presentations makes it impractical to perform the task manually.

The development of computer directed drawing devices has taken several directions.

COMPUTER DIRECTED TRACER SYSTEM

With the complexity of products increasing, the need for technical handbooks and

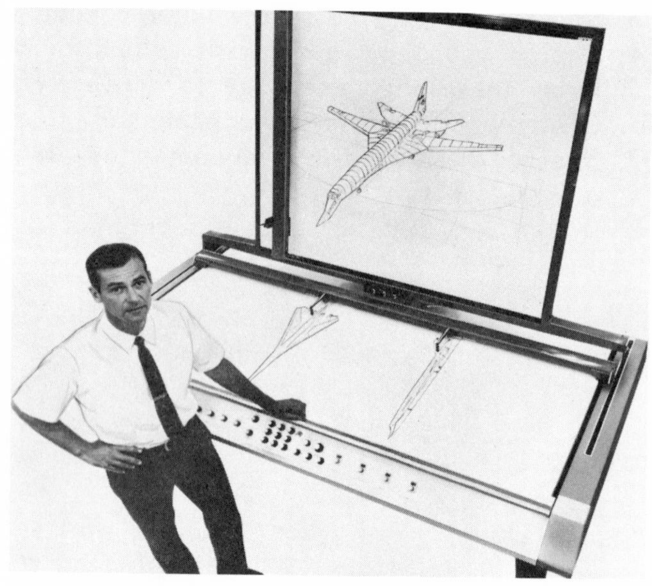

Fig. 22-2. Computer directed Illustromat 1100 is capable of making accurate perspective drawings. (Perspective Inc.)

Fig. 22-1. A typical computer center. (California Computer Products, Inc.)

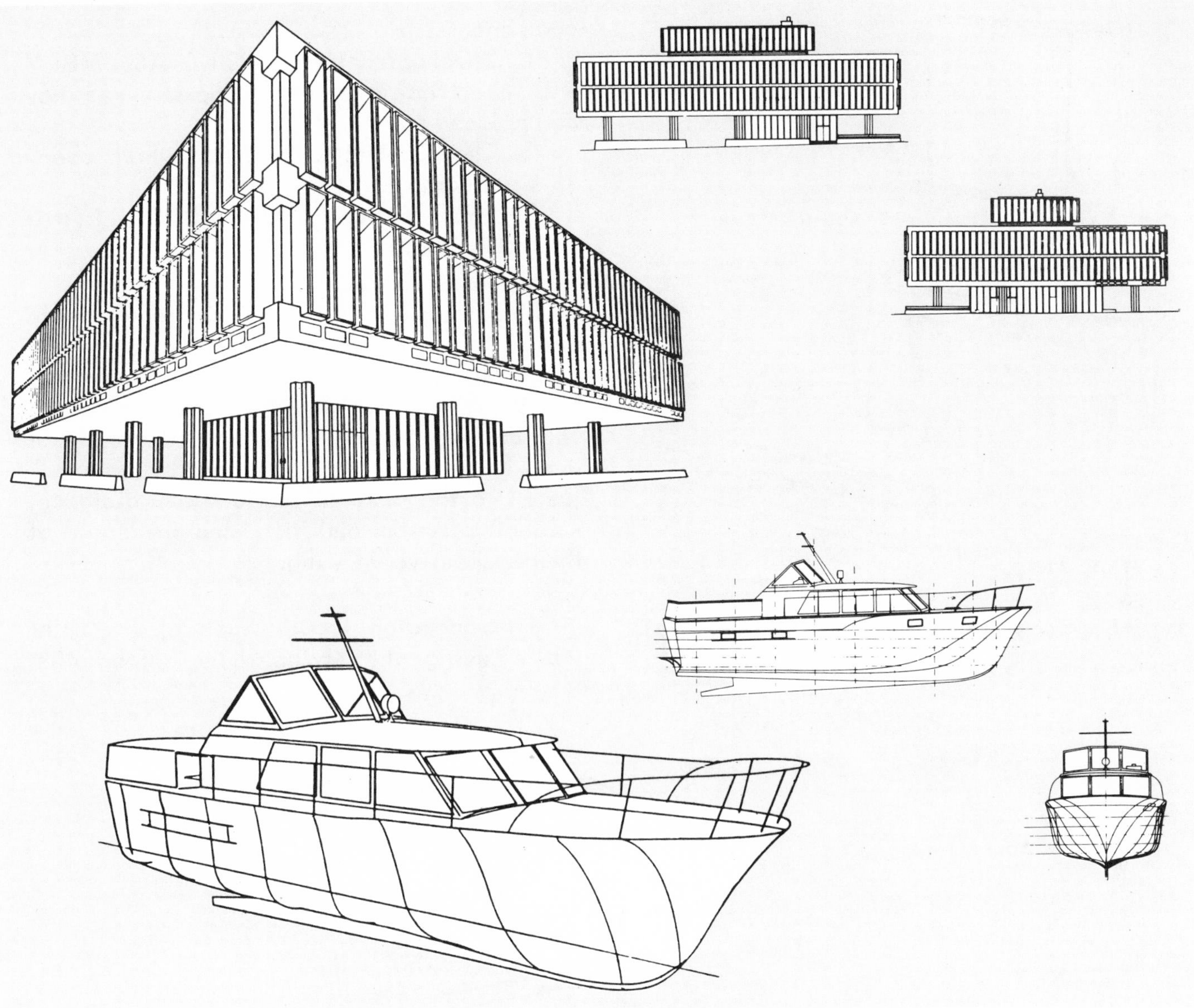

Fig. 22-3. Perspective drawings made by computer directed instrument from two views of an engineering drawing.

manuals becomes much greater. As an example, twenty-one different manuals are required to provide information on the operation and maintenance of one of our modern jet aircraft. The manuals would make a stack seven feet high if they were piled one on another.

To meet the demand for pictorial drawings such as this, the development of automated equipment has been necessary.

The computer directed instrument shown in Fig. 22-2 is one such device. This instrument can draw mathematically and mechanically accurate perspective, isometric, and stereoscopic 3-D views.

The finished view can be given any tilt and rotation from 0 to 360 degrees. Scaled enlargements or reductions can be made, and the missing third view can be produced from a drawing having only two views.

A special preliminary drawing need not be made. The operator sets up the machine and traces two views (side and end views) from the engineering drawing. The instrument completes the perspective. See Fig. 22-3. The unit consists of three primary

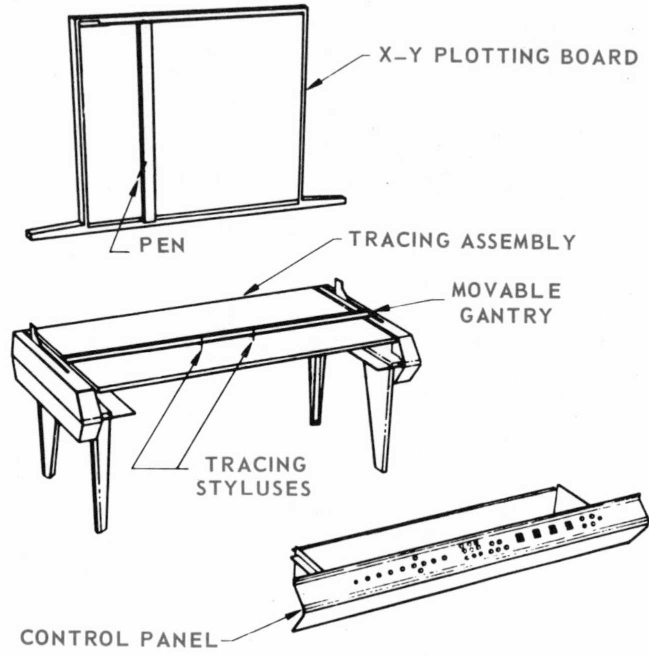

Fig. 22-4. Basic components of the Illustromat 1100. (Perspective Inc.)

components, Fig. 22-4:

1. The horizontal TRACING ASSEMBLY with two tracing styluses. The styluses may be replaced by pencils.
2. The CONTROL PANEL which operates the computer.
3. The vertically-mounted PLOTTING BOARD with a motorized pen.

Machine setup is rapid and comparatively simple. Two perpendicular (blueprint) views of the subject are placed on the horizontal board. A sheet of paper is attached to the vertical board on which the three-dimensional view is to be drawn. The controls are adjusted to the desired station point distance, subject rotation and tilt, and the scale of the perspective drawing.

The operator traces each of the print views using the stylus. Line information

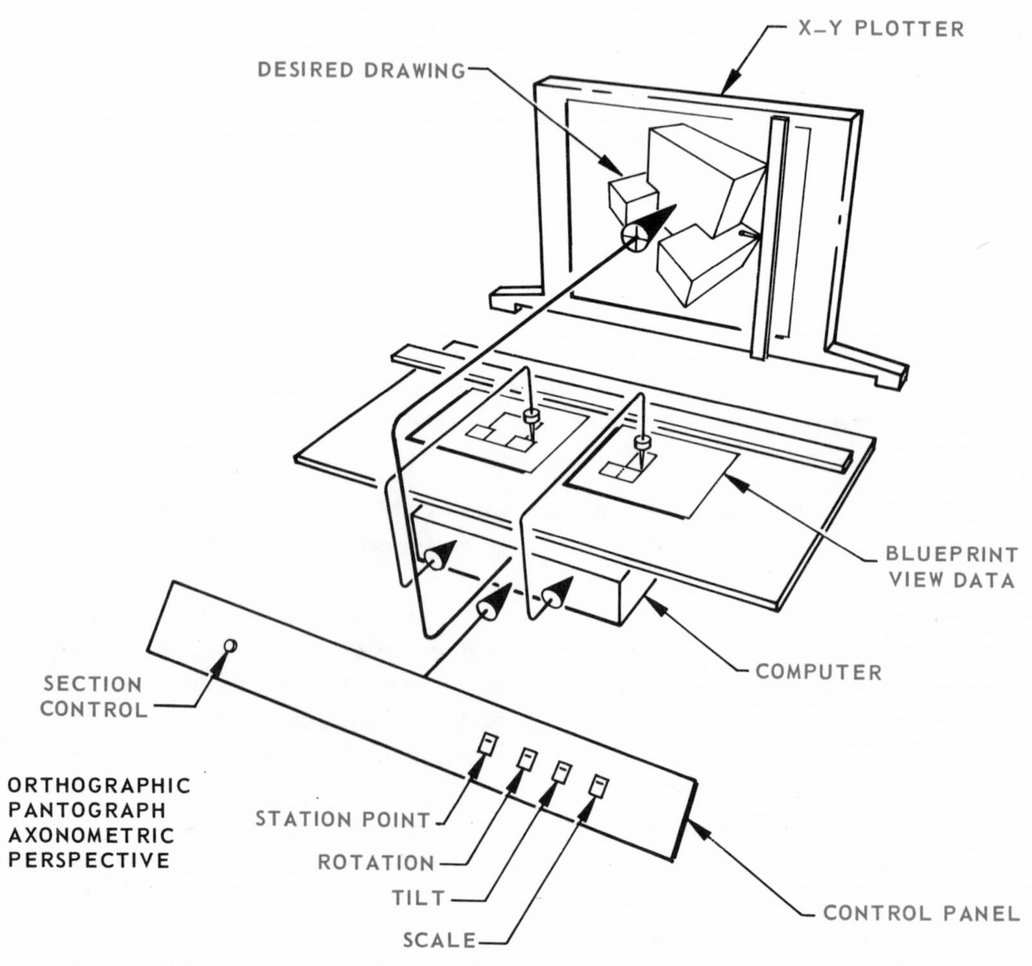

Fig. 22-5. Operation of the computer directed Illustromat 1100.

from the two print views of the subject is fed into the computer. The computer converts the two-dimensional data to three-dimensional data; then directs the pen on the vertical plotting board to make the drawing, Fig. 22-5.

Using this setup, accurate drawings (perspective, assembly, exploded view, etc.) can be completed quickly. See Fig. 22-6.

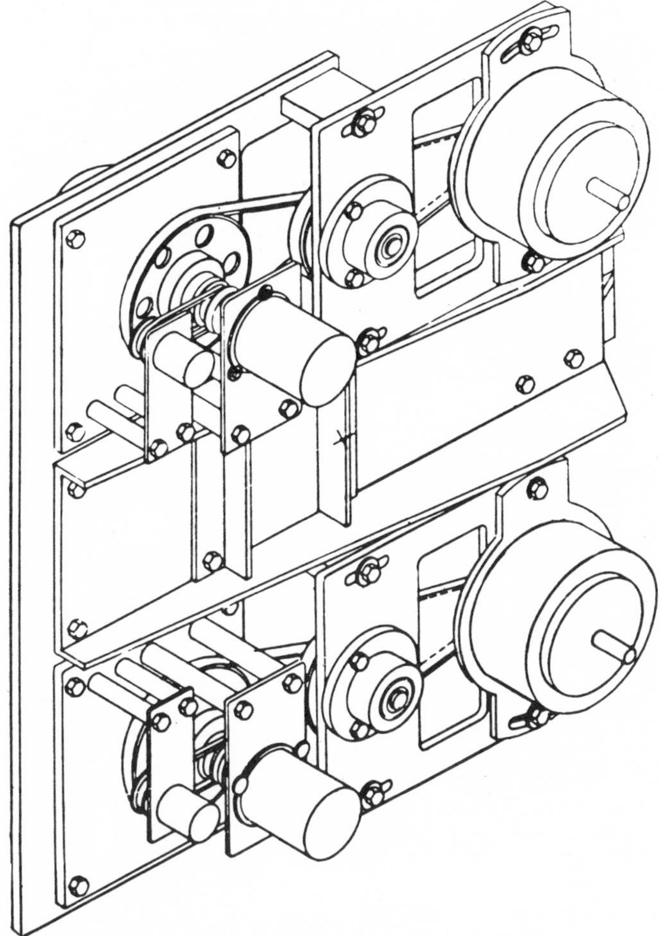

Fig. 22-6. Drawing made by computer directed device in 1.9 hours. (Perspective Inc.)

PHOTO-COMPOSING SYSTEM OF AUTOMATED DRAFTING

Another approach to automated drafting makes use of a photo-composing system, Fig. 22-7. Typical applications for this system include the repeated use of symbols such as schematics, printed circuits, flow charts, industrial piping, computer wiring, and hydraulic diagrams.

Fig. 22-7. The Mergenthaler DIAGRAMMER. (Mergenthaler Linotype Co.)

To operate the machine, one of the 256 symbol buttons on the SELECTOR UNIT, Fig. 22-8, is selected by the operator and pushed. A full scale projection of the symbol is projected onto the VIEWING SCREEN. By working the controls, the image can be magnified, rotated to any angular position, and moved horizontally, vertically or diagonally over the VIEWING SCREEN.

Fig. 22-8. Partial view of the 256 push buttons on the Selector Unit which causes the symbols to be projected onto the viewing screen. (Mergenthaler Linotype Co.)

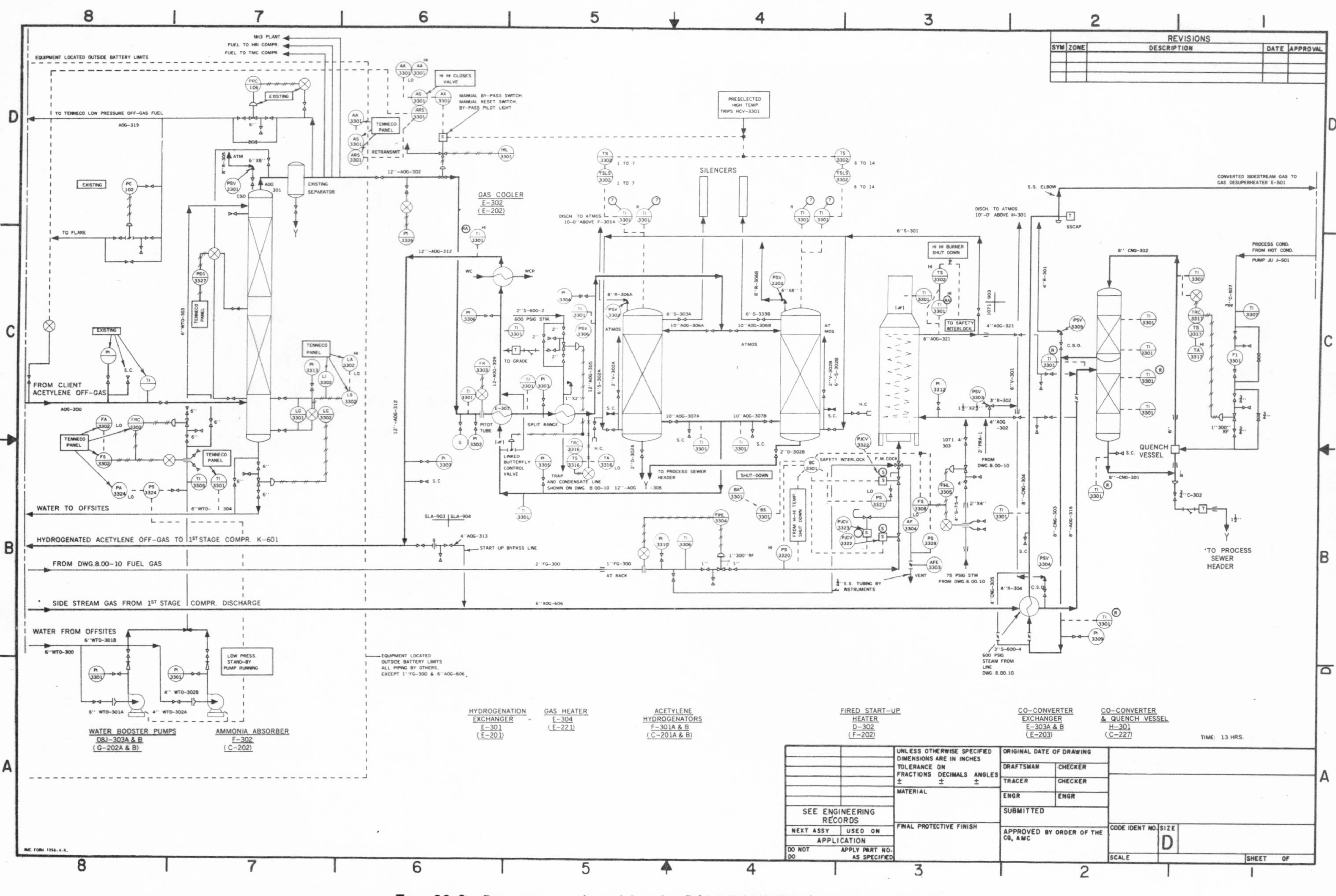

Fig. 22-9. Drawing produced by the ***DIAGRAMMER*** *(greatly reduced).*
(Mergenthaler Linotype Co.)

Fig. 22-10. System in which user establishes two-way communication with computer by using a light pen, alphabet letters and numbers on push buttons to present problem on cathode ray tube. (Control Data Corp.)

The symbol or line (the unit can produce seven basic types of lines) is recorded on film when the operator touches the EXPOSE BUTTON. Exposed images can be maintained on the film throughout the drawing procedure for progress reference. A rough sketch is frequently used as a guide.

Lettering is placed on the film by pushing appropriate buttons. The lettering may be run up, down, to the right or to the left. A space bar is provided for word spacing.

When all elements have been selected, positioned and exposed, the film is removed from the operating unit for normal photographic processing.

The drawing in Fig. 22-9 (reduced in size) was produced on the unit.

COMPUTERIZED GRAPHIC DISPLAY SYSTEMS

These systems are among the most versatile of computer graphic techniques. They enable the user to establish two-way communication with the computer in alphabet letters or numbers, Fig. 22-10.

The basic elements of such a system are:

1. The computer.
2. A "software" system (program) that provides contact between the user and the computer.
3. A cathode ray tube (vacuum tube in which slender beams are projected onto the surface) (CRT) display scope.
4. A hand-held pen, used for drawing and activating controls on the scope.
5. A keyboard with symbol push buttons.

By using the light pen and the push buttons, the problem is shown on the scope in graphic symbols - alphabet and number characters, points and lines. Curves are represented by a sequence of dots or short straight lines. The computer provides an immediate display of the information. Displayed data may be analyzed and modified if needed. Engineering changes on one drawing can be made on thousands of drawings in a computer file. For example: all hex head cap screws can be changed to socket head cap screws by a single entry.

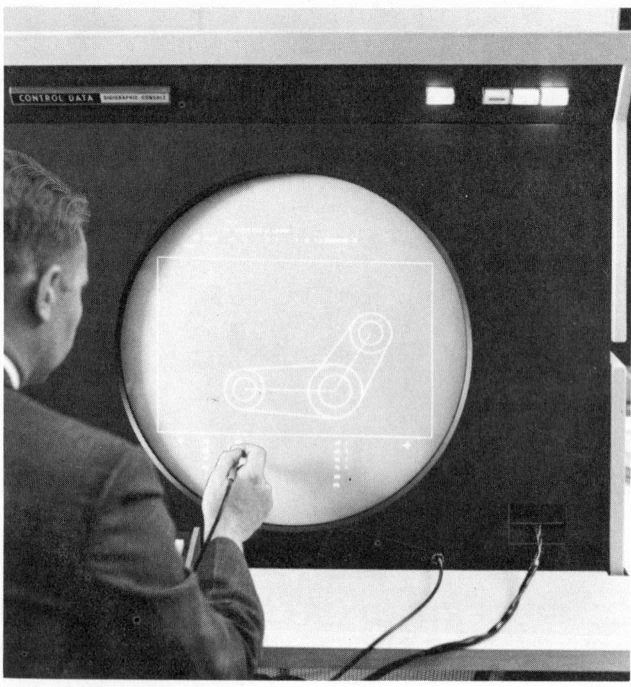

Fig. 22-11. Hand-held light pen is used to trace path of tool cutter around part. This activates computer which automatically punches out tape that controls operation of machine tool.

The light pen, when pointed at any specific item (character, point or line) displayed on the screen identifies the displayed information to the computer for proper action. Action may also be initiated by using the appropriate push button.

It is possible to write a program to enable the computer to smooth and straighten hand-drawn lines and to draw lines of requested lengths or desired angles.

Any section of the diagram can be viewed at a reduced or enlarged size. Different views of the same figure can be viewed at one time, and a geometric figure can be rotated to any desired viewing angle.

The system can be adapted to virtually any situation that involves calculation and manipulation of data or symbols.

NUMERICALLY CONTROLLED MACHINE TOOLS

Preparing tapes which operate numerically controlled machines is another important application of computer graphics.

Fig. 22-12. An electrical field in the "glass sandwich" that makes up the writing surface of the Data Tablet senses the location of the stylus 200 times a second. Depending on the electronics involved, the movement of the pen is converted into signals which perform a wide range of functions. (Sylvania Electric Products Inc.)

In using the system shown in Fig. 22-11, a description of the part to be produced is entered on the cathode ray tube from a print of the finished part. The path of the machine cutting tool is traced around the part, using a hand-held light pen. The computer then punches out a control tape that will enable the machine to make the required cuts.

SYLVANIA DATA TABLET[1]

The SYLVANIA DATA TABLET, Fig. 22-12, is a unique computer directed system. With it, data processing is joined with basic pen-and-pencil methods of processing and transmitting information. The unit enables the computer and other electronic equipment to accept handwritten information. It can add to or replace conventional input techniques--punched cards and paper or magnetic tapes.

The system consists of an electronic stylus and a writing surface composed of a transparent conductive film sandwiched between protective layers of glass, Fig. 27-12. Direct current voltages proportional to position are impressed on the conductive layer in such a manner that coordinate position ("X" - horizontal axis, "Y" - vertical axis) can be sensed by the stylus as it moves over the tablet surface. The position of the stylus is recorded or sampled 200 times a second providing a precision of image. Accuracy of line recording is less than one percent deviation from the actual drawing.

A small cabinet contains the associated solid state electronics that permits the unit to be used with almost any computer.

As the writing surface is transparent, the tablet can be placed over the image on the CRT and used, for example, for designating positions on a military map, Fig. 22-13, and permit combat commanders to be kept up to date on the latest developments on the battle field.

The unit can also be used in the field of publishing. Copy readers can display portions of a manuscript and make alterations and corrections with the stylus alone or with keyboard entries.

An important feature of the Data Tablet is its ability to simplify and expedite the flow of information into computers. It is possible to feed information directly into a computer, ask questions, obtain drawings, and make revisions merely by drawing symbols or schematics instead of having to write new computer programs and to prepare new punch cards or tapes.

[1]Developed by the Applied Research Laboratory of Sylvania Electric Products Inc., Waltham, Massachusetts.

Fig. 22-13. The Data Tablet can be used at a desk, console or as a transparent overlay for a CRT. (Sylvania Electric Products Inc.)

GLOSSARY OF COMPUTER GRAPHIC TERMS

ADDRESS: A number, name or label identifying a specific location within the computer's memory apparatus.

ALPHANUMERIC: Pretaining to a set of characters that contains both letters and numbers.

ANALOG: Denotes the use of physical variables - distance, rotation or voltage, to represent and correspond with numerical variables that occur in computation.

ANALOG COMPUTER: A computer that operates on analog data by preforming physical processes on these data.

CATHODE RAY TUBE (CRT): A vacuum tube in which cathode rays are used to produce luminous spots on its surface. Similar to the picture tube in a TV set.

COMPUTER: A device capable of solving problems by accepting data, preforming prescribed operations on the data, and supplying the results of these operations.

CONSOLE: Location of computer controls, as well as various lights and the cathode ray tube display.

COORDINATES: The positions or locations of points or planes.

DATA: Facts or information taken in, acted upon or emitted by a machine used for handling information.

DIGITAL COMPUTER: A computer which produces results from numeric information only, and performs operations by means of counting, rather than measuring as in analog computers.

GRAPHIC: Written or drawn.

HARDWARE: The computer and its accessories.

KEYBOARD: Part of a device that punches holes in a card or tape to represent data, or a device that communicates directly with a computer.

LIGHT PEN: A hand held pen-like device containing a photocell or photo multiplier, used for the generation of lines on a display.

MACHINE LANGUAGE: Instructions written as binary (base two) codes.

MEMORY: A term referring to the equipment and media used for storing information (data and instructions) in machine language in electrical or magnetic form.

NUMERICAL CONTROL SYSTEM: A system in which actions are controlled by the direct insertion of numerical data.

OPTICAL SCANNER: A device that optically scans (examines) printed or written data and generates electrical representation for input to the computer.

PLOTTER: Automatic drawing equipment controlled by a tape or directly by a computer.

PROGRAM: A set of instructions for the computer that defines a desired sequence of conditions for a process or function, and the operations required between these conditions.

PUNCHED CARDS: Cards of uniform size and shape, suitable for punching in a meaningful pattern and for mechanical handling. The punched holes are usually sensed electrically or mechanically.

PUNCHED TAPE: Paper or plastic tape into which a pattern of holes has been punched to convey information.

REAL TIME: The time the computer needs to respond with a solution.

SOFTWARE: The means of communicating with the machine. The program.

TIME-SHARING: Process in which the computer switches rapidly from one problem to another, giving to each of a number of human users the illusion of working upon his problem all of the time.

TEST YOUR KNOWLEDGE - UNIT 22

1. In computer graphics (automated drafting) drawing devices are used in connection with a computer to make __________ and __________ presentations.
2. Computer directed instruments can draw mathematically and mechanically accurate perspective, isometric and stereoscopic 3D views. True or False?
3. In the photo composing system of automated drafting, typical applications include the repeated use of __________.
4. In a system in which the user establishes a two-way communication with a computer, the operator must use a __________, __________ letters and __________ buttons.
5. In preparing tapes which operate numerically controlled machines, a description of the part to be produced is entered on the __________ tube from a print of the finished part.
6. A cathode ray tube is a vacuum tube in which cathode rays are used to produce __________ on its surface.

Industry photo. The architect who designed this building communicates, through his drafting ability, with these carpenters by means of blueprints.

Fig. 23-1. Ship models at the Naval Ship Research and Development Center used to determine a proposed ship's seaworthiness, power requirements, resistance to water, and stability characteristics.

Unit 23
MODELS, MOCKUPS, AND PROTOTYPES

To many people, modelmaking is an interesting hobby. You probably have made models of famous planes, boats or cars from wood or plastic.

Industry makes extensive use of three modelmaking activities - MODELS, MOCKUPS and PROTOTYPES. These are used for engineering, educational and planning purposes. Models have proved to be very helpful in solving design problems and to check the workability of a design or idea before it is put into production.

What is industry's definition of a model, mockup and prototype? In general, the following applies:

MODEL. A scale replica of a planned or existing object, Fig. 23-1. The model may be constructed to see how the product will look, to check out scientific theory, demonstrate ideas or for training or advertising purposes.

MOCKUP. A full size three-dimensional copy of an object, Fig. 23-2. This is usually made of plywood, plaster, clay, fiber glass, plaster or a combination of materials.

*Fig. 23-2. Mockup of the **ASTRO III** a turbine powered three wheel "dream" car. Ideas from this car may be used in future auto production. (Chevrolet)*

*Fig. 23-3. The **TURBOTRAIN** is the prototype of trains that will furnish high speed transportation. It is powered by a gas turbine and is capable of reaching speeds up to 170 mph. (Sikorsky Aircraft)*

PROTOTYPE. A full size operating model of the production item, Fig. 23-3. It is usually handcrafted to check out and eliminate possible design and production "bugs."

HOW MODELS, MOCKUPS AND PROTOTYPES ARE USED

Many industries employ models, mockups and prototypes for design tools, Fig. 23-4. A few of the more important applications are:

AUTOMOTIVE INDUSTRY

The automotive industry places great importance on the use of models, mockups and prototypes. Mistakes can be very costly.

Fig. 23-4. An automobile design studio is a busy place, with various members of the design team working simultaneously on different stages of the design process. In the foreground a designer works out a detail idea in sketches. In the background, the shape of an experimental GM–X is taking form in a full-size clay model at the hands of modelmakers, working under the direction of the chief designer. (General Motors)

Fig. 23-5. The automotive industry uses sketches to develop ideas. Scale models are made of promising ideas for further study. (Oldsmobile)

An automobile starts "life" as a series of sketches, Fig. 23-5. These are usually developed around specifications supplied by management. Promising sketches are usually drawn full size for additional evaluation.

Clay models are used for three-dimensional studies. Upon the completion of further design development, a fiber glass prototype is usually constructed, Fig. 23-6.

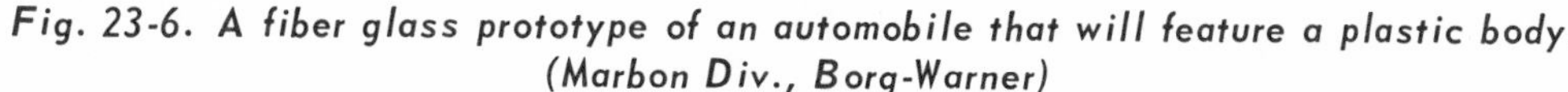

Fig. 23-6. A fiber glass prototype of an automobile that will feature a plastic body. (Marbon Div., Borg-Warner)

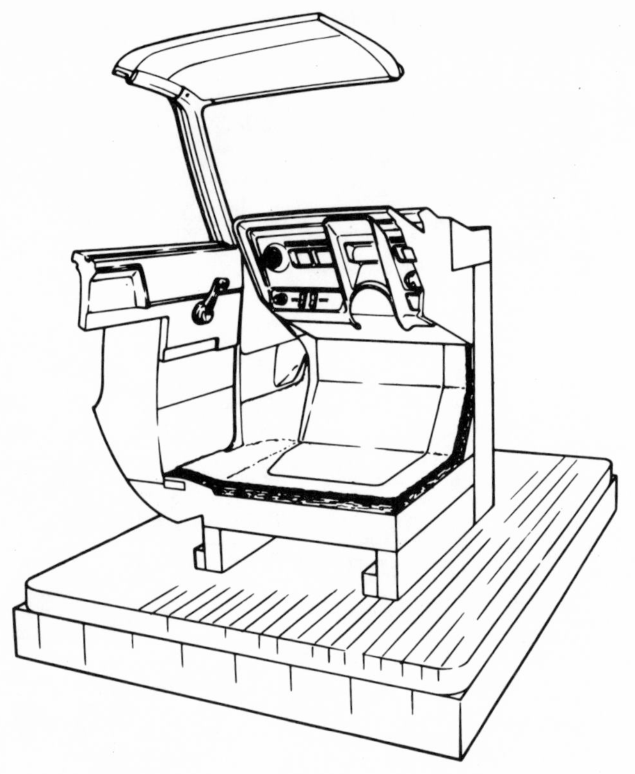

Fig. 23-7. Wood models (called "bucks") are used to prove the accuracy of molding and trim designs. (General Motors)

Production fixtures (devices to hold body panels and other parts while they are welded together) and other tools needed are developed from accurate full size wood and plaster models, Fig. 23-7.

After many months of development and production planning, manufacturers are ready to make new model automobiles available to the public.

AEROSPACE APPLICATIONS

The tremendous cost of aerospace vehicles makes it mandatory that they first be developed in model and mockup form, Fig. 23-8.

Fig. 23-8. A full-scale mockup of the supersonic Concorde jet airplane. The truck in the background will give some indication of its size. Using mockups such as this, ideas can be checked without the great expense of reworking the actual plane. (Trans World Airlines)

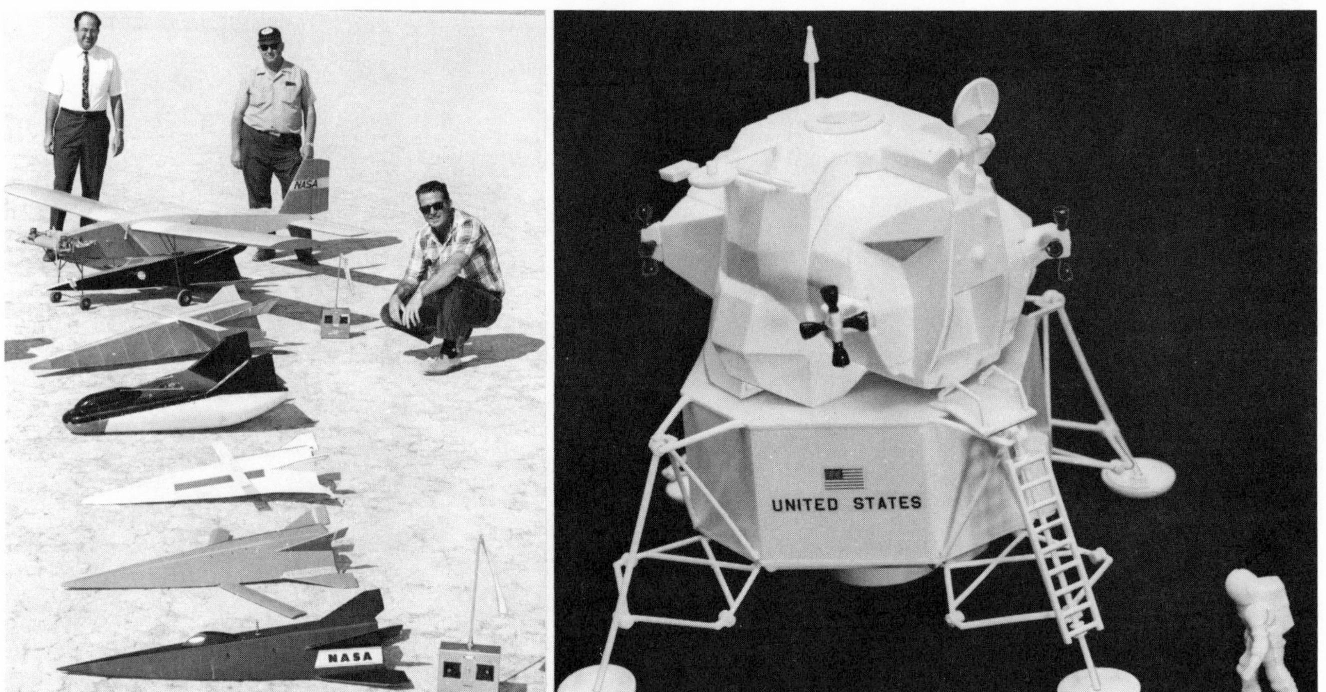

Fig. 23-9. Left. A few of the many shuttle re-entry vehicle design ideas tested in model form. To test full size aircraft would have cost millions of dollars and endangered the lives of many people. (NASA)
Fig. 23-10. Right. Models were used when planning moon landings.

Flight characteristics can be determined with considerable accuracy without endangering human life, by testing complicated models in a wind tunnel or in free flight, Fig. 23-9.

Exploration of the moon was first planned with models, Fig. 23-10.

Prototype aircraft, Fig. 23-11, are the first two or three preproduction planes (usually handcrafted) that are flown to check the data obtained from wind tunnel research and to secure a license for that particular type plane.

Fig. 23-11. Prototype of the 747. It was used to flight prove design data. (Boeing)

Fig. 23-12. Architectual model of the Bel Air, Maryland, high school (school where the author teaches).

ARCHITECTURE

You have seen photos of proposed buildings in the real estate section of your Sunday paper. Some of these illustrations were of models that were very accurate miniature replicas of the proposed buildings, Fig. 23-12.

Many people use models when planning a new home, Fig. 23-13. This helps them to visualize how the house will look when completed. The model enables the owner to see the completed design in three dimensions and also how paint colors and shrubbery plantings will look.

SHIP BUILDING

Ship hulls are tested in model form, Fig. 23-14, before designs are finalized and construction starts. Specially designed equipment tows the model hull through the water in the

Fig. 23-13. Many potential home builders make models such as this one. They can see how their home will look before starting to build. (Taylor Made Models)

Fig. 23-14. Left. Model of a U. S. Coast Guard Vessel in seaworthiness test in the 140 ft. basin at the Naval Ship Research and Development Center. (U. S. Navy) Fig. 23-15. Right. Highly detailed model of a power generating plant. Note the clear plastic walls. The model was used to train operating personnel.

test basin. The model hull behaves like the full size ship so design faults can be located and corrected.

CITY PLANNING

Most large cities use scale models to show city officials and planners how proposed changes and future developments will look. Models, while rather costly, permit intelligent decisions to be made before large sums of money are spent acquiring land and existing buildings are torn down.

CONSTRUCTION ENGINEERING

Many construction projects are designed from carefully constructed models, Fig. 23-15. By working from models, engineers can see how space can best be utilized. In some instances, they can determine how the proposed project will affect the surrounding community, Fig. 23-16. This helps to minimize field problems and changes during construction.

Models may also be used to train personnel in plant operation.

Fig. 23-16. Model of a portion of the Susquehanna River in Maryland to test how the proposed atomic power generating plant will affect the river and its ecology (branch of biology dealing with relations between animal and plant life and their environment). (Philadelphia Electric Co./Alden Research Lab)

CONSTRUCTION MODELS

Model making materials are readily available commercially, Fig. 23-17. Many products made for the model railroader are ideally suited for making architectural models. Kits are available for the small home builder who wants to design his own home.

A professional touch can be added to models by using accurately scaled furniture, automobiles and figures that can be purchased at toy and hobby shops.

Preprinted sheets of brick and stone can be glued to a suitable thickness of balsa wood for walls and partitions. Various types of abrasive paper are suitable for roofing, driveways and walkways. Simulated window glass can be made from transparent plastic sheet. Several different scale sizes of window and door frames are available molded in plastic. Most model shops can supply bushes and trees in various types and sizes.

Fig. 23-17. A few of the model making materials available at hobby shops.

Other types of models - autos, planes and boats, are made from bass wood, mahogany, balsa wood, metal, plaster and various kinds of plastics.

Regular model making paint is produced in hundreds of colors and is ideal for painting all types of models. However, care must be exercised when painting models that have plastic in their construction. Be sure the paints used are designed for plastics. If you are not sure, paint a small portion of the plastic that is hidden from view to determine whether the paint is compatible with the material.

Models may be assembled with model airplane cement or the white glues . . . Elmer's, Titebond, etc.

TEST YOUR KNOWLEDGE - UNIT 23

1. Industry uses models, mockups and prototypes for what purposes?
2. A MODEL is ______________________.
3. A MOCKUP is ______________________.
4. A PROTOTYPE is ______________________.
5. List four industries that employ models, mockups and prototypes as design tools. Briefly describe how each industry listed uses them.
6. Walls and partitions in a model home are usually constructed of ______________ wood.
7. Roofs, driveways and walkways can be made from different kinds and grades of ________.
8. Models of cars, planes and boats may be made from what materials?

OUTSIDE ACTIVITIES

1. Visit a model or hobby shop, then prepare a list of available materials which may be used for building architectural models.

2. Review technical magazines and clip illustrations that show models, mockups and prototypes being used for engineering, educational, planning or other purposes. (Do not cut up library copies.)
3. Make a collection of materials suitable for building model homes.
4. Visit a professional model maker in your community. With his permission, make a series of slides showing how he makes models and of the models he has made. Give a talk to your class on professional modelmaking.
5. Make a scale model of your drafting room. Discuss alternate layouts.

Fig. 23-18. Plastic model helicopter to be used as a basis for constructing a hanger.

6. Construct a model helicopter similar to the one shown in Fig. 23-18. Design and construct from balsa wood a minimum building that would protect the helicopter from the elements.
7. Design and construct a full size model of a disposable waste basket, using corrugated cardboard.

Model showing the details of the tip of Manhattan and a portion of Brooklyn. This model will be used for planning purposes. Many draftsmen will be needed to plan the future expansion of the city. (Lester Associates, Inc.)

Unit 24
DRAFTING AIDS

DRAFTING AIDS are devices and techniques that help the draftsman do his work better and/or in less time. Typical aids commonly used by industry are included.

TEMPLATES

TEMPLATES, Fig. 24-1, are probably the most widely used of the drafting aids. These convenient, time saving drafting tools are available in almost an unlimited range of standard symbols and figures. See Fig. 24-2.

Made of transparent plastic, these templates enable the draftsman to do normally time consuming jobs with ease and accuracy.

Fig. 24-1. Templates enable the draftsman to do normally time consuming jobs with ease and accuracy.

Fig. 24-2. A few of the many different types of templates used in the modern drafting room.

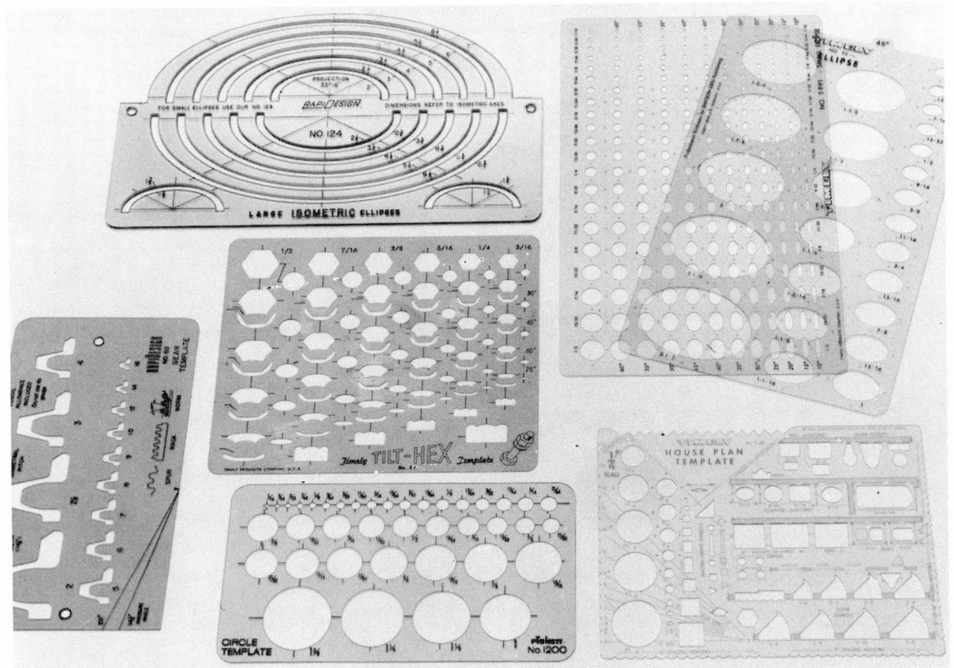

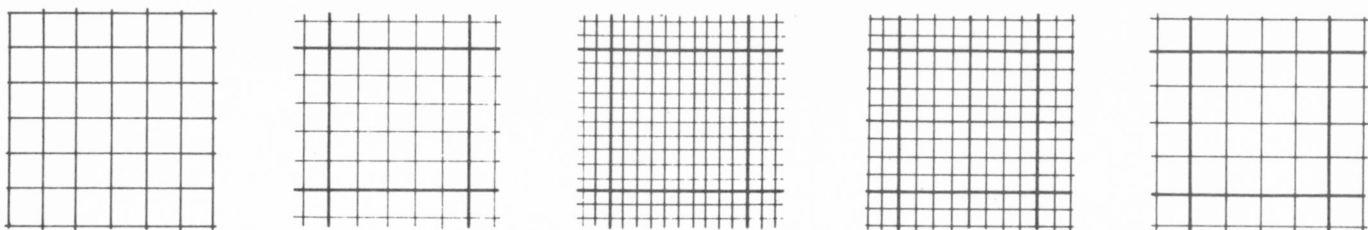

Fig. 24-3. Graph paper is manufactured in many different grid sizes.

GRIDS FOR GRAPHS AND SCALE DRAWINGS

A graph is a picture of a collection of facts. Many graphs are plotted on the familiar square or cross-section grid paper (See Unit 16).

In many instances, drawing the grid (cross-sectioned background) is more time consuming than plotting and drawing in the graph or chart. Many types and sizes of grid patters are available commercially in the form of prepunched loose leaf sheets (8 1/2 in. by 11 in.) and in other dimensions. See Fig. 24-3. The grid lines are usually printed in green ink. However, some grid patterns are printed in orange, pale blue or black.

Designers, draftsmen, surveyors, engineers and architects also make considerable use of commercially prepared graph sheets to make preliminary design studies, Fig. 24-4. For example, the isometric grid makes it very easy to convert an orthographic drawing into an isometric drawing, Fig. 24-5.

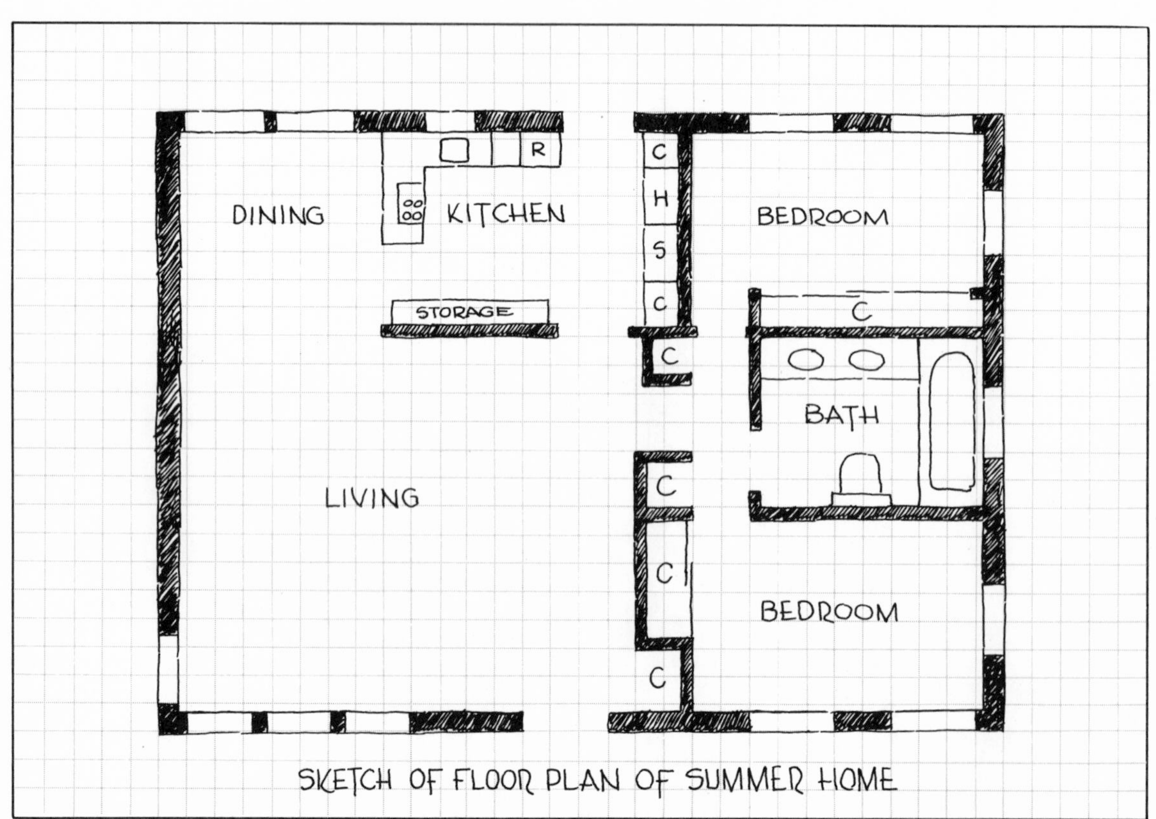

Fig. 24-4. A floor plan sketched on graph paper. Changes can be made easily on the grid. Each square equals one foot.

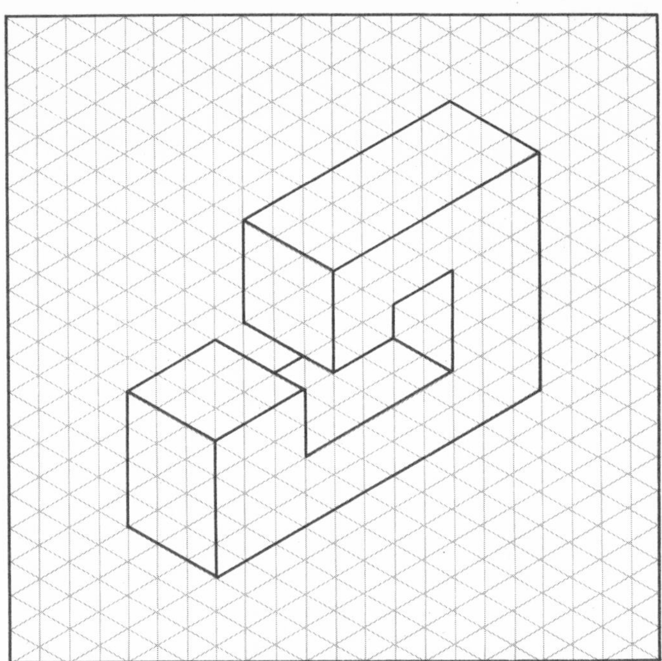

Fig. 24-5. An isometric drawing made on graph paper designed for that purpose.

PREPRINTED MATERIALS

Preprinted materials are another means of simplifying work and saving time for the draftsman. Drafting aids that fall into this category are:

1. Preprinted DRAWING SHEETS, Fig. 24-6. Imprinted drawing forms come in standard drawing sheet sizes and most types of drawing media.

2. TITLE BLOCKS preprinted on acetate or Mylar sheet are widely used, Fig. 24-7. One side of the plastic sheet is coated with a pressure sensitive adhesive making it easy to apply to the drawing sheet. Title blocks are printed to order.

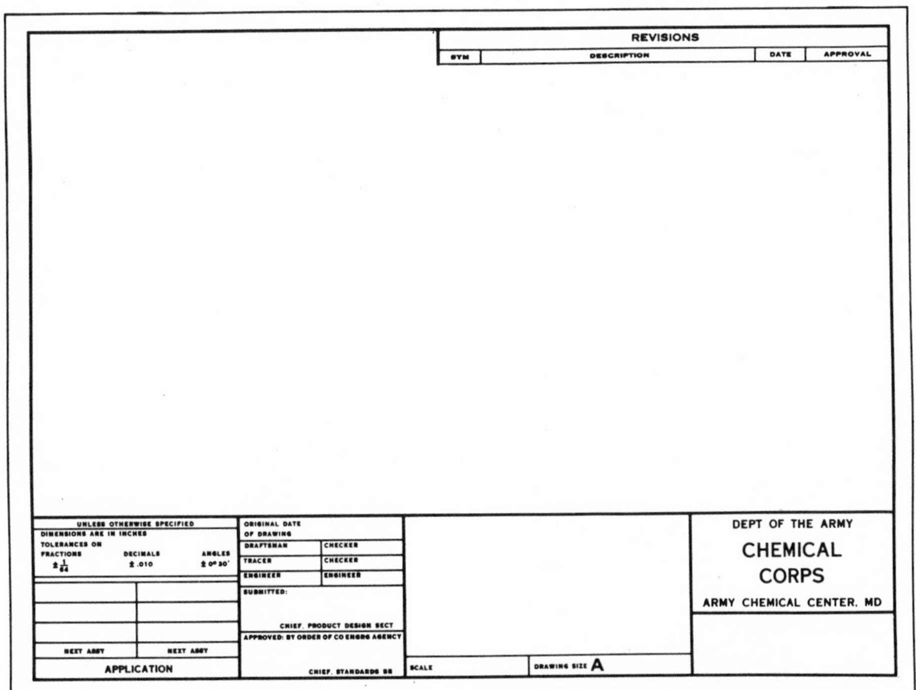

REVISIONS

SYM	DESCRIPTION	DATE	APPROVAL

UNLESS OTHERWISE SPECIFIED
DIMENSIONS ARE IN INCHES
TOLERANCES ON
FRACTIONS ± 1/64 DECIMALS ± .010 ANGLES ± 0° 30'

NEXT ASSY | NEXT ASSY
APPLICATION

ORIGINAL DATE OF DRAWING
DRAFTSMAN | CHECKER
TRACER | CHECKER
ENGINEER | ENGINEER
SUBMITTED:
CHIEF, PRODUCT DESIGN SECT
APPROVED: BY ORDER OF CO ENGRG AGENCY
CHIEF, STANDARDS BR

SCALE | DRAWING SIZE A

DEPT OF THE ARMY
CHEMICAL CORPS
ARMY CHEMICAL CENTER, MD

Fig. 24-6. Above. Preprinted drawing sheets save the draftsman's time. All sheets are identical in format. Fig. 24-7. Below. Preprinted title block.

UNLESS OTHERWISE SPECIFIED DIMENSIONS ARE IN INCHES TOLERANCES ON FRACTIONS ± 1/64 DECIMALS ± 0.010 ANGLES ± 1°	DRAWN BY	WALKER INDUSTRIES	
	DATE	TITLE	
	CHK'D		
MATERIAL	HEAT TREATMENT	SCALE	DRAWING NO.
		SHEET	

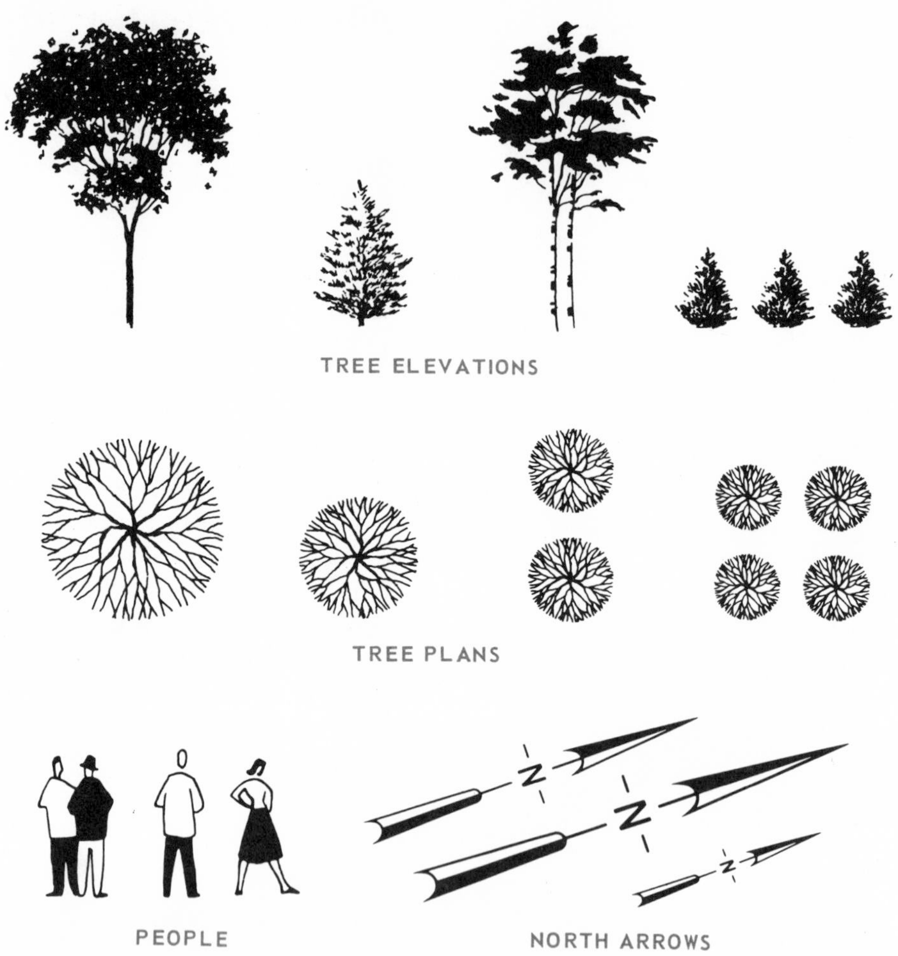

Fig. 24-8. Graphic symbols simplify repetitive work. Many designs are available.

Fig. 24-9. Applying a graphic symbol to an architectural drawing by burnishing (rubbing) it into position.

3. GRAPHIC SYMBOLS, Fig. 24-8, are used to simplify repetitive work. These are applied to the drawing sheet by means of pressure sensitive adhesive or by the transfer method. The symbol sheet is placed in position on the drawing and the symbol is transferred by burnishing (rubbing) the backing sheet with a smooth burnishing tool, Fig. 24-9. The transferred symbol may be removed with a soft pencil eraser.

4. Preprinted LETTERING which is available in hundreds of styles and sizes, enables students to do professional type lettering, Fig. 24-10.

The material is manufactured in TRANSFER LETTERING (the letters are transferred to the drawing by burnishing the back of the sheet holding the letters), Fig. 24-11, and CUT-OUT LETTERING (the individual letters are cut out and held in place by a pressure sensitive adhesive), Fig. 24-12.

LEADERSHIP IN THE
18 leadership in the creat
LEADERSHIP IN
24 leadership in the
LEADERSHIP
24 leadership in
LEADERS
30 leadership

Leaders
Leadershi
leadership in
Leadership In T

LEADERSHIP IN THE
24 leadership in the cre
LEADERSHIP IN
36 leadership in
LEADERSHIP IN THE
24 leadership in the
LEADERSHIP

Fig. 24-10. Preprinted lettering. Many styles and sizes are produced. (Formatt)

Fig. 24-11. Left. Transfer letters are applied by placing the letter in position and burnishing the back of the sheet with a smooth object. Right. Removing the lettering sheet after the letter has ben applied to the sheet. The guide line is erased after the entire line of lettering has been applied.

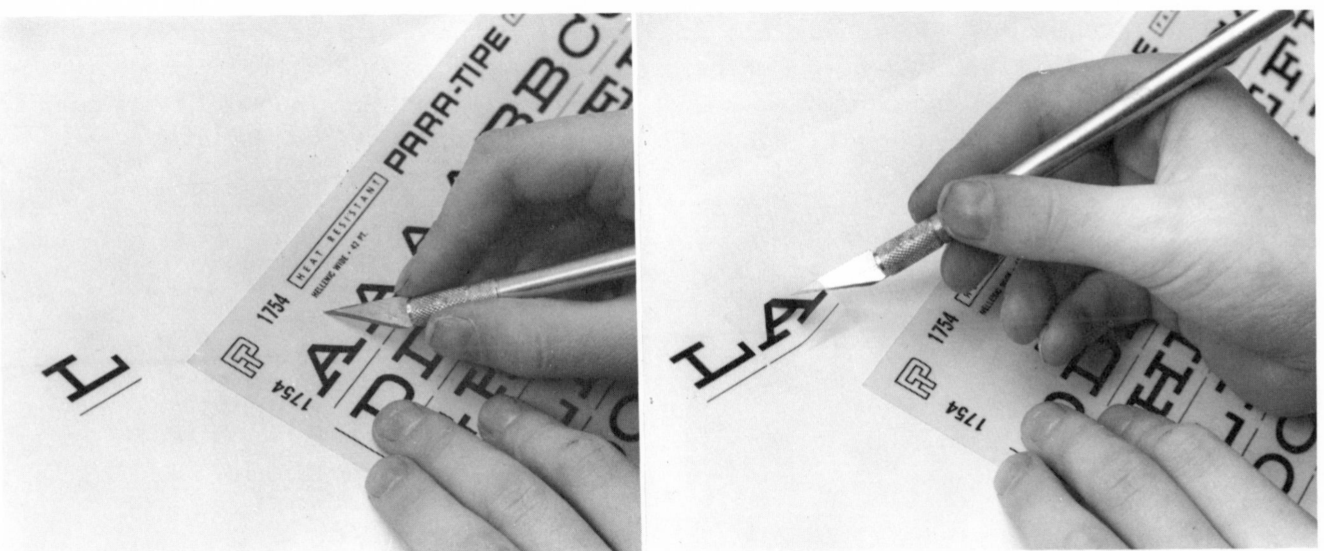

Fig. 24-12. Left. Removing a **CUT-OUT LETTER** *from the lettering sheet. Right. Applying the letter to the sheet. The guide line is cut away after the entire line of lettering has been set into position.*

Fig. 24-13. Shading mediums. Many patterns are available.

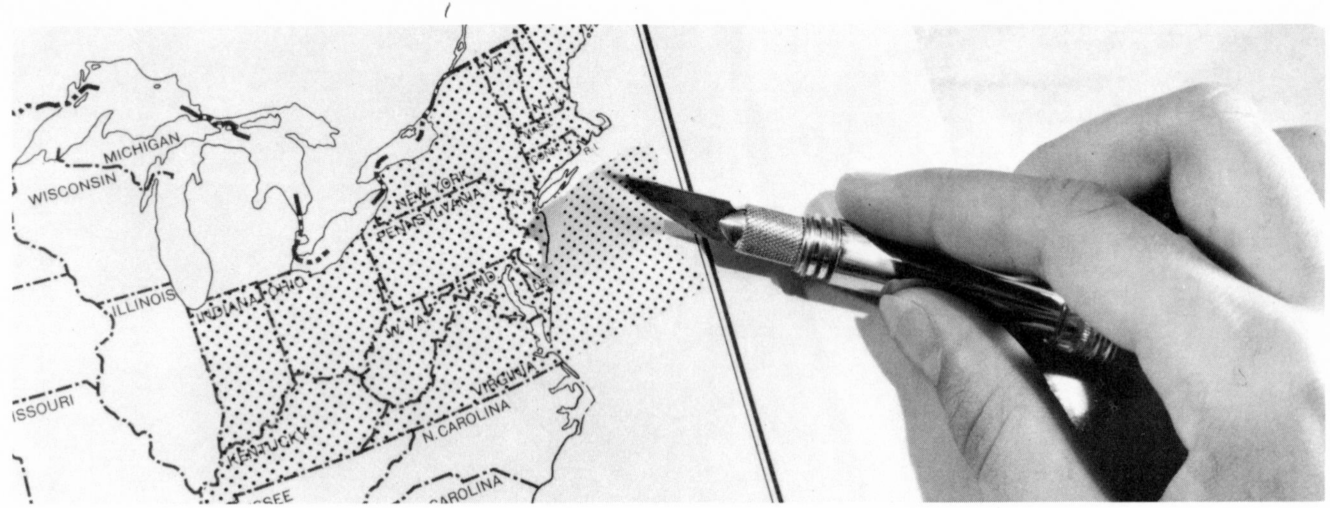

Fig. 24-14. Shading media being used to differentiate between areas on a map. (Formatt)

5. SHADING MEDIUMS, Fig. 24-13, make it possible to show differences in areas in mapping, Fig. 24-14, and to highlight areas. There are patterns for indicating swamps, forests and other areas.

6. PRESSURE SENSITIVE TAPES, Fig. 24-15, supplement conventional drafting instruments, for making charts, graphs, electronic diagrams, etc. Many different colors, styles and widths are available.

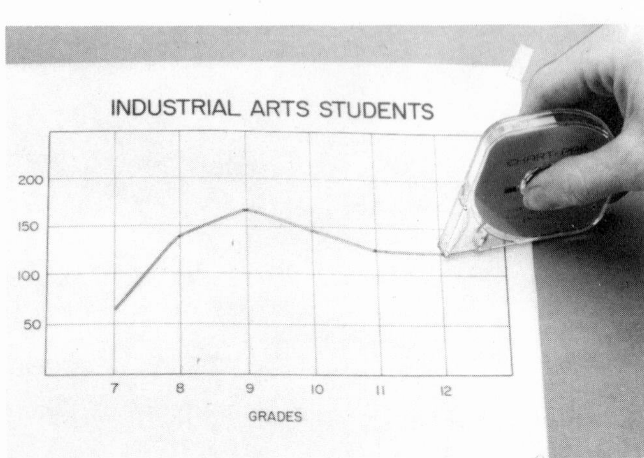

Fig. 24-15. Graphs are made easily using pressure sensitive tapes. Different color tapes may be used to provide different bits of information.

TEST YOUR KNOWLEDGE - UNIT 24

1. Drafting aids are ____________________.
2. Templates are:
   a. Convenient and time saving drafting tools.
   b. Available in a great number of standard styles and symbols.
   c. Tools which enable the draftsman to do normally time consuming jobs with ease and accuracy.
   d. All of the above.
3. Match these words with the descriptions below.
   a. ________ Grids.
   b. ________ Preprinted title blocks.
   c. ________ Preprinted drawing sheets.
   d. ________ Graphic symbols.
   e. ________ Preprinted lettering.
   f. ________ Shading mediums.
   g. ________ Pressure sensitive tape.

   A. Used to simply repetitive work.
   B. Supplement conventional drafting tools.
   C. Enables students to do professional type lettering.
   D. Preprinted on Mylar or acetate sheet.
   E. Available in all standard drawing sheet sizes.
   F. Used to show difference in areas in mapping, or to highlight areas on a drawing.
   G. Often used to plot charts and graphs.
   H. Made of transparent plastic and are available in an almost unlimited range of standard symbols and figures.

OUTSIDE ACTIVITIES

1. Secure samples of the following:
   a. Graph paper.
   b. Pressure sensitive tapes.
   c. Preprinted symbols.
   d. Preprinted lettering.
   e. Preprinted title blocks.
   f. Preprinted drawing sheets.
   g. Shading mediums.
2. Get samples of work done on a preprinted drawing sheet.
3. Prepare one of the graphing problems in Unit 16 using pressure sensitive tapes.
4. Contact manufacturers of pressure sensitive drafting aids for catalogs to be kept in your drafting room library.
5. Demonstrate the proper use of templates.

Unit 25
DESIGN IN INDUSTRIAL ARTS

In many ways, good design may be considered a carefully thought out plan for a direct solution to a problem. This is sometimes called CREATIVE PLANNING. The development of the plan includes creating, inventing and research. Usually many ideas are studied, tried-out, analyzed and then either incorporated into the design or discarded.

DESIGN GUIDELINES

Though design is not an exact science, good design is characterized by certain qualities that are easily recognized:

FUNCTION. How well does the design fit the purpose for which it was planned? Does it fulfill a need? Function often dictates the product's appearance, Fig. 25-1.

Fig. 25-1. The job for which this all-terrain (it operates on rough ground, water and snow) TERRA TIGER was designed pretty well dictated how it should look. (Allis-Chalmers)

HONESTY. Is there an honest use of material? The qualities of the material (strength, texture, etc.) should be emphasized to the fullest. The product should be able to do the job it was designed to do, Fig. 25-2.

Fig. 25-2. There is an honest use of materials in this speed boat. The fiber glass hull is light, of high strength, and is not affected by salt water. Color is molded in the hull, and it is easy to clean. (Evinrude Motors)

APPEARANCE. Do the individual parts of the design create interest when they are brought together, Fig. 25-3. Are the proportions in balance and do the components belong together? Is the product pleasing in appearance?

Fig. 25-3. This water pitcher, a copy of a colonial American pitcher, is a fine example of good design. Its proportions are in balance and it is pleasing in appearance. (Shirley Pewter Shop, Williamsburg)

CRAFTSMANSHIP. Craftsmanship is an inherent part of good design. Quality must be built into the product. It cannot be added on.

Simply stated, good design is distinguished by certain recognizable qualities. All of the qualities are necessary. Overlooking or eliminating one may destroy the entire design.

DESIGNING A PROJECT

How should YOU go about designing a project? The easiest approach is to follow a pattern similar to that used by a professional designer. Think of the project as a DESIGN PROBLEM:

1. STATE THE PROBLEM. What is the purpose of the project? How will it be used?
2. THINK THROUGH THE PROBLEM. What must the project do? How can it be done? What are the limitations that must be considered? How have others solved similar problems? Study comparable products for ideas, Fig. 25-4.

Fig. 25-4. This young man wants to design a particular kind of model airplane. He is studying examples of models that others have designed and built. He is thinking through the design problem.

Fig. 25-5. Making sketches of ideas.

3. DEVELOP YOUR IDEAS. Next, make sketches of your ideas, Fig. 25-5. Develop the ideas and have them criticized by your teacher and fellow students. Determine what materials will be best suited for the project.

4. MAKE MODELS. Put your best ideas into model form. Develop fabrication techniques for your ideas.

5. PREPARE WORKING DRAWINGS. After you are satisfied that you have solved the problem to the best of your ability, prepare working drawings for the project.

6. Construct the project to the best of your ability. Do a job that you will be proud to show to others. See Fig. 25-6.

The development of a well designed product (project) takes time. Do not be discouraged if your first attempts at designing fall short of your goal. It takes time to acquire the skill and ability to solve design problems. You are urged to read and study the many fine publications on design to acquaint yourself with numerous well designed products that are available.

Keep a notebook of your ideas and ideas you have clipped from newspapers, advertising folders and magazines (but not those in the library). Include photos of projects you have designed and constructed to make it easier to evaluate your work and to help you see your improvement.

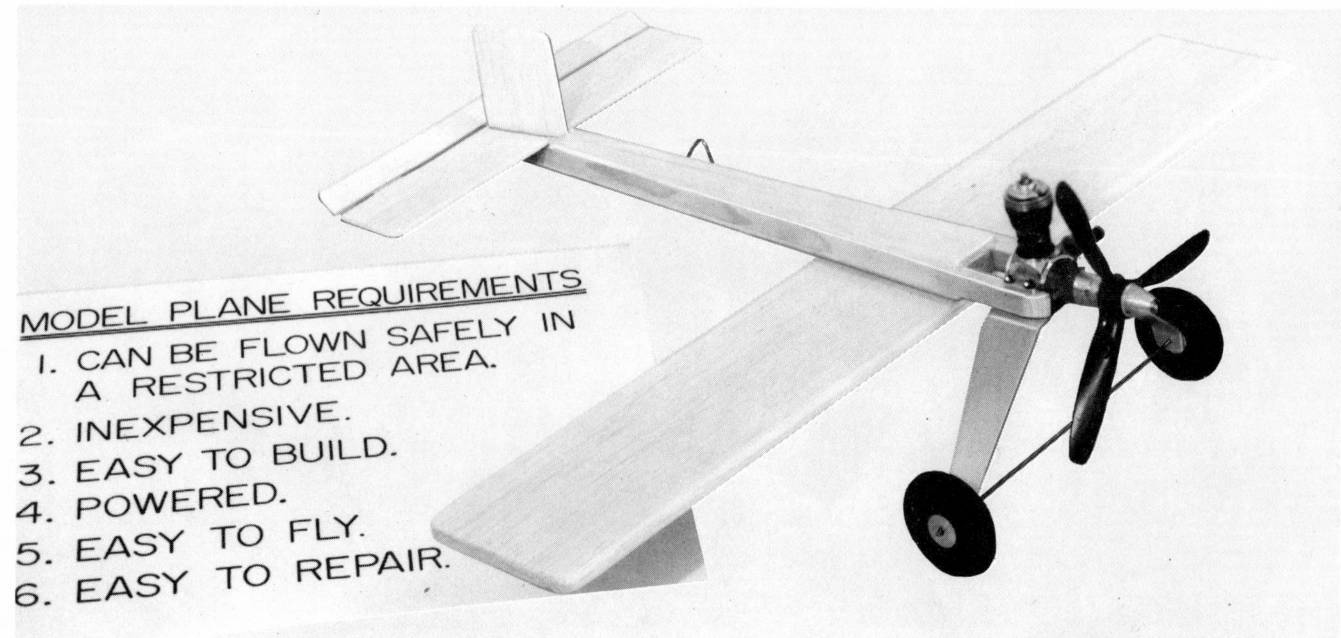

Fig. 25-6. The completed model ready for painting, trimming and test flying. The model meets design requirements.

TEST YOUR KNOWLEDGE - UNIT 25

1. A well-designed product should serve several purposes:
   a. ____________________.
   b. ____________________.
   c. ____________________.
2. Good design is hard to describe. Why?
3. Good design is characterized by certain qualities or guidelines. Describe the guidelines briefly.
4. When solving a design problem a series of steps should be followed. List and briefly describe the steps.

OUTSIDE ACTIVITIES

1. The design problems listed below are presented to get you started. You may select one and develop it, or you may originate a problem of your own.

   a. Shoeshine box.
   b. Coffee table.
   c. Table lamp.
   d. Hand launched glider.
   e. Jet propelled (CO_2 cartridge or Jetex Rocket engine) model automobile.
   f. Disposable waste basket (made from corrugated cardboard).
   g. Clock (movement may be purchased from School Products Co., 312 East 23rd Street, New York).
   h. Book rack.
   i. Tray.
   j. Wall shelf.
   k. Salad server (laminated wood, wood, plastic, metal).
   l. Tool box.
   m. Turned lamp.
   n. Turned bowl.
   o. Plastic soft food spreader.
   p. Bird house.
   q. Fiber glass model boat.
   r. Cutting board.
   s. Metal lamp.
   t. Sheet metal tool box.

Unit 26
DRAFTING AND DESIGN OCCUPATIONS

If you like drawing and designing as described in this book why not consider a job in one of the drafting and design occupations?

There is a great need for drawings. For example, more than 27,000 drawings are needed to manufacture an automobile. The field of drafting is providing employment for over one million men and women. The work of other millions requires them to be able to read and interpret drawings. See Fig. 26-1.

DRAFTING OCCUPATIONS

The draftsman makes working plans and detailed drawings. He prepares these from the specifications and information received verbally and from sketches and notes.

Fig. 26-1. Craftsmen that build airplanes must be able to understand drawings. (Cessna Aircraft Corp.)

The draftsman usually starts out as a TRAINEE DRAFTSMAN where he redraws or repairs damaged drawings. He may revise engineering drawings or make simple detail drawings under the direct supervision of a senior draftsman.

During his period of training, the trainee draftsman may be enrolled in formal shop and classroom courses within the company or at a community college or technical school.

Upon completion of the training period, the trainee draftsman advances to JUNIOR DRAFTSMAN. Similar positions are known as: DETAILER, DETAIL DRAFTSMAN and ASSISTANT DRAFTSMAN.

In this position he will prepare detailed and working drawings of machine parts, electronic/electrical devices, structures, etc., from rough design drawings. He may also prepare simple assembly drawings, charts and graphs and make simple calculations according to established drafting room procedures.

From junior draftsman he progresses to DRAFTSMAN. The draftsman applies independent judgment in the preparation of original layouts with intricate details. He must have an understanding of machine shop practices, the proper use of materials and be able to make extensive use of reference books and handbooks.

With experience the draftsman will become a SENIOR DRAFTSMAN where he will be expected to do complex original work.

In time he may become LEAD or CHIEF DRAFTSMAN where he will be responsible for the work of his department or section.

Most draftsmen specialize in a particular field of technical drawing - aerospace, architecture, structural, etc. Regardless of the field of specialization, the draftsman must be able to draw rapidly, with accuracy and neatness. He must also have a thorough understanding of the materials and manufacturing methods found in his field of specialization. In addition, the draftsman must have a working knowledge of mathematics, science, English, materials and manufacturing processes.

The manufacturing industries employ large numbers of draftsmen. Others are employed by architectural and engineering firms and our government.

INDUSTRIAL DESIGNER

The work of the INDUSTRIAL DESIGNER has considerable influence on virtually every item used in our daily living whether it is the design of a small tape player or of a giant jet plane. Many top designers are on the staffs of major automobile manufacturers, Fig. 26-2.

*Fig. 26-2. An **INDUSTRIAL DESIGNER** employed by an automobile manufacture. He is preparing preliminary sketches of an experimental car. (Oldsmobile)*

In general, the chief function of the industrial designer is to simplify and improve the operation and appearance of industrial products. Design simplification usually means fewer parts to wear or malfunction. Appearance plays an important role in the sale of a product. The industrial designer must be aware of changing customs and tastes, and he must know why people buy and use different products.

Fig. 26-3. Teaching is a challenging profession that offers considerable freedom. (Bob Meyers)

It is recommended that the prospective designer have an engineering degree and have a working knowledge of engineering and manufacturing techniques and materials and their properties. He should be a good draftsman and have artistic ability.

TOOL DESIGNER

Tools and devices needed for the manufacture of industrial products are designed by the TOOL DESIGNER.

He originates the designs for cutting tools, special holding devices (fixtures), jigs, dies and machine attachments that are needed to manufacture the product. He may be responsible for making the necessary drawings, or supervise others in making them.

The tool designer must be familiar with machine shop practices, be an accomplished draftsman and have a working knowledge of algebra, geometry and trigonometry.

As industrial technology expands and more automated machinery is introduced, there will be a constantly increasing demand for competent tool designers.

TEACHERS

TEACHING, Fig. 26-3, is a satisfying profession which is often overlooked by students. Teachers of industrial arts, vocational and technical education are in a fortunate position. They work in a challenging profession that offers a freedom not found in many other professions.

Four years of college training are needed and while industrial experience is ordinarily not required it is highly recommended.

ENGINEERS

The ENGINEER usually specializes in one of the many branches of the profession. There are at least 25 engineering specialties recognized - aeronautical, industrial, chemical, structural, civil, electrical, metallurgy to name but a few. See Fig. 26-4.

Fig. 26-4. Engineers tend to specialize. The structural engineer is responsible for the design of structures such as this bridge in Philadelphia. (Bethlehem Steel Co.)

The engineer provides the technical, and in many instances the managerial leadership in industry and government. Depending upon his specialization, the engineer may be responsible for the design and development of new products and processes, plan structures and highways, or work out new ways of transforming raw materials into saleable products.

Laws in all 50 states and the District of Columbia provide for the licensing of engineers whose work may effect life, health and property.

A professional engineering license usually requires graduation from an approved engineering college, four years of experience and passing an examination. Some states will accept experience in place of a college degree.

ARCHITECTS

In general, the ARCHITECT plans and designs all kinds of structures. However, he may specialize in specific fields of architecture - private homes, industrial buildings, schools, etc.

When planning a structure, the architect first consults the client on the purpose of the building, its size, location, cost range and other requirements. Upon completion and approval of preliminary drawings, detailed working drawings and specification sheets are prepared. As construction progresses the architect usually makes periodic inspections to determine whether the plans are being followed and construction details are to specifications.

Architects must be licensed in most states. This requires graduation from an approved college program, several years of experience (similar to the internship of a physician) and the passing of a special examination.

Several years of experience are acceptable in a few states in place of graduation from an architectural program.

MODELMAKER

Industry makes extensive use of models, mockups and prototypes for engineering, educational and planning purposes. Their preparation is often the responsibility of the engineering drafting group although professional MODELMAKERS are frequently used.

The modelmaker makes scale models of the proposed building or product to show the client what it will look like and, in some cases, how it will work.

In order to interpret the designer's or engineer's plans accurately, the modelmaker must be able to read and understand drawings. He must also be able to prepare accurate drawings.

TECHNICAL ILLUSTRATOR

Technical illustration is a process of preparing art work for industry. The TECHNICAL ILLUSTRATOR prepares pictorial matter for engineering and educational purposes. See Fig. 26-5. Technical illustrators must have a combination of technical and artistic abilities. The technical ability is necessary to understand the mechanical aspects of the job, and the artistic ability to show the building or product in three dimensions. He must be able to make accurate sketches and finish the drawing according to standards established by industry.

Fig. 26-5. After the ideas for a new car are carried through full size line drawings, a full size colored rendering is made to enable the designers to evaluate the shape as it would actually appear in paint and chrome. Here a technical illustrator uses an airbrush to put the finishing touches on a rendering of an experimental small car. (General Motors Corp.)

TEST YOUR KNOWLEDGE - UNIT 26

1. Draftsmen make ______________________.
2. The draftsman usually starts as a ______ and advances to ___________ and ______ on his way to becoming lead or chief draftsman.
3. A good draftsman is _________, ________ and ______________.
4. The industrial designer's main job is to ____________________________________.
5. ____________ is a profession that offers a freedom not usually found in other professions.
6. ____________ in general, plan and design all kinds of structures and buildings. Most states require that he be ________.
7. The engineer usually specializes in one of the many branches of the profession. List four kinds of engineering:
   a. ____________________.
   b. ____________________.
   c. ____________________.
   d. ____________________.
8. The modelmaker makes ______________ to show the client ______________.
9. The ______________ must have a combination of technical and artistic abilities.

OUTSIDE ACTIVITIES

1. Invite a representative from the local Government Employment Service Office to discuss, with your class, employment opportunities in the drafting occupations.
2. Make a study of the HELP WANTED COLUMNS in your daily newspaper for a period of two weeks. Prepare a list of drafting and related jobs available, salaries offered and the minimum requirements for securing the jobs. How often are additional benefits such as insurance, hospitalization, etc., mentioned in the ads.
3. Summarize the information on the drafting occupations given in the OCCUPATIONAL OUTLOOK HANDBOOK (a Government publication) and make it available to the class.

Unit 27
METRICATION IN DRAFTING

More than one hundred years ago a federal law was passed that said in part-

"It shall be lawful throughout the United States of America to employ the weights and measures of the metric system. . ."

This law made the metric system of weights and measures legal in the United States. However, Congress did NOT make the metric system obligatory and, as the people at that time were not acquainted with the metric system, they retained the more familiar inch-pound system based on English practice.

Fig. 27-1. This four-cylinder, 2300 cubic-centimetre (2.3 litre) engine was the first power plant designed, developed and manufactured in the United States entirely to metric standards. (Ford Motor Co.)

Today, because of world trade, the United States is being urged to adopt the INTERNATIONAL SYSTEM OF UNITS (the metric system). Already, some large firms are starting to convert, Fig. 27-1.

The International System of Units (Systéme International d'Unités) has officially been abbreviated to SI.

The conversion by the U.S. from the inch-pound system to SI units will be very costly and take many years.

DRAFTING AND THE CONVERSION TO THE METRIC SYSTEM

Drafting will play a very important role in the conversion process. Until the inch disappears altogether, draftsmen will find themselves making drawings to three different specifications.

1. Drawings will be made using the inch unit of measure.

2. Drawings classified as DUAL DIMENSIONED drawings list dimensions in both inches (usually decimal inches) and in the metric system (usually in millimetres), Fig. 27-2. Metre is the approved spelling.

CAUTION: Dual dimensioning is applicable where interchangeable parts are to be manufactured in BOTH U.S. customary (inch) and SI (metric) units of measurements. IT IS NOT RECOMMENDED AS A PRACTICE FOR IMPLEMENTING THE TRANSITION FROM INCH TO METRIC UNITS.

3. Drawings utilizing SI (metric) units and symbols, Fig. 27-3.

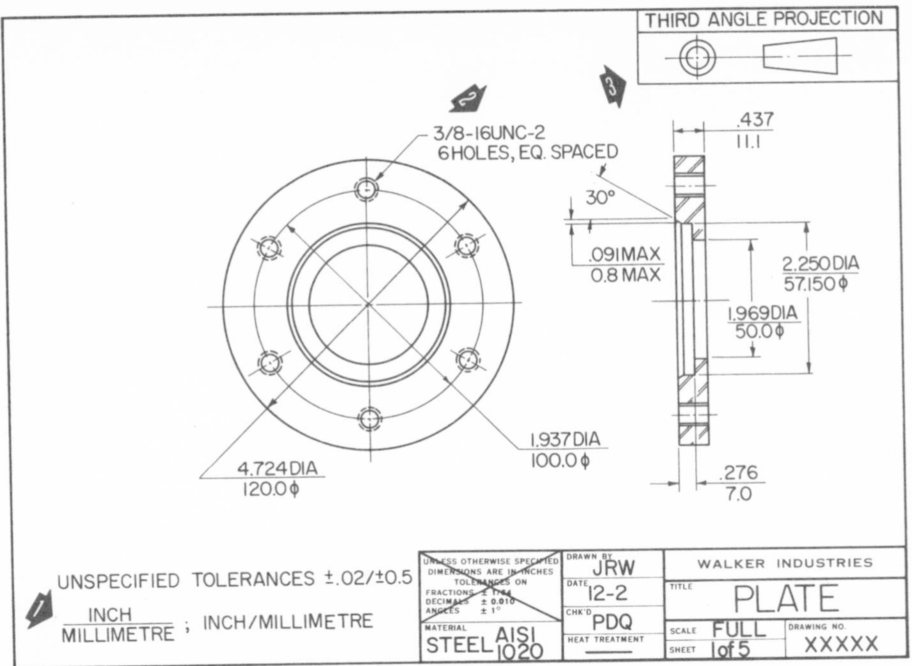

Fig. 27-2. A typical ***DUAL DIMENSIONED*** *drawing. Dual dimensioning is applicable where interchangeable parts are to be manufactured in both U.S. customary (inch) and SI (metric) units of measurements. Notice the important points (arrows) on the drawing. 1. Note indicating how the inch and millimetre dimensions can be identified. 2. Thread size is* ***NOT*** *given in metric measurement because there is* ***NO*** *metric thread this size. 3. It is recommended that all dual dimensioned drawings indicate the* ***ANGLE OF PROJECTION*** *used, to eliminate any confusion when used in different countries.*

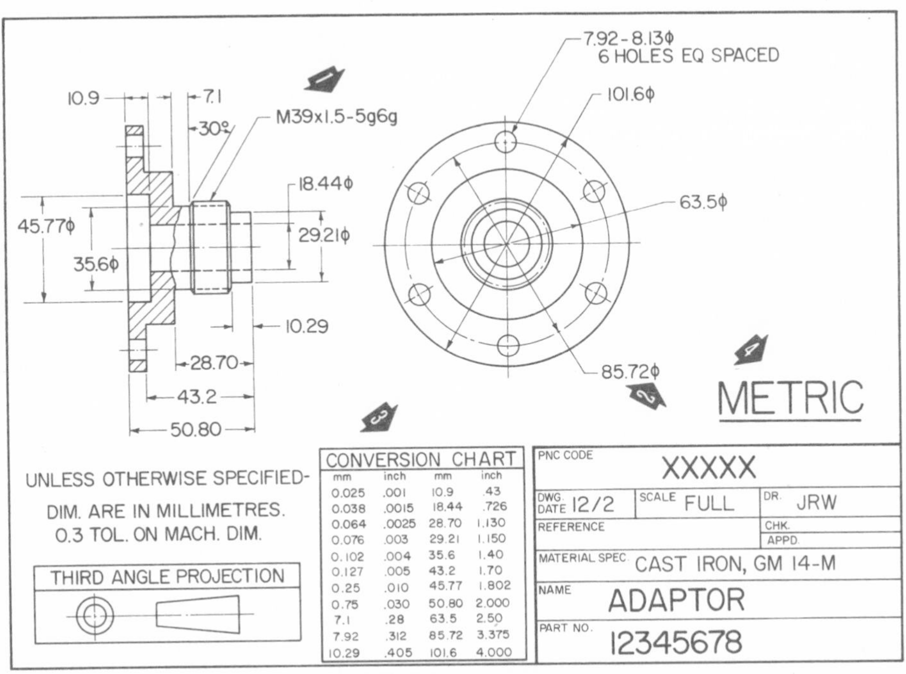

CONVERSION CHART

mm	inch	mm	inch
0.025	.001	10.9	.43
0.038	.0015	18.44	.726
0.064	.0025	28.70	1.130
0.076	.003	29.21	1.150
0.102	.004	35.6	1.40
0.127	.005	43.2	1.70
0.25	.010	45.77	1.802
0.75	.030	50.80	2.000
7.1	.28	63.5	2.50
7.92	.312	85.72	3.375
10.29	.405	101.6	4.000

Fig. 27-3. A drawing utilizing SI (metric) units of measurement. Until it is no longer considered necessary, equivalent inch values ***MAY*** *be shown on metric drawings. The metric dimensioning will be used for manufacturing with the inch equivalent given for reference only. Many U.S. companies have adopted this technique. Important points on the drawing are indicated by numbered arrows. 1. Metric threads are specified differently from the familiar NC and NF threads. The M denotes that it is a metric thread. The 39 denotes the nominal thread diameter in millimetres. The 1.5 indicates the thread pitch in millimetres and the 5g6g indicate tolerance class designation. 2. Symbol used to indicate diameter. 3. Conversion chart (for reference only). 4. The term* ***METRIC*** *is lettered on the drawing to alert the user that metric dimensions are used. It will be used on drawings until it is no longer considered necessary. Some drawings may have* ***METRIC*** *stamped at an angle to better catch the eye.*

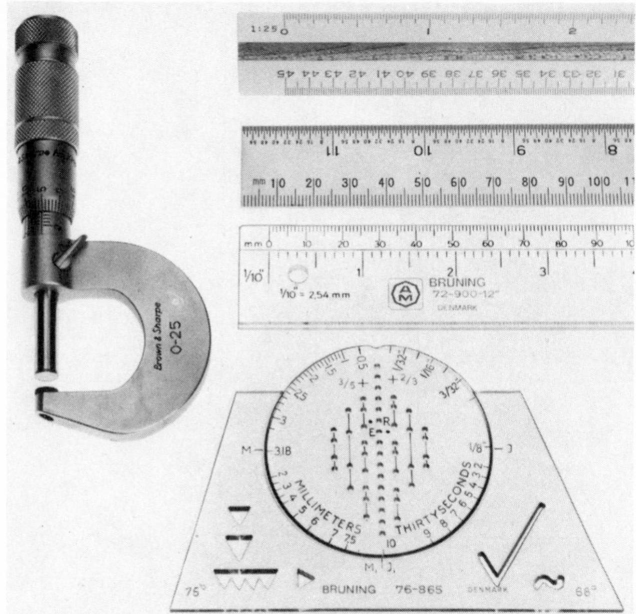

Fig. 27-4. A few of the many metric drafting tools with which the draftsman must become familiar.

PROBLEM AREAS

Designing and drawing products in SI units involves a great deal more than simply converting inches to millimetres. For example: materials and tools will be manufactured to SI units. You as a draftsman or designer cannot specify a 9/16 in. bolt by merely listing the metric equivalent of 9/16 in. (14.287 5mm) because THERE ARE NO METRIC BOLTS THAT CORRESPOND TO THIS DIAMETER. The same problem would occur if a 1 in. dia. shaft were specified as 25.4mm diameter. There is NO standard SI unit shaft this diameter. The closest size would be 25.0mm diameter. To get the 25.4 diameter shaft, a very expensive machining operation would be required to turn a larger shaft to 25.4mm in diameter.

UNDERSTAND THE METRIC SYSTEM

To best understand the metric system, the engineer, designer, draftsman or craftsman will have to THINK in SI units. They will also have to learn to read and use metric measuring tools, Fig. 27-4.

In addition, they must remember that the symbols employed in the metric system are all important and inaccuracies will result if they are used carelessly. Lower-case mm means millimetre (0.001 metre) while Mm means megametre (1 000 000 metres), a very great difference in size.

SI DIMENSIONING PRACTICES

It is not possible to include the entire metric dimensioning system in a text of this limited nature. The material included will serve as an introduction to SI standards recommended for use by the industries of the U.S.

The millimetre (mm) is the basic unit for linear dimensions on drawings used by the manufacturing industries. In building construction, the metre (m) is recommended. Drawings made for topographic purposes (surveying, mapping, etc.) where considerable distances are depicted should use the kilometre (k) and a sizable reduction will result.

Angular dimensions are specified the SAME in both inch and metric systems.

30°

Whenever possible, specify dimensions in even millimetres.

Millimetre values less than one should be shown with a zero to the LEFT of the decimal point.

0.2 not .2

A zero is shown to the RIGHT of the decimal point where the millimetre value is a whole untoleranced number.

25.0 not 25

The decimal point for millimetre dimensions shall be the SAME as that used for inch dimensions (some European countries use a comma as a decimal point).

Commas SHALL NOT be employed to denote thousands. A space shall be used in place of a comma. Also a space shall be added after every third digit to the RIGHT of the decimal point.

1 234.567 8 not 1,234.5678

Dual dimensioning is a procedure where both U.S. customary units (inch) and SI units (metric) of measurement are shown on the same drawing. Two methods have been designated to display dual dimensions on a drawing, Fig. 27-5. Use either the POSITION or BRACKET method, but avoid using both methods on the same drawing.

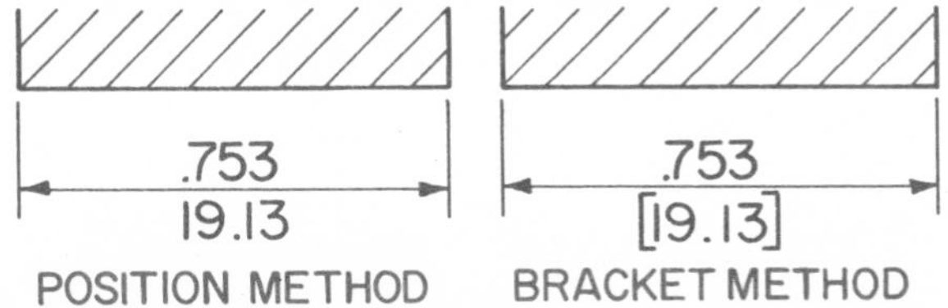

Fig. 27-5. Methods of indicating inches and millimetres on a dual dimensioned drawing. Avoid using BOTH on the same drawing.

Drawings employing the position method and used primarily in the United States will display the inch dimensions above or to the left of the millimetre dimension. The bracket method encloses the metric dimension in square brackets. Drawings used primarily in Europe will have the inch dimension enclosed in square brackets.

A note must be placed on the drawing to show how the inch and millimetre dimension can be identified, Fig. 27-6.

INCH / MILLIMETRE ; INCH/MILLIMETRE

DIMENSIONS IN [ ] ARE MILLIMETRES

Fig. 27-6. Type of note placed on drawing to identify inch and millimetre dimensions.

Hole diameter or shaft diameter is indicated on SI unit drawings with the number and symbol ϕ.

Various metric references can be found in Figs. 27-7, 27-8, and 27-9.

COMMON LINEAR UNITS IN THE METRIC SYSTEM

One of the advantages of the metric system is the ability to decimalize all of the units.

Unit	Symbol		Equals
kilometre	k	=	1 000 metres (thousands)
hectometre*	h	=	100 metres (hundreds)
dekametre*	da	=	10 metres (tens)
metre	m	=	1 metre (unit of linear measure)
decimetre*	d	=	0.1 metre (tenths)
centimetre*	c	=	0.01 metre (hundredths)
millimetre	mm	=	0.001 metre (thousandths)
micrometer	μ	=	0.000 001 metre (millionths)

* To be avoided where possible.

FOR EXAMPLE: 1000 metres = 1 kilometre
0.001 metre = 1 millimetre

USEFUL APPROXIMATIONS

A metre is about 10 percent longer than a yard.
A litre has 5 percent more volume than a quart.
Subtract a tenth of the pounds and divide by two to get kilograms.
To calculate kilometres (or km/h), add half and then one tenth of miles (or mph).
Remember that water freezes at 0°C. and boils at 100°C. Your body temperature is about 37°C.

Fig. 27-7. Metric linear units and useful approximations.

SI UNITS & CONVERSIONS

PROPERTY	UNIT	SYMBOL	EXACT CONVERSION			APPROXIMATE EQUIVALENCY
			FROM	TO	MULTIPLY BY	
length	metre	m	inch	mm	2.540×10	25mm = 1 in.
	centimetre	cm	inch	cm	2.540	300mm = 1 ft.
	millimetre	mm	foot	mm	3.048×10^{-4}	
mass	kilogram	kg	ounce	g	2.835×10	2.8g - 1 oz.
	gram	g	pound	kg	4.536×10^{-1}	kg = 2.2 lbs. = 35 oz.
	tonne (megagram)	t	ton (2000 lb)	kg	9.072×10^{2}	1t = 2200 lbs.
density	kilogram per cub. metre	kg/m^3	pounds per cu. ft.	kg/m^3	1.602×10	$16kg/M^3 = 1 lb./ft^3$
temperature	deg. Celsius	^oC	deg. Fahr.	^oC	$(^oF-32) \times 5/9$	$0^oC = 32^oF$
						$100^oC = 212^oF$
area	square metre	m^2	sq. inch	mm^2	6.452×10^{2}	$645mm^2 = 1 in.^2$
	square millimetre	mm^2	sq. ft.	m^2	9.290×10^{-2}	$1m^2 = 11 ft.^2$
volume	cubic metre	m^3	cu. in.	mm^3	1.639×10^{4}	$16400mm^3 = 1 in.^3$
	cubic centimetre	cm^3	cu. ft.	m^3	2.832×10^{-2}	$1m^3 = 35 ft.^3$
	cubic millimetre	mm^3	cu. yd.	m^3	7.645×10^{-1}	$1m^3 = 1.3 yd.^3$
force	newton	N	ounce (Force)	N	2.780×10^{-1}	1N = 3.6 oz.
	kilonewton	kN	pound (Force)	kN	4.448×10^{-3}	4.4N = 1 lb.
	meganewton	MN	Kip	MN	4.448	1kN = 225 lb.
stress	megapascal	MPa	$pound/in^2$ (psi)	MPa	6.895×10^{-3}	1MPa = 145 psi
			Kip/in^2 (ksi)	MPa	6.895	7MPa = 1 ksi
torque	newton-metres	N.m	in-ounce	N.m	7.062×10^{3}	1N.m = 140 in.oz.
			in.pound	N.m	1.130×10^{-1}	1N.m = 9 in.lb.
			ft pound	N.m	1.356	1N.m =.75 ft.lb.
						1.4N.m = 1 ft.lb.

Fig. 27-8. Metric units and conversions.

TWIST DRILL DATA

METRIC DRILL SIZES (mm)[1] Preferred	Available	Decimal Equivalent in Inches (Ref)
	.40	.0157
	.42	.0165
	.45	.0177
	.48	.0189
.50		.0197
	.52	.0205
.55		.0217
	.58	.0228
.60		.0236
	.62	.0244
.65		.0256
	.68	.0268
.70		.0276
	.72	.0283
.75		.0295
	.78	.0307
.80		.0315
	.82	.0323
.85		.0335
	.88	.0346
.90		.0354
	.92	.0362
.95		.0374
	.98	.0386
1.00		.0394
	1.03	.0406
1.05		.0413
	1.08	.0425
1.10		.0433
	1.15	.0453
1.20		.0472
1.25		.0492
1.30		.0512
	1.35	.0531
1.40		.0551
	1.45	.0571
1.50		.0591
	1.55	.0610
1.60		.0630
	1.65	.0650

METRIC DRILL SIZES (mm)[1] Preferred	Available	Decimal Equivalent in Inches (Ref)
1.70		.0669
	1.75	.0689
1.80		.0709
	1.85	.0728
1.90		.0748
	1.95	.0768
2.00		.0787
	2.05	.0807
2.10		.0827
	2.15	.0846
2.20		.0866
	2.30	.0906
2.40		.0945
2.50		.0984
2.60		.1024
	2.70	.1063
2.80		.1102
	2.90	.1142
3.00		.1181
	3.10	.1220
3.20		.1260
	3.30	.1299
3.40		.1339
	3.50	.1378
3.60		.1417
	3.70	.1457
3.80		.1496
	3.90	.1535
4.00		.1575
	4.10	.1614
4.20		.1654
	4.40	.1732
4.50		.1772
	4.60	.1811
4.80		.1890
5.00		.1969
	5.20	.2047
5.30		.2087
	5.40	.2126
5.60		.2205
	5.80	.2283

1 Metric drill sizes listed in the "Preferred" column are based on the R'40 series of preferred numbers shown in the ISO Standard R497. Those listed in the "Available" column are based on the R80 series from the same document.

TWIST DRILL DATA (CONTINUED)

METRIC DRILL SIZES (mm)[1] Preferred	Available	Decimal Equivalent in Inches (Ref)
6.00		.2362
	6.20	.2441
6.30		.2480
	6.50	.2559
6.70		.2638
	6.80[2]	.2677
	6.90	.2717
7.10		.2795
	7.30	.2874
7.50		.2953
	7.80	.3071
8.00		.3150
	8.20	.3228
8.50[2]		.3346
	8.80	.3465
9.00		.3543
	9.20	.3622
9.50		.3740
	9.80	.3858
10.00		.3937
	10.30	.4055
10.50		.4134
	10.80	.4252
11.00		.4331
	11.50	.4528
12.00		.4724
12.50		.4921
13.00		.5118
	13.50	.5315
14.00		.5512
	14.50	.5709
15.00		.5906
	15.50	.6102
16.00		.6299
	16.50	.6496
17.00		.6693
	17.50	.6890
18.00		.7087
	18.50	.7283
19.00		.7480

METRIC DRILL SIZES (mm)[1] Preferred	Available	Decimal Equivalent in Inches (Ref)
	19.50	.7677
20.00		.7874
	20.50	.8071
21.00		.8268
	21.50	.8465
22.00		.8661
	23.00	.9055
24.00		.9449
25.00		.9843
26.00		1.0236
	27.00	1.0630
28.00		1.1024
	29.00	1.1417
30.00		1.1811
	31.00	1.2205
32.00		1.2598
	33.00	1.2992
34.00		1.3386
	35.00	1.3780
36.00		1.4173
	37.00	1.4567
38.00		1.4961
	39.00	1.5354
40.00		1.5748
	41.00	1.6142
42.00		1.6535
	43.50	1.7126
45.00		1.7717
	46.50	1.8307
48.00		1.8898
50.00		1.9685
	51.50	2.0276
53.00		2.0866
	54.00	2.1260
56.00		2.2047
	58.00	2.2835
60.00		2.3622

1 Metric drill sizes listed in the "Preferred" column are based on the R'40 series of preferred numbers shown in the ISO Standard R497. Those listed in the "Available" column are based on the R80 series from the same document.

2 Recommended only for use as a tap drill size.

Fig. 27-9. Metric twist drill data. (General Motors)

GLOSSARY

ACCURATE: Made within the tolerances allowed.

ACOUSTIC TILE: Tile made of sound absorbing materials.

ACUTE ANGLE: An angle less than 90 deg.

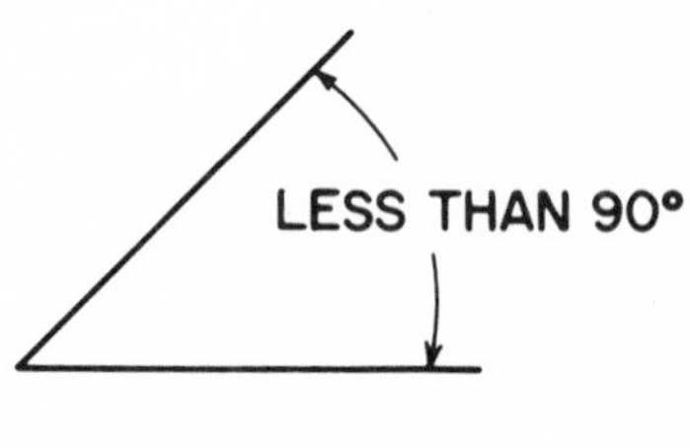

ACUTE ANGLE

AIR BRUSH: A device used to spray paint by means of compressed air.

AIR CONDITIONING: The control of the temperature, humidity, motion, dust and distribution of the air within a structure.

ALLOY: A mixture of two or more metals fused or melted together to form a new metal.

ALLOWANCE: The limits permitted for satisfactory performance of machined parts.

ANGLE: The figure formed by two lines coming together to a point.

ANNEALING: The process of heating metal to a given temperature (the exact temperature and period the temperature is held depends upon the composition of the metal being annealed) and cooling it slowly to remove stresses and induce softness.

ARC: Portion of a circle.

ASPHALT ROOFING: A roofing material made by saturating an asbestos or felt with asphalt.

ASSEMBLY: A unit fitted together from manufactured parts.

ATTIC: The part of a building just below the roof.

AXES: The plural of axis.

AXIS: The center line of a view or of a geometric figure.

BASEMENT: The base story of a structure. It is usually underground.

BEAM: A term used to describe joists, rafters, and girders.

BEAM COMPASS: Compass used to draw large circles and arcs.

BEVEL: The angle that is formed by a line or a surface that is not at right angles to another line or surface.

BLOWHOLE: A hole produced in a casting when gasses are entrapped during the pouring operation.

BLUEPRINT: A reproduction of a drawing that has a bright background with white lines.

BRAZING: Joining metals by the fusion of nonferrous alloys that have melting temperatures above 800 deg. F. but lower than the metals being joined.

BUSHING: A bearing for a revolving shaft. A hardened steel tube used on jigs to guide drills and reamers.

CASTING: An object made by pouring molten metal into a mold.

CAULK: To fill cracks and seams with caulking material.

CASEHARDEN: A process of surface hardening iron base alloys so that the surface layer or case is made substantially harder than the interior or core of the metal.

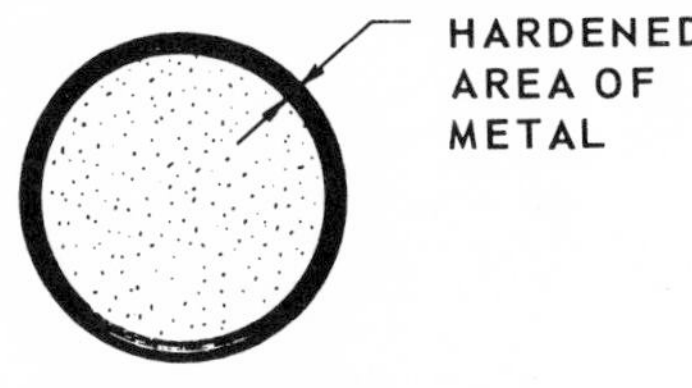

CASEHARDENING

CENTER LINE SYMBOL.

CHAMFER: See BEVEL.
CIRCUMFERENCE: The perimeter of a circle.
CIRCUMSCRIBE: To draw a line around.
CLEARANCE: The distance by which one part clears another part.
CLOCKWISE: From left to right in a circular motion. The direction clock hands move.

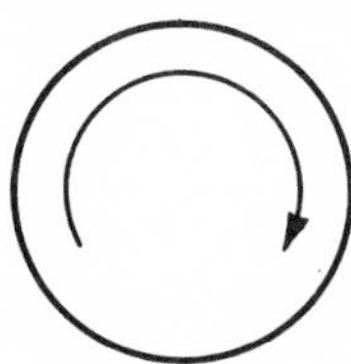

CLOCKWISE

COMPUTER: An electronic calculator performing a sequence of computations.
CONCAVE SURFACE: A curved depression in the surface of an object.

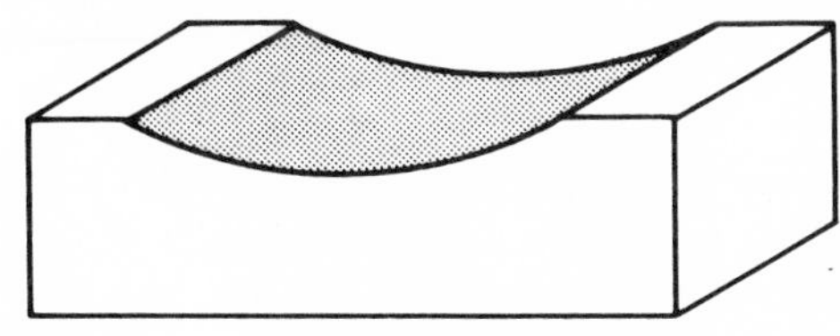

CONCAVE SURFACE

CONCENTRIC: Having a common center.

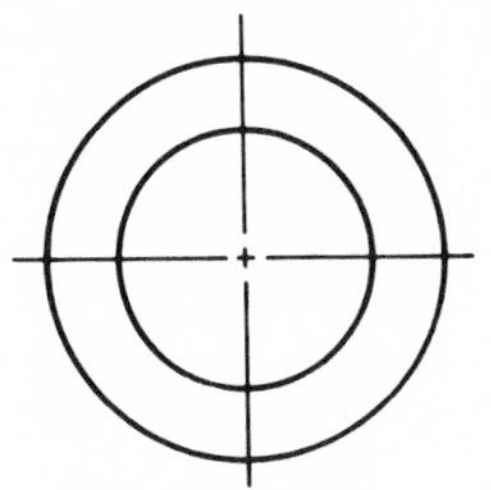

CONCENTRIC CIRCLES

CONCRETE: A mixture of portland cement, sand, gravel and water.
CONICAL: Shaped like a cone.
CONTOUR: The outline of an object.
CONVENTIONAL: Not original; customary, or traditional.
CONVEX SURFACE: A rounded surface on an object.

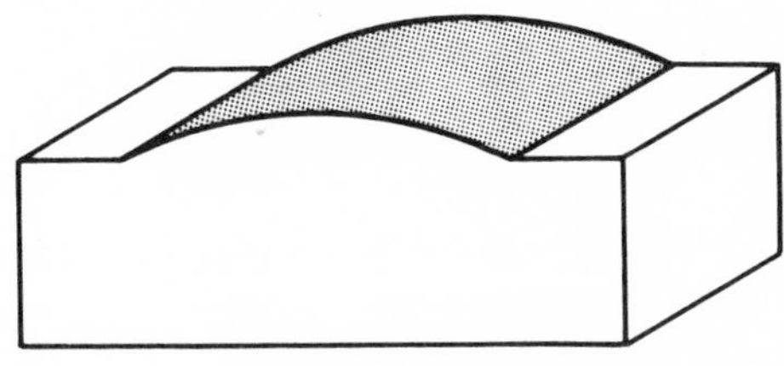

CONVEX SURFACE

CORE: A body of sand or other material that is formed to a desired shape and placed in a mold to produce a cavity or opening in a casting.
COUNTERBORE: To enlarge a hole to a given depth.

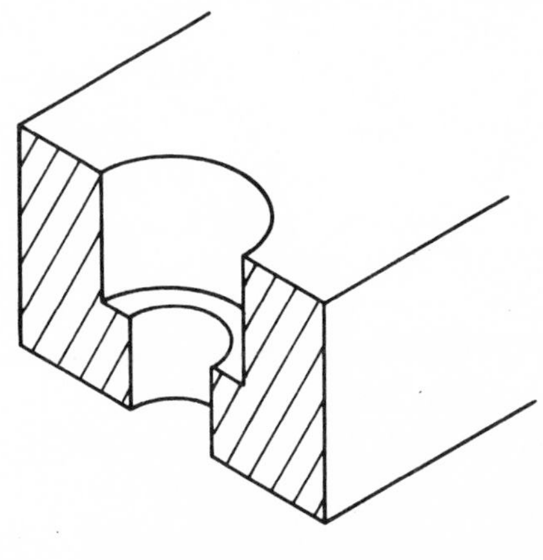

COUNTERBORED HOLE

COUNTERCLOCKWISE: From right to left in a circular motion.

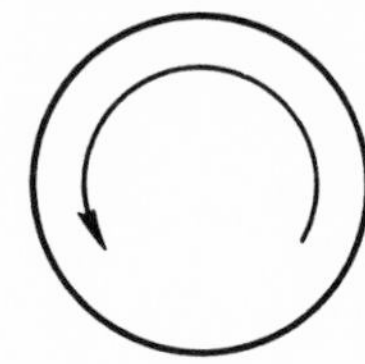

COUNTERCLOCKWISE

COUNTERSINK: To chamfer a hole to receive a flat or fillister head fastener.

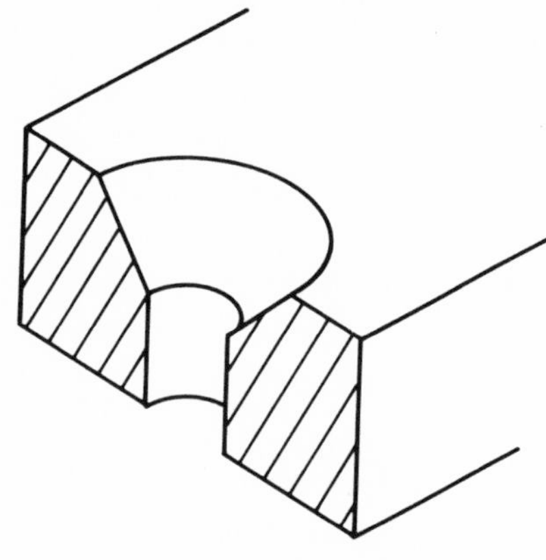

COUNTERSUNK HOLE

CURVED LINE: A line of which no part is straight.

CYLINDER: A geometric figure with a uniform circular cross-section through its entire length.

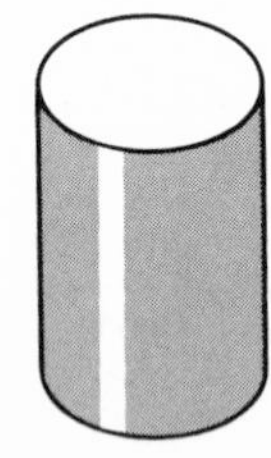

CYLINDER

DIAGONAL: A line running across from corner to corner.

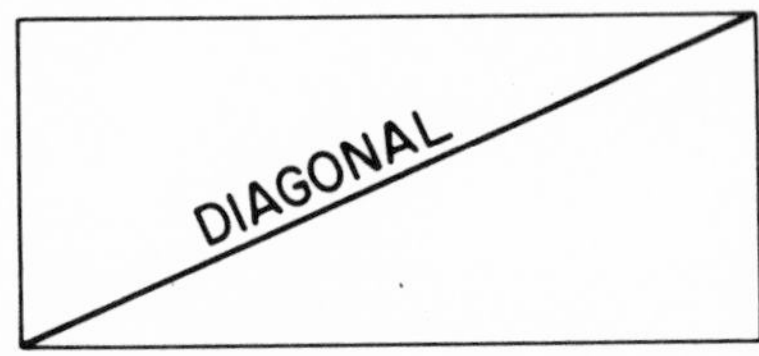

DIAMETER: The length of a straight line running through the center of a circle.

DIE: The tool used to cut external threads. A tool used to shape materials.

DIE CASTING: A method of casting metal under pressure by injecting it into metal dies of a die casting machine.

DRAFT: The clearance on a pattern or mold that allows easy withdrawal of the pattern from the mold.

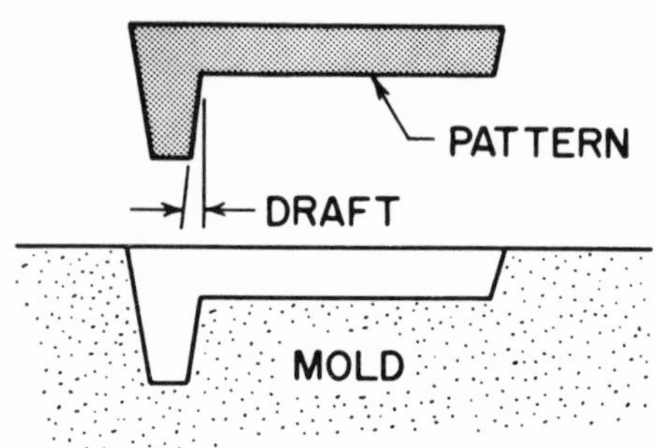

DRILLING: Cutting round holes by use of a cutting tool called a drill.

DRILL ROD: Accurately ground and polished tool steel rods.

DRIVE FIT: Using force or pressure to fit two pieces together. One of several classes of fits.

DUCT: A sheet metal pipe or passageway used for conveying air.

ECCENTRIC: Not on a common center.

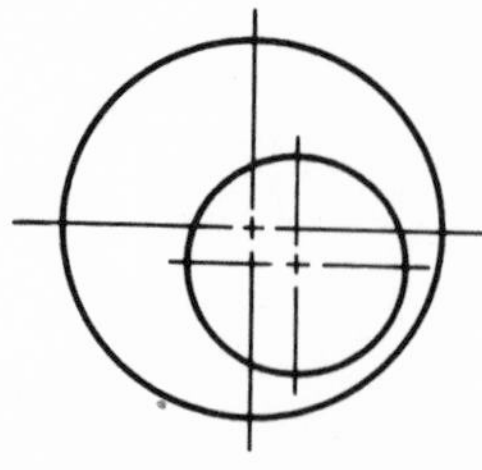

ECCENTRIC CIRCLES

ELIMINATE: To do away with.

ELLIPSE: A closed curve in the form of a symmetrical oval.

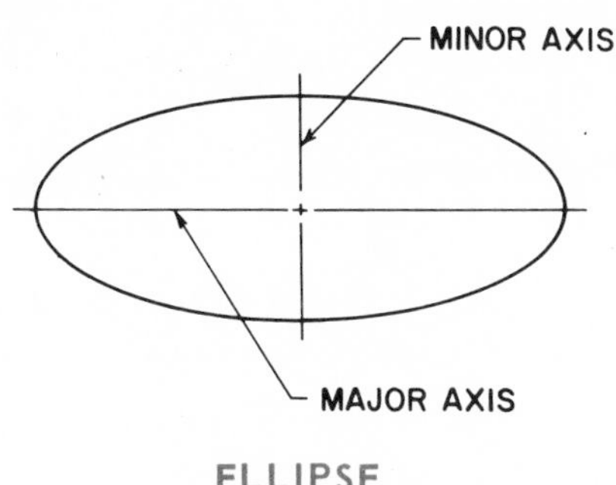

ELLIPSE

ENGINEERING DRAWING: Also known as a technical drawing or drafting, is the graphic language of the engineer.

EQUAL: The same.

EQUILATERAL: A figure having equal length sides.

EXPANSION FIT: The reverse of shrink fit. The piece to be fitted is placed in liquid nitrogen or dry ice until it shrinks enough to fit into the mating piece. Interference develops between the fitted pieces as the cooled piece expands to normal size.

FERROUS METAL: A metal that contains iron as its major ingredient.

FILLET: The curved surface connecting two surfaces that form an angle.

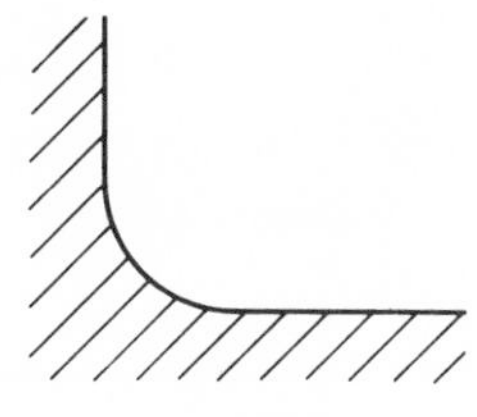

FILLET

FIXTURE: A device for holding metal while it is being machined.

FORGE: To form material using heat and pressure.

FLASH: A thin fin of metal formed at the parting line of a forging or casting where a small portion of metal is forced out between the edges of the die.

FLASK: A frame of wood or metal consisting of a cope (the top portion) and a drag (the bottom portion) used to hold sand that forms the mold used in the foundry.

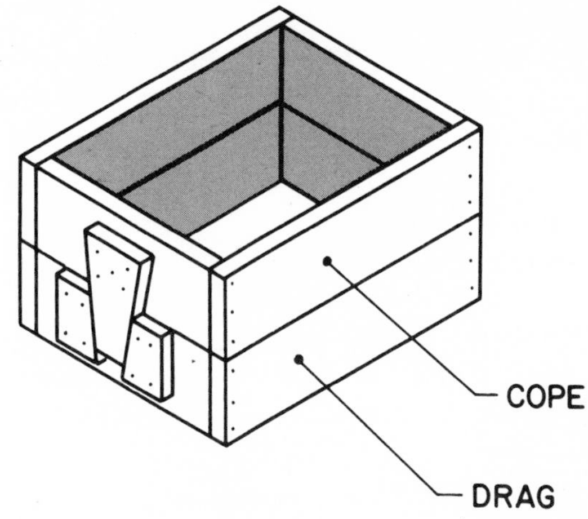

FLASK

FLOOR PLAN: The drawing that shows the exact shape, dimensions, and arrangements of the rooms of a building.

FORCE FIT: The interference between two mating parts sufficient to require force to press the pieces together. The joined pieces are considered permanently assembled.

FREE FIT: Used when tolerances are liberal. Clearance is sufficient to permit a shaft to turn freely without binding or overheating when properly lubricated.

FRUSTUM: The figure formed by cutting off a portion of a cone or pyramid parallel to its base.

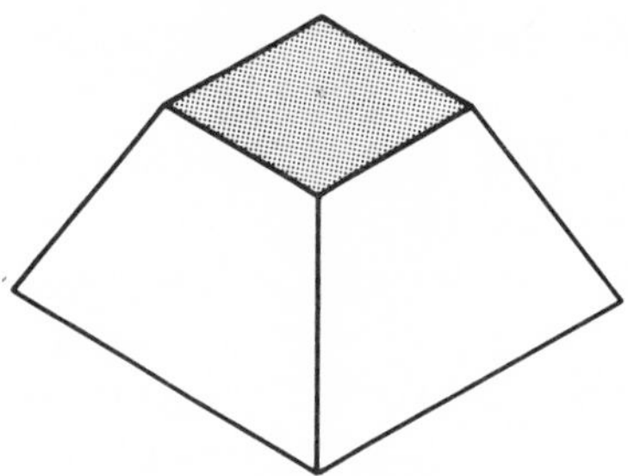

FRUSTUM

GATE: The opening that guides the molten metal into the cavity that forms the mold. See SPRUE.

GEARS: Toothed wheels that transmit rotary motion from one shaft without slippage.

HARDENING: The heating and quenching of certain iron-base alloys for the purpose of producing a hardness superior to that of the

untreated metal.
HEAT TREATMENT: The careful application of a combination of heating and cooling cycles to a metal or alloy to bring about certain desirable conditions such as hardness and toughness.
HEXAGON: A six-sided figure with each side forming a 60 deg. angle.
INCLINED: Making an angle with another line or plane.
INSCRIBE: To draw one figure within another figure.
INSPECTION: The measuring and checking of finished parts to determine whether they have been made to specifications.
INTERCHANGEABLE: Refers to a part that has been made to specific dimensions and tolerances and is capable of being fitted in a mechanism in place of a similarly made part.
INVESTMENT CASTING: A process that involves making a wax, plastic or a frozen mercury pattern, surrounding it with a wet refractory material, melting or burning the pattern out after the investment material has dried and set, and finally pouring molten metal into the cavity.
JAMB: The vertical parts of a door frame.
JIG: A device that holds the work in position and guides the cutting tool.
KEY: A small piece of metal imbedded partially in the shaft and partially in the hub to prevent rotation of a gear or pulley on the shaft.
KEYWAY: The slot or recess in the shaft that holds the key.
LAY OUT: To locate and scribe points for machining and forming operations.
LINE CONVENTIONS: Symbols that furnish a means of representing or describing some part of an object. It is expressed by a combination of line weight and appearance.
MACHINE TOOL: The name given to that class of machines which, taken as a group, can reproduce themselves.
MAJOR DIAMETER: The largest diameter of a thread measured perpendicular to the axis.
MESH: To engage gears to a working contact.
MICROINCH: One-millionth of an inch.
MILL: To remove metal with a rotating cutter on a milling machine.
MINOR DIAMETER: The smallest diameter on a screw thread measured across the root of the thread and perpendicular to the axis. Also known as the "root diameter."
MOLD: The material that forms the cavity into which molten metal is poured.

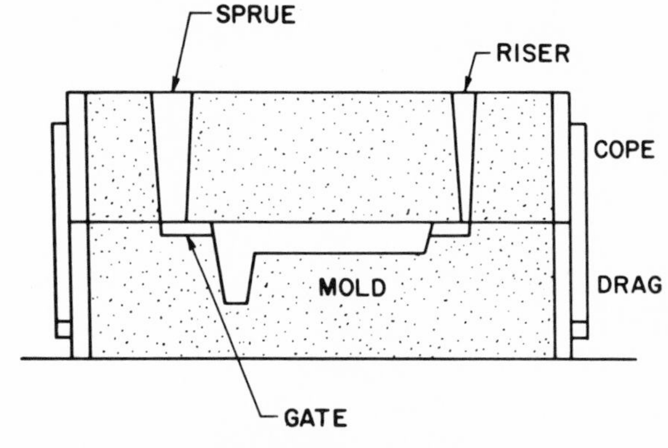

PARTS OF A MOLD

NATIONAL COURSE (NC): The coarse series of the American National thread series.
NATIONAL FINE (NF): The fine series of the American National thread series.
NC: Abbreviation for the National Coarse series of threads.
NF: Abbreviation for the National Fine series of threads.
OBTUSE ANGLE: An angle more than 90 deg.

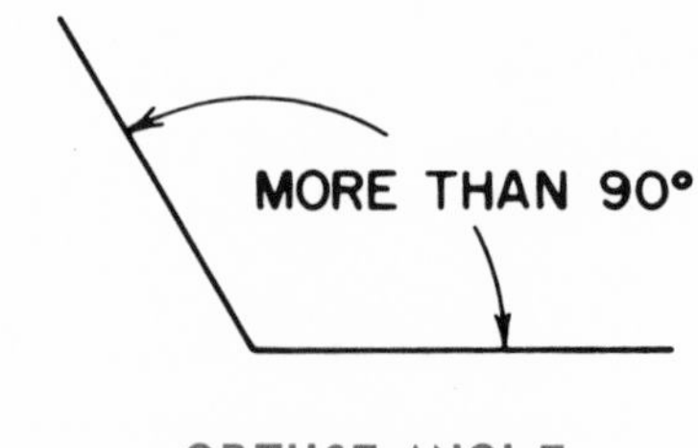

OBTUSE ANGLE

OCTAGON: An eight-sided geometric figure with each side forming a 45 deg. angle.
OD: Abbreviation for outside diameter.
PENTAGON: A five-sided geometric figure with each side forming a 72 deg. angle.
PERIMETER: The boundry of a geometric figure.
PERMANENT MOLD: Mold ordinarily made of metal that is used for the repeated production of similar castings.
PERPENDICULAR: A line at right angles to

a given line.

PHOTO DRAWING: A drawing prepared using a photograph on which dimensions, notes and specifications have been added.

ROUGH LAYOUT: A rough pencil plan that arranges lines and symbols so that they have a pleasing relation to one another.

PITCH: The distance from a point on one thread to a corresponding point on the next thread.

PLATE: Another name given to a drawing.

PROFILE: The outline of an object.

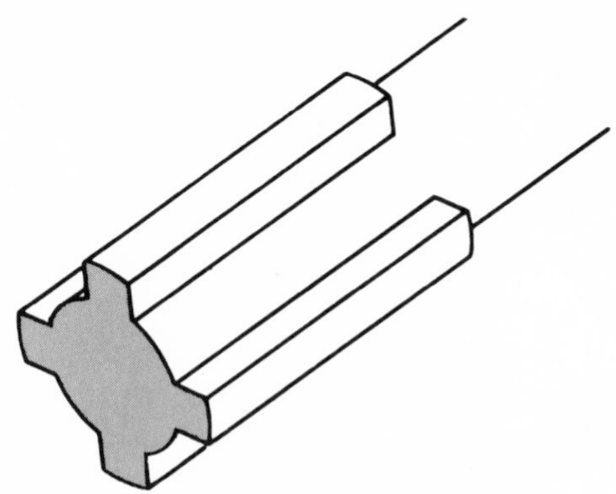

PROFILE

PROJECT: To extend from.

RACK: A flat strip with teeth designed to mesh with teeth on a gear. Used to change rotary motion to a reciprocating motion.

RADIUS: The length of a straight line running from the center of a circle to the perimeter of the circle.

RADII: The plural of radius.

REAM: To finish a drilled hole to exact size with a reamer.

RECTANGLE: A geometric figure with opposite sides equal in length and each corner forming a 90 deg. angle.

RIGHT ANGLE: A 90 deg. angle. The angle that is formed by a line that is perpendicular to another line.

RISER: An opening in the mold that permits the gasses to escape. The gasses are formed when molten metal is poured into the mold.

ROTATE: To turn or revolve around a point.

SEAM: The line formed where two edges are joined together.

SEGMENT: Any part of a divided line.

SKETCH: To draw without the aid of drafting instruments.

SLIDE RULE: A portable calculating device.

SPLINE: A series of grooves, cut lengthwise, around a shaft or hole.

SPOTFACE: To machine a circular spot on the surface of a casting to furnish a bearing surface for the head of a bolt or a nut.

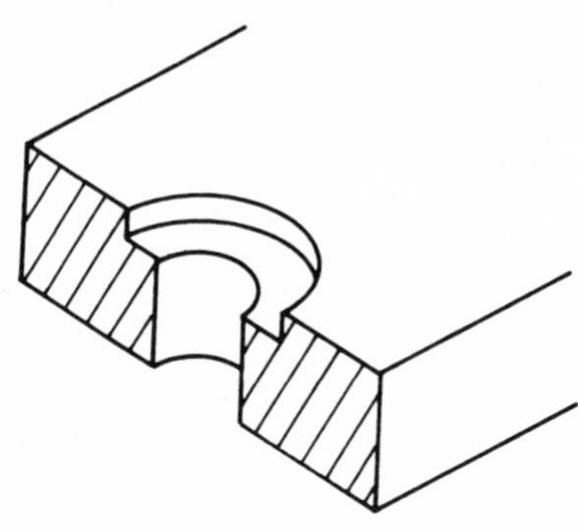

SPOTFACED HOLE

SPRUE: The opening in a mold that leads to the gate, which in turn leads to the cavity into which molten metal is poured.

SQUARE: To machine or cut at right angles. A geometric figure with four equal length sides and four right (90 deg.) angles.

SYMBOL: A figure or character used in place of a word or group of words.

TAP: The tool used to cut internal threads.

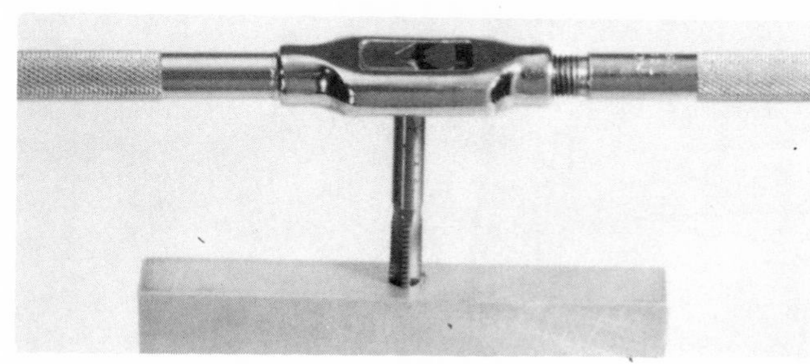

TAPER: A piece that increases or decreases in size at a uniform rate to assume a wedge or conical shape.

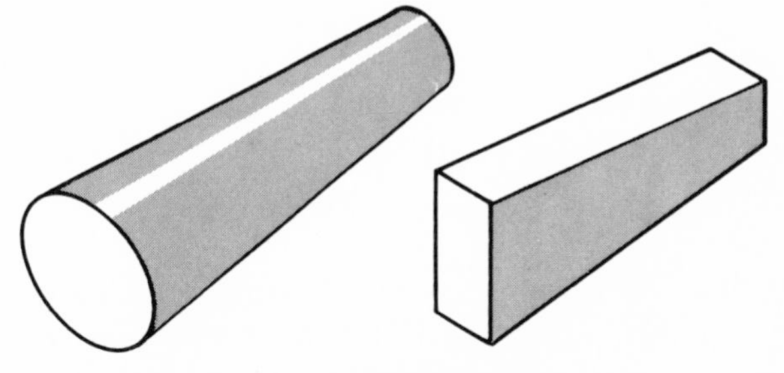

TAPERS

TEMPLATE: A pattern or guide.

THREAD: The act of cutting a screw thread.

THREADS

TRAIN: A series of meshed gears.

TRIANGLE: A three-sided geometric figure.

TRUNCATE: To cut off a geometric solid at an angle to its base.

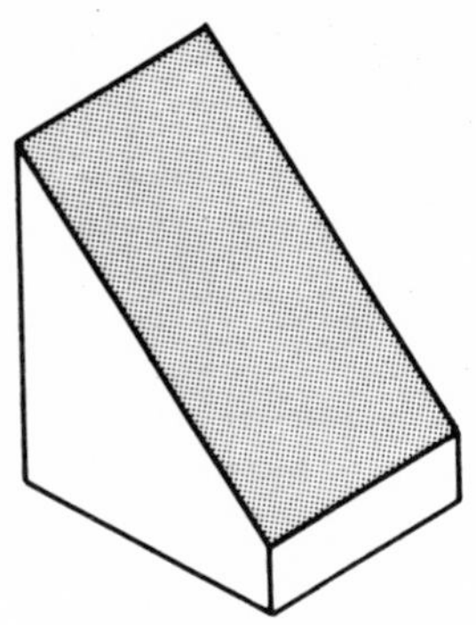

TRUNCATED PRISM

UNIFIED THREADS: A series of screw threads that have been adopted by the United States, Canada, and Great Britain to attain interchangeability of certain screw threads.

VERTICAL: At right angles to a horizontal line or plane.

WORKING DRAWING: A drawing or drawings that give the craftsman the necessary information to make and assemble a product.

ACKNOWLEDGMENTS

While it would be a most pleasant task, it would be impossible for one person to develop the material included in this text by visiting the various industries represented and observing, studying and taking the photos first hand.

My sincere thanks to those who helped in the gathering of the necessary material, information and photographs. Their cooperation was most appreciated.

John R. Walker
Bel Air, Maryland

TABLES

METALS WE USE

SHAPES	LENGTH	HOW MEASURED	* HOW PURCHASED
Sheet less than 1/4 in. thick	to 144 in.	Thickness x width, widths to 72 in.	Weight, foot or piece
Plate more than 1/4 in. thick	to 20 ft.	Thickness x width	Weight, foot or piece
Band	to 20 ft.	Thickness x width	Weight, or piece
Rod	12 to 20 ft.	Diameter	Weight, foot or piece
Square	12 to 20 ft.	Width	Weight, foot or piece
Flats	Hot rolled 20–22 ft. Cold finished	Thickness x width	Weight, foot or piece
Hexagon	12 to 20 ft.	Distance across flats	Weight, foot or piece
Octagon	12 to 20 ft.	Distance across flats	Weight, foot or piece
Angle	Lengths to 40 ft.	Leg length x leg length x thickness of legs	Weight, foot or piece
Channel	Lengths to 60 ft.	Depth x web thickness x flange width	Weight, foot or piece
I–beam	Lengths to 60 ft.	Height x web thickness x flange width	Weight, foot or piece

* Charge made for cutting to other than standard lengths.

PHYSICAL PROPERTIES OF METALS

METAL	SYMBOL	SPECIFIC GRAVITY	SPECIFIC HEAT	MELTING POINT*		LBS. PER CUBIC INCH
				DEG. C	DEG. F.	
Aluminum (Cast)	Al	2.56	.2185	658	1217	.0924
Aluminum (Rolled).	Al	2.71	–	–	–	.0978
Antimony	Sb	6.71	.051	630	1166	.2424
Bismuth	Bi	9.80	.031	271	520	.3540
Boron.	B	2.30	.3091	2300	4172	.0831
Brass.	–	8.51	.094	–	–	.3075
Cadmium.	Cd	8.60	.057	321	610	.3107
Calcium	Ca	1.57	.170	810	1490	.0567
Carbon	C	2.22	.165	–	–	.0802
Chromium	Cr	6.80	.120	1510	2750	.2457
Cobalt	Co	8.50	.110	1490	2714	.3071
Copper.	Cu	8.89	.094	1083	1982	.3212
Columbium . . .	Cb	8.57	–	1950	3542	.3096
Gold	Au	19.32	.032	1063	1945	.6979
Iridium.	Ir	22.42	.033	2300	4170	.8099
Iron	Fe	7.86	.110	1520	2768	.2634
Iron (Cast) . . .	Fe	7.218	.1298	1375	2507	.2605
Iron (Wrought) .	Fe	7.70	.1138	1500–1600	2732–2912	.2779
Lead	Pb	11.37	.031	327	621	.4108
Lithium	Li	.057	.941	186	367	.0213
Magnesium . . .	Mg	1.74	.250	651	1204	.0629
Manganese . . .	Mn	8.00	.120	1225	2237	.2890
Mercury	Hg	13.59	.032	38.7	37.7	.4909
Molybdenum. . .	Mo	10.2	.0647	2620	4748	.368
Monel Metal. . .	–	8.87	.127	1360	2480	.320
Nickel	Ni	8.80	.130	1452	2646	.319
Phosphorus . . .	P	1.82	.177	43	111.4	.0657
Platinum.	Pt	21.50	.033	1755	3191	.7767
Potassium. . . .	K	0.87	.170	62	144	.0314
Selenium.	Se	4.81	.084	220	428	.174
Silicon	Si	2.40	.1762	1427	2600	.087
Silver.	Ag	10.53	.056	961	1761	.3805
Sodium	Na	0.97	.290	97	207	.0350
Steel	–	7.858	.1175	1330–1378	2372–2532	.2839
Strontium	Sr	2.54	.074	–	–	.0918
Sulphur.	S	2.07	.175	115	235.4	.075
Tantalum	Ta	10.80	–	2850	5160	.3902
Tin	Sn	7.29	.056	232	450	.2634
Titanium.	Ti	5.3	.130	1900	3450	.1915
Tungsten	W	19.10	.033	3000	5432	.6900
Uranium	U	18.70	–	–	–	.6755
Vanadium	V	5.50	–	1730	3146	.1987
Zinc	Zn	7.19	.094	419	786	.2598

* Circular of the Bureau of Standards No. 35, Department of Commerce and Labor.

MEASUREMENT SYSTEMS

ENGLISH SYSTEM

MEASURES OF TIME

60 sec. = 1 min.
60 min. = 1 hr.
24 hr. = 1 day
365 dy. = 1 common yr.
366 dy. = 1 leap yr.

DRY MEASURES

2 pt. = 1 qt.
8 qt. = 1 pk.
4 pk. = 1 bu.
2150.42 cu. in. = 1 bu.

MEASURES OF LENGTH

12 in. = 1 ft.
3 ft. = 1 yd.
5 1/2 yd. = 1 rod
320 rods = 1 mile
5,280 ft. = 1 mile
1,760 yd. = 1 mile
6,080 ft. = 1 knot

LIQUID MEASURES

16 fluid oz. = 1 pt.
2 pt. = 1 qt.
32 fl. oz. = 1 qt.
4 qt. = 1 gal.
31 1/2 gal. = 1 bbl.
231 cu. in. = 1 gal.
7 1/2 gal. = 1 cu. ft.

MEASURES OF AREA

144 sq. in. = 1 sq. ft.
9 sq. ft. = 1 sq. yd.
30 1/4 sq. yd. = 1 sq. rod
160 sq. rods = 1 acre
640 acres = 1 sq. mile

MEASURES OF WEIGHT (Avoirdupois)

7,000 grains (gr.) = 1 lb.
16 oz. = 1 lb.
100 lb. = 1 cwt.
2,000 lb. = 1 short ton
2,240 lb. = 1 long ton

MEASURES OF VOLUME

1,728 cu. in. = 1 cu. ft.
27 cu. ft. = 1 cu. yd.
128 cu. ft. = 1 cord

METRIC SYSTEM

The basic unit of the metric system is the meter (m). The meter is exactly 39.37 in. long. This is 3.37 in. longer than the English yard. Units that are multiples or fractional parts of the meter are designated as such by prefixes to the word "meter". For example:

1 millimeter (mm.) = 0.001 meter or 1/1000 meter
1 centimeter (cm.) = 0.01 meter or 1/100 meter
1 decimeter (dm.) = 0.1 meter or 1/10 meter
1 meter (m.)
1 decameter (dkm.) = 10 meters
1 hectometer (hm.) = 100 meters
1 kilometer (km.) = 1000 meters

These prefixes may be applied to any unit of length, weight, volume, etc. The meter is adopted as the basic unit of length, the gram for mass, and the liter for volume.

In the metric system, area is measured in square kilometers (sq. km. or km.^{2}), square centimeters (sq. cm. or cm.^{2}), etc. Volume is commonly measured in cubic centimeters, etc. One liter (1) is equal to 1,000 cubic centimeters.

The metric measurements in most common use are shown in the following tables:

MEASURES OF LENGTH

10 millimeters = 1 centimeter
10 centimeters = 1 decimeter
10 decimeters = 1 meter
1000 meters = 1 kilometer

MEASURES OF WEIGHT

100 milligrams = 1 gram
1000 grams = 1 kilogram
1000 kilograms = 1 metric ton

MEASURES OF VOLUME

1000 cubic centimeters = 1 liter
100 liters = 1 hectoliter

CONVERSION TABLES

TO REDUCE	MULTIPLY BY	TO REDUCE	MULTIPLY BY
LENGTH			
miles to km.	1.61	km. to miles	0.62
miles to m.	1609.35	m. to miles	0.00062
yd. to m.	0.9144	m. to yd.	1.0936
in. to cm.	2.54	cm. to in.	0.3937
in. to mm.	25.4	mm. to in.	0.03937
VOLUME			
cu. in. to cc. or ml.	16.387	cc. to cu. in.	0.061
cu. in. to l.	0.0164	l to cu. in.	61.024
gal. to l.	3.785	l to gal.	0.264
WEIGHT			
lb. to kg.	0.4536	kg. to lb.	2.2
oz. to gm.	28.35	gm. to oz.	0.0353
gr. to gm.	0.0648	gm. to gr.	15.432

DECIMAL EQUIVALENTS NUMBER SIZE DRILLS

NO.	SIZE OF DRILL IN INCHES	NO.	SIZE OF DRILL IN INCHES	NO.	SIZE OF DRILL IN INCHES	NO.	SIZE OF DRILL IN INCHES
1	.2280	21	.1590	41	.0960	61	.0390
2	.2210	22	.1570	42	.0935	62	.0380
3	.2130	23	.1540	43	.0890	63	.0370
4	.2090	24	.1520	44	.0860	64	.0360
5	.2055	25	.1495	45	.0820	65	.0350
6	.2040	26	.1470	46	.0810	66	.0330
7	.2010	27	.1440	47	.0785	67	.0320
8	.1990	28	.1405	48	.0760	68	.0310
9	.1960	29	.1360	49	.0730	69	.0292
10	.1935	30	.1285	50	.0700	70	.0280
11	.1910	31	.1200	51	.0670	71	.0260
12	.1890	32	.1160	52	.0635	72	.0250
13	.1850	33	.1130	53	.0595	73	.0240
14	.1820	34	.1110	54	.0550	74	.0225
15	.1800	35	.1100	55	.0520	75	.0210
16	.1770	36	.1065	56	.0465	76	.0200
17	.1730	37	.1040	57	.0430	77	.0180
18	.1695	38	.1015	58	.0420	78	.0160
19	.1660	39	.0995	59	.0410	79	.0145
20	.1610	40	.0980	60	.0400	80	.0135

NATIONAL COARSE AND NATIONAL FINE THREADS AND TAP DRILLS

SIZE	THREADS PER INCH	MAJOR DIA.	MINOR DIA.	PITCH DIA.	TAP DRILL 75 PERCENT THREAD	DECIMAL EQUIVALENT	CLEARANCE DRILL	DECIMAL EQUIVALENT
2	56	.0860	.0628	.0744	50	.0700	42	.0935
	64	.0860	.0657	.0759	50	.0700	42	.0935
3	48	.099	.0719	.0855	47	.0785	36	.1065
	56	.099	.0758	.0874	45	.0820	36	.1065
4	40	.112	.0795	.0958	43	.0890	31	.1200
	48	.112	.0849	.0985	42	.0935	31	.1200
6	32	.138	.0974	.1177	36	.1065	26	.1470
	40	.138	.1055	.1218	33	.1130	26	.1470
8	32	.164	.1234	.1437	29	.1360	17	.1730
	36	.164	.1279	.1460	29	.1360	17	.1730
10	24	.190	.1359	.1629	25	.1495	8	.1990
	32	.190	.1494	.1697	21	.1590	8	.1990
12	24	.216	.1619	.1889	16	.1770	1	.2280
	28	.216	.1696	.1928	14	.1820	2	.2210
1/4	20	.250	.1850	.2175	7	.2010	G	.2610
	28	.250	.2036	.2268	3	.2130	G	.2610
5/16	18	.3125	.2403	.2764	F	.2570	21/64	.3281
	24	.3125	.2584	.2854	I	.2720	21/64	.3281
3/8	16	.3750	.2938	.3344	5/16	.3125	25/64	.3906
	24	.3750	.3209	.3479	Q	.3320	25/64	.3906
7/16	14	.4375	.3447	.3911	U	.3680	15/32	.4687
	20	.4375	.3725	.4050	25/64	.3906	29/64	.4531
1/2	13	.5000	.4001	.4500	27/64	.4219	17/32	.5312
	20	.5000	.4350	.4675	29/64	.4531	33/64	.5156
9/16	12	.5625	.4542	.5084	31/64	.4844	19/32	.5937
	18	.5625	.4903	.5264	33/64	.5156	37/64	.5781
5/8	11	.6250	.5069	.5660	17/32	.5312	21/32	.6562
	18	.6250	.5528	.5889	37/64	.5781	41/64	.6406
3/4	10	.7500	.6201	.6850	21/32	.6562	25/32	.7812
	16	.7500	.6688	.7094	11/16	.6875	49/64	.7656
7/8	9	.8750	.7307	.8028	49/64	.7656	29/32	.9062
	14	.8750	.7822	.8286	13/16	.8125	57/64	.8906
1	8	1.0000	.8376	.9188	7/8	.8750	1-1/32	1.0312
	14	1.0000	.9072	.9536	15/16	.9375	1-1/64	1.0156
1-1/8	7	1.1250	.9394	1.0322	63/64	.9844	1-5/32	1.1562
	12	1.1250	1.0167	1.0709	1-3/64	1.0469	1-5/32	1.1562
1-1/4	7	1.2500	1.0644	1.1572	1-7/64	1.1094	1-9/32	1.2812
	12	1.2500	1.1417	1.1959	1-11/64	1.1719	1-9/32	1.2812
1-1/2	6	1.5000	1.2835	1.3917	1-11/32	1.3437	1-17/32	1.5312
	12	1.5000	1.3917	1.4459	1-27/64	1.4219	1-17/32	1.5312

SCREW THREAD ELEMENTS FOR UNIFIED AND NATIONAL FORM OF THREAD

THREADS PER INCH (n)	PITCH (p) $p = \frac{1}{n}$	SINGLE HEIGHT — SUBTRACT FROM BASIC MAJOR DIAMETER TO GET BASIC PITCH DIAMETER	DOUBLE HEIGHT — SUBTRACT FROM BASIC MAJOR DIAMETER TO GET BASIC MINOR DIAMETER	83 1/3 PERCENT DOUBLE HEIGHT — SUBTRACT FROM BASIC MAJOR DIAMETER TO GET MINOR DIAMETER OF RING GAGE	BASIC WIDTH OF CREST AND ROOT FLAT $\frac{p}{8}$	CONSTANT FOR BEST SIZE WIRE ALSO SINGLE HEIGHT OF 60 DEG. V-THREAD	DIAMETER OF BEST SIZE WIRE
3	.333333	.216506	.43301	.36084	.0417	.28868	.19245
3 1/4	.307692	.199852	.39970	.33309	.0385	.26647	.17765
3 1/2	.285714	.185577	.37115	.30929	.0357	.24744	.16496
4	.250000	.162379	.32476	.27063	.0312	.21651	.14434
4 1/2	.222222	.144337	.28867	.24056	.0278	.19245	.12830
5	.200000	.129903	.25981	.21650	.0250	.17321	.11547
5 1/2	.181818	.118093	.23619	.19682	.0227	.15746	.10497
6	.166666	.108253	.21651	.18042	.0208	.14434	.09623
7	.142857	.092788	.18558	.15465	.0179	.12372	.08248
8	.125000	.081189	.16238	.13531	.0156	.10825	.07217
9	.111111	.072168	.14434	.12028	.0139	.09623	.06415
10	.100000	.064952	.12990	.10825	.0125	.08660	.05774
11	.090909	.059046	.11809	.09841	.0114	.07873	.05249
11 1/2	.086956	.056480	.11296	.09413	.0109	.07531	.05020
12	.083333	.054127	.10826	.09021	.0104	.07217	.04811
13	.076923	.049963	.09993	.08327	.0096	.06662	.04441
14	.071428	.046394	.09279	.07732	.0089	.06186	.04124
16	.062500	.040595	.08119	.06766	.0078	.05413	.03608
18	.055555	.036086	.07217	.06014	.0069	.04811	.03208
20	.050000	.032475	.06495	.05412	.0062	.04330	.02887
22	.045454	.029523	.05905	.04920	.0057	.03936	.02624
24	.041666	.027063	.05413	.04510	.0052	.03608	.02406
27	.037037	.024056	.04811	.04009	.0046	.03208	.02138
28	.035714	.023197	.04639	.03866	.0045	.03093	.02062
30	.033333	.021651	.04330	.03608	.0042	.02887	.01925
32	.031250	.020297	.04059	.03383	.0039	.02706	.01804
36	.027777	.018042	.03608	.03007	.0035	.02406	.01604
40	.025000	.016237	.03247	.02706	.0031	.02165	.01443
44	.022727	.014761	.02952	.02460	.0028	.01968	.01312
48	.020833	.013531	.02706	.02255	.0026	.01804	.01203
50	.020000	.012990	.02598	.02165	.0025	.01732	.01155
56	.017857	.011598	.02320	.01933	.0022	.01546	.01031
60	.016666	.010825	.02165	.01804	.0021	.01443	.00962
64	.015625	.010148	.02030	.01691	.0020	.01353	.00902
72	.013888	.009021	.01804	.01503	.0017	.01203	.00802
80	.012500	.008118	.01624	.01353	.0016	.01083	.00722
90	.011111	.007217	.01443	.01202	.0014	.00962	.00642
96	.010417	.006766	.01353	.01127	.0013	.00902	.00601
100	.010000	.006495	.01299	.01082	.0012	.00866	.00577
120	.008333	.005413	.01083	.00902	.0010	.00722	.00481

Using the Best Size Wires, the measurement over three wires minus the Constant for Best Size Wire equals the Pitch Diameter.

LETTER SIZE DRILLS

A	0.234	J	0.277	S	0.348
B	0.238	K	0.281	T	0.358
C	0.242	L	0.290	U	0.368
D	0.246	M	0.295	V	0.377
E	0.250	N	0.302	W	0.386
F	0.257	O	0.316	X	0.397
G	0.261	P	0.323	Y	0.404
H	0.266	Q	0.332	Z	0.413
I	0.272	R	0.339		

WOOD SCREW TABLE

LENGTH	GAUGE STEEL SCREW	GAUGE BRASS SCREW	GAUGE NO.	DECIMAL	APPROX. FRACT.	DRILL SIZE A	DRILL SIZE B	DRILL SIZE C
1/4	0 to 4	0 to 4	0	.060	1/16	1/16		
3/8	0 to 8	0 to 6	1	.073	5/64	3/32		
1/2	1 to 10	1 to 8	2	.086	5/64	3/32	1/16	3/16
5/8	2 to 12	2 to 10	3	.099	3/32	1/8	1/16	1/4
3/4	2 to 14	2 to 12	4	.112	7/64	1/8	1/16	1/4
7/8	3 to 14	4 to 12	5	.125	1/8	1/8	3/32	1/4
1	3 to 16	4 to 14	6	.138	9/64	5/32	3/32	5/16
1 1/4	4 to 18	6 to 14	7	.151	5/32	5/32	1/8	5/16
1 1/2	4 to 20	6 to 14	8	.164	5/32	3/16	1/8	3/8
1 3/4	6 to 20	8 to 14	9	.177	11/64	3/16	1/8	3/8
2	6 to 20	8 to 18	10	.190	3/16	3/16	1/8	3/8
2 1/4	6 to 20	10 to 18	11	.203	13/64	7/32	5/32	7/16
2 1/2	8 to 20	10 to 18	12	.216	7/32	7/32	5/32	7/16
2 3/4	8 to 20	8 to 20	14	.242	15/64	1/4	3/16	1/2
3	8 to 24	12 to 18	16	.268	17/64	9/32	7/32	9/16
3 1/2	10 to 24	12 to 18	18	.294	19/64	5/16	1/4	5/8
4	12 to 24	12 to 24	20	.320	21/64	11/32	9/32	11/16
4 1/2	14 to 24	14 to 24	24	.372	3/8	3/8	5/16	3/4
5	14 to 24	14 to 24						

SCREWS ALSO AVAILABLE WITH PHILLIPS HEAD

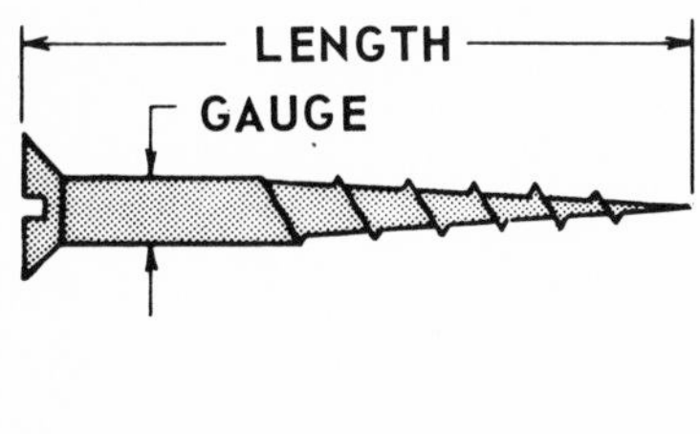

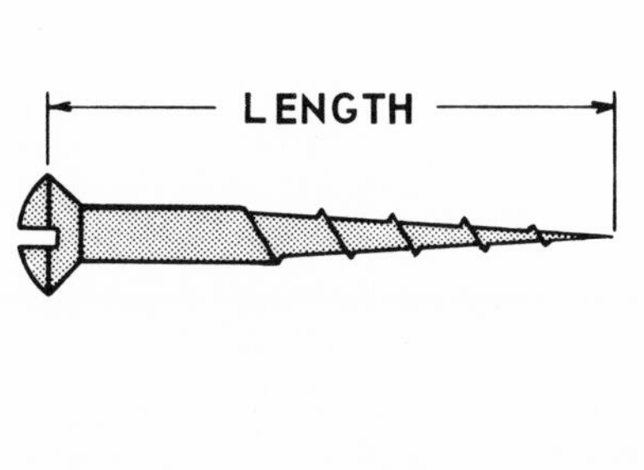

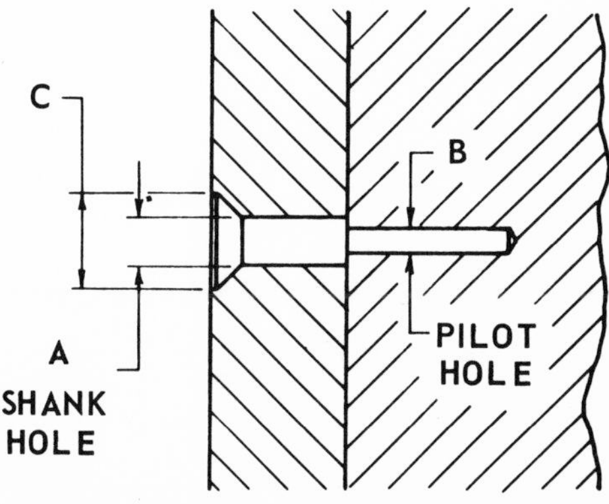

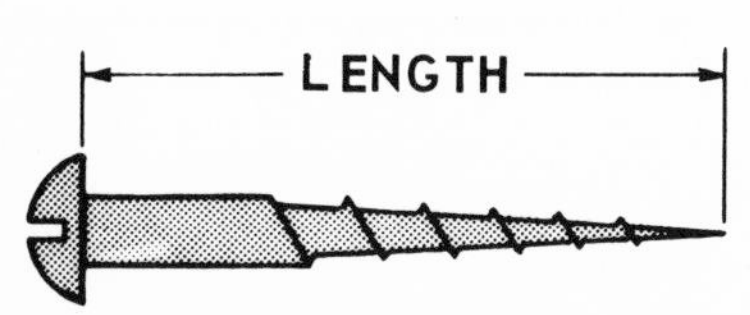

MACHINE SCREW AND CAP SCREW HEADS

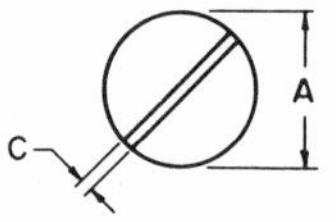

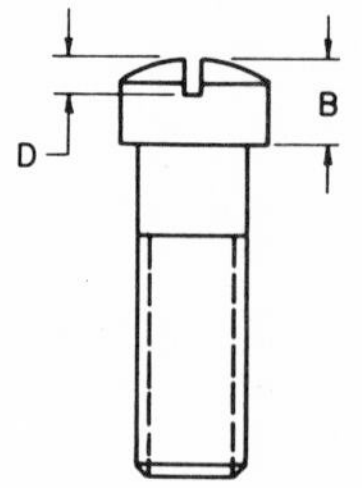

FILLISTER HEAD

SIZE	A	B	C	D
#8	.260	.141	.042	.060
#10	.302	.164	.048	.072
1/4	3/8	.205	.064	.087
5/16	7/16	.242	.077	.102
3/8	9/16	.300	.086	.125
1/2	3/4	.394	.102	.168
5/8	7/8	.500	.128	.215
3/4	1	.590	.144	.258
1	1 5/16	.774	.182	.352

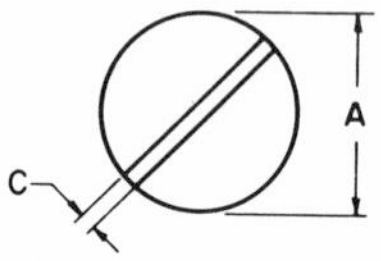

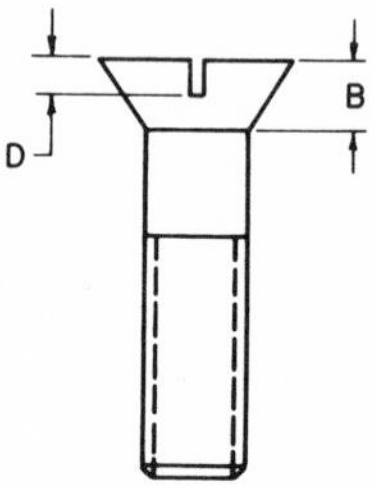

FLAT HEAD

SIZE	A	B	C	D
#8	.320	.092	.043	.037
#10	.372	.107	.048	.044
1/4	1/2	.146	.064	.063
5/16	5/8	.183	.072	.078
3/8	3/4	.220	.081	.095
1/2	7/8	.220	.102	.090
5/8	1 1/8	.293	.128	.125
3/4	1 3/8	.366	.144	.153

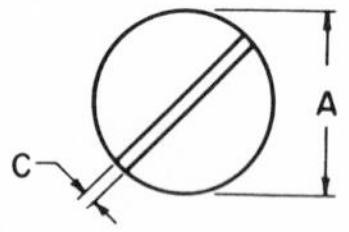

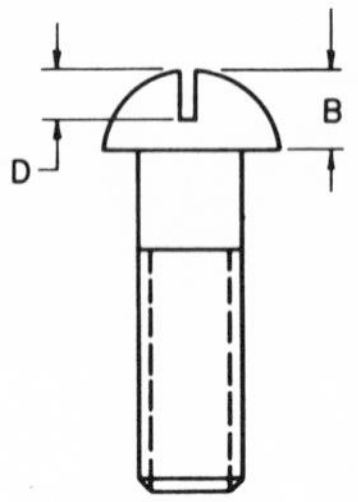

ROUND HEAD

SIZE	A	B	C	D
#8	.297	.113	.044	.067
#10	.346	.130	.048	.073
1/4	7/16	.1831	.064	.107
5/16	9/16	.236	.072	.150
3/8	5/8	.262	.081	.160
1/2	13/16	.340	.102	.200
5/8	1	.422	.128	.255
3/4	1 1/4	.526	.144	.320

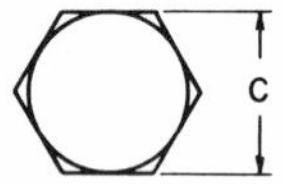

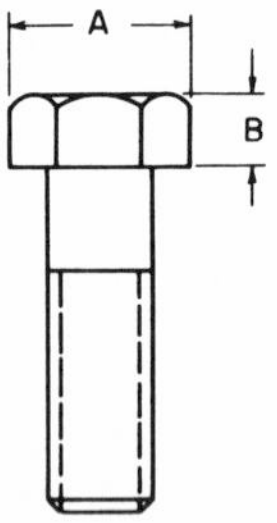

HEXAGON HEAD

SIZE	A	B	C
1/4	.494	.170	7/16
5/16	.564	.215	1/2
3/8	.635	.246	9/16
1/2	.846	.333	3/4
5/8	1.058	.411	15/16
3/4	1.270	.490	1 1/8
7/8	1.482	.566	1 5/16
1	1.693	.640	1 1/2

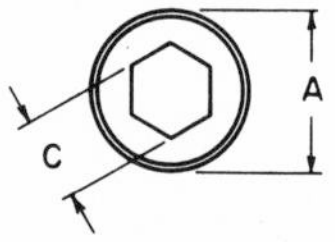

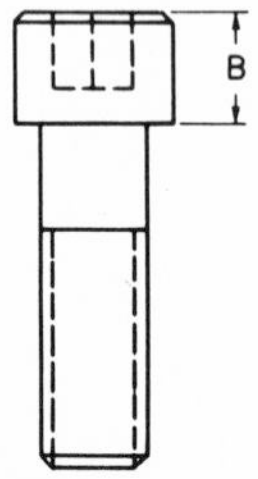

SOCKET HEAD

SIZE	A	B	C
#8	.265	.164	1/8
#10	5/16	.190	5/32
1/4	3/8	1/4	3/16
5/16	7/16	5/16	7/32
3/8	9/16	3/8	5/16
7/16	5/8	7/16	5/16
1/2	3/4	1/2	3/8
5/8	7/8	5/8	1/2
3/4	1	3/4	9/16
7/8	1 1/8	7/8	9/16
1	1 5/16	1	5/8

FRACTIONAL INCHES INTO DECIMALS AND MILLIMETERS

INCH	DECIMAL INCH	MILLIMETER	INCH	DECIMAL INCH	MILLIMETER
1/64	0.0156	0.3967	33/64	0.5162	13.0968
1/32	0.0312	0.7937	17/32	0.5312	13.4937
3/64	0.0468	1.1906	35/64	0.5468	13.8906
1/16	0.0625	1.5875	9/16	0.5625	14.2875
5/64	0.0781	1.9843	37/64	0.5781	14.6843
3/32	0.0937	2.3812	19/32	0.5937	15.0812
7/64	0.1093	2.7781	39/64	0.6093	15.4781
1/8	0.125	3.175	5/8	0.625	15.875
9/64	0.1406	3.5718	41/64	0.6406	16.2718
5/32	0.1562	3.9687	21/32	0.6562	16.6687
11/64	0.1718	4.3656	43/64	0.6718	17.0656
3/16	0.1875	4.7625	11/16	0.6875	17.4625
13/64	0.2031	5.1593	45/64	0.7031	17.8593
7/32	0.2187	5.5562	23/32	0.7187	18.2562
15/64	0.2343	5.9531	47/64	0.7343	18.6531
1/4	0.25	6.5	3/4	0.75	19.05
17/64	0.2656	6.7468	49/64	0.7656	19.4468
9/32	0.2812	7.1437	25/32	0.7812	19.8437
19/64	0.2968	7.5406	51/64	0.7968	20.2406
5/16	0.3125	7.9375	13/16	0.8125	20.6375
21/64	0.3281	8.3343	53/64	0.8281	21.0343
11/32	0.3437	8.7312	27/32	0.8437	21.4312
23/64	0.3593	9.1281	55/64	0.8593	21.8281
3/8	0.375	9.525	7/8	0.875	22.225
25/64	0.3906	9.9218	57/64	0.8906	22.6218
13/32	0.4062	10.3187	29/32	0.9062	23.0187
27/64	0.4218	10.7156	59/64	0.9218	23.4156
7/16	0.4375	11.1125	15/16	0.9375	23.8125
29/64	0.4531	11.5093	61/64	0.9531	24.2093
15/32	0.4687	11.9062	31/32	0.9687	24.6062
31/64	0.4843	12.3031	63/64	0.9843	25.0031
1/2	0.50	12.7	1	1.0000	25.4

INDEX

T

U

V

W